ADVANCES IN MOLTEN SALT CHEMISTRY
Volume 3

CONTRIBUTORS TO THIS VOLUME

Carlos E. Bamberger
Chemistry Division
Oak Ridge National Laboratory
Oak Ridge, Tennessee

Paul E. Field
Department of Chemistry
Virginia Polytechnic Institute and State University
Blacksburg, Virginia

H. Lloyd Jones
Department of Chemistry
Colorado State University
Fort Collins, Colorado

D. H. Kerridge
Department of Chemistry
The University
Southampton, England

Robert A. Osteryoung
Department of Chemistry
Colorado State University
Fort Collins, Colorado

Roy E. Thoma
Oak Ridge National Laboratory
Oak Ridge, Tennessee

L. V. Woodcock
Department of Physical Chemistry
Cambridge University
Cambridge, England

A Continuation Order Plan is available for this series. A continuation order will bring delivery of each new volume immediately upon publication. Volumes are billed only upon actual shipment. For further information please contact the publisher.

ADVANCES IN MOLTEN SALT CHEMISTRY
Volume 3

Edited by
J. BRAUNSTEIN

Oak Ridge National Laboratory
Oak Ridge, Tennessee

GLEB MAMANTOV

The University of Tennessee
Knoxville, Tennessee

and
G. P. SMITH

Oak Ridge National Laboratory
Oak Ridge, Tennessee

PLENUM PRESS • NEW YORK AND LONDON

The Library of Congress cataloged the first volume of this title as follows:

Advances in molten salt chemistry. v. 1–
1971–
New York, Plenum Press.

v. illus. 24 cm.

1. Fused salts—Collected works.

QD189.A33   546'.34   78–131884

Library of Congress   71 ₍4₎

Library of Congress Catalog Card Number 78-131884

ISBN-13: 978-1-4615-8272-4   e-ISBN-13: 978-1-4615-8270-0
DOI: 10.1007/978-1-4615-8270-0

© 1975 Plenum Press, New York
Softcover reprint of the hardcover 1st edition 1975

A Division of Plenum Publishing Corporation
227 West 17th Street, New York, N.Y. 10011

United Kingdom edition published by Plenum Press, London
A Division of Plenum Publishing Company, Ltd.
Davis House (4th Floor), 8 Scrubs Lane, Harlesden, London, NW10 6SE, England

PREFACE

The first chapter of this volume deals with computer simulation of molten salt behavior by molecular dynamics calculations. The next four chapters are reviews of experimental work: Chapter 2 deals with the solubility of nonreactive gases in molten salts, Chapter 3 with various types of organic reactions in molten tetrachloroaluminates, Chapter 4 with techniques for the study of molten fluorides, and Chapter 5 with the physical and chemical properties of thiocyanate melts. The last chapter is a collection of phase diagrams for binary and ternary fluoride systems.

J. B., G. M., G. P. S.

CONTENTS

Chapter 2

GAS SOLUBILITY IN MOLTEN SALTS

P. Field

Chapter 3

ORGANIC REACTIONS IN MOLTEN TETRACHLOROALUMINATE SOLVENTS

H. L. Jones and R. A. Osteryoung

Chapter 4

EXPERIMENTAL TECHNIQUES IN MOLTEN FLUORIDE CHEMISTRY

C. E. Bamberger

Chapter 5

THE CHEMISTRY OF THIOCYANATE MELTS

D. H. Kerridge

Chapter 6

PHASE DIAGRAMS OF BINARY AND TERNARY FLUORIDE SYSTEMS

R. E. Thoma

Chapter 1

MOLECULAR DYNAMICS CALCULATIONS ON MOLTEN IONIC SALTS

L. V. Woodcock*

School of Chemistry
Leeds University

1. INTRODUCTION

The advent of computers in recent years has given considerable impetus to progress in many areas of physical science. This article reviews some aspects of the application of modern computer simulation techniques to studies of the physical chemistry of molten salts.

Inherent in fundamental chemistry and the complex nature of matter associated with the interpretation of experimental observables ·are certain problems which are mathematical rather than conceptual, and which are therefore amenable to numerical solutions. These mathematical obstacles, which may involve the collective wave mechanics of many elementary particles, or cooperative motions of large numbers of molecules condensed together, are many-body problems which could have a foreseeable solution but which would constitute a formidable computational exercise. The application of computers to solve problems of this type arising in real context is generally described as numerical simulation. Although we are here primarily concerned with the computer simulation of molten salt systems on a molecular level, it is instructive to briefly review the levels of

* Ramsay Memorial (British) Fellow 1971–1973. Present address: Department of Physical Chemistry, Cambridge University.

simulation in physical chemistry and the relationships of molecular dynamics calculations thereto.

Computational chemistry as a subject may be subdivided so as to incorporate three distinct levels for the interpretation of experimental observables. These are phenomenological, molecular, and electronic. Each level of interpretation may be associated with a general dynamic many-body problem and corresponds to a particular level of simulation.

The deepest level of understanding for any physicochemical phenomenon is in terms of the elementary particles, i.e., nuclei and electrons, of the constituent atoms, molecules, or ions. Simulation at the electronic level begins with the specification of a particular set of electrons and nuclei and an approximate solution to Schrödinger's wave equation. It is presently only possible to obtain an exact analytical solution for two-particle systems, such as the hydrogen atom, but essentially quantitative results may now be obtained for systems containing many electrons. The accuracy depends on the complexity of the system, the number of electrons, and the level of approximation. As a consequence of the enhanced power of modern computers, these *ab initio* computations are becoming increasingly accurate, and the prediction of certain molecular parameters, to within 5 or 10%, say, for systems containing relatively large numbers of electrons is now practicable. The molecular properties which are computed in these circumstances may be intramolecular energy levels, dipole moments, polarizabilities, etc., or intermolecular potentials between two or more molecules. Potential-energy surfaces govern the motions of molecules in condensed phases and the mechanisms of chemical reactions. These are the parameters in terms of which the phenomenological properties of macroscopic systems are to be interpreted on a molecular level.

Once the intermolecular potential-energy surfaces are determined, in order to calculate the macroscopic experimental observables of a condensed state, such as thermodynamic functions or transport coefficients, it is necessary to appeal to statistical mechanics for tractable expressions relating the molecular parameters to bulk properties. Here again, even in the case of simple classical mechanical many-body systems, it is only possible to proceed to fruition through exact analytical derivation for certain idealized states. These include very dilute gases where only binary encounters need to be considered, and low-temperature crystals which may be represented as a system of independent harmonic oscillators. Simulation on the molecular level (MD calculations) is applied to compute physical properties from molecular parameters in those condensed states of matter where approximate analytical solutions to the many-body problem are inadequate. This includes all real liquids, dense gases, anharmonic crystals, and many other physical molecular systems such as solutions and interfaces.

In molecular dynamics computations the classical Newtonian equations of motion for representative numbers of molecules are solved numerically. Unlike the *ab initio* simulation of molecular parameters outlined above, MD calculations of bulk properties are essentially exact in that there are no conceptual approximations involved. The input model consists of a set of atomic masses and intermolecular potentials which may be obtained from *ab initio* simulation or determined semiempirically from physical measurements and parametrization. The bulk physical properties which are computed by these methods (described in Secs. 3 and 4) may be considered as the "experimental" properties of the mathematical model for the molecular interactions. Hence, MD calculations are often referred to as computer experiments. While there remain some difficulties in the general and quantum mechanical simulation of many-particle motion, the remaining problems associated with the interpretation of macroscopic properties on a molecular level are either technical or involve specification of the potential surfaces. We shall see in Sec. 3.3, however, that those two factors impose considerable limitations on the extent of applicability.

Application of computer simulation at the phenomenological level of interpretation is extensively employed in all branches of physical science. Here the parameters describing the input model may correspond to the bulk properties of a thermodynamically unstable system, the objective of a simulation program being to follow the progress of the system toward equilibrium. Generally, the laws of hydrodynamics, irreversible thermodynamics, or reaction kinetics are solved numerically in order to predict the future of a rate process, or to test a proposed mechanism when the rate constants or transport coefficients are known. Molecular dynamics calculations may eventually be complementary to phenomenological simulation techniques since it can be applied to calculate the unknown macroscopic variables.

The fundamental objective of molecular theories of dense fluids is to quantitatively interpret physical properties in terms of molecular parameters. Equilibrium theories of the liquid state, in general, have so far been largely unsuccessful in attempts to account analytically for thermodynamic functions in terms of intermolecular pair potentials for even the simplest systems. Transport theories which relate linear transport coefficients to intermolecular forces and masses have not yet been formulated because the problem here is even more complex. Despite these difficulties many less fundamental theoretical approaches to the liquid state have been proposed at varying levels of approximation, some of which have been applied to the interpretation of molten salt physical properties.

Computer simulation results may be utilized in a number of roles in relationship to approximate theoretical approaches. When a theory fails to agree with experiment, it is often not possible to determine whether the disagreement is due to the inadequacy of the model or the mathematical approximations in the implementation of the theory. Computer "experiments" effectively measure the properties of the mathematical model, and thus the ambiguity arising from comparisons of theory with real experiment may be resolved. When a theory agrees with experiment there is always the possibility of a cancellation of discrepancies since two or more independent approximations are inevitably involved. A comparison with MD results would clearly indicate if this were the case.

In those theories based upon simplified structural models of the liquid state, such as significant structures theory, cell or cellular-hole theories, the results of computer simulation may be applied in a different role. When macroscopic properties are related to molecular parameters which are adjusted to fit certain other bulk properties, the MD results can be used to test the physical reality of the underlying assumptions or to provide independent estimates of theoretically important parameters. Some theories involve experimental observables which are not accurately known; for example, radial distribution functions are required in some transport theories of liquids. The computations can provide quasi-experimental data such as temporal or spatial distribution functions which are required to implement theories.

In Sec. 5 we shall see that the application of computer simulation enables many of the shortcomings of approximate analytical theory to be elucidated; it gives rise to more conclusive comparison between theory and experiment; it furnishes the structural and dynamic details which indicate the direction of improved approximations and leads to a more comprehensive understanding of physical properties on a molecular level.

2. INTERMOLECULAR FORCES IN MOLTEN SALTS

2.1. True and Effective Pair Potentials

The interaction energy between a pair of molecules or ions generally depends upon the medium in which they exist. The total potential energy of a group of N ions may not therefore, a priori, be obtained simply as a sum of isolated pair potentials but must be expressed as a series, i.e.,

$$\Phi(r_1 \ldots r_N)$$

$$= \frac{1}{2} \sum_i \sum_j{}' \phi_{ij} + \frac{1}{3!} \sum_i \sum_j{}' \sum_k{}' \phi_{ijk} \cdots \frac{1}{n!} \sum_i \sum_j{}' \sum_k{}' \cdots \sum_m{}' \phi_{ijk\cdots m} \quad (1)$$

where the first term in the expression represents the contribution of the isolated pair potentials to the total energy. When the total potential energy is defined relative to a standard state of infinitely dispersed ions, the pair potential is correspondingly defined as the energy of a pair of ions at a particular distance apart relative to the energy of the pair of ions at infinite separation. It is important to bear in mind that this is not generally the lowest potential energy curve; at larger separations the atomic state is more stable than the ionic state. Successive terms in Eq. (1) represent the sums of contributions from three-, four-, n-body potentials. The primes on the summation signs indicate that contributions to the sums for which any two ions have the same label one are to be omitted and the dependence of $\phi_{ij\ldots m}$ on $r_i, r_j \ldots r_m$ is implicit.

Nonadditive (many-body) contributions to Φ are troublesome first because little is known theoretically or experimentally regarding the functional forms and magnitudes involved, and second because of the difficulties encountered in the explicit incorporation of many-body potentials into statistical mechanics or numerical work and, in the present context, into molecular dynamics computations. Despite these complexities, there is strong evidence that in simple ionic liquids (excluding liquid metals), as in simple atomic liquids, the pair potential is predominant. This evidence arises from two sources: (1) from the theoretical predictions by *ab initio* calculations of potential energy surfaces[1] and (2) from the comparisons between semiempirical pair potentials obtained using gas phase[2] and crystalline state[3] data. In molten ionic salts, moreover, it is thought that the only appreciable contributions to many-body potentials arise from internal electrostatic polarization of the ions.[4]

The limitations imposed on statistical mechanical treatments and numerical applications by the restriction to pairwise additive systems have given rise to the concept of effective pair potential as distinct from the isolated pair potentials. An effective pair potential may be defined to reproduce a particular property of a system. For example, if this is the potential energy, we have

$$\Phi(r_1 \ldots r_N, \varrho, T) = \left\langle \frac{1}{2} \sum_i \sum_j{}' \phi^{\text{eff}}(r_{ij}, \varrho, T) \right\rangle \quad (2)$$

to define the corresponding effective pair potential for a thermal time or ensemble average (denoted by angular brackets). Many-body potentials are geometry dependent, and the net contribution to ϕ^{eff} depends upon the state of the system. Parametrized pair potentials which are obtained by fitting experimental data, unless relating to the dilute gaseous state, are thus properly described as semiempirical effective pair potentials. The extent to which an effective pair potential is state and property dependent is some measure of the importance of many-body potentials.

Quantitative experimental information on pair potentials between monatomic ions has been obtained largely from two main sources, i.e., from spectroscopic studies of low-density vapors and from thermodynamic properties of crystals. When parameters entering plausible functional forms are determined so as to reproduce empirical crystal data, using essentially exact relationships, effective pair potentials are obtained which are found to be property insensitive. For example, two- or three-parameter pair potentials can satisfactorily account for the lattice energy, density, compressibility, expansivity, and heat capacities of ionic crystals. These effective pair potentials, moreover, correspond very closely to semiempirical pair potentials obtained by fitting curves to spectroscopic data for isolated ion pairs in the gas phase. The closeness of these two types for the alkali halides is direct evidence for the absence of appreciable many-body effects. Theoretical arguments leading to the conclusion that polarization forces represent the only important nonpairwise additive contribution in molten ionic salts have been propounded by Stillinger.[4]

2.2. Semiempirical Models

The predominant long-range contribution to the isolated ion–ion pair potential arises from the interaction of the charges on each ion according to Coulomb's law. The electrostatic contribution to the pair potential is therefore simply $Z_i Z_j e^2/r$, where the charge distribution of an ion is considered as a point charge centered on the nucleus. At short distances, in accord with the Pauli exclusion principle, repulsive forces operate whenever the electron clouds appreciably overlap. The crudest approximation to this effect is the hard-core potential which is defined such that

$$\phi = Z_i Z_j e^2/r \qquad \text{for} \qquad r > \sigma$$

and

$$\phi = \infty \qquad \text{for} \qquad r < \sigma$$

σ being the collision diameter. More realistic representations are continuous functions with a steeply increasing gradient over the appropriate distance range such as the inverse power potential or the exponential form which more closely conforms with the theoretically predicted functional form for short-range repulsion. These may be written as

$$\phi = Z_i Z_j e^2 / r + b r^{-n} \tag{3}$$

$$\phi = Z_i Z_j e^2 / r + B \exp(-Ar) \tag{4}$$

respectively. The former is often referred to as the Pauling form,[5] and the latter is usually described as the Born–Mayer potential,[6] after those who first attempted to parametrize alkali halide crystal data on the basis of the Born model of ionic crystals. There have been many determinations of the parameters entering Eqs. (3) and (4) for pairs of ions in alkali halides from a variety of experimental data.[7]

A further elaboration is the inclusion of terms for van der Waals dispersion forces which arise from the mutual response of fluctuating electronic charge distributions. Dispersion potentials may be expressed as multipole expansions of the form[1]

$$\phi^D = C r^{-6} + D r^{-8} + \cdots \tag{5}$$

in which the first term corresponds to London's dipole-induced dipole dispersion energy, and further terms to interactions between higher moments of the multipole expansion. To a first approximation, terms beyond $C r^{-6}$ are often neglected. As a consequence of the predominance of the Coulombic contribution to the interionic pair potential, dispersion potentials in ionic liquids are relatively unimportant.

A number of attempts have been made to approximate the pair potential curves of isolated alkali halide molecules using semiempirical functions which assume that the molecule is constituted of ions. Rittner[8] used a sophisticated expression which includes terms corresponding to Eqs. (4) and (5) and also terms for mutual polarization of the ions. The Rittner potential has the form

$$\phi = \frac{Z_i Z_j e^2}{r} + B \exp(-Ar) + \frac{C}{r^6} + \frac{e^2(\alpha_i + \alpha_j)}{2r^4} + \frac{2e^2 \alpha_i \alpha_j}{r^7} \tag{6}$$

in which α is the linear coefficient of electrostatic polarization of an ion. In Rittner's analysis of the alkali halide vapor phase data, C was calculated from the London formula

$$C = \tfrac{3}{2}\alpha_1\alpha_2[I_2E/(I_2 + E)] \tag{7}$$

where I_2 is the second ionization potential of the alkali metal atom and E is the electron affinity of the halogen atom. The semiempirical method of determining the dispersion potential coefficients of Mayer,[9] however, is probably more accurate. Here the coefficient C can be expressed as

$$C_{ij} = 3h(\alpha_i\alpha_j\nu_i\nu_j)/2(\nu_i + \nu_j) \tag{8}$$

where h is Planck's constant and ν_i and ν_j are the series limit frequencies of the discrete optical spectra of the ions i and j. Values of C and D obtained for the alkali halides are tabulated by Mayer in the literature. Having fixed all the parameters in Eq. (12) in accord with the polarizable ion model, Rittner determined the repulsive parameter B and A such that

$$d\phi/dr = 0 \qquad \text{at} \qquad r = r_0 \tag{9}$$

the equilibrium internuclear distance (from microwave spectra) and

$$d^2\phi/dr^2 = p \tag{10}$$

the force constant (from vibrational spectra). The above potential has been found to give satisfactory agreement with experiment for the binding energy, the dipole moment, and the fundamental vibrational frequency.

Varshni and Shukla[10] have carried out an independent analysis of the parameters entering three different pair potentials for alkali halide dimers. These are the Born–Mayer potential [Eq. (4)], the Rittner potential [Eq. (12)], and a Gaussian form, i.e.,

$$\phi = Z_iZ_je^2/r + B\exp(-Ar^2) \tag{11}$$

The authors also used the two criteria [Eqs. (9) and (10)] to fix the two repulsive parameters in each case. Although the Rittner potential was found to be generally superior, the Gaussian expression appears to be an improvement on the simple exponential form without additional adjustable parameters.

One disadvantage of using data on gaseous ion pairs as a source of semiempirical cation–anion pair potentials is that there is no information on the short-range part of the pair potential between like ions. This is perhaps not too serious, however, since the Coulombic repulsion in itself ensures that unlike ions do not approach very closely in the molten salt at ordinary temperatures (see, for example, Sec. 5.2). Although this source of data on pair potentials has not yet been utilized in molecular dynamics calculations, it seems feasible that it would provide an alternative to crystal data. These data may be particularly valuable for more complex melts such as mixtures, or melts with divalent cations, where the predominant species in the melt are geometrically akin to those found in the vapor phase. It is also worth noting that the most probable cation–anion nearest neighbor distance in the molten salt corresponds more closely to the isolated ion-pair equilibrium internuclear distance than to the crystalline-phase lattice spacing.

An interesting and possibly useful connection between the potential-energy functions of two rare gas atoms and the two corresponding isoelectronic alkali halide ions has been demonstrated.[11] When an electrostatic term is simply added to pair potentials for two rare gas atoms, the similarity with semiempirical effective pair potentials extracted from crystal data seems quite remarkable. This approach is limited, however, since NaCl and KF, for example, would have the same set of pair potentials despite substantial differences in their respective thermodynamic properties.

By far the major source of information on pair potentials in the alkali and alkaline earth halides, and many other inorganic salts, is the thermodynamic properties of the crystalline state, usually at ordinary temperatures. Analysis of the kind whereby parameters entering semiempirical forms are tailored to reproduce two or more properties in a given approximation is the subject of a very extensive literature.[7] Here we will only mention one such analysis for the alkali halides in any detail, and the reader is referred to review articles in the literature for more detailed accounts and further references. Most of the computer-simulation studies reported so far for the alkali halides have been based upon the set of semiempirical effective pair potentials obtained by Tosi and Fumi from an analysis of density data on the 17 alkali halides with the rock salt structure.[12,13]

Tosi and Fumi[13] employed the traditional Born model of an ionic crystal, wherein the lattice energy Φ^L is pairwise additive and given by the Madelung electrostatic energy plus short-range repulsion and possibly also dispersion energy. As part of an extensive analysis, the generalized Huggins–Mayer form for the short-range repulsion was adopted. The form of the pair potential may be written as

$$\phi_{ij} = Z_i Z_j e^2/r + B c_{ij} \exp[A(\sigma_i + \sigma_j - r)] + C_{ij}r^{-6} + D_{ij}r^{-8} \qquad (12)$$

in which the parameters B, A, σ_i, and σ_j are empirically adjustable. In the Tosi–Fumi analysis, values of C and D were taken from Mayer, and c_{ij} values from Pauling. The quantities σ_i and σ_j are distance parameters related to each "ionic radius," A is a constant for each salt, and B is a constant for all the salts. These parameters have been determined by Tosi and Fumi from the equations of state of the alkali halide crystals at 298°K. In the classical quasi-harmonic approximation for the kinetic energy, we have, for the first two derivations of the lattice energy with respect to volume,

$$d\Phi^L/dV = \beta T/\varkappa \tag{13}$$

and

$$\frac{d^2\Phi^L}{dV^2} = \frac{1}{V\varkappa}\left\{1 + \frac{T}{\varkappa}\left[\left(\frac{\partial \varkappa}{\partial T}\right)_p + \frac{\beta}{\varkappa}\left(\frac{\partial \varkappa}{\partial p}\right)_T\right]\right\} \tag{14}$$

where $\varkappa$ is the isothermal compressibility and β is the coefficient of thermal expansion. The quantity Φ^L is substituted in terms of the pair potentials in Eq. (12), and an iterative least-squared minimization computational procedure is used to determine B and the various A and σ values which best conform with all the resultant simultaneous equations. The important conclusion of the Tosi–Fumi analysis is that the set of effective pair potentials thus obtained are found to reproduce the experimental cohesive energies of the alkali halides to within 1% with the exception of Li salts.

In Table I the dissociation energy (D) and internuclear distance at the minimum (r_0) in the Tosi–Fumi potentials are compared with vapor-phase data. The close agreement for most alkali halides indicates (1) that nonadditive potentials are comparatively small in magnitude and (2) that spectroscopic data on vapor-phase clusters is a reliable source of interionic potentials for use in condensed phases. Also given in Table I are values of the ion-induced dipole (column 5) and induced-dipole–induced dipole energies (column 6) calculated from the Rittner formula using Pauling's polarizabilities[14] for distances corresponding to r_0. If polarization effects are negligible in the crystal, it is evident that the simple assumptions associated with the derivation of the Rittner expression are probably invalid. These approximate considerations, do, however, show the induced-dipole–induced-dipole contribution to be only a few percent of the ion-induced dipole contribution. In condensed phases of salts in which only the anions have appreciable induced dipoles, since this energy term has an r^{-7} dependence, and like ions do not approach closely, it appears to be reasonable to neglect contributions to the potential beyond the ion-induced dipole.

TABLE I. Alkali Halide Pair Potential Parameters from Vapor-Phase[a] and Crystal Data[b] Together with Contributions from Polarization Potentials Estimated Using Rittner Formulas [Equation (6)][c]

	r_0 (Å)		D (ergs $\times 10^{-12}$)		Polarization potentials at r_0 (ergs $\times 10^{-12}$)	
	Vapor	TF potential	Vapor	TF potential	ID	DD
LiF	1.55	1.34($-$0.21)	-12.53	$-13.26(-0.73)$	-2.17	-0.07
LiCl	2.02	1.84($-$0.18)	-10.42	$-10.21(+0.21)$	-2.57	-0.04
LiBr	2.17	2.00($-$0.17)	-10.00	$-9.49(+0.51)$	-2.53	-0.03
LiI	2.39	1.97($-$0.42)	-9.38	$-9.11(+0.27)$	-2.53	-0.02
NaF	1.84	1.56($-$0.28)	-10.49	$-11.52(-1.03)$	-1.24	-0.12
NaCl	2.36	2.28($-$0.08)	-9.03	$-8.71(+0.32)$	-1.44	-0.08
NaBr	2.50	2.39($-$0.11)	-8.61	$-8.29(+0.32)$	-1.48	-0.07
NaI	2.71	2.49($-$0.22)	-8.16	$-7.82(+0.34)$	-1.57	-0.06
KF	2.13	1.97($-$0.16)	-9.45	$-9.67(-0.22)$	-1.06	-0.21
KCl	2.67	2.59($-$0.08)	-7.99	$-7.78(+0.21)$	-1.03	-0.15
KBr	2.82	2.78($-$0.04)	-7.64	$-7.34(+0.30)$	-1.03	-0.13
KI	3.08	2.97($-$0.11)	-7.22	$-6.87(+0.35)$	-1.03	-0.11
RbF	2.27	2.20($-$0.07)	-9.10	$-8.93(+0.17)$	-1.08	-0.22
RbCl	2.79	2.80($+$0.01)	-7.64	$-7.35(+0.29)$	-0.98	-0.19
RbBr	2.95	2.93($-$0.02)	-7.33	$-7.03(+0.30)$	-0.96	-0.17
RbI	3.18	3.17($-$0.01)	-6.91	$-6.55(+0.36)$	-0.97	-0.14
CsF	2.35	2.55($+$0.20)	-8.89	$-8.11(+0.78)$	-1.33	-0.30
CsCl	2.91	2.91($+$0.00)	-7.64	$-7.08(+0.56)$	-0.99	-0.24
CsBr	3.07	3.09($+$0.02)	-7.36	$-6.74(+0.62)$	-0.94	-0.21
CsI	3.32	3.33($+$0.01)	-6.88	$-6.30(+0.58)$	-0.92	-0.18

[a] L. Brewer and E. Brackett, *Chem. Rev.* **66**:425 (1961).
[b] M. P. Tosi and F. G. Fumi, *J. Phys. Chem. Solids* **25**:45 (1964).
[c] r_0 is the equilibrium distance, D is the corresponding potential energy, and the last two columns represent the ion-induced dipole and induced dipole–induced dipole potentials.

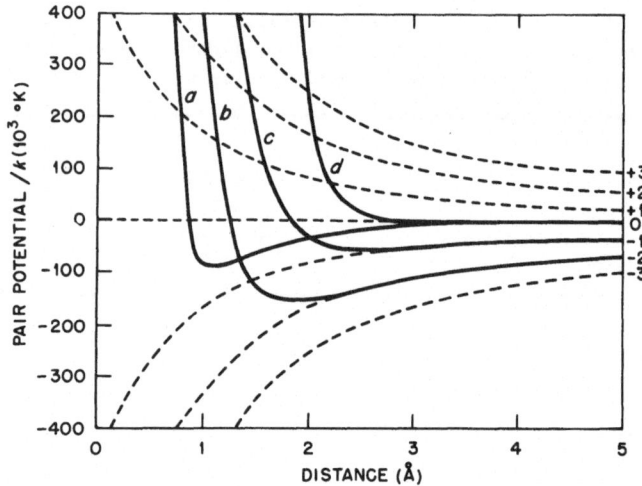

Fig. 1. Comparisons of interionic, interatomic, and intramolecular pair potentials. (a) intramolecular Morse potential for nitrogen; (b) interionic pair potential for $Ca^{++}-F^-$ (Reitz *et al.*); (c) interionic pair potential for K^+-Cl^- (Tosi and Fumi), (d) interatomic Lennard-Jones potential for argon. Faint lines represent the electrostatic potential $Z_i Z_j e^2/r$ for integral values of $Z_i Z_j$ as given on the right.

The pair potential between K^+ and Cl^- ions from Tosi and Fumi is shown in Fig. 1 together with a typical intramolecular potential (the Morse potential for nitrogen), a typical interatomic potential (the Lennard-Jones potential for argon), and the pair potential obtained for CaF_2 by Reitz *et al.* from data on the elastic constants of the crystal.[15]

There are a number of important features of the potential-energy surfaces in ionic systems which are evident from Fig. 1. Firstly, even up to molten salt temperatures, the depths of the potential wells are orders of magnitude greater than kT. (One can imagine the like ion pair potential as being simply the Coulombic repulsion.) Secondly, unlike atom–atom intramolecular potentials, the range of the ionic pair potential is extremely great. For example, at 1000°K the KCl pair potential only becomes equal to kT at around 170 Å, or 60 ion diameters. Finally, the restoring force of the potential becomes larger as the cation–anion distance becomes smaller and/or the cation increases its charge. In a later section, we will examine these observations in the light of MD results for particular systems and compare them with Fajans rules,[16] based upon polarization effects, which are traditionally used for interpreting tendencies of groups IIa and IIa halides to behave like molecular compounds.

3. COMPUTATIONAL TECHNIQUES

3.1. Molecular Dynamics Simulation

The most straightforward method of computing the physical properties of a model system is by direct numerical integration of the classical Newtonian equations of motion. The parameters entering the model for an ionic melt will be a prescribed set of interionic potentials, the ionic masses and two external thermodynamic variables, in this case the energy and the density. A molecular dynamics calculation begins with an initial arbitrary set of positions and momenta for a small representative number of molecules and solves the equations of motion so that the molecular trajectories are determined as a function of time. Only equilibrium situations are being simulated, and the computations inevitably involve an initial equilibration period which has to be discounted.

In a system containing N ions, interacting through model pair potentials of the type described in the previous section, let m_i denote the mass of the ith ion, r_i its position vector, and F_i the resultant force on the ion due to interactions with the other $N-1$ ions. In the absence of external forces, Newton's classical equation of motion is

$$m_i(d^2 r_i/dt^2) = F_i \tag{15}$$

For a system of N ions, therefore, in three dimensions, there are $3N$ coupled second-order partial differential equations to be solved numerically. As a consequence of the limitations of modern computers, it is presently only practicable to simulate the motions of small numbers of ions, i.e., up to 10^3. Thus, in order to eliminate the surface effects associated with very small systems, the N ions are subjected to periodic boundary conditions. Each ion in the parent cube (the volume of which is determined by the prescribed density) has images of itself at all combinations of unit translations by the cube along its edges. If, during the course of a molecular dynamics simulation, an ion moves out of the basic cube, a corresponding image moves into the cube at the opposite face and becomes the parent ion. Because the entire system is isolated, the total internal energy, momentum, and density are conserved and, consequently, the sequence of configurations generated during a molecular dynamics simulation corresponds to a petit (finite) microcanonical ensemble in statistical thermodynamics.

There are a number of different algorithms available for the numerical solution of coupled partial differential equations. Rahman,[17] in his original simulation of liquid argon, transformed the $3N$ second-order equations into $6N$ first-order finite difference formulas which are then solved using a predictor–corrector procedure. If X_i represents a component of the position vector, v_i^x a velocity component, and Δt is a small increment in time, the finite difference formulas can be written for the velocity and position, respectively:

$$v_i(t + \Delta t) = v_i^x(t - \Delta t) + 2\Delta t F_i^x(t)/m_i + \theta(\Delta t^3)$$

$$v_i(t = \Delta t) = v_i^x(t) + \tfrac{1}{2}\Delta t[F_i^x(t) + F_i^x(t + \Delta t)]/m_i + \theta(\Delta t^3)$$

$$X_i(t + \Delta t) = X_i(t - \Delta t) + 2\Delta t v_i^x(t) + \theta(\Delta t^3)$$

$$X_i(t + \Delta t) = X_i(t) + \tfrac{1}{2}\Delta t[v_i^x(t) + v_i^x(t + \Delta t)] + \theta(\Delta t^3)$$

The force on an ion is obtained as a summation over pair forces by

$$F_i^x = \sum_{\substack{j=1 \\ j \neq i}}^{N} \frac{(X_i - X_j)}{r} \times \frac{d\phi_{ij}}{dr} \tag{16}$$

Any number of corrector cycles could in fact be incorporated into the algorithm, but it is clear that each additional cycle requires an extra summation over pair forces per time increment. The quantity $\theta(\Delta t^3)$ implies that the error associated with the finite time increment is proportional to Δt^3. The general objective is to maximize the yield of molecular dynamics in real time relative to that of machine time within acceptable limits for conservation of energy and momentum. Whereas increasing Δt will enhance the former, additional cycles in the algorithm serve to increase the computational time per time increment. More sophisticated algorithms, however, have a higher stability limit or permit a larger value of Δt to be used at a given level of energy conservation. Verlet noted that since repulsive forces between colliding molecules are steep, a small value of Δt has to be used anyhow, and he introduced a simple finite-difference formula for the second-order derivative,[18] i.e.,

$$X_i(t + \Delta t) = 2X_i(t) - X_i(t - \Delta t) + \Delta t^3 F_i^x(t)/m_i + \theta(\Delta t^4)$$

and calculated the velocity from

$$v_i^x(t) = [X_i(t + \Delta t) - X_i(t - \Delta t)]/2\Delta t + \theta(\Delta t^2)$$

One of the disadvantages of dynamic simulation is that it may only rigorously be applied to conservative systems at constant energy and density. Experiments on real systems, however, are usually carried out under conditions of constant temperature and pressure. A heuristic modification of Verlet's algorithm[19] has been found to simulate the equilibrium properties of constant temperature conditions and to reproduce the first and second moments of the energy and the virial in the canonical ensemble. If the requisite mean temperature is T^0, the corresponding mean-squared velocity of species of a component α with mass m_α is given by

$$\overline{(v_i^2)}_\alpha^0 = 3kT^0/m \tag{17}$$

and the algorithm simply becomes

$$X_i(t + \Delta t) = X_i(t) + f_\alpha(t)[X_i(t) - X_i(t - \Delta t)] + \Delta t^2 F_i^x(t)/m_i$$

where

$$f_\alpha^2(t) = (v_i^2)_\alpha^0 \Big/ \left[N_\alpha^{-1} \sum_{j=1}^{N_\alpha} v_i^2(t) \right] \tag{18}$$

While this approach must be used with caution for computing time-dependent properties of small systems, results for 216 ion molten salt systems reveal no systematic discrepancies with constant energy runs with the same mean temperature for both equilibrium and some time-dependent properties.

A severe limitation of the present techniques is the excessive demands upon computational facilities. Typically, several hours of central processing time on the most powerful computers presently available are required to integrate the equations of motion of, for instance, 512 ions during 10^{-10} sec. Owing to the N^2 dependence of mill-time and the large memory requirements, it is only possible to handle up to around 10^3 ions. Molecular dynamics calculations may only therefore be applied to study time-dependent phenomena with a characteristic relaxation time less than the duration of the machine experiment, or to frequency-dependent properties where the wavelength is within the limit of the constraint imposed on the wavelengths of fluctuations by the periodic boundary. Despite these limitations, for simple liquids and molten salts the thermodynamic properties and high-frequency time-dependent properties can be computed with precision.

3.2. The Monte Carlo Method

An alternative numerical means whereby the equilibrium properties of a model system may be computed is the Monte Carlo approach. In statistical mechanics the sequence of configurations for a dynamic system is represented by an ensemble of static configurations so that the time-average of a fluctuating dynamic variable is equated to an ensemble average. Thus, if f is any dynamic variable of the model system of N ions

$$\frac{1}{\tau} \int_0^{\tau} f(\mathbf{r}_1 \cdots \mathbf{r}_N, t)\, dt = \int_{\tau \to \infty} \cdots \int f(\mathbf{r}_1 \ldots \mathbf{r}_N) P(\mathbf{r}_1 \ldots \mathbf{r}_N)\, d\mathbf{r}_1 \ldots d\mathbf{r}_N$$

(19)

where $\mathbf{r}_i$ is the position vector of the ith ion, $P(\mathbf{r}_1 \cdots \mathbf{r}_N)$ is a prescribed probability function for observing the configuration $\mathbf{r}_1 \cdots \mathbf{r}_n$, and the $3N$-dimensional integral on the right side extends over all phase space. The form of $P(\mathbf{r}_1 \cdots \mathbf{r}_N)$ depends upon the independent state variables used to define the ensemble. In the most widely used case of the isochoric-isothermal (canonical) ensemble it is

$$P(\mathbf{r}_1 \ldots \mathbf{r}_N) = \frac{\exp[-\Phi(\mathbf{r}_1 \ldots \mathbf{r}_N)/kT]}{\int \ldots \int \exp[-\Phi(\mathbf{r}_1 \ldots \mathbf{r}_N)/kT]\, d\mathbf{r}_1 \ldots d\mathbf{r}_N}$$

(20)

where k is Boltzmann's constant, T is the absolute temperature, and Φ is the total potential energy of the configuration.

In order to compute thermodynamic averages of the model system, the rather formidable $3N$-dimensional phase integral has to be evaluated numerically. The straightforward numerical procedure would be to systematically integrate, by small stepwise increments, over all possible configurations. Alternatively, the crudest Monte Carlo approach would simply select random sets of $3N$ coordinates, compute the Boltzmann factor for each configuration, i.e., its probability, and weight each configuration accordingly. Most of the computing expenditure involves a summation over pairs of ions to obtain $\Phi(\mathbf{r}_1 \cdots \mathbf{r}_N)$ and hence $P(\mathbf{r}_1 \cdots \mathbf{r}_N)$. It is immediately clear that both the direct numerical approach and the crude Monte Carlo approach would be impractical for condensed phases because high-energy configurations with negligible Boltzmann factors would be selected in the overwhelming majority of trials.

A more sophisticated Monte Carlo method which is suitable for condensed molecular systems is that of Metropolis et al.[20] The essential feature of this method is that configurations are chosen with a probability of selection proportional to $P(\mathbf{r}_1 \cdots \mathbf{r}_N)$ and then all the configurations thus chosen are weighted evenly in the calculation of the averages. Starting from any arbitrary initial configuration the sequence of steps in the algorithm may be listed:

1. Randomly select an ion i.
2. Displace i randomly from $\mathbf{r}_i$ to $\mathbf{r}_i + \Delta\mathbf{r}$.
3. Compute the change in potential energy $\Delta\Phi$.
4. If $\Delta\Phi > 0$, compute $\exp(-\Delta\Phi/kT)$.
5. If $\exp(-\Delta\Phi/kT) < \text{RAND}$, return i to $\mathbf{r}_i$.
6. In either event accumulate current configurational properties.
7. Return to step 1.

The quantity $\Delta\mathbf{r}$ is a random displacement from a uniform distribution of probabilities within a preselected volume, usually cubic. RAND is a pseudo-random number from a rectangular distribution between 0 and 1. When the two conditions are met, namely, ergodicity, i.e., all configurations having the same energy are equally accessible, and steady state, i.e., there exists an equilibrium situation, then it can be proved that this procedure generates a petit (finite) canonical ensemble in the limit of a sufficiently large number of cycles.

MC calculations can also be carried out for systems in the isobaric–isothermal (N–p–T) ensemble.[21] The prescribed probability function in this case is given by

$$P(\mathbf{r}_1 \cdots \mathbf{r}_N, V) = \exp[-(\Phi_N + pV)/kT]Q_{NpT}^{-1} \tag{21}$$

where $Q(N$–p–$T)$ is the normalizing phase integral. The algorithm proceeds in a manner similar to the N–V–T system. At stage 2, in addition to the displacement of an ion, the volume would also be given a small random displacement by scaling all the interionic distances. The pseudo-Boltzmann factor for a phase point now becomes

$$V^N \exp[-(\Phi_q + pV_q/kT)]$$

and at stage 4 the ratio of new to old Boltzmann factors, i.e.,

$$\exp[-(\Delta\Phi + p\Delta V)/kT + N\ln(V'/V)]$$

in which V' is the provisional new volume, is compared to the random number. These N–p–T computations have so far only been applied to hard disks and simple liquid mixtures, but mention has been included because of the advantages to be gained in applying this approach to molten salts and their mixtures; comparisons with experiment are facilitated when the computer results can directly relate to the same external conditions.

Monte Carlo methods suffer from the disadvantage that the information regarding the time-dependent properties is lost in the replacement of a dynamic sequence by a statistical ensemble. On the other hand, there is the advantage that Monte Carlo methods, unlike molecular dynamics, can be rigorously applied to any kind of statistical mechanical ensemble corresponding to the external conditions of the experimental data. Systematic comparisons between the two methods for computing equilibrium properties of molten ionic salts have not been reported, but, on overall balance, the molecular dynamics method seems to be more popular mainly because time-dependent ·properties are also computed.

3.3. Electrostatic Summations

Both molecular dynamics and Monte Carlo calculations on ionic systems involve computation of the electrostatic potential energy of each ion in the configuration. In MD calculations the electrostatic force components acting on each ion must also be evaluated. Owing to the limitations discussed in the previous sections regarding the excessive demands on computational facilities, it is necessary to consider carefully the implications of the long-range Coulombic forces with respect to economics and practicalities.

Consider the appropriate system of a neutral assembly of N ions confined within a basic cubic cell of side L with infinite cubic periodicity. Let $\mathbf{n}$ represent a real space vector index over all the image cubes, i.e., extending from $-\infty$ to $+\infty$, and let the basic cube correspond to $\mathbf{n} = 0, 0, 0$. If the charge on an ion in the cell is denoted by q, then its electrostatic potential energy of interaction with the other ions is given by

$$\Phi_i^E(\mathbf{r}_1 \ldots \mathbf{r}_N) = q_i \sum_{j=1}^{N} \sum_{\mathbf{n}}' q_j \mathbf{r}_{ij}^{-1} \tag{22}$$

The prime on the second summation indicates that when $\mathbf{n} = 0, 0, 0$ the term corresponding to $j = i$ is omitted, and the dependence of Φ_i^E on the configuration $(\mathbf{r}_1 \cdots \mathbf{r}_N)$ will henceforth be implicit. The negative gradient of Φ_i^E at $\mathbf{r}_i$ is the electrostatic force vector acting on the ion, which may be written as

$$\mathbf{F}_i^E = -\nabla \mathbf{r}_i \Phi_i^E \tag{23}$$

where the operator $\nabla \mathbf{r}_i$ denotes $(d/dx_i, d/dy_i, d/dz_i)$.

The major difficulties in extending MD computations to ionic systems arise from the summation of very slowly convergent series implied by

Eqs. (22) and (23) for each ion in the basic cube. With the possible exception of some systems at very high temperatures or low charge density, it may be deduced immediately that the usual method of truncating the pair potential beyond contributions arising from within a prescribed cut-off distance would be impractical. Any criterion for establishing a truncation level in these computations must consider the error associated with Φ_i^E relative to kT. Whereas in the simple van der Waals liquids, for example, truncation after several atomic diameters leads to errors an order of magnitude less than kT, for molten ionic salts there is no indication that the errors would ever converge to a small fraction of kT using this mode of truncation. This is because the summation series in Eq. (22) is unconditionally nonconvergent. Even if the series were to converge satisfactorily in particular configurations, the magnitude of the Coulombic pair potential relative to kT implies that MC or MD computations on ionic systems using straightforward spherical truncation are inconceivable.

The problem at hand is analogous to the well-known Madelung problem of evaluating the electrostatic contributions to thermodynamic lattice energies of ionic crystals. One could regard the basic cube of the system as the unit cell of an infinite simple cubic ionic crystal in which the ions may be disordered. As a consequence of many extensive studies of the Madelung problem, the methods which have hitherto been applied to computer simulation of ionic systems are originally derived from crystal theory.

A very simple procedure involving only direct summation was described by Evjen.[22] The technique leads to a reasonably fast-converging series for the Madelung constant of simple cubic static ionic crystal lattices, and the simplicity of the approach is particularly appealing for adaptation to MD and MC computations. Regretfully, preliminary considerations indicate that this method leads to very serious discrepancies when applied to disordered systems of high effective charge density.[23,24] Very briefly, the most straightforward adaptation to the present context translates the origin of the coordinates for the entire system so that, for a particular ion i, $\mathbf{r}_i$ corresponds to the center of $\mathbf{n} = 0, 0, 0$ and has the vector components $L/2, L/2, L/2$. The electrostatic potential is then evaluated exactly according to Eq. (22) with truncation being invoked at a particular value of $\mathbf{n}$, usually $\mathbf{n} = 0, 0, 0$ in machine calculations. The magnitudes of the errors arising from truncation at $\mathbf{n} = 0, 0, 0$ for small disordered ionic systems leads one to suspect inadequacies in the method irrespective of the level of truncation. This has indeed been confirmed, and it has further been shown that the Evjen method leads to instability during computer simulation of ionic crystals also.[24] In systems of low effective charge density, such as the

Debye–Hückel model for dilute electrolytes, however, there is some evidence that the Evjen method is satisfactory.[25]

A rigorous and much more sophisticated procedure for enhancing the convergence of Madelung sums in ionic lattices is that due to Ewald.[26] Prior to its adaptation for computer simulation of molten ionic salts, the Ewald method was first employed by Barker in a Monte Carlo study of a hydrogenous plasma,[27] and later by Brush et al. in a similar study of a one-component plasma.[28] Although the Ewald method represents a purely mathematical transformation of the direct summation series of Eq. (22), it is instructive to consider the various contributions to the potential from a physical point of view, particularly in the derivation. First, the electrostatic potential at r_i is expressed as a sum of the potentials at that point due to $N - 1$ infinite cubic lattices plus the self-potential of the lattice images of ion i, as expressed by Eq. (22). Then each point charge lattice is neutralized by the superposition of a uniform background of opposite charge and, since we are dealing with a neutral system, this has no net physical effect. Thus, the contribution to the charge density at r_i of a single neutralized lattice may be considered. The following trick employed by Ewald transforms the slowly convergent direct summation for the potential due to a point-charge lattice in a uniform neutralizing background into an expression consisting of two provisionally convergent series.

Component 1 of the charge density is defined as an infinite cubic lattice of normalized Gaussian charge distributions opposite in sign to, and centered on, the point charges. Component 2 comprises the uniform neutralizing background superimposed on a cubic lattice of normalized Gaussian distributions which are opposite in sign to those in component 1. Clearly, when the two components are taken together, the respective Gaussian charge distributions cancel, as do the neutralizing backgrounds in the final summation over all the lattices corresponding to each ion in the basic MD cube.

If we denote the half-width of the Gaussian distribution by η, the contribution of component 1 to the charge density at a point r within a single cubic lattice may be written as

$$\varrho^1(\mathbf{r}) = \sum_n [\delta \mid \mathbf{r} - \mathbf{r}_j \mid - \eta^{-3}\pi^{-3/2} \exp(- \mid \mathbf{r} - \mathbf{r}_j \mid \eta^{-2})] \qquad (24)$$

where the point charges are represented as delta functions. The contribution to the charge density from the second component is

$$\varrho^2(\mathbf{r}) = \sum_n \eta^{-3}\pi^{-3/2} \exp(- \mid \mathbf{r} - \mathbf{r}_j \mid \eta^{-2}) - L^{-3} \qquad (25)$$

In these equations j is an index over each ion in the lattice and corresponds to a value of $\mathbf{n}$. Component 2 is then expanded into a Fourier series with the periodicity of the lattice

$$\varrho^2(\mathbf{r}) = L^{-3} \sum_{\mathbf{h}}{}' \exp(-\pi^2\eta^2\mathbf{h}^2 + 2\pi i\mathbf{h} \cdot \mathbf{r}_j) \tag{26}$$

in which $\mathbf{h}$ denotes the reciprocal lattice position vectors corresponding to the real cell vectors $\mathbf{n}$, and the prime on the summation again indicates that terms corresponding to $\mathbf{h} = 0, 0, 0$ are to be omitted. The electrostatic potentials are obtained after substituting for ϱ into the Poisson equation and integrating

$$\phi^1(\mathbf{r}) = \sum_{\mathbf{n}} [1 - \mathrm{erf}(|\,\mathbf{r} - \mathbf{r}_j\,|\,\eta^{-1})](\mathbf{r} - \mathbf{r}_j)^{-1} \pm (\pi\eta^2 L^{-3}) \tag{28}$$

and

$$\phi^2(\mathbf{r}) = \pi^{-1}L^{-3} \sum_{\mathbf{h}}{}' \mathbf{h} \exp(-\pi^2\eta^2\mathbf{h}^2 + 2\pi i\mathbf{h} \cdot \mathbf{r}_j) \tag{27}$$

[erf is the usual error function defined by $\mathrm{erf}(x) = (2/\sqrt{\pi})\int_0^x \exp(-u^2)du$]. The last term in Eq. (27) is a constant of integration and in fact cancels in the final summation [Eq. (34) below] and can thus be neglected. If we now consider the contribution to the potential at a point $\mathbf{r}_i$ arising from the image lattice of ion i itself, then an additional term is included by virtue of the absence of component 1. Using Eqs. (27) and (28), this is given as

$$\phi^0(\mathbf{r}_i) = \phi^1(\mathbf{r}_i) + \phi^2(\mathbf{r}_i) + \pi^{1/2}\eta^{-1} \tag{29}$$

The electrostatic potential energy of an ion in a configuration $\mathbf{r}_i \ldots \mathbf{r}_N$ may then be expressed as

$$\Phi_i^E(\mathbf{r}_i) = q_i \sum_{j=1}^{N} q_j[\phi^1(\mathbf{r}_{ij}) + \phi^2(\mathbf{r}_{ij})] - q_i^2\phi^0(i) \tag{30}$$

where j is an index over all the ions in the cell. Thus, for the total electrostatic potential energy of an ion, we have the three components

$$\Phi_i^{E_1} = \sum_{j=1}^{N} \sum_{\mathbf{n}} \frac{q_iq_j[1 - \mathrm{erf}(|\,r_{ij,\mathbf{n}}\,|\,\eta^{-1})]}{|\,r_{ij,\mathbf{n}}\,|} \tag{31}$$

$$\Phi_i^{E_2} = \frac{1}{\pi L} \sum_{j=1}^{N} \sum_{\mathbf{h}\neq 0} q_iq_j \frac{\exp(-\pi^2\,|\,\mathbf{h}\,|^2\,\eta^2)}{|\,\mathbf{h}\,|^2} \cos\left(\frac{2\pi}{L} \cdot \mathbf{h} \cdot r_{ij,\mathbf{n}}\right) \tag{32}$$

$$\Phi_i^{E_0} = -q_i^2\eta^{-1}\pi^{-1/2} \tag{33}$$

and the total electrostatic potential energy of the configuration is

$$\Phi^E = \frac{1}{2} \sum_{i=1}^{N} (\Phi_i^{E_0} + \Phi_i^{E_1} + \Phi_i^{E_2})$$ (34)

Differentiation of Eqs. (31) and (32) leads to the expressions for the two Ewald contributions to the components of the electrostatic force acting on an ion

$$F_x^{(1)} = q_i \sum_{j \neq i} q_j \sum_{\mathbf{n}} \frac{x_{ij,\mathbf{n}}}{r_{ij,\mathbf{n}}^3} \left\{ 1 - \mathrm{erf}\left(\frac{r_{ij,\mathbf{n}}}{\eta}\right) + \frac{2r_{ij,\mathbf{n}}}{\pi^{1/2}\eta} \exp\left[-\left(\frac{r_{ij,\mathbf{n}}}{\eta}\right)^2\right]\right\}$$ (35)

$$F_x^{(2)} = \frac{2q_i}{L^2} \sum_{j \neq i} q_j \sum_{\mathbf{h}} \frac{h_x}{|\mathbf{h}|^2} \exp\left(\frac{-\pi^2 |\mathbf{h}|^2}{\eta^2}\right) \sin\left(\frac{2\pi}{L} \mathbf{h} \cdot \mathbf{r}_{ij}\right)$$ (36)

so that the total force on an ion is

$$\mathbf{F}_i = \mathbf{F}^{(1)} + \mathbf{F}^{(2)}$$ (37)

Implementation of the Ewald method raises some difficulties regarding the computing requirements. Although, in theory, reasonable convergence of both the Ewald sums may be obtained for a range of values, considerable care is required in the selection of an optimum value which leads to acceptable truncation levels and which is computationally feasible. In molecular dynamics simulation, each time increment in the numerical integration requires approximately $N^2/2$ computations of the pairwise additive force components of Eqs. (35) and (36). Even when full use is made of tabulation techniques, calculations on systems with 216 ions are only marginally practicable on present computers.

In the earlier Monte Carlo studies,[29,30] preliminary computations were carried out to find that value of η which gives convergence to three figures in Φ^1 when the series [Eq. (31)] is truncated at $|\mathbf{r}_i - \mathbf{r}_{j,n}| - \mathbf{n} > L/2$. When the electrostatic energy is evaluated, it is found that $\Phi^{(1)} + \Phi^{(0)}$ contribute more than 99% of the total, and only those terms in the first series corresponding to $\mathbf{h} = 1$ need to be included at a modest level of accuracy. For a 216 ion system this may be achieved with a reciprocal half-width of $5.714L^{-1}$. In order to obtain at least three-figure convergence in $\Phi^{(2)}$, however, around $150L$ vectors are required with this choice of η. It is worth noting that it is possible to obtain a value of η for which the Fourier sum [Eq. (32)] may be completely neglected, such that a set of effective pair potentials are obtained which reproduce both the thermodynamic energy and the pressure of the system. For a 216-ion alkali halide system this turns out to be around $\alpha = 5$.

The situation regarding convergence of the electrostatic forces in ionic dynamics calculations is much less favorable. Here, truncation at small values of $\mathbf{h}$ can lead to errors of as much as 50% in the components of the force on an ion. Calculation of $F_x^{(1)}$ [Eq. (36)] involves sums over pairs of ions of terms which depend only on distance and therefore presents no more problems than ordinary short-range contributions to the force. Calculation of $F_x^{(2)}$ in Eq. (37), on the other hand, involves sums over pairs which depend both on distance and orientation. This problem may be generally overcome by tabulating, in advance, contributions from all possible distances and orientations on a three-dimensional grid.[31]

If one is to resort to tabulation on a three-dimensional grid, however, the best approach is simply to use the dimensionless function

$$f(\Delta x, \Delta y, \Delta z) = \sum_{j=1}^{\infty} \frac{(X_i - X_j)}{r^3} \tag{38}$$

from which the component of the force on an ion in an infinite cubic array of point charges may be obtained. Unlike the expression for the potential, the summation in Eq. (38) is convergent. The function $f(\Delta x, \Delta y, \Delta z)$ may be parametrized in the form of a simple three-dimensional polynomial in inverse powers of Δx, Δy, and Δz, the displacement of the ion from the nearest lattice point. Only the forces in molecular dynamics calculations need to be rigorously computed. The Evjen method may then be used to compute the potential energy to within three or four figures since the errors involved in truncation cannot influence the ionic trajectories.

4. CALCULATION OF PHYSICAL PROPERTIES

4.1. Equilibrium Properties

Thermodynamic properties may be calculated from either molecular dynamics or Monte Carlo data. The configurational averages from which these properties are obtained are accumulated during the course of the simulation, discounting the atypical configurations from the initial equilibration period. Although the MD and MC techniques yield essentially exact results in the long run, residual errors inevitably arise due to the finite duration of either the time integration in the MD method or the number of trial displacements in the MC method. The magnitudes of the statistical errors can be assessed by taking the standard deviations of independent subaverages.

The internal energy is calculated by adding the translational kinetic energy to the mean potential energy which is obtained as the sum over pair potentials

$$U = \tfrac{3}{2}NkT + \left\langle \sum_{i=1}^{N} \sum_{j>i}^{N} \phi_{ij}(r) \right\rangle \tag{39}$$

and the pressure is obtained from the virial theorem[32]

$$\frac{pV}{NkT} = 1 - \left\langle \sum_{i=1}^{N} \sum_{j>i}^{N} r\, \frac{d\phi_{ij}(r)}{dr} \right\rangle \tag{40}$$

(In these expressions, angular brackets are used to denote either a time average or an ensemble average.) The average temperature of a molecular dynamic simulation is given by the mean kinetic energy

$$T = \frac{1}{3Nk} \left\langle \sum_{i=1}^{N} m_i \mathbf{v}_i^2 \right\rangle \tag{41}$$

Second-order thermodynamic properties, such as the derivatives of internal energy or pressure with respect to either volume or temperature, may be calculated using two different approaches. When sufficient computing facilities are available, simulations are carried out for a number of V–T points to obtain the $U(V, T)$ and $p(V, T)$ equations of state. The internal pressure $(\partial U/\partial V)_T$, isochoric heat capacity $(\partial U/\partial T)_V$, isothermal compressibility $-(\partial V/\partial p)_T/V$, and the thermal pressure coefficient $(\partial p/\partial T)_V$ are subsequently all determined by numerical differentiation.

When it is desirable to compute these derived properties from a single V–T simulation, statistical mechanical relationships for fluctuations in dependent state variables may be utilized.[33] The isochoric heat capacity (C_V), for example, is simply related to fluctuations around the mean total potential energy and is computed from

$$C_V = \frac{\langle \Phi^2 \rangle - \langle \Phi \rangle^2}{kT^2} + \tfrac{3}{2}Nk \tag{42}$$

Fluctuations in the total virial (Ψ) lead to a statistical mechanical expression for the isothermal compressibility $(\varkappa_T)$

$$\varkappa_T = p^{-1} + \frac{\langle W \rangle}{kT} - \frac{(\langle \psi^2 \rangle - \langle \psi \rangle^2)}{kTV} \tag{43}$$

where

$$\psi = \sum_{i=1}^{N} \sum_{j>i}^{N} r \frac{d\phi_{ij}(r)}{dr} \tag{44}$$

and

$$W = \frac{1}{9} \sum_{i=1}^{N} \sum_{j>i}^{N} \left(\frac{rd\phi_{ij}(r)}{dr} + \frac{r^2 d^2\phi_{ij}(r)}{dr^2} \right) \tag{45}$$

The covariance between the potential energy and the virial is related to the thermal pressure coefficient (γ_V)

$$\gamma_V = \frac{Nk}{V} + \frac{\langle \Phi\psi \rangle - \langle \Phi \rangle \langle \psi \rangle}{kT^2 V} \tag{46}$$

The internal pressure (p_i) is simply related to the thermal pressure coefficient

$$p_i = T\gamma_V + p \tag{47}$$

Isobaric properties can be obtained similarly from well-known thermodynamic relationships. For example, the thermal expansivity [$\alpha_p = (\partial V/\partial T)_p/V$] is given by

$$\alpha_p = \frac{1}{T} \left[1 - V\left(\frac{\partial H}{\partial p}\right)_T \right] \tag{48}$$

where the enthalpy H is given by

$$H = U + pV \tag{49}$$

and for the heat capacity at constant pressure [$(\partial H/\partial T)_p$] we have

$$C_p = C_V + VT\alpha p\gamma_V \tag{50}$$

The liquid-crystal freezing transition is an especially important property of molten salts, and it can be calculated from MD or MC data. In order to locate a coexistence point between two competing phases, the free-energy difference must be calculated as a function of the external variables, usually temperature and density with pressure constant. The transition temperature corresponds to a zero free energy difference. The entropic contribution to the free energy of a dynamic system, however, is not rigorously related to any configurational average and therefore is not quite as easy to compute as the energetic contribution. By performing a range of simulations along isotherms, isochores, or isobars, and utilizing the thermodynamic equations for entropy changes,

$$\Delta S_T = \frac{1}{T} \left[\int_{V_1}^{V_2} p \, dV + \Delta U \right] \tag{51}$$

$$\Delta S_V = \int_{T_1}^{T_2} (C_V/T) \, dT \tag{52}$$

$$\Delta S_p = \int_{T_1}^{T_2} (C_p/T) \, dT \tag{53}$$

we can assess the entropies and therefore free energies of gaseous and crystalline states. The problem is more difficult for the liquid phase because it is bounded by discontinuities in state functions at the crystal–liquid and liquid–vapor transitions. A number of methods have been proposed for computing the entropy or free energy in the liquid state.

Hoover and co-workers have developed two exact methods for the free energy. In a study of the hard-sphere solid–fluid transition,[34] an artificial thermodynamically reversible path was devised to link the two phases. The crystalline phase is stabilized by an external periodic field, whereby each particle is confined to its cell, such that at low densities the only difference in entropy between the solid and fluid is Nk, the communal entropy. The configurational entropy difference is then given by Eq. (51). An alternative approach[35] involves the introduction of a coupling parameter. The Helmoltz chemical potential $(\partial A/\partial N)_{V,T}$ may also be written as $A(N, V, T) - A(N - 1, V, T)$ in the limit of large N. The problem reduces, therefore, to constructing a reversible path connecting the two states N, V, T and $N - 1, V, T$ in order to carry out the integral over the work required to introduce or remove one particle. In the computer model a labeled particle has its pair potentials with the other $N - 1$ particles multiplied by the coupling parameter λ. When $\lambda = 1$ the particle is entirely present, and when $\lambda = 0$ the particle is absent. Computations are carried out for a range of λ values which permits the chemical potential to be computed from

$$(\partial A/\partial N)_{V,T} = \int_0^1 \langle \Phi_i \rangle \, d\lambda \tag{54}$$

Neither of these two methods has yet been applied to the study of molten salts, but their adaptation to the freezing transitions of ionic liquids would appear to be practicable, although perhaps computationally expensive.

No one has yet proposed a generally applicable method for the determination of free energy from a single V–T simulation, but there are approximations which are useful in certain circumstances. In calculating the melting transition of KCl,[30] use has been made of the approximate formulas arising in the rigorous treatment of the cell theory of liquids due to Kirkwood.[36] This approximation has been elaborated upon in the literature and discussed in more depth with reference to MC calculations.[37,38]

4.2. Transport Coefficients

In recent years developments in the field of nonequilibrium statistical mechanics have resulted in considerable progress in the computation and interpretation of transport properties of liquids.[39] These advancements are based upon the so-called "fluctuation dissipation theorem" which relates linear transport coefficients to time correlation functions for locally fluctuating dynamic variables in liquids at equilibrium. In a molten ionic salt at equilibrium, for example, the thermal actions of the ions alone give rise to local charge fluctuations, the dissipation of which is related to the coefficient for Ohmic conductance in the presence of an external electric field.

Consider first the self-diffusion of an ion in a molten salt. The corresponding transport coefficient may be defined in terms of the flux of a particular dynamic variable, in this case the position of the ion. Thus,

$$D = \frac{1}{6\tau} \langle [\mathbf{r}_i(t) - \mathbf{r}_i(t + \tau)]^2 \rangle \tag{55}$$

where τ is a time increment large compared to the relaxation time for translational diffusion and the angular brackets denote an ensemble average of time origins. Equation (55) may alternatively be expressed as

$$D = \frac{1}{6} \frac{d}{d\tau} \langle [\mathbf{r}_i(t) - \mathbf{r}_i(t + \tau)]^2 \rangle \tag{56}$$

where τ is again large. Equation (56) is generally termed as an Einstein expression for the transport coefficient. Now, if the displacement of the ion in time τ is replaced by an integral over its velocity during that interval, we have

$$\Delta \mathbf{r}_i = \int_0^\tau \mathbf{v}_i(t) \, dt \tag{57}$$

and substituting into Eq. (55), we have

$$D = \frac{1}{6\tau} \left\langle \left[\int_0^\tau \mathbf{v}_i(t) \, dt \right]^2 \right\rangle \tag{58}$$

or, since at equilibrium the integral is independent of the time origin,

$$D = \frac{1}{3} \int_0^\tau \left(1 - \frac{t}{\tau}\right) \langle \mathbf{v}_i(t) \cdot \mathbf{v}_i(t + \tau) \rangle \, d\tau \tag{59}$$

and since $\tau \to \infty$ can be reduced to

$$D = \frac{1}{3} \int_0^\alpha \langle \mathbf{v}_i(0) \cdot \mathbf{v}_i(t) \rangle \, dt \tag{60}$$

where each origin of time is conveniently put at $t = 0$. The time-dependent ensemble average in Eq. (60) is the velocity autocorrelation function. At $t = 0$ it has the value of the mean-squared velocity $(3kT/m_i)$, and at long times it approaches zero as the correlations are destroyed by random fluctuations in $\mathbf{v}_i$.

Thus, so far, there are three possible methods for computing the self-diffusion coefficient from molecular dynamics. These are (1) from the actual mean-squared displacement over a sufficiently large time increment [Eq. (55)], (2) from the long-time slope of the mean-squared displacement against time curve [Eq. (56)], and (3) from the area under the velocity autocorrelation function [Eq. (60)]. A recent derivation[40] has further shown that D is also related to the autocorrelation function of the force (or alternatively the acceleration)

$$D = \frac{-1}{6m^2} \int_0^\alpha \mathbf{F}_i(0) \cdot \mathbf{F}_i(t) t^2 \, dt \tag{61}$$

It is important to note that additional information regarding the process of self-diffusion is acquired in the progressive representation of D by methods (1)–(4) above. The curves representing D by Eqs. (56), (60), and (67) for K^+ ions in molten KCl are shown in Figs. 2, 3, and 4. The correlation functions in Figs. 3 and 4 have been normalized to unity at $t = 0$ by dividing by the mean-squared value.

Linear coefficients for bulk transport coefficients, such as the electrical and thermal conductivities or the shear and bulk viscosities, are also related to time correlation functions of appropriate dynamic variables which may be obtained from MD data. In the case of electrical conductivity, the dynamic variable in question is the electrical charge of a number of ions in a subvolume of the melt, and the flux of charge $\mathbf{J}$ is

$$\mathbf{J} = \sum_i e_i \mathbf{v}_i \tag{62}$$

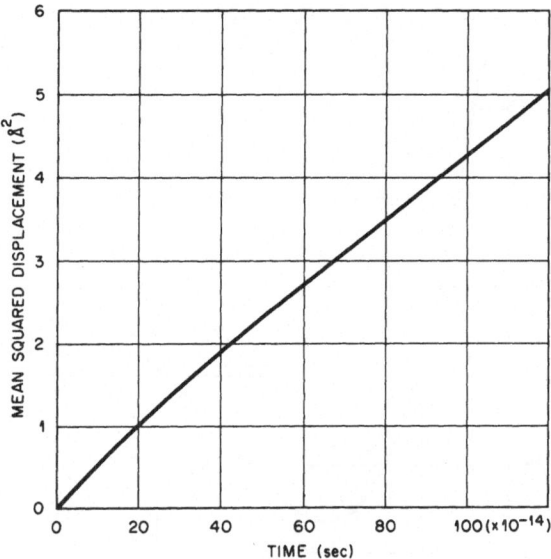

Fig. 2. Mean-squared displacement of K^+ ions in molten
KCl at T_f from MD data.

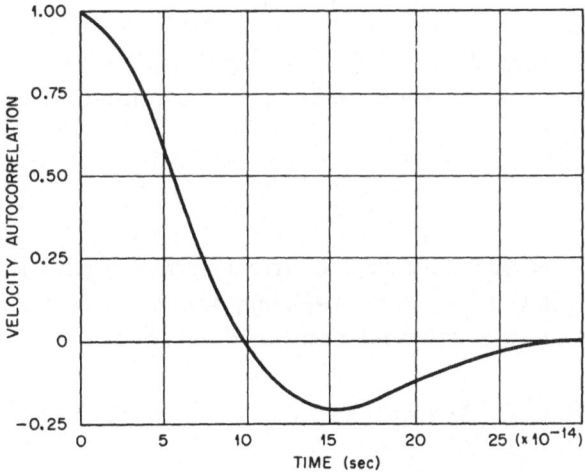

Fig. 3. Normalized velocity autocorrelation function for
K^+ ions in molten KCl at T_f from MD data.

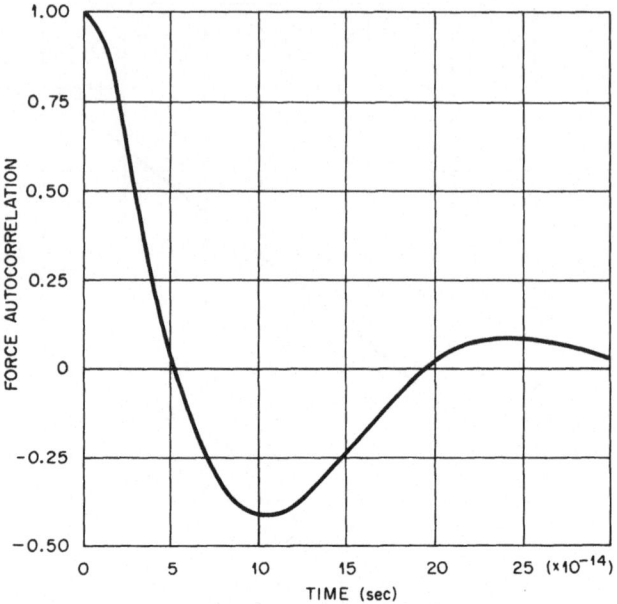

Fig. 4. Normalized force autocorrelation function for K$^+$
ions in molten KCl at T_f from MD data.

where e is the electronic charge. At zero frequency, the electrical conductivity σ of an isotropic system may be defined by

$$\mathbf{J} = \sigma\mathbf{E} \tag{63}$$

where $\mathbf{E}$ is the applied electric field. The expression for the electrical conductivity in terms of the autocorrelation of the current in the equilibrium
situation is then

$$\sigma = \frac{1}{kT} \int_0^\alpha \langle \mathbf{J}(0) \cdot \mathbf{J}(t) \rangle \, dt \tag{64}$$

Since the total charge is additive, the current–current N-particle autocorrelation function can be resolved into two components, i.e., into an autocorrelation function and a cross-correlation function. Thus,

$$\langle \mathbf{J}(0) \cdot \mathbf{J}(t) \rangle = \left\langle \sum_{i=1}^N e_i\mathbf{v}_i(0) \cdot e_i\mathbf{v}_i(t) \right\rangle + \left\langle \sum_{\substack{i,j \\ j \neq i}}^N e_i\mathbf{v}_i(0) \cdot e_j\mathbf{v}_j(t) \right\rangle \tag{65}$$

which, when compared with Eq. (60) for D leads to an interesting observation regarding deviations from the Nernst–Einstein relationship in ionic

liquids. This may be written

$$D^+/kT + D^-/kT = (\sigma/Ne^2)(1 - \alpha) \qquad (66)$$

whereupon the deviation parameter α is related to the cross-correlations by

$$\alpha\sigma = \frac{1}{kT} \int_0^\alpha \left\langle \sum_{\substack{i,j \\ j\neq i}}^N e_i \mathbf{v}_i(0) \cdot e_j \mathbf{v}_j(t) \right\rangle dt \qquad (67)$$

In a manner analogous to Eq. (61) for D, σ can also be computed from the autocorrelation of the force[40]

$$\sigma = -\frac{1}{6m^2kT} \int_0^\alpha \left\langle \sum_{i,j}^N e_i \mathbf{F}_i(0) \cdot e_i \mathbf{F}_j(t) \right\rangle t^2 \, dt \qquad (68)$$

The force correlation may also be expressed as a sum of auto- and cross-terms and, moreover, may be resolved into correlations arising from different components of the interionic pair potential. When electromigration data for D^+, D^- and σ is expressed in terms of phenomenological friction coefficients, the use of Eqs. (61) and (68) with MD data should provide a detailed quantitative insight into both interionic frictional forces and deviations from the Nernst–Einstein relationship.

Linear coefficients for the transport of energy and momentum may also be calculated in terms of correlation functions of N-ion dynamic variables which include auto- and cross-components. For the thermal conductivity, the fluctuating dynamic variable is the energy displacement, so that the heat current entering the correlation function is given by

$$\mathbf{K}(t) = V \sum_{i=1}^N (\Phi_i + \tfrac{1}{2}m_i \mathbf{v}_i^2)\mathbf{r}_i \qquad (69)$$

and the coefficient of thermal conductivity $\varkappa$ is

$$\varkappa = \frac{1}{3VkT^2} \int_0^\alpha \langle \mathbf{K}(0) \cdot \mathbf{K}(t) \rangle \, dt \qquad (70)$$

The shear and bulk viscosity coefficients are calculated from the diagonal and off-diagonal elements of the microscopic stress tensor $\mathbf{G}$, respectively. This is

$$G_{xy} = m_i \sum_{i=1}^N v_i^x v_j^y + \sum_{i=1}^N F_{ij}^x (Y_i - Y_j) \qquad (71)$$

The shear viscosity (η) is given by

$$\eta = \frac{1}{VkT} \int_0^\alpha \langle G_{xy}(0)G_{xy}(t)\rangle \, dt \tag{72}$$

and the bulk viscosity (ν) is

$$\nu = \frac{1}{VkT} \int_0^\alpha \langle G_{xx}(0)G_{xx}(t)\rangle \, dt \tag{73}$$

Although the equations have not been included, corresponding mean-squared displacement formulas could equally be employed to represent and compute σ, $\varkappa$, η, and ν in a manner analogous to Eqs. (55) and (56) for D.

Very little progress has in fact been made in the actual computation of transport coefficients other than molecular diffusion. The quantity D is easier to calculate because one can immediately average over N time origins for each molecule (or ion) in the system. Since approximately 10^5 time origins are required for 2 or 3% accuracy, the equivalent computation of bulk transport coefficients is computationally a more expensive undertaking. In practice, the requisite dynamic variables are recorded, usually on magnetic tape, as the MD simulation proceeds, and separate programs are used to compute the transport coefficients. Such computations for hard spheres and simple liquids have been reported,[41] but no one has yet reported any computations of transport data for ionic liquids apart from diffusion coefficients.

4.3. Spectroscopic Properties

Certain absorption and scattering properties of liquids, such as IR and Raman intensity profiles, dielectric relaxation, NMR spin-lattice relaxation times, and other related phenomena may be computed or quantitatively interpreted using MD data. By analyzing the response of statistical mechanical systems to weak time-dependent external fields, the experimental power spectra are related to time-correlation functions of dynamic variables by Fourier transformation. These may be computed from MD data and then Fourier-inverted to corresponding power spectra for comparison with experiment, or vice versa.

Gordon[42] has shown that the analysis of spectra in the Heisenberg representation of quantum mechanics leads to an expression for IR absorption profiles which is a Fourier transformation of the autocorrelation iunction of the dipole moment operator $\mathbf{M}$ for N absorbing molecules or fons. The frequency-dependent intensity is

$$I(\omega) \propto \frac{1}{2\pi} \int_{-\infty}^{\infty} e^{-i\omega t} \langle \mathbf{M}(0) \cdot \mathbf{M}(t) \rangle \, dt \tag{74}$$

If $\boldsymbol{\mu}_i$ is the molecular dipole moment and $\mathbf{M} = \sum_{i=1}^{N} \boldsymbol{\mu}_i$, and we take only the real part of the transform, the positive time-correlation reduces to

$$I(\omega) \propto \frac{1}{2\pi} \int_{0}^{\infty} \sum_{i,j} \boldsymbol{\mu}_i(0) \cdot \boldsymbol{\mu}_j(t) \cos(\omega t) \, dt \tag{75}$$

In the liquid state cross-correlations of the type $\langle \boldsymbol{\mu}_i(0) \cdot \boldsymbol{\mu}_j(t) \rangle_{j \neq i}$ may become important. Gordon has further shown[43] that for linear or, more generally, symmetric top molecules, the absorption intensity at a displacement ω' from the band origin may be expressed as

$$I^{\mathrm{a}}(\omega') \propto \frac{1}{2\pi} \int_{0}^{\infty} \langle \cos \theta(0) \cos \theta(t) \rangle_i \cos(\omega t) \, dt \tag{76}$$

where θ is the orientational displacement. The correlation functions in Eqs. (75) and (76) can be computed from MD data and used, *inter alia*, to interpret the experimental band shapes in terms of the intermolecular forces and torques.

In Raman spectroscopy the intensity of scattered light is related to correlation functions of the polarizability tensor.[43] The differential scattering cross section for a frequency range $d\omega$ and a solid angle element $d\Omega$ is

$$\frac{d^2\sigma}{d\Omega \, d\omega} = \lambda^{-4} \sum |\langle i\boldsymbol{\epsilon}_i \mathbf{a} \boldsymbol{\epsilon}_s \rangle|^2 \varrho_i \delta(\omega - \omega_i) \tag{77}$$

where the wavelength of the scattered light is $2\pi\lambda$, $\boldsymbol{\epsilon}_i$ and $\boldsymbol{\epsilon}_s$ are unit vectors along directions of incident and scattered radiation, and $\mathbf{a}$ is the polarizability tensor of a group of interacting molecules or ions. This may be converted into the form of a correlation function

$$I^s(\omega) = \frac{1}{2\pi} \int_{-\infty}^{\infty} e^{i\omega t} \langle (\boldsymbol{\epsilon}_i \mathbf{a}(0) \boldsymbol{\epsilon}_s) \boldsymbol{\epsilon}_i \mathbf{a}(t) \boldsymbol{\epsilon}_s \rangle \, dt \tag{78}$$

where $\omega = \omega_i - \omega_s$. Experimentally, where the wave vectors of the incident and scattered light are perpendicular, only a depolarized component of scattered light appears; but when they are parallel, both polarized and depolarized components are obtained. Thus, the total polarized intensity may be obtained from

$$I^{\text{pol}}(\omega) = I_\perp(\omega) - \tfrac{4}{3}I_\parallel(\omega) \tag{79}$$

The polarizability tensor may be diagonalized for nonoptically active scatterers. The average of the diagonal components gives the trace $\bar{\alpha}$

$$\bar{\alpha} = \tfrac{1}{3}(\alpha_{xx} + \alpha_{yy} + \alpha_{zz}) \tag{80}$$

and its anisotropy is defined by

$$\beta_{xy} = \alpha_{xy} - \bar{\alpha} \tag{81}$$

For polarized and depolarized scattering intensities, respectively, we have therefore

$$I^{\text{pol}}(\omega) \propto \frac{1}{2\pi} \int_0^\infty \langle \bar{\alpha}(0)\bar{\alpha}(t) \rangle \cos(\omega t)\, dt \tag{82}$$

and

$$I^{\text{depol}}(\omega) \propto \frac{1}{2\pi} \int_0^\infty \langle \boldsymbol{\beta}(0)\boldsymbol{\beta}(t) \rangle \cos(\omega t)\, dt \tag{83}$$

The actual computation of Raman bandshapes from MD data requires expressions for $\bar{\alpha}$ and $\boldsymbol{\beta}$ in terms of the intermolecular forces plus the field of the light. While this is generally a complex problem, it can be shown that for symmetric top molecules expression (82) for the polarized intensity reduces to[43]

$$I^{\text{pol}}(\omega) \propto \frac{1}{2\pi} \int_0^\infty \langle [\cos^2\theta(0) - \tfrac{1}{3}][\cos^2\theta(t) - \tfrac{1}{3}] \rangle \cos(\omega t)\, dt \tag{84}$$

where θ is the orientational displacement.

Only a few calculations have so far been reported for angular correlation functions. In Fig. 5 some preliminary results of angular correlation functions from MD data on molten KCN are shown.[44] The $P_1[\cos\theta]$ and $P_2[\cos\theta]$ correlation functions are those corresponding to Eqs. (76) and (84). The angular momentum autocorrelation function and the torque autocorrelation function are related to the coefficient of rotational diffusion by expressions analogous to Eqs. (60) and (61) for the translational diffusion coefficient. The area under the angular momentum autocorrelation function corresponds to the relaxation time measured in NMR spin–lattice relaxation experiments.[43]

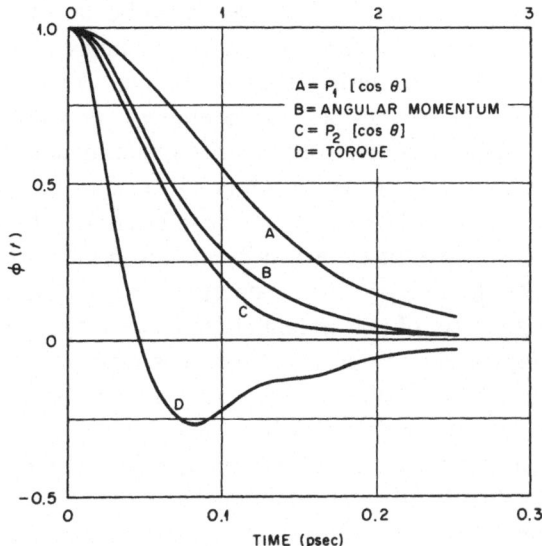

Fig. 5. Normalized angular correlation functions $\phi(t)$ for CN^- ions in molten KCN at 1000°K from MD data (J. H. R. Clarke, S. Miller, and L. V. Woodcock).

5. APPLICATIONS

5.1. Studies of Interionic Forces

Computer simulation may be considered as an experimental measurement of the properties of the mathematical model, defined by its interionic pair potentials, ionic masses, and thermodynamic state. Comparisons between computed properties and experimental data therefore lead to deductions regarding the true nature of the intermolecular forces. The general criterion for the adequacy of a mathematical model for a particular liquid is its ability to reproduce the various thermodynamic, transport, and spectroscopic properties. In making these comparisons, account must be carefully taken of the experimental uncertainties and the limitations of the simulation techniques discussed in Sec. 3.

Each individual liquid state property will, in general, have a characteristic sensitivity to particular components of the interionic potentials. For example, the density and compressibility depend largely upon the short-range repulsive parts, whereas the internal energy, on the other hand, is determined more by the depth of the potential energy surfaces. Both thermodynamic and transport properties may be largely insensitive to polarization forces in ionic media, but these are manifested in spectroscopic properties such as light-scattering-intensity profiles. Thus, the magnitude and directions of discrepancies between computed and experimental quantities can lead to definite conclusions regarding interionic potentials.

When quantitative agreement is observed, within the bounds of both experimental and statistical errors, it may be inferred that the pair potentials employed are satisfactory effective pair potentials in the particular thermodynamic state for the one or more properties for which agreement is obtained. When a given set of pair potentials can account for many properties over a wide temperature or density range, it is probable that these are a representation of the actual interactions and, moreover, are also the isolated pair potentials. This is the situation here, mainly as a consequence of the predominance of the electrostatic term in the pair potential for molten ionic salts. It is this feature that enables the molten alkali halides to be studied numerically as a fundamentally important class of liquids from a theoretical standpoint.

A series of Monte Carlo computations have been carried out for molten KCl[30] and a number of other alkali halides[45] based upon the Tosi–Fumi model interionic potentials.[13] It has since been pointed out that the criterion used in terminating the second Ewald component of the potential [Eq. (32)] may lead to serious truncation errors.[31] More recent MC computations, however,[46] indicate that the errors in the electrostatic energy and virial are less than 1%. These preliminary data would appear to be within the bounds of experimental uncertainties anyhow, and comparisons with experiment are of interest.

In Table II the calculated data on internal energies and densities for nine alkali halides are compared with experiment. At 298°K, as has already been mentioned in Sec. 2.2, the Tosi–Fumi potentials adequately represent the density and its derivatives with respect to temperature and pressure and, as shown in Table II, also for the lattice energy U to within 1%, with the exception of Li salts and CsF where the discrepancy in U is as much as 5%.

When one compares absolute values, the overall agreement for both the high-temperature crystal and the molten salt is good. A notable feature of the discrepancies (given in brackets in Table II) for the liquid state, when compared with those obtained for the crystal at its melting point, is that they remain more or less constant on fusion (KF is an exception). Thus, the calculated heats of fusion agree well with the experimental values. There appears to be a systematic discrepancy for the computed heat capacities of

TABLE II. Comparisons of Computed and Experimental Thermodynamic Properties for Some Alkali Halide Melts

| | T_f(°K) | Molar internal energy (KJ) | | | | | | Molar volume (cm³) | | | |
| | | Crystal (298°K) | | Crystal (T_f) | | Liquid (T_f) | | Crystal (T_f) | | Liquid (T_f) | |
		Expt.[a]	Calc.[b]	Expt.[a,c]	Calc.	Expt.[a,c,d]	Calc.	Expt.[e]	Calc.	Expt.[e]	Calc.
LiCl	883	−832.2	−828.4(+3.8)	−800.4	−795.6(+4.8)	−780.5	−781.1(−0.6)	22.3	22.3(+0.0)	28.2	29.1(+0.9)
LiI	742	−743.5	−704.6(+38.9)	[g]	[g]	−702.7	−674.9(+27.8)	[g]	[g]	43.1	47.5(+4.4)
NaCl	1073	−764.0	−762.3(+1.7)	−719.7	−719.3(+0.4)	−691.7	−691.9(−0.2)	30.1	31.2(+1.1)	37.6	39.1(+1.5)
KF	1130	−794.1	−798.7(−4.6)	−748.0	−738.2(+9.8)	−719.8	−731.2(−11.4)	25.9	30.6(+4.7)	30.4	35.1(+4.7)
KCl	1045	−693.7	−693.3(+0.4)	−651.2	−654.3(−3.1)	−625.0	−628.0(−3.0)	40.3	41.8(+1.5)	48.8	50.2(+1.4)
KBr	1007	−663.2	−664.4(−1.2)	−622.6	−622.3(+0.3)	−596.1	−595.8(+0.3)	47.9	49.5(+1.6)	55.9	60.9(+5.0)
KI	955	−627.2	−623.4(+3.8)	−589.2	−586.3(+2.9)	−565.2	−562.4(+2.8)	58.6	59.7(+1.1)	67.9	71.1(+3.2)
RbCl	995	− 66.5	−673.6(−7.1)	−627.7	−633.7(−6.0)	−604.0	−609.4(−5.4)	47.2	48.0(+0.8)	53.9	56.6(+2.7)
CsF	976	−721.7	−746.0(−24.3)	−684.8	−709.8(−25.0)	−662.1	−676.8(−14.7)	33.6	36.0(+2.4)	41.3	43.9(+2.6)

[a] M. P. Tosi, J. Phys. Chem. Solids 24:965 (1963).
[b] M. P. Tosi and F. G. Fumi, J. Phys. Chem. Solids 25:45 (1964) (Table 2, p. 49, 1st set of data).
[c] K. K. Kelley, U.S. Bur. Mines Bull. 584 (1960) (values of C_p for LiCl, LiI, RbCl, and CsF crystals were estimated by Kelley).
[d] A. S. Dworkin and M. A. Bredig, J. Phys. Chem. 64:269 (1960) (value for ΔH_f for KF as given by Kelley).
[e] N.S.R.D.S.–NBS 15, Molten Salts, Vol. 1, ed. G. J. Janz (1968).
[g] The computer model for LiI is found to be unstable in the crystalline phase at this temperature, and spontaneous fusion occurs.

the crystalline states somewhere between 298°K and the normal melting temperatures.

Also given in Table II are comparisons of the computed and experimental molar volumes. The MC computations are actually performed in the N–V–T ensemble, and the zero-pressure results are obtained by interpolation between two runs for each salt. Considering the large experimental uncertainties, the overall agreement for the absolute molar volumes is satisfactory. In this case, however, the discrepancies appear to be as much due to the changes on fusion as to any contributions from differences in crystalline state thermal expansivities. The computed molar volumes for NaCl, KCl, KBr, KI, and RbCl are probably within the region of uncertainty associated with the experimental values, which may be as high as 5%. An intriguing feature of the comparisons as a whole is that the volume discrepancies are all in a direction which might be expected to arise from the explicit neglect of polarization potentials in the computer model. Discrepancies in the internal energies, on the other hand, are, with the exception of LiI, in the opposite direction. Until more accurate computational and experimental data are available, however, it would seem unjustifiable to draw any conclusions from the comparatively small differences between calculated and experimental values.

On balance, the evidence from Table II suggests that, within the bounds of experimental errors, for first-order thermodynamic properties, polarization potentials are not important. Nevertheless, it is of interest to have at least an approximate estimate of contributions from ion-induced dipole forces. Although no one has yet reported any computer simulation results with the explicit incorporation of a provision for ion polarization in the model, these estimates have been obtained using computer-generated configurations for the Tosi–Fumi potentials. If the dipole moment induced in an ion by the internal electrostatic field of the melt is given by

$$\boldsymbol{\mu}_i = \alpha_i \mathbf{F}_i^E \tag{85}$$

the corresponding contribution to the total potential energy of a configuration may be written as

$$\Phi^P = \frac{1}{2} \sum_{i=1}^{N} \sum_{\substack{j=1 \\ j \neq i}}^{N} \frac{Z_i e \alpha_j \mathbf{F}_j^E \cdot (\mathbf{r}_i - \mathbf{r}_j)}{r_{ij}^3} \tag{86}$$

Configurations (200 in each case for 216 ions) of LiCl, NaCl, and KCl have been utilized to compute averages of Φ^P from Eq. (86). The results are given in Table III. It can be observed that the contributions of Φ^P to

TABLE III. Polarization Energies of Molten Alkali Chlorides Estimated from Computer-Generated Configurations for the Tosi–Fumi Potentials Using Equation (86)

	Φ^P (KJ-mole^{-1})	$100 \times \Phi^P/U$
LiCl	−40.2	4.9%
NaCl	−19.8	2.6%
KCl	−14.4	2.2%

U for the alkali metal chlorides are small, even for LiCl. The development of sophisticated effective pair potentials incorporating polarization potentials, of the type derived by Stillinger, for example,[4] would not appear to be warranted in the case of simple alkali halides.

Larsen *et al.* have carried out a Monte Carlo computation of the excess thermodynamic properties of mixing NaCl with KCl using modified Tosi–Fumi potentials.[47] The excess properties in this case are small and difficult to calculate accurately relative to the absolute values. Their results (Table IV), however, are promising and indicate that the extension of this type of computation to other mixtures should yield valuable information regarding interionic potentials and possibly also polarization.

Self-diffusion and bulk transport data could, in principle, be used to obtain information on the intermolecular potentials. The situation at present, however, is that the experimental data available, even for simple ionic melts, is not sufficiently accurate to lead to meaningful comparisons. In Table V some recently obtained[48] accurate self-diffusion data on the Tosi–Fumi model for NaCl is compared with experimental data. While the 10% or so discrepancy is well outside the bounds of the calculation errors, it is probably around the same magnitude as the experimental uncertainty.

TABLE IV. Some Results of a Monte Carlo Calculation by Larsen *et al.* for the Excess Thermodynamics Properties of Mixing NaCl and KCl (1:1) at 1 atm and 1083°K (Reference 47)

	ΔV (cm^3-mole^{-1})	ΔU (KJ-mole^{-1})	ΔS (KJ-mole^{-1})	ΔG (KJ-mole^{-1})
MC	1.2 ± 0.4	3.3 ± 0.4	5.8	−3.3
Expt.	0.24 ± 0.08	−0.54 ± 0.04	4.32	−6.53

TABLE V. Self-Diffusion Coefficients for Molten Sodium Chloride Calculated from MD Data by J. W. Lewis[a] Compared with Experimental Values of Bockris et al.[b,c]

T	ϱ(MD)	MD		Expt.	
		D_+	D_-	D_+	D_-
1100	1.46187	8.60	7.89	8.15	6.45
1200	1.41463	10.27	9.47	10.8	8.70
1300	1.37767	12.23	11.28	13.6	10.9

[a] J. W. Lewis (to be published).
[b] J. O'M. Bockris et al., J. Phys. Chem. **69**:1629 (1965).
[c] D in units of 10^{-5} cm^2-sec^{-1}. ϱ_{MD} is density in g-cm^{-3}.

The general conclusions regarding the adequacy of the Tosi–Fumi potentials, based upon the available data, is that where agreement is obtained in the crystal at 298°K, agreement is also obtained for the melt within the bounds of experimental errors for the various properties considered.

5.2. Microstructure and Mechanisms

The equilibrium molten salt microstructure is a fundamental quantity which is required for both qualitative and quantitative interpretation of thermodynamic behavior. Mechanisms for time-dependent phenomena such as electromigration or spectral bandshapes are also intimately related to the equilibrium microstructure. Prior to the advent of computer simulation, the conventional experimental probes, x-ray and neutron diffraction, were the only appreciable source of quantitative microscopic structural data on molten salts.[49] Until very recently, the structural information which was forthcoming from these sources has been comparatively meager and very often ambiguous in its detailed interpretation. Current developments in neutron scattering techniques, however, using isotopic samples, may enhance this experimental approach for molten salts in the future.

It is a comparatively trivial task to extract from computer simulation studies structural information to practically any requisite degree of detail. The problem is that of reducing all the data obtained as computer-generated configurations into forms which are experimentally, theoretically, or conceptually meaningful. A complete description of liquid-state microstructure would require the determination of a large number of n-body distribution functions. At present, only the two-body distributions have any experimental or theoretical significance in the case of molten salts. So far, only data relating to the pair distribution have therefore been determined for simple molten salts.

Some qualitative insight into local geometries may be gained from studies of three-dimensional models. Figure 6 depicts a model of computer-generated instantaneous configuration of ions in molten KCl.[50] This is helpful in the qualitative reappraisal of those theories of molten salts which are based upon postulates of simplified physical models. While these theories, with suitably determined parameters, may always reproduce certain time or ensemble averages of dynamic variables, it is evident from Fig. 6 that they grossly oversimplify a very complex situation. Quasi-crystalline theories, in particular, do not conform with the computed structure. The model illustrates that although, on average, each ion is surrounded by ions of opposite charge, there are situations in which like ions approach quite closely. There is a distinct tendency toward ion-pair

Fig. 6. Model of the geometry of a computer-generated configuration of ions in molten KCl at T_f.

formation and clustering amidst large unoccupied voids. There remain strong electrostatic fields within these voids, and the term "free volume" is not appropriate in this context. Fluctuations in local microstructure are seen to be large, implying that many-event mechanisms involving cooperative motions will be required to describe self-diffusion or electromigration.

The two-body radial distribution function (RDF) is the most important quantity in the description of equilibrium microstructure. The RDF may be defined as

$$g_{\alpha\beta}(r) = (d\bar{n}_\beta/dr)/(4\pi r^2 \varrho_\beta) \tag{87}$$

where $\bar{n}_\beta$ is the mean number of ions of type β around an ion of type α in a sphere of radius r, and ϱ_β is the number density of ions of type β. For a pure molten alkali halide there will generally be three discriminate two-body distribution functions to consider, i.e., $g_{++}(r)$, $g_{+-}(r)$, and $g_{--}(r)$, where the subscripts refer to positive and negative ions. The indiscriminate RDF, denoted by the subscript m, is then

$$g_m(r) = \tfrac{1}{2}[g_{+-}(r) + \tfrac{1}{2}(g_{++}(r) + g_{--}(r)] \tag{88}$$

Discriminate RDFs for crystalline and molten NaCl at the melting temperature[45] are shown in Figs. 7 and 8. RDFs for molten KCl have been tabulated and discussed in detail.[51] In the case of KCl and other salts with similarly sized anions and cations, the differences between $g_{++}(r)$ and $g_{--}(r)$ are small compared to those obtained for binary mixtures of non-ionic, nonpolar molecules. This is largely a consequence of the predominance of the repulsive Coulombic interaction in determining the distribution between similarly charged ions. The latter observation should be a general

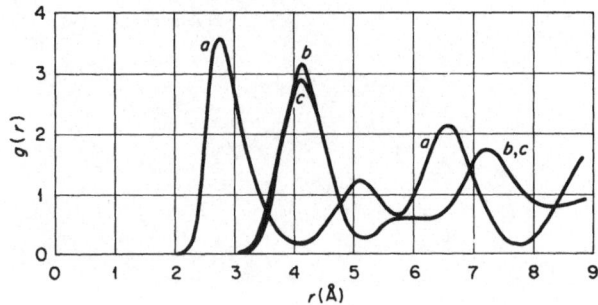

Fig. 7. Radial distribution functions for crystalline NaCl at T_f. (a) $g_{+-}(r)$; (b) $g_{--}(r)$; (c) $g_{++}(r)$.

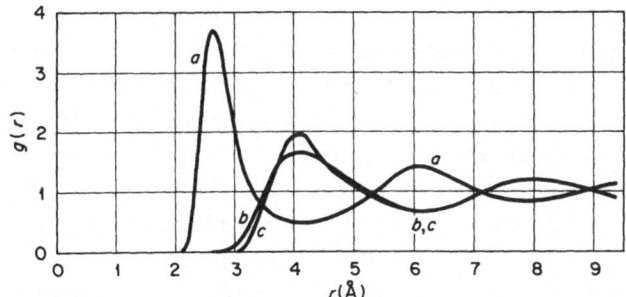

Fig. 8. Radial distribution functions for molten NaCl at T_f.
(a) $g_{+-}(r)$; (b) $g_{--}(r)$; (c) $g_{++}(r)$.

feature of ionic melt microstructure and is particularly evident from the RDFs for molten NaCl (Fig. 7) and also even LiCl.[19] Many other aspects of these RDFs have been previously discussed.[30,51]

The experimental equilibrium microstructure, i.e., the RDFs obtained from x-ray and neutron diffraction provide additional criteria for testing the adequacy of the computer model and obtaining information on interionic forces. A striking example of this application is given in a recent experimental investigation of molten CuCl.[52] When experimental curves were compared with RDFs from an MC calculation for a simple ionic model, large discrepancies were observed. The simple ionic model was shown to be highly unsatisfactory. It appears that in group IIb halides association occurs and is due, in part, to angle-dependent covalent forces reflecting the electronic structure of the copper ion.

The RDF only contains information on the average number of ions in spherical shells, and further information about local structure may be obtained by studying the corresponding radial fluctuation function (RFF) which may be defined as

$$W_{\alpha\beta}(r) = [\bar{n}_\beta^2(r) - \bar{n}_\beta^2(r)]/(\tfrac{4}{3}\pi r^3 \varrho_\beta) \qquad (89)$$

and which measures fluctuations around the radial occupation function $\bar{n}_\beta(r)$. Computed RFF curves have so far only been reported for simple Lennard-Jones liquids[53] (argon) and for molten KCl.[51] A striking difference in $W(r)$ for these two types of liquid is observed at short range. In ionic liquids fluctuations are suppressed due to ion pairing and the electroneutrality tendency.

The actual distribution of radial occupation numbers in molten KCl has been computed for significant distances and found to be unexpectedly broad. Figure 9 shows the actual probability distribution $P(n_\beta, r)$ for radial occupation numbers of both like and unlike ions corresponding to the first minima in $g_m(r)$ and $g_{+-}(r)$. The range of these distributions shows, for instance, that during the course of the simulation there are situations in which as many as three like ions are to be found inside the distance usually considered to define the first "coordination" shell. A value of 0 for $P(n_i, g_m^{min})$ is, however, the most probable. The broadness of these distributions support the contention that the microstructures of simple ionic melts are, by far, too complex for simple qualitative generalizations.

The angular distribution function (ADF) is closely related to the RFF, and both these functions contain information regarding the three-body distribution. Larsen *et al.* in their study of the NaCl/KCl mixture[47] have obtained the various ADFs for both the pure liquids and the mixture. The ADF is defined as

$$a(r, \theta) = dn(r, \theta)/(n_0 2\pi^2 \sin\theta \, dr \, d\theta) \qquad (90)$$

whereupon $a(r, \theta)2\pi^2 \sin\theta \, dr \, d\theta$ represents the probability of finding a third ion k in a volume element $2\pi^2 \sin\theta \, dr \, d\theta$, where θ denotes the angle at $\mathbf{r}_i$ in a triangle $\mathbf{r}_i, \mathbf{r}_j, \mathbf{r}_k$, and r is the distance between i and j. The discriminate ADFs obtained by Larsen *et al.* for molten NaCl at 1073°K are

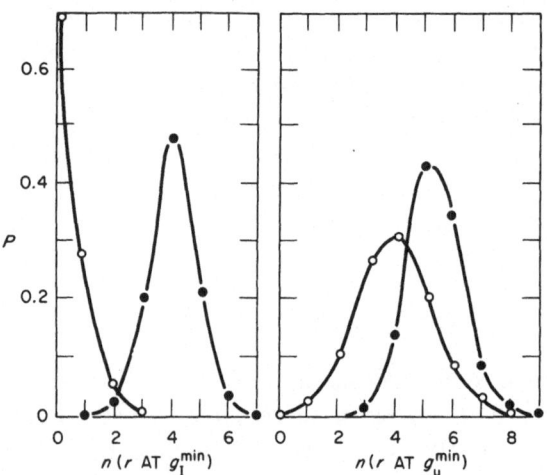

Fig. 9. Normalized probabilities of radial occupation numbers for like and unlike ions in molten KCl (T_f) at distances corresponding to the first minimum in $g_m(r)$ (3.7 Å) and $g_u(r)$ (4.5 Å), respectively. Solid circles, $P_u(n, r)$; open circles, $P_l(n, r)$.

shown in Fig. 10. These functions further illustrate the ineffectiveness of short-range repulsions between like ions in determining local structure. The Na–Na–k and Cl–Cl–k areas are all very similar and exhibit little ordering after the "collisional" angle. The Na–Cl–k curves, on the other hand, have peaks around 90° and 180°, reflecting a high degree of symmetry in the distribution of oppositely charged neighbors round an ion.

Preliminary reports of calculations of structure in molten salts containing divalent cations have been published. The results of these computations may be important in understanding chemical binding in a wide range of ionic-covalent compounds. According to Fajan's rules,[16] the "covalent character" of a chemical bond between a metal ion and a halide ion increases with the anion/cation radius ratio and also with the cationic charge, i.e., with increase in polarization. Evidence in support of Fajan's ideas is prevalent in standard physical and inorganic textbooks, e.g., (1) the diagonal relationship, (2) orders of melting points in the alkaline earth halides, and (3) comparisons between properties of compounds in series such as $LiCl–BeCl_2–AlCl_3$ and ordinary covalent compounds.

Potential-energy surfaces in the simple ionic model (see Fig. 1), however, indicate that many of the above observations are amenable to interpretation without invoking either electrostatic polarization or angle-dependent covalent forces. From Fig. 1 it is clear that as the cation charge increases, or as the cation/anion radius ratio decreases, the ionic pair potential increases in depth and restoring force. While bulk properties will inevitably be modified by polarization effects, the need to invoke angle-dependent covalent forces is questionable. Figure 1 shows that in the ionic model, when the cation is small and multiply charged, the depth of the pair potential is many orders of magnitude greater than kT. Since the force constants are also large, one then expects behavior similar to covalent molecular compounds. Two recent computer simulation MD studies of K_2MgCl_4 and BeF_2 illustrate these points.

There has in the past been some uncertainty about the interpretation of observed Raman bands in group IIa/Ia halide melt mixtures in terms of "complex" ion formation.[54] Whereas the geometry and coordination numbers of transition metal complexes are the result, in part, of angle-dependent covalent interactions between metal ion and ligand, with the central ion reflecting its electronic structure, group IIa ions, or Mg^{++} in particular, would not be expected to show angle-dependent interactions with ligands. It has been found from an MD computation on K_2MgCl_4[55] that a simple ionic model can give rise to symmetric clusters sufficiently

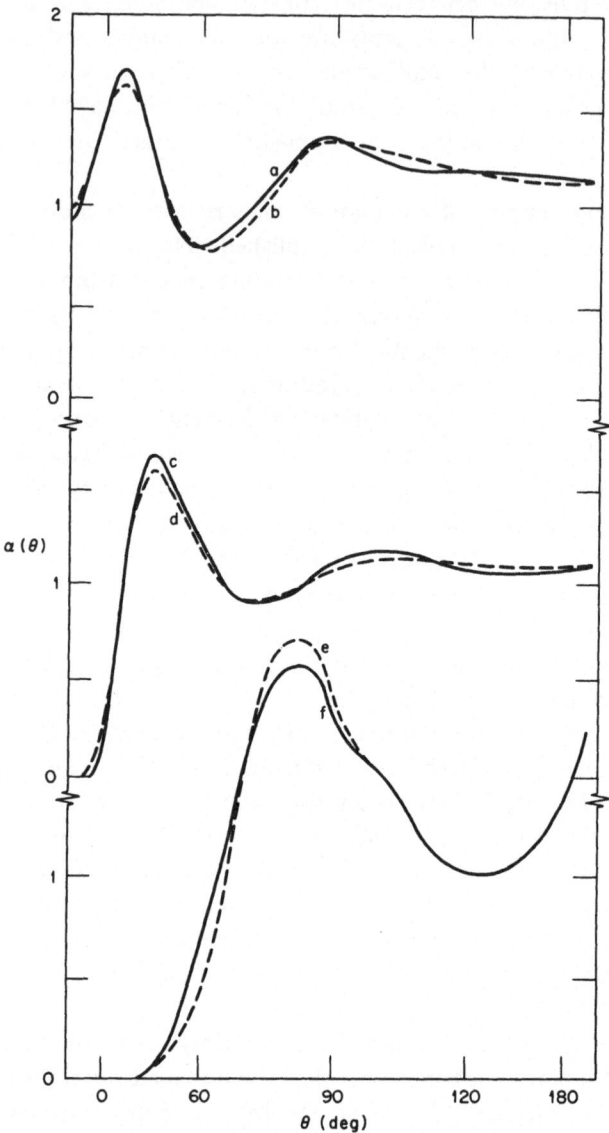

Fig. 10. Angular distribution functions $\alpha(\theta)$ for molten NaCl
(1083°K) from MC data by Larsen, Førland, and Singer.
(a) Cl⁻Na⁺Cl⁻ (i.e., i, j, k in text); (b) Na⁺Na⁺Cl⁻; (c)
Cl⁻Cl⁻Cl⁻; (d) Na⁺Na⁺Na⁺; (e) Na⁺Cl⁻Cl⁻; (f) Cl⁻Na⁺Na⁺.

long-lived to account for the Raman bands. A complete analysis of the computed RDFs for K_2MgCl_4 showed the structure to be trigonal $MgCl_3^-$ clusters; KCl ion pairs and K^+ ions tended to be on the exterior of the trigonal clusters. The disagreement with experiment (tetrahedral symmetry is observed experimentally) probably reflects the inadequacy of the Mg–Cl pair potential. Minor alterations could, however, lead to tetrahedral symmetry exclusively since the differences in potential energies between competing clusters are many times greater than kT for T less than $\sim 1000°K$. The average lifetime of each $MgCl_3^-$ cluster between exchange of one Cl^- is 5×10^{-12} sec and the mechanism proceeds via a highly asymmetric transient $MgCl_4^=$ cluster.

In a calculation on molten BeF_2,[56] it has been shown that a simple ionic model gives a liquid structure which, in the short range, is tetrahedrally crosslinked. In molten ionic salts and their mixtures, generally, when there are insufficient ligands (anions) to satisfy the preferred coordination number of the most attractive cation, then either a lower coordination number must be accepted or crosslinking occurs. At liquid densities and ordinary molten salt temperatures entropy considerations will favor crosslinking. The computed structure of BeF_2, shown in Fig. 11, was found to be in excellent agreement with available diffraction and spectroscopic data on the real liquid. The high symmetry and apparent angular dependence of the pair potentials in certain ionic systems is a consequence, not primarily of covalence or polarization, but of simple Coulombic repulsions between the "ligands" themselves.

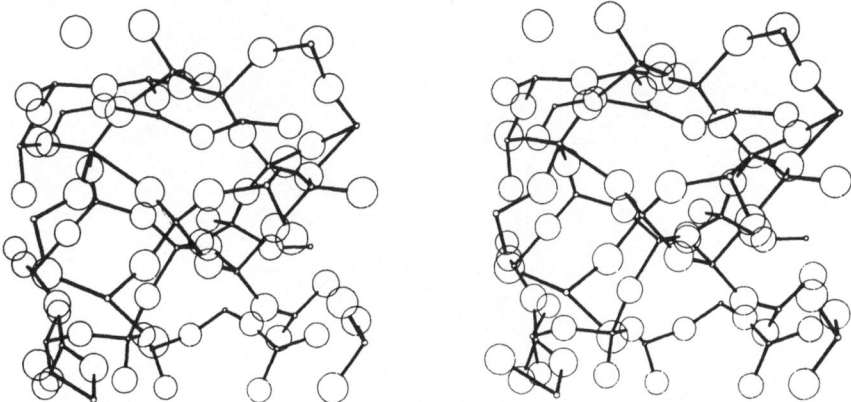

Fig. 11. Stereoscopic view of a typical instantaneous MD configuration of molten BeF_2 1153°K computed by A. Rahman, R. H. Fowler, and A. H. Narten.

All the necessary information concerning the time dependence of the microstructure is available from MD calculations, but there is difficulty in presenting the necessary distance and time dependence together. Some progress is being made by J. W. E. Lewis (Figs. 12–14) in the application of computer graphics to the three-dimensional representation of displacement–time probability functions.

The minimum information of any value is the mean-squared displacement $\overline{\Delta r^2}(t)$ curves shown in Fig. 2. This may be expanded by considering the probability that an ion will be displaced a distance r in time t such that

$$\overline{\Delta r^2}(t) = \int r^2 P_i(r, t)\, dr$$

where $P_i(r, t)$ is the single-ion-displacement correlation function. These functions are shown for Na$^+$ and Cl$^-$ ions in molten NaCl in Figs. 12 and 13,

Fig. 12. Displacement–time self-correlation function for Na$^+$ ions in molten NaCl. (Unpublished data reproduced by permission of Dr. J. W. E. Lewis.)

Fig. 13. Displacement–time self-correlation function for Cl^- ions in molten NaCl. (Unpublished data reproduced by permission of Dr. J. W. E. Lewis.)

respectively. Despite the ionic mass and size difference here, both functions are very similar. This is probably a consequence, to some extent, of the pairing tendency. It is also noticeable that there are no indications of preferred "jump distances" or "jump times" which would be presumably manifested in subpeaks.

The double-ion-displacement correlation function, sometimes referred to as the Van Hove distinct pair correlation function, $g_{\alpha\beta}(r, t)$, is simply the time-dependent RDF defined by a manner analogous to Eq. (87) except that $\bar{n}_\beta$ now becomes $\bar{n}_\beta(t)$ which denotes the number of ions of type β in a sphere centered on α at a time t that were also present at $t = 0$. Thus $g_{\alpha\beta}(r) = g_{\alpha\beta}(r, t = 0)$. In Fig. 14 the Na^+–Na^+ function is shown. The ordering between these pairs decays within approximately 10^{-12} sec. When further correlation functions of this type are available, we will have a much better basis for improving the current models for mass transport (see Sec.

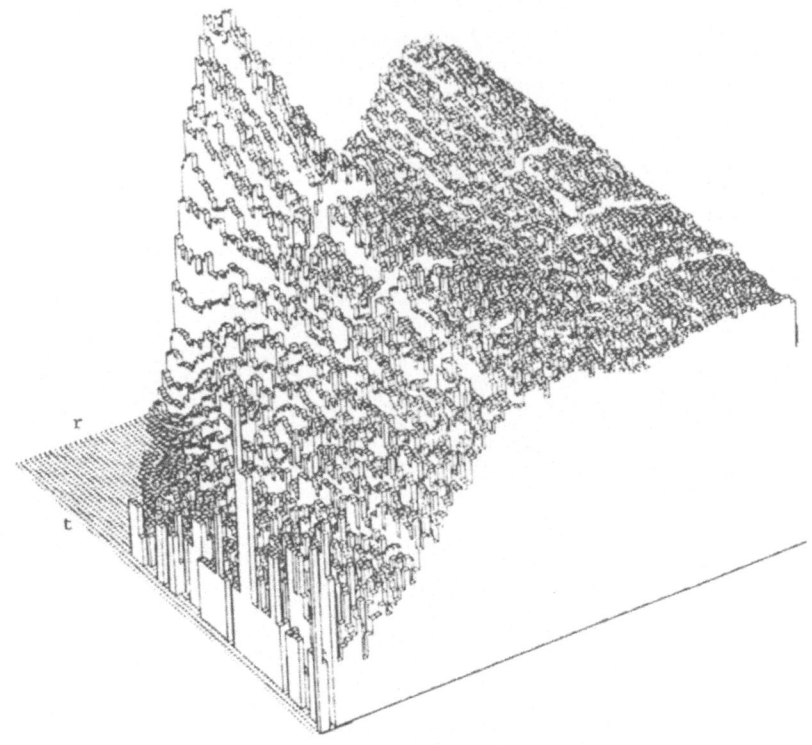

Fig. 14. Van Hove distinct pair correlation function $G_d(r, t)$ for Na$^+$–Na$^+$ pairs in molten NaCl. (Unpublished data reproduced by permission of Dr. J. W. E. Lewis.)

5.4) and for understanding the time dependence of ion pairs, symmetric clusters, or crosslinked structures.

5.3. Interpretation of Experimental Observables

5.3.1. Thermodynamic Properties

Computer simulation can be employed to furnish the information necessary to understand phenomenological thermodynamic behavior in terms of theoretically meaningful molecular-level parameters. This is perhaps the most important application from the point of view of the experimental molten-salt physical chemist. Thermodynamic properties may be resolved into contributions from either the different types of ion pairs or from the various components of the interionic pair potentials. For pairwise additive potentials, these data may further be represented as an integral over the appropriate pair distribution function.

In Monte Carlo computations for molten KCl[30] and some other alkali halides,[45] the 'internal potential energy has been resolved into electrostatic, overlap repulsion, and dispersion energy components (Table VI). This resolution corresponds to the Tosi–Fumi form of the interionic pair potentials [Eq. (12)]. Contributions to the virial, and hence the pressure, may be obtained from the electrostatic and dispersion terms directly, and, for the short-range repulsion, from the pressure data in Table II, using Eq. (40), after subtraction of the other contributions. The short-range overlap repulsion energy has been further resolved into contributions from each of the three different types of ion pairs.

The electrostatic energy is the predominant cohesive energy term. The mean Madelung "constants" may be calculated from the total electrostatic energy by

$$\bar{M} = \Phi^E r_0 / N_0 e^2 \tag{91}$$

where r_0 is the near-neighbor lattice parameter corresponding to a simple cubic lattice at the same density as the molten salt, e is the electronic charge, and N_0 is Avogadro's number. Calculated values for all the alkali halides are

TABLE VI. Resolution of the Potential Energies of Some Molten Alkali Halides at T_f into Contributions from Electrostatic (Φ^E), Overlap Repulsion (Φ^R), Dipole–Dipole Dispersion (Φ^{DD}), and Dipole–Quadrupole Dispersion (Φ^{DQ}) Terms in the Tosi–Fumi Potentials [Equation (12)][a]

	Φ^E	Φ^R_{++}	Φ^R_{+-}	Φ^R_{--}	Φ^{DD}	Φ^{DQ}
LiCl	−926.3	0.5	127.2	14.4	−15.9	−2.5
LiI	−817.6	0.5	124.9	21.5	−20.6	−3.2
NaCl	−802.1	0.8	95.2	5.4	−15.3	−2.3
KF	−870.7	7.7	131.1	1.1	−25.1	−4.5
KCl	−722.2	1.8	90.7	2.5	−23.2	−3.3
KBr	−679.9	1.4	78.4	1.1	−21.0	−3.0
KI	−641.0	0.6	75.7	4.0	−23.0	−3.3
RbCl	−688.7	2.1	82.1	1.4	−27.1	−3.8
CsF	−759.8	8.0	81.8	0.2	−31.7	−4.9

[a] Data for KCl from L. V. Woodcock and K. Singer, *Trans. Faraday Soc.* 67:12 (1972); data for other alkali halides from J. W. Lewis, K. Singer, and L. V. Woodcock (unpublished results). Energy units: KJ-mole⁻¹.

slightly greater than the crystal constant (1.74756), contrary to a number of rather uninspired predictions.[57] For molten KCl, $\bar{M}$ has been computed over a wide V–T range and found to increase linearly with volume at constant temperature.[30]

The total overlap repulsion term decreases the cohesive energy by around 15% for all the salts. The contributions from interactions between like-ion neighbors are small. Even in the extreme case of LiI, the I^-–I^- overlap energy is only -2.5% of the electrostatic energy. The use of model pair potentials from vapor phase data which incorporate only the repulsive electrostatic term between like ions therefore appears justified. These observations also explain, to some extent, the success of corresponding state treatments for correlation of thermodynamic data on some molten alkali halides.[58]

The dipole–dipole and dipole–quadrupole dispersion terms contribute only 3% and 0.5% (approximately) of the electrostatic potential. The major part of this energy, furthermore, arises from very short-range interactions. Because of the complexity of the underlying theoretical reasons for this type of resolution, and neglect of other terms such as polarization, it hardly seems justifiable to include dispersion forces, particularly the dipole–quadrupole, in these computations using semiempirical effective pair potentials. If dispersion terms are accounted for in the determination of the parameters in the overlap repulsion term, as in the Tosi–Fumi analysis, then they have also to be used in the molten salt computations of course.

The effective range of the Coulombic potential in molten ionic salts has some intrinsic interest both with regard to theoretical models and to technical aspects of MC computations. The electrostatic potential of an ion may be expressed as an integral over the radial charge density function, i.e.,

$$\Phi_i^E = -4\pi e^2 \varrho_0 \int_0^\infty r\varrho_i^E(r)\, dr \tag{92}$$

where ϱ_0 is the ion number density. The radial charge density functions $\varrho_i^E(r)$ are obtained by subtracting the discriminate pair distribution functions

$$\varrho_+^E(r) = g_{++}(r) - g_{+-}(r) \tag{93}$$

and

$$\varrho_-^E(r) = g_{+-}(r) - g_{--}(r) \tag{94}$$

The total charge density surrounding an ion is determined by the electroneutrality condition, so we have

$$Z_i = -\varrho_0 \int 4\pi r^2 \varrho_i{}^E(r) \, dr \qquad (95)$$

$\varrho_+{}^E(r)$ for molten KCl has been calculated from MD data and is shown in Fig. 15. The neutralizing charge density surrounding an ion extends up to several ionic diameters as an oscillatory function with alternating charge. It is the function $r\varrho_i{}^E(r)$ of Eq. (92), however, that determines the radially resolved contributions to the electrostatic potential. When this is plotted (Fig. 16), the convergence of the integral is seen to be remarkably slow, but, in contrast to ionic lattices, it is definite.

5.3.2. Phase Transitions

The V–T diagram of the condensed phases of KCl at negligible pressures calculated from MD data is shown in Fig. 17. It is likely that the application of computer "experiments" to supercooled liquids and the vitreous state will become increasingly important, particularly when more complex real glass-forming salts are simulated. It is worthwhile, therefore, to consider some of the conceptual aspects of applying computer simulation to supercooled liquids and the vitreous state.

There are certain advantageous factors which render the computer model more amenable to the attainment of metastable states. First, surface interactions are eliminated by the imposition of periodic boundary conditions. Second, there are no impurities in the system. Therefore, the two effects which are conducive to heterogeneous nucleation are absent. Furthermore, the fluctuations that can lead to homogeneous nucleation in macroscopic systems are not observed for small numbers of ions during microscopic time intervals. Thus, the actual study of time-dependent aspects of macroscopic phase transitions by molecular dynamics computations is presently impractical. When a typical computer-simulation liquid model is "cooled" below the normal melting temperature, crystallization is not observed and, eventually, liquid-state degrees of freedom are frozen out and an amorphous solid phase is obtained.

In the use of MD calculations to investigate glass transformation phenomena, however, there is a serious limitation in that processes with characteristic relaxation times exceeding around 10^{-10} sec cannot be simulated. The real experimentalist works in the time domain of the order of seconds upwards, whereas the simulationist may only implement state changes over time intervals ranging from around 10^{-10} sec to instantaneous.

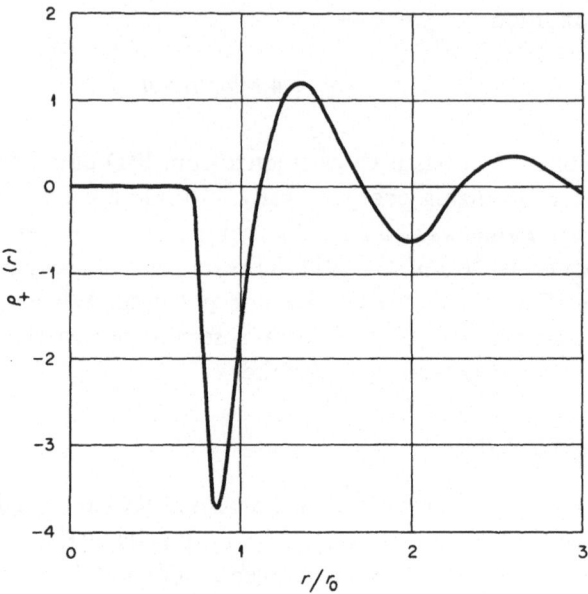

Fig. 15. Charge density radial distribution function $\varrho_+{}^E(r)$
for molten KCl at T_f.

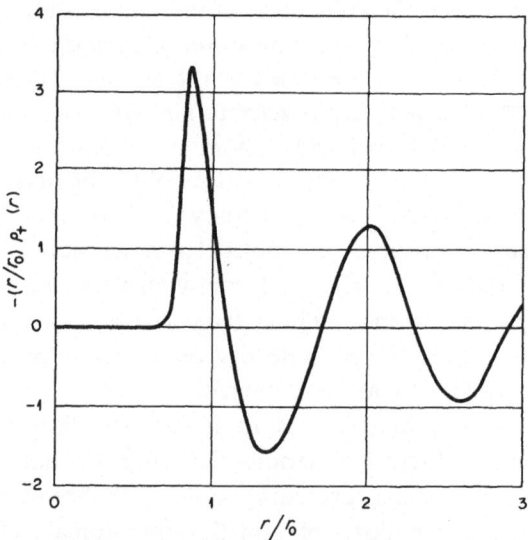

Fig. 16. Radially resolved mean electrostatic potential
energy of K^+ ions in molten KCl at T_f.

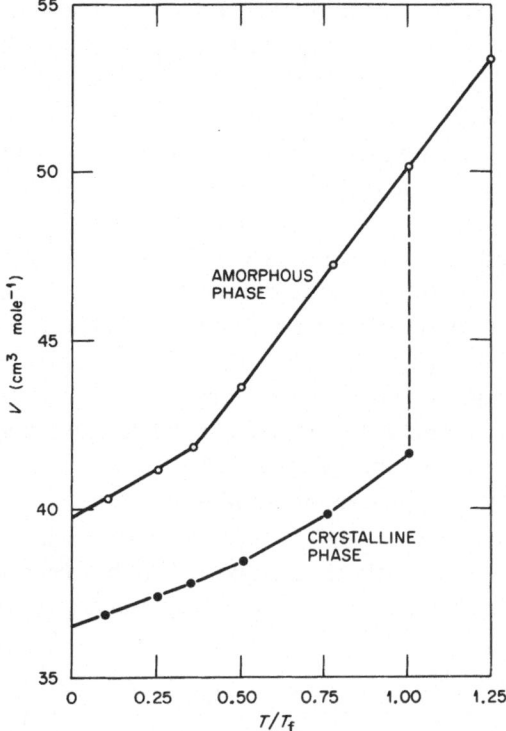

Fig. 17. Temperature dependence of the zerobaric molar volume of KCl in amorphous and crystalline phases from MD data.

Changes of state variables can be accomplished by simply adjusting either the coordinates for density changes or the momenta for temperature changes. When a real amorphous material is obtained in a metastable solid form after freezing out certain degrees of freedom, the thermodynamic state obtained may not correspond to a simulation of that process if the characteristic relaxation time exceeds the duration limit of the computer "experiment." While it may always be possible to obtain an amorphous solid from a liquid configuration with sufficient computational "compression," it is questionable whether or not the reequilibrated state would correspond to a real experimental state obtained with a cooling rate many orders of magnitude slower.

A series of MD computations on supercooled liquid KCl, using a number of widely different irreversible vitrification paths, have been carried out.[59] The computer model has been found to undergo a "glass" transition at around $0.3T_f$. This is evidenced by (1) discontinuities in the heat capacity and thermal expansivity (see, e.g., Fig. 17), (2) on approach to zero of the calculated coefficients of self-diffusion, as shown in Fig. 18, and (3) on absence of communal contributions to the entropy.

The diffusion data are best illustrated by the mean-squared displacement versus time curves shown in Fig. 19 and by the velocity autocorrelation functions for the "glass" compared with the crystal at the same temperature and with the liquid at T_f (Fig. 20). The amorphous solid phase was further found to exhibit first-order thermodynamic properties and RDFs which were independent of the thermal and mechanical stress history of the systems for all the vitrification paths considered. For KCl, the available equilibration time in the machine experiment, 10^{-11} sec, exceeds the relaxation time for translational diffusion. An amorphous solid phase with a well-defined free-energy minimum appears to be formed.

Location of the liquid–crystal equilibrium melting transition requires computation of the free energy of each phase (see Sec. 4.1) or the free-energy difference between the two phases. This has been carried out for KCl using an approximate method of entropy determination based upon

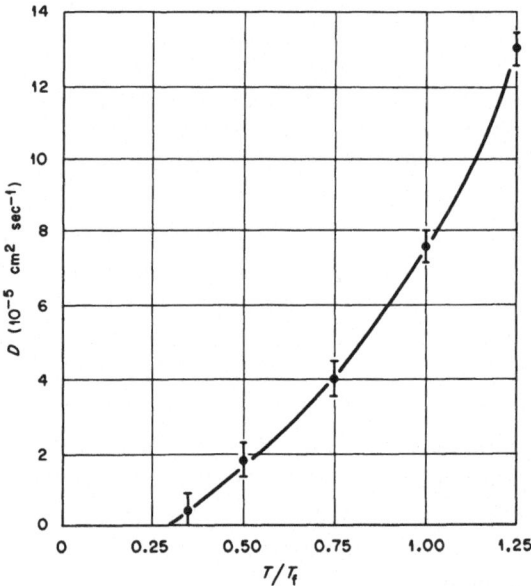

Fig. 18. Temperature dependence of the self-diffusion coefficient of K$^+$ ions in amorphous (supercooled liquid) KCl from MD data.

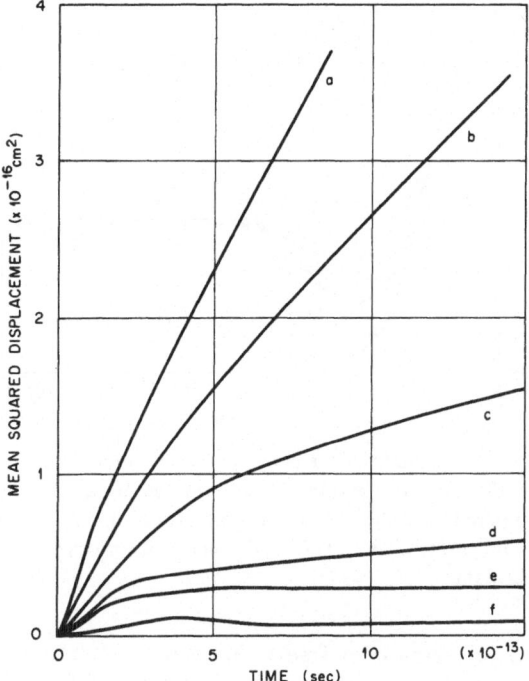

Fig. 19. Mean-squared displacement curves for K^+ ions in molten KCl and the vitreous form from MD data. (a) $T/T_f = 1.0$; (b) $T/T_f = 0.75$; (c) $T/T_f = 0.5$; (d) $T/T_f = 0.35$; (e) $T/T_f = 0.25$; (f) $T/T_f = 0.1$.

the cell theory of condensed phases.[30] The N-ion configurational integral

$$Q^\dagger = \int_v \cdots \int_v \exp[-\Phi(\mathbf{r}_1 \ldots \mathbf{r}_N)/kT] \, d\mathbf{r}_1 \ldots d\mathbf{r}_N \qquad (96)$$

which determines the configurational contribution to the free energy according to

$$A^\dagger = kT \log_e Q^\dagger \qquad (97)$$

is approximated by the N-th power of the mean one-ion configurational integral, so that

$$Q^\dagger \sim \left\langle \left(\int_v \exp(-\Phi(\mathbf{r}_1)/2kT)_{N-1} \right\rangle^N \qquad (98)$$

and the configuration entropy, defined by

$$S^\dagger = (\langle \Phi \rangle - A^\dagger)/T \qquad (99)$$

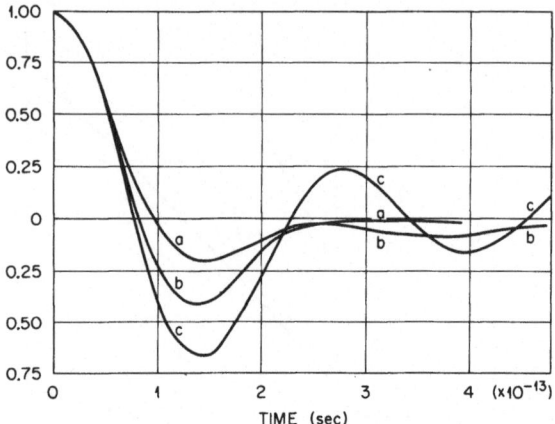

Fig. 20. Normalized velocity autocorrelation function for K^+ ions in "vitreous KCl" (curve b) at $0.25T_f$, compared with the crystalline state also at $0.25T_f$ (curve c) and molten KCl at T_f. All curves are plotted from MD data.

can be expressed in terms of a free volume by

$$S^\dagger = Nk \log_e v \tag{100}$$

which, by appropriate manipulation, can be obtained from MC data.[37,38,60] The calculated melting point of KCl using this method has been found to agree well with the experimental value.

In the interpretation of experimentally observed transition points, the computer simulation data can be resolved so that a better insight into the reasons can be obtained. In this case, the free-energy difference can be resolved into its energy and entropy terms, as can the thermochemical experimental data, but the potential energies and entropies from simulation studies can be further resolved. Some results of such a resolution of the free-energy difference between amorphous and crystalline KCl are given in Table VII. The errors in these small differences may be as high as 1 or 2 kJ-mole^{-1}, but the data contain some significant features.

From Table VII it can be seen that by far the major contribution to the enthalpy difference arises from the electrostatic term. In the crystal this will be proportional to the lattice spacing, or approximately so. Thus, crystals with a close packed lattice such as the alkali halides will have a larger enthalpic contribution to the free-energy difference than, say, alkali metal nitrates where the volume of the crystal relative to that of the liquid

TABLE VII. Computed Free-Energy Difference between Amorphous and Crystalline Phases of KCl at Zero Pressure and Resolution into Energetic and Entropic Components[a]

T/T_f	ΔG	ΔH	$\Delta \Phi^E$	$\Delta \Phi^R$	$\Delta \Phi^D$	$T \Delta S$	$NkT \ln(v_f^L/v_f^e)$
0.0	16.3	16.3	1.0	−4.9	2.2	0	0
0.1	16.2	17.3	20.0	−4.9	2.2	1.2	1.2
0.25	14.3	17.9	20.4	−4.8	2.3	3.6	3.6
0.35	13.2	18.2	20.6	−4.7	2.3	5.0	4.9
0.5	10.7	20.4	21.1	−4.2	2.4	9.7	7.7
0.75	5.0	24.3	23.5	−3.6	2.8	19.3	10.9
1.0	0.0	26.3	25.4	−2.8	3.7	26.3	9.0

[a] Energy units: KJ-mole^{-1}.

is greater. This accounts for the higher melting points of the alkali halides. It is also interesting to note that the free-volume contribution to the entropy difference increases up to around the glass transition and then begins to decrease again. This may be a reflection of the clustering tendency in ionic liquids which increases with the volume.

5.3.3. Transport Coefficients

Important questions in the interpretation of transport data concern the effect upon transport coefficients of the different masses of the ions, the various interionic potentials, and the complex interdependence. Up to now, experimental investigations of mass effects in transport have been confined to the study of isotopic samples which differ only slightly in mass. In computer "experiments," masses can be varied without difficulty up to large extremes within the classical limit. This facility, together with available methods for resolving the forces and potentials, should provide the detailed information required for the understanding and interpretation of transport data in ionic melts.

Only linear coefficients of translational diffusion have so far been successfully computed from MD data. The various methods of both the translational and rotational diffusion coefficients have been discussed in Sec. 4.2.

Some exploratory computations have been carried out into the possibility of computing the linear coefficient of electrical conductance[81] from Eq. (64). Current–current correlation functions obtained for molten

KCl at two different pressures are shown in Fig. 21. It should be emphasized that these results are not accurate, particularly at longer times, but, nevertheless, some of the difficulties to be encountered in conductance calculations are already evident. Firstly, there is the difficulty of the constraint upon cross-correlation contributions imposed by the conservation of momentum within the basic MD cube. Ideally, larger systems would be employed, and the fluctuations and their corresponding time correlations could then be averaged over a number of subvolume elements. Secondly, the amount of computing time for the requisite number of time origins for, say, 1% accuracy at long times appears to be prohibitively large. Experimental values of deviations from the Nernst–Einstein relationship are often small, and the accuracy of the cross-correlations is then all the more important. Finally, it is evident from Fig. 21 that the actual integration involves a small difference between two large areas; the negative correlation is appreciable, particularly at the higher pressure. Thus, only qualitative features of these correlation functions are really significant.

The current–current correlation function computed for molten LiCl, compared with the single-ion velocity autocorrelation functions, is shown

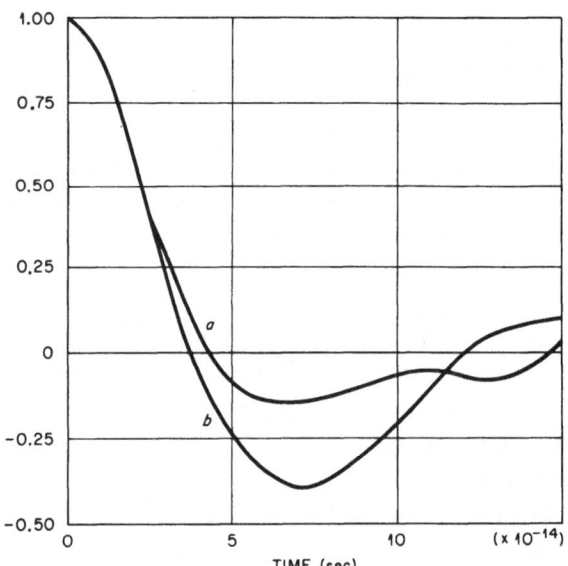

Fig. 21. Normalized current–current correlation function for molten KCl at T_f and two different pressures from MD data. Molar volumes: (a) 51.0; (b) 45.0 cm³-mole⁻¹. [$Vm_{p=0}$ is 50.2 cm³-mole⁻¹.]

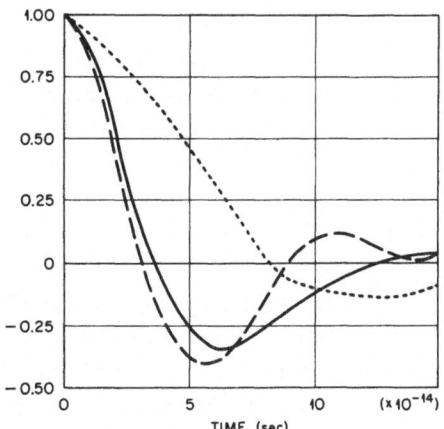

Fig. 22. Normalized current–current correlation function for molten LiCl at 900°K from MD data (solid line) compared with the velocity autocorrelation functions of Li^+ (dashed line) and Cl^- (dotted line).

in Fig. 22. The high transport number of the Li^+ ion is reflected in the closeness of its velocity autocorrelation function with the current autocorrelation function.

Computations of the viscosity or thermal conductivity of ionic liquids, from either the correlation functions [Eqs. (70)–(73)] or their corresponding Einstein formulas, have so far not been reported. Reasonably promising results, however, have been reported for hard spheres and simple Lennard–Jones liquids,[41] and it seems likely that these calculations will be extended to ionic liquids in the near future.

5.3.4. Spectral Data

MD computations of high-frequency spectral data, such as IR absorption and Raman scattering of light, can be applied in a number of capacities in the study of either interionic forces or in the quantitative interpretation of the optical interaction itself. Whereas thermodynamic and transport properties may be largely insensitive to the orientation dependence of interionic potentials, or polarization potentials, these are manifested in certain spectroscopic properties. Comparisons between computed and experimental spectra, therefore, could lead to deductions regarding the nature of the potential-energy surfaces in the real liquid. MD computations can also be used to test proposed mechanisms for interactions between the scattering media and the optical field. Some preliminary results have been obtained for the application of MD computations to the interpretation of high-frequency interionic light scattering[62] and intraionic rotational-vibrational Raman intensity profiles in molten ionic salts.[23]

In the quantitative theoretical interpretation of light scattering from liquids, there are generally two types of approximation. The first is in the relationship between the polarizability tensor and the interionic forces plus the field of the light, and the second is in the ionic motions which give rise to the time dependence of the polarizability anisotropy. By utilizing MD computations to eliminate the second type of approximation, proposed models for the induced anisotropy may be assessed by comparing calculated and experimental spectral densities.

Simple molten alkali halides, like KCl for instance, do not exhibit Raman vibrational bands, but a quasi-exponential Rayleigh wing extending up to ~ 100 cm^{-1} is observed (Fig. 23). The dipole-induced dipole model which seems to be the most plausible interpretation of intermolecular light scattering from simple nonionic nonpolar liquids is inapplicable here since the spectrum from KCl is predominantly polarized. Moreover, in ionic liquids, the internal electrostatic field is very much greater than the optical field. An alternative model based upon internal charge-induced anisotropy has been suggested.

The polarizability of a spherical ion may be expanded in a Taylor

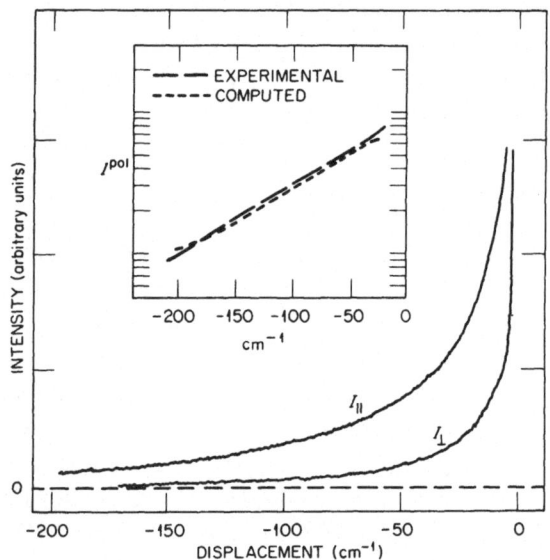

Fig. 23. Interionic light-scattering spectra for molten KCl at 1073°K compared with a density-of-states computation of the electrostatic field at an anion. $I_{\text{pol}}(\omega)$ is the trace scattering given by $I_{\text{pol}}(\omega) = I_{\parallel}(\omega) - I_{\perp}(\omega)$. (J. H. R. Clarke and L. V. Woodcock.)

series, whereupon the optically induced dipole moment is given by

$$\mu(t) = (\alpha_i + \tfrac{1}{2}\gamma_i\psi_i^2(t) + \cdots)\mathbf{E}_0(t) \tag{101}$$

where ψ_i is the internal electrostatic field at $\mathbf{r}_i$, $\mathbf{E}_0(t)$ is the time-dependent optical field, and α_i and γ_i are the linear and hyperpolarizabilities of the ion, respectively. Thus, the polarized intensity profile should be related to the Fourier transform of $\langle Tr\psi_i^2(0)\psi_i^2(t)\rangle$ and, furthermore, the scattering intensity should be proportional to the square of the hyperpolarizability. Relative intensity data, however, suggest that the hyperpolarizability model is inadequate and the total intensity for a number of alkali halides is in fact proportional to α_i of the anions.

An alternative implementation of the CIA mechanism is to equate the time dependence of the magnitude of the induced anisotropy with that of the internal electrostatic field. A polarized spectrum which was computed on this basis (shown in the inset to Fig. 23) has a profile coincident with the experimental data. The high symmetric component in the motions of small clusters of ions, shown by computer simulation to be a characteristic feature of ionic liquid microstructure (Sec. 5.2), would be expected to give rise to appreciable polarized intensity.

In Fig. 24 the orientational correlation function $[\phi_{P_2}(t)]$ is compared with the Fourier transform of the Raman spectra for molten KCN. The calculated and experimental correlation functions are in good agreement. This being so, the other correlation functions descriptive of rotational diffusion (shown in Fig. 24) can be used to interpret the features of the spectra in terms of the equilibrium structure, the potential-energy surfaces, and the ionic motions. Approximate theoretical models for angular motions may also be tested.

The small-angle Brownian diffusion model predicts that the ratio of relaxation times τ_{p_1}/τ_{p_2} should be 3, and that $\phi_J(t)$ should decay much faster than either $\phi_{p_1}(t)$ or $\phi_{p_2}(t)$. For CN^- ions in molten KCN, from the MD data, $\phi_{p_2}(t)$ is everywhere close to $\phi_J(t)$, implying a predominance of "large-angle jumps." Cross-correlations of the form

$$\left\langle \sum_{i,j}^{N} [\cos^2\theta_i(0) - \tfrac{1}{3}][\cos^2\theta_j(t) - \tfrac{1}{3}] \right\rangle$$

were also computed for $j \neq i$ and found to be around 20% of the $j = i$ variance at $t = 0$, and negative. This reflects the presence, on average, of K^+ ions directly between neighboring CN^- ions as the Coulombic field tends to align neighboring CN^- ions in opposite directions. The corre-

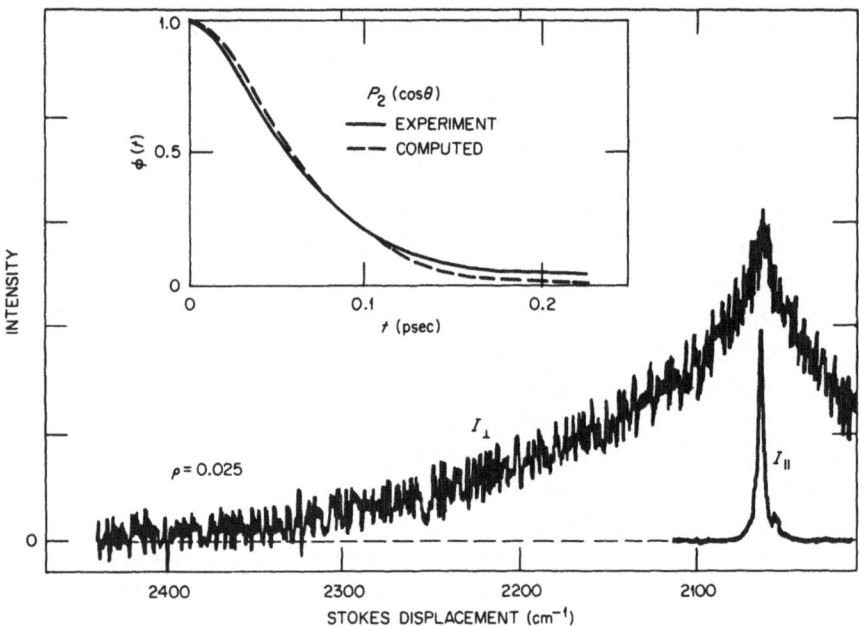

Fig. 24. Raman scattering from the 2065 cm⁻¹ CN⁻ stretching vibration in molten KCN at 1000°K. Inset shows a comparison of the Fourier transform of the orientational contribution with the MD-computed $\phi_{P_2}(t)$ correlation function for CN⁻. (J. H. R. Clarke, S. Miller, and L. V. Woodcock.)

lation functions in Fig. 5 broadly conform to a picture of the CN⁻ ions rotating relatively freely in a direction perpendicular to their nearest K⁺ "partners."

5.4. Reappraisal of Molten Salt Theories

5.4.1. Equilibrium Models

Various approaches to the equilibrium theory of molten ionic salt properties are reviewed in articles by Stillinger[4] and by Bloom and Bockris.[64] In the interpretation of thermodynamic data on molten salts, the most widely employed approach has been the physical type of approximation (as distinct from rigorous representation and mathematical truncation) whereby a simple structural model is postulated, certain properties corresponding to which can be computed with mathematical rigor. Here we will briefly examine some of the fundamental simplifying assumptions of these models in the light of computer simulation results.

The *Lennard-Jones–Devonshire lattice model* of liquids has been applied to the thermodynamic equation of state of molten ionic salts.[4,65] This involves two basic approximations. The configurational integral for N ions is replaced by the N-th power of the mean single-ion configurational integral [Eqs. (96)–(98)], and then the "smearing" approximation is applied. In the latter approximation the free volume of an ion is represented as an integral over a spherically symmetric Boltzmann weighted potential energy surface, and thus we have [Ref. 65, Eq. (7)]

$$v_f = 4\pi \int_{\text{cell}} \exp[(-\phi(r) - \phi(0)]/kT]r^2 \, dr \qquad (102)$$

where $\phi(0)$ is the potential at the lattice origin. MC computations[30] of the entropy of both crystalline and molten KCl at T_f indicate that the first approximation is almost exact in these high density states. In Fig. 25 a radially resolved free-volume function obtained from MC data is shown. The MC free volume may be written as

$$v_f^{\text{MC}} = \varrho_0 \int_0^\infty \cdot 4\pi\lambda(r)r^2 \, dr \qquad (103)$$

where

$$\lambda(r) = \langle \exp\{[\Phi_i(r) - \Phi_i^{\text{min}}]/kT\} \rangle \qquad (104)$$

Figure 25 indicates that the entire thermodynamic free volume arises within

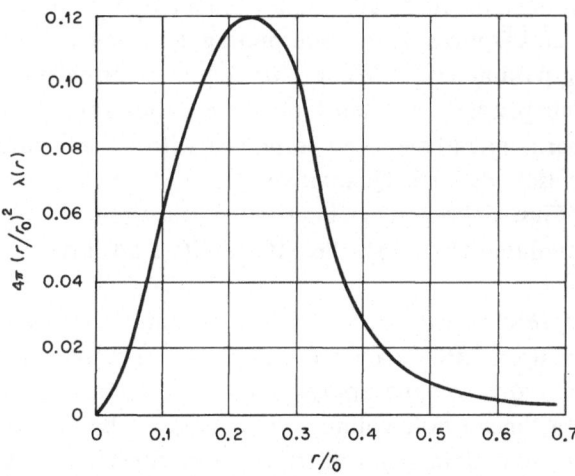

Fig. 25. Radially resolved thermodynamic free volume of
K+ ions in molten KCl at T_f from MC data.

around $0.8r_0$. The exponential tail of $\lambda(r)$, however, would appear to be inconsistent with the hard-sphere model for $\phi(r)$ used previously. It should be remembered that approximations such as Eq. (102) may always reproduce one or more thermodynamic averages with the correct choice of semi-empirical adjustable parameters. McQuarrie,[65] for example, obtained reasonable agreement for some properties of alkali halides using a Madelung constant for molten KCl of 1.1 and a mean free volume of 0.028 Å³. The corresponding MC results for these two quantities are 1.80 for $\bar{M}$ and 1.13 Å³ for v_f^{MC}.

The *cellular-hole model* (not to be confused with the hole theory mentioned below) is an extension of the lattice model whereby allowance is made for the presence of vacant sites.[4,66] The concentration of vacant sites is related to the expansion on fusion. This modification conforms with the experimentally observed decrease in "mean coordination number" on fusion from 6 to around 4. That $\bar{M}$ is found to increase slightly on fusion, however, suggests this model is not a realistic improvement. The radial free volume function (Fig. 25), moreover, shows no indication of any appreciable contributions beyond $\sim 0.8r_0$.

The *significant liquid structures model of Eyring* has been applied to molten salts by a number of authors.[67-69] In this model it is assumed that the equilibrium configuration can be divided into two types of ion, some with "crystalline degrees of freedom," and others "gas-like." A number of ions equal to the fractional expansion on fusion are deemed to have "gas-like" degrees of freedom. Because of the nature of the two empirically adjustable parameters, there are some fundamental doubts as to the applicability of this approach for interpreting experimental observables in terms of ionic parameters.[5] MD or MC data, nevertheless could be used to provide independent estimates of these adjustable parameters. Unfortunately, parameters have been employed with no indication as to how they were precisely defined. Qualitatively, in view of the absence of quasi-crystallinity (Figs. 6–9) and of long-range contributions to the thermodynamic free volume (Fig. 25), the MC results lend little support for this approach either.

The *hole theory*, extended by Bockris and co-workers to molten salts,[64,69-71] is not a molecular theory in the usual sense of the term, but a continuum or hydrodynamic approximation. The essential feature of the model is that a "mean hole volume" may be given by the reversible work required to create a Boltzmann distribution of spherical cavities against a macroscopic surface tension. The hole concentration is determined by the assumption that the expansion volume on fusion is due to the appearance of holes. This would appear from MC data to be a more realistic qualitative simple picture than the lattice models, but it is inconceivable that cavities or voids of no more than ionic-size dimensions could be treated realistically with continuum thermodynamics. The vacancies in ionic melt microstructure, moreover, do not contribute to the thermodynamic free volume and, therefore, do not give rise to translation diffusion by any "jump mechanism."

The *random mixing theory* of binary molten salt mixtures[72] has been tested for the 1:1 NaCl–KCl mixture with the angular distribution functions from MC calculations. Even the small size difference between Na^+ and K^+ leads to a significant discrepancy between the discriminate ADFs for each ion (Fig. 10), indicating a deviation from the random-mixture theory for this favorable example. If one considers, instead of "ionic radii" or "polarization parameters," the relevant pair potentials, the large difference in depth between the Na^+–Cl^- and K^+–Cl^- pair potentials (Table I), i.e., 6 times greater than kT at 1100°K, the result is understandable. The Cl^- ions would overwhelmingly prefer Na^+ "partners," thus leading to deviations from the random-mixture model.

5.4.2. Transport Theory

The above-mentioned cellular-hole and hole models of fused salts have been extensively employed to interpret linear coefficients of mass transport.[74] The unifying concept in all these model-based approaches is that the transport of ions proceeds via a single-event mechanism with a characteristic activation energy. Even prior to the advent of MD computations, suspicions against this type of activated-state mechanism, in ionic liquids and indeed in liquids generally, were strongly expressed.[73] The MD results, however, confirm these earlier misgivings and should dispense once and, hopefully, for all time, with the "jump diffusion" description of mass transport in simple ionic liquids. Consequently, the terms "activation energy" and "activation volume" should be regarded only as macroscopic variables, which are sometimes useful in the presentation of transport data, without mechanistic significance.

The evidence against this type of approach may be summarized as (1) the equilibrium microstructure does not conform with the requisite picture for a single-event mechanism (Figs. 6–10); (2) the thermodynamic free volume of an ion is confined to its immediate vicinity, implying that diffusion must be highly cooperative (Fig. 25); (3) Van Hove self-correlation functions do not exhibit peaks at preferred jump distances or times; they decay monotonically (Figs. 12 and 13); and finally (4) printouts of simple ion trajectories show that positional displacements do not follow any simple pattern; there are no observed instances where ions are suddenly displaced by distances of the order of r_0. In fact, the time dependence is very complex, as are the potential-energy surfaces on which the ions move.

The Rice–Allnatt approach to the molecular theory of transport in liquids has been extended to molten ionic salts.[74–76] On the basis of a previously derived approximation for the molecular-friction coefficient, Rice observed that electrostatic forces should have a negligible contribution to the frictional retarding forces in ionic liquids.[74] The Rice–Allnatt model for momentum transport is essentially a two-event mechanism which was postulated as an extension to the single-event Brownian mechanism, or small-step diffusion model, proposed earlier by Kirkwood.[77] Both of these models have been examined in the light of force and velocity autocorrelation functions for molten KCl from MD data.[78]

In the small-step diffusion model the velocity autocorrelation function decays exponentially and the corresponding force autocorrelation exhibits a fast initial decay with a long negative tail. A further consistency requirement of this model is that the force autocorrelation has a characteristic relaxation time much less than m/ζ, where m is the ionic mass and ζ is the friction coefficient (kT/D). From Figs. 4 and 5 it is evident that there are significantly large deviations from this model. The velocity autocorrelation function has a distinct non-Gaussian decay with a negative minimum and a long tail. The force autocorrelation function decays slower than that required by the model and, furthermore, is oscillatory at longer times. The correlation time corresponding to the minimum in the force autocorrelation function is 10.3×10^{-14} sec which, in fact, turns out to be greater than m/ζ $(3.35 \times 10^{-14}$ sec$)$.

The two-event mechanism postulated in the Rice–Allnatt extension of the Kirkwood model is that there exists a basic dynamic event, i.e., a "strongly repulsive binary encounter," followed by a quasi-Brownian destruction of correlation by the "rapidly fluctuating soft-force field" between "collisions." This implies that there are two independent mechanisms for momentum transfer. In the implementation of this model, in order to obtain expressions for transport coefficients, it has been assumed that the electrostatic potential does not contribute to the dissipation of energy, i.e., to the transport processes.

In Fig. 26 the force autocorrelation, shown in Fig. 4, has been resolved into autocorrelations of F^E (the electrostatic force), F^R (the overlap repulsive

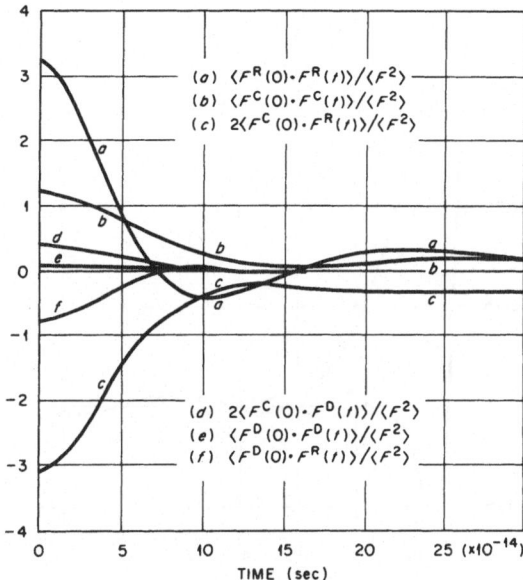

Fig. 26. Resolution of the total force autocorrelation function, shown in Fig. 4, into electrostatic, overlap repulsion, and dispersion potential auto- and cross-correlations. (S. I. Smedley and L. V. Woodcock.)

force), and F^D (the dispersion force) and the three cross-correlation functions. The slow decay of $\langle F^E(0)F^E(t)\rangle$ supports the original contention of Rice that electrostatic forces cause only slow dissipation of momentum. The large contribution of the electrostatic term to the mean-squared force and the very large negative cross-correlation contribution, with a comparatively slow decay, however, indicate that the electrostatic forces have a large indirect contribution to energy dissipation mainly through cross correlations with the short-range repulsion. It is unrealistic to regard the dissipation of momentum as occurring on two distinct independent time scales. The dispersion forces have a negligible effect, but it is noticeable that the cross correlations involving F^D are large compared to $\langle F_D(0)F_D(t)\rangle$.

Recently reported computations[79] of the velocity autocorrelation functions of molten LiI (Fig. 27) are of interest in the present context. The decay of the I$^-$ curve appears to conform to a Brownian-type model. It would be interesting to examine the force autocorrelations in this system and, in particular, to determine the extent to which this behavior is due to the size of the I$^-$ ion, or to its large mass.

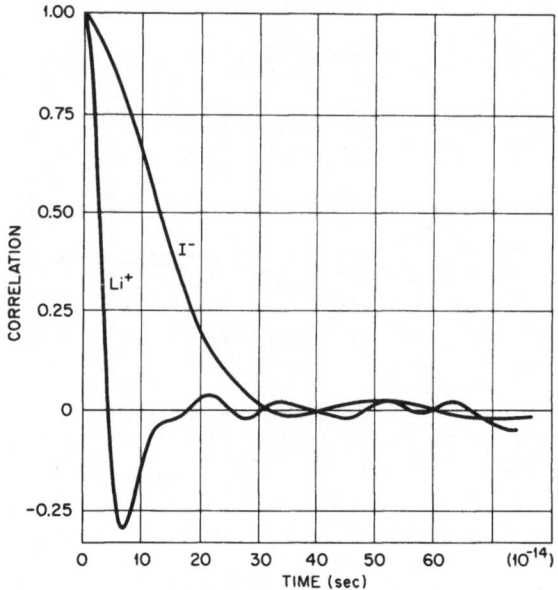

Fig. 27. Normalized velocity autocorrelation functions for molten LiI (1100°K) from MD data by J. W. E. Lewis.

6. CONCLUSIONS

Computer simulation may be considered as an established tool for interpreting the physical chemistry of molten ionic salts. A diversity of preliminary Monte Carlo and molecular dynamics results for a number of molten salts have been reviewed, and three distinct topics for reflection have emerged. These are (1) the success of the simple ionic model, (2) the part played by electrostatic forces in determining the equilibrium micro-structure and hence physical properties, and (3) the inadequacy of current theoretical models for both equilibrium and transport properties in simple ionic liquids.

The simple ionic model is generally a satisfactory representation of the interactions in molten alkali halides. Neither polarization potentials nor short-range overlap potentials between like ions contribute significantly to the ionic dynamics. When parameters derived from crystal data are used in the ionic model, agreement with experiment is less satisfactory when the anion/cation radius ratio differs appreciably from unity. The observed tendency to form ion pairs suggests that parameters obtained from vapor-phase ion-pair data may be a better representation. The ionic model can further account for the occurrence of symmetric clusters and crosslinked structures in group IIa halides and Ia/IIa halide mixtures. More extensive computations based on ionic models for Ia, IIa, and IIIa halides and mixtures should yield valuable information on the nature of chemical binding and electrovalency in particular.

Electrostatic forces have a profound symmetrizing effect on the equilibrium microstructure. The enormous depths of the ionic pair potentials relative to kT implies that there is no chance of finding an ion without at least one oppositely charged neighbor at a distance close to the minimum in the pair potential. This effect in the molten alkali halides gives rise to a complex assemblage of ion pairs and clusters. Electromigration is only a very slight perturbation on self-diffusion, and must therefore have the same mechanism. This involves cooperative motions of clusters and chain-like exchange of partners, but certainly no "single-ion jumps." In mixtures of molten ionic salts with divalent cations present, the anions stick to the attractive cation, resulting in more well-defined and longer-lived symmetric clusters. At higher divalent cation concentrations, or for pure group IIa halides (and probably also IIb halides to some extent), when there are insufficient anions to satisfy the preferred coordination number, crosslinking or ligand sharing occurs. Hence the highly symmetric lattice-like structure of molten BeF_2, for example. The symmetry in many of these ionic melts may be considered as a secondary structuring effect of the electrostatic forces through Coulombic repulsions between neighboring anions. The actual symmetry type depends upon the composition of the melt, the details of the potential-energy surfaces, including polarization effects, and entropy considerations.

All the theoretical models which have so far been postulated for simple ionic liquids are a gross oversimplification of the computer simulation results. It seems unlikely that a generally acceptable picture that could lead to satisfactory analytic expressions for a number of bulk properties will evolve. A basic application of MD calculation is to indicate the direction of improved approximations. Unfortunately, the situation with regard to simple ionic liquids is that no such single structural feature would suffice. Similarly, a many-event, highly cross-correlated mechanism would be required for a realistic model of mass transport. In the *quantitative* interpretation of molten ionic salt properties, as with other simple liquids, computer simulation seems to be the only presently acceptable approach.

ACKNOWLEDGMENTS

I wish to thank Dr. J. W. E. Lewis for permission to discuss his unpublished recent MD results.

Financial assistance from the Ramsay Memorial Trust during the tenure of a General British Fellowship at Southampton University (1971–1972) and Leeds University (1972–1973) and the hospitality of these host institutions are gratefully acknowledged.

7. REFERENCES

1. H. Margenau and N. R. Kestner, *Theory of Intermolecular Forces*, Pergamon Press, Oxford (1971).
2. S. H. Bauer and R. F. Porter, in: *Molten Salt Chemistry* (M. Blander, ed.), Interscience Publishers, New York (1964), pp. 607–680.
3. M. P. Tosi, *Advan. Solid State Phys.* **16**:1 (1964).
4. F. H. Stillinger, in: *Molten Salt Chemistry* (M. Blander, ed.), Interscience Publishers, New York (1964), pp. 1–105.
5. L. Pauling, *J. Am. Chem. Soc.* **49**:765 (1927); see also L. Pauling, *Nature of the Chemical Bond*, Cornell University Press, Ithaca (1960), Sec. 13.2.
6. M. Born and J. E. Mayer, *Z. Physik* **75**:1 (1932).
7. M. P. Tosi, *Advan. Solid State Phys.* **16**:1 (1964); J. C. Slater, *Quantum Theory of Solids*, McGraw-Hill, New York (1965), Vol. 3, Chap. 9, pp. 206–232; T. C. Waddington, *Advan. Inorg. Chem. Radiochem.* **1**:157 (1959); E. A. Guggenheim, *Applications of Statistical Mechanics*, Clarendon Press, Oxford (1965), Chap. 5.
8. E. S. Rittner, *J. Chem. Phys.* **19**:1030 (1951).
9. J. E. Mayer, *J. Chem. Phys.* **1**:270 (1933).
10. Y. P. Varshni and R. C. Shukla, *J. Chem. Phys.* **35**:582 (1961).
11. A. A. Frost and J. H. Woodson, *J. Am. Chem. Soc.* **80**:2615 (1958).
12. F. G. Fumi and M. P. Tosi, *J. Phys. Chem. Solids* **21**:31 (1964).
13. M. P. Tosi and F. G. Fumi, *J. Phys. Chem. Solids* **21**:45 (1964).
14. L. Pauling, *Proc. Roy. Soc.* (London) **A114**:191 (1927).
15. J. R. Reitz, R. N. Seitz, and R. W. Genberg, *J. Phys. Chem. Solids* **19**:73 (1961).
16. F. A. Cotton and G. Wilkinson, *Advanced Inorganic Chemistry*, Interscience Publishers, London (1962), pp. 157–159.
17. A. Rahman, *Phys. Rev.* **136A**:405 (1964).
18. L. Verlet, *Phys. Rev.* **159**:98 (1967).
19. L. V. Woodcock, *Chem. Phys. Lett.* **10**:257 (1971).
20. N. Metropolis, A. W. Rosenbluth, M. N. Rosenbluth, A. H. Teller, and E. Teller, *J. Chem. Phys.* **21**:1P87 (1953); see also W. W. Wood and F. R. Parker, *J. Chem. Phys.* **27**:720 (1957).
21. W. W. Wood, *J. Chem. Phys.* **48**:415 (1968); also I. R. McDonald, *Chem. Phys. Lett.* **3**:241 (1969).
22. H. M. Evjen, *Phys. Rev.* **39**:675 (1932).
23. J. Krogh-Moe, J. Østvold, and T. Førland, *Acta Chem. Scand.* **23**:2421 (1969).

24. L. V. Woodcock, Ph.D. thesis, University of London (1970).
25. S. Card and J. P. Valleau, *J. Chem. Phys.* **52**:6232 (1970).
26. P. P. Ewald, *Ann. Phys.* (*Paris*) **21**:1087 (1921).
27. A. A. Barker, *Australian J. Phys.* **18**:119 (1965).
28. S. G. Brush, H. L. Sahlin, and E. Teller, *J. Chem. Phys.* **45**:2102 (1966).
29. L. V. Woodcock and K. Singer, *Proc. Culham Conf. Computational Physics*, paper No. 25, H.M.S.O. (1969).
30. L. V. Woodcock and K. Singer, *Trans. Faraday Soc.* **67**:12 (1971).
31. M. J. L. Sangster and M. Dixon, to be published.
32. J. O. Hirchfelder, C. F. Curtiss, and R. B. Bird, *Molecular Theory of Gases and Liquids*, Wiley, New York (1954), pp. 134–136.
33. T. L. Hill, *Statistical Mechanics*, McGraw-Hill, New York (1956), Chap. 4, pp. 97–121.
34. W. G. Hoover and F. H. Ree, *J. Chem. Phys.* **49**:3609 (1968).
35. D. R. Squire and W. G. Hoover, *J. Chem. Phys.* **50**:701 (1969).
36. J. G. Kirkwood, *J. Chem. Phys.* **18**:380 (1950).
37. E. Gosling and K. Singer, *J. Chem. Soc. Faraday Trans. Part II* **69**:1004 (1973).
38. J. P. Valleau and W. Whittington, *J. Chem. Soc. Faraday Trans. Part II* **69**:1009 (1973).
39. R. Zwanzig, *Ann. Rev. Phys. Chem.* **16**:67 (1965).
40. S. Harris, *Mol. Phys.* **23**:861 (1972).
41. B. J. Alder, D. M. Gass, and T. E. Wainwright, *J. Chem. Phys.* **53**:3813 (1970); L. Verlet in *Molecular Motions in Liquids*, J. Lascombe, ed., Reidel Publ. Co., Dordrecht, Holland (1973).
42. R. G. Gordon, *J. Chem. Phys.* **43**:1307 (1965).
43. R. G. Gordon, *Advan. Mag. Res.* **3**:1 (1968).
44. J. H. R. Clarke, S. Miller, and L. V. Woodcock, Proc. Conf. Molecular Motions in Liquids, University of Paris, Orsay, July 1973 (to be published).
45. J. W. E. Lewis, K. Singer, and L. V. Woodcock, *J. Chem. Soc. Faraday Trans. II* (in press).
46. D. Adams and K. Singer (private communication).
47. B. Larsen, T. Førland, and K. Singer (in press).
48. J. W. E. Lewis (unpublished results).
49. H. A. Levy and M. D. Danford, in: *Molten Salt Chemistry* (M. Blander, ed.), Interscience Publishers, New York (1964), pp. 109–125.
50. L. V. Woodcock, *Nature* (*Phys. Lett.*) **232**:63 (1971).
51. L. V. Woodcock, *Proc. Roy. Soc.* (*London*) **A328**:83 (1972).
52. D. I. Page and K. Mika, *J. Phys.* (*C*) **4**:3034 (1971).
53. L. V. Woodcock, *Z. Naturforsch.* **26a**:287 (1971).
54. V. A. Maroni, E. J. Hathaway, and E. J. Cairns, *J. Chem. Phys.* **75**:155 (1971); V. A. Maroni, *J. Chem. Phys.* **75**:4789 (1971); see also M. Bredig, in: *Structure and Properties of Molten Salts* (G. Mamantov, ed.), Marcel Dekker, New York (1967).
55. B. R. Sundheim and L. V. Woodcock, *Chem. Phys. Lett.* **15**:191 (1972).
56. A. Rahman, R. H. Fowler, and A. H. Narten, *J. Chem. Phys.* **57**:3010 (1972).
57. G. E. Blomgren, *Ann. N.Y. Acad. Sci.* **79**:781 (1960); D. A. McQuarrie, *J. Phys. Chem.* **66**:1508 (1962); H. Bloom and J. O'M. Bockris, in: *Fused Salts* (B. R. Sundheim, ed.), McGraw-Hill, New York (1964), p. 14.
58. H. Reiss, S. W. Mayer, and J. L. Katz, *J. Chem. Phys.* **35**:820 (1961); M. Blander, *Advan. Chem. Phys.* **11**:83 (1967).

59. L. V. Woodcock, Proc. Conf. Computer Simulation, San Diego, 1972, Sec. IV, Physical Science, Vol. 1, pp. 847–852.
60. T. L. Hill, *Statistical Mechanics*, McGraw-Hill, New York (1956), Chap. 8, pp. 354–392.
61. L. V. Woodcock (unpublished results).
62. J. H. R. Clarke and L. V. Woodcock, *J. Chem. Phys.* **57**:1006 (1972).
63. S. Miller, Thesis, University of Southampton (1974).
64. H. Bloom and J. O'M. Bockris, in: *Fused Salts* (B. R. Sundheim, ed.), McGraw-Hill, New York (1964), pp. 1–62.
65. D. A. McQuarrie, *J. Phys. Chem.* **66**:1508 (1962).
66. I. G. Mungulesur and G. H. Vasu, *Rev. Roumaine Chem.* **11**:681 (1966).
67. G. E. Blomgren, *Ann. N.Y. Acad. Sci.* **79**:781 (1960).
68. C. M. Carlson, H. Eyring, and T. Ree, *Proc. Nat. Acad. Sci. U.S.* **46**:333 (1960).
69. R. Vilcu and C. Misdolea, *J. Chem. Phys.* **46**:906 (1967).
70. J. O'M. Bockris and N. E. Richards, *Proc. Roy. Soc.* (*London*) **A24a**:44 (1957).
71. J. O'M. Bockris and G. W. Hooper, *Disc. Faraday Soc.* **32**:218 (1961).
72. T. Førland, in: *Fused Salts* (B. R. Sundheim, ed.), McGraw-Hill, New York (1964), pp. 73–78.
73. B. R. Sundheim, in: *Fused Salts* (B. R. Sundheim, ed.), McGraw-Hill, New York (1964), pp. 233–247.
74. S. A. Rice, *Trans. Farad. Soc.* **58**:499 (1962).
75. B. Berne and S. A. Rice, *J. Chem. Phys.* **40**:1347 (1964).
76. S. A. Rice and P. Gray, *Statistical Mechanics of Simple Liquids*, Interscience Publishers, London (1965), Chap. 6.
77. J. G. Kirkwood, *J. Chem. Phys.* **14**:180 (1946).
78. S. E. Smedley and L. V. Woodcock, *J. Chem. Soc. Faraday Trans. II* **69**:955 (1974).
79. J. W. Lewis (unpublished data).

Chapter 2

GAS SOLUBILITY IN MOLTEN SALTS

Paul E. Field

Department of Chemistry
Virginia Polytechnic Institute and State University
Blacksburg, Virginia

1. INTRODUCTION

Interest in the solubilities of nonreactive gases in molten salts stemmed originally from the technological need developed as a consequence of gaseous fission products of radiated molten salts. A secondary technological development associated with industrial uses of molten salts as heat treatment baths, electrolytes, and metallurgical slags led to other studies of the influence of various gases and vapors on the physical and chemical properties of the fused salts. Systematic research of gaseous solutions of molten salts has lagged behind technological studies and lies considerably more dormant than analogous studies in water and organic solvents. Theoretical studies of the liquid state and solution theory, on the other hand, have been pursued which provide applications to ionic liquid solvents such as the simpler molten salts.

This review covers a survey of the literature through 1972 and includes a summary of experimental results of studies of 14 gaseous solutes in various pure and mixed metal salts of seven anions. Because of the diversity of solvents in molten salt systems, solvents have been classified by the anion. The summary of the experimental studies is listed by reference number in Table I. With the exception of water in some of the lower melting salts such as $LiClO_4$ and $LiNO_3$, the solubilities included in this survey are at temperatures above the critical temperature of the solute.

TABLE Ia. Temperature Dependence Studies of Rare Gases in Ionic Melts

Gas	F^-	Cl^-	Br^-	I^-	NO_3^-	ClO_4^-	$CO_3^=$
He	Li–Na–K[35]				Li[9]		
	Li–Be[36]				Na[9,10]		
	Na–Be[37]				Na–K[7]		
	Na–Zr[3]						
	Na–Zr–U[3]						
Ne	Li–Na–K[35]						
	Li–Be[36]						
	Na–Zr[3]						
Ar	Li–Na–K[35]	K[29]	K[29]	K[29]	Li[9,46]		
	Li–Be[36]				Na[9,10]		
	Na–Be[37]				K[46]		
	Na–Zr[3]				Rb[9]		
					Na–K[7]		
Xe	Li–Be[36]						
	Na–Be[37]						
	Na–Zr[3]						
	Na–Zr–U[2]						

TABLE Ib. Temperature Dependence Studies of Nonpolar Gases in Ionic Melts

Gas	F^-	Cl^-	Br^-	I^-	NO_3^-	ClO_4^-	$CO_3^=$
N_2					Li[9,46]		
					Na[9,10]		
					K[46]		
					Na–K[7]		
O_2					Na–K[6,7]		Li–Na[27]
							Li–Na–K[27]
Cl_2		Na[30]					
		K[30]					
		Mg[30]					
		Na–K[30]					
		Na–Mg[30]					
		K–Mg[30]					
		K–Pb[31]					
CO_2	Na–Be[38]	Na[2]	K[2]	K[2]	Na[10]		
	Na–Al[2]	K[2]			K[46]		
CH_4					Na–K[7]		

TABLE Ic. Temperature Dependence Studies of Polar Gases in Ionic Melts

Gas	F⁻	Cl⁻	Br⁻	I⁻	NO₃⁻	ClO₄⁻	CO₃⁻
HF (DF)	Li–Be[18,40] Na–Be[40] Na–Zr[41]						
HCl		Na[15-17] K[16,17] Rb, Cs[16] Mg[17] Ca, Sr, Ba[34] Na–K[17] K–Mg[17]					
NH₃					Li[26] Li–K[26] Li–Na–K[26]	Li–K[26]	
H₂O		Li–K[32]			Li[13,22] Na[22] Li–Na[13] Li–K[13,21] Na–K[24] Li–Na–K[24] K–NaNO₂[25]	Li–K[24]	
CO		Na–K[33]					

It is not the intent of the author to present a unified analysis but to examine the available data in the light of the various semiempirical and theoretical models available.

The value of gases as solutes in the study of solution behavior rests primarily on their slight solubility. This in turn guarantees both negligible solute–solute interaction and effect on solvent–solvent interaction, and permits interpretation of the solution properties in terms of the solvent–solute interaction. Molten salts as solvents offer the advantage of strong classical forces of interaction, i.e., Coulombic forces, thus enhancing the interaction potential between solute and solvent beyond that found in the weaker interactions of nonelectrolyte solvents or the more complicated interactions of hydrogen bonding found in aqueous solutions. In this respect, the lower-melting ionic salts offer the additional advantage of not overwhelming the potential terms due to the larger kinetic terms. The disadvantage of these salts is their lower geometrical symmetry, e.g., alkali metal nitrates.

Studies in ionic solvents are not seriously hampered by the occurrence of dissimilar species, i.e., cation and anion, in solution since systematic investigation permits the determination of the separate effects by studies of series of cations having a common anion, and vice versa. Moreover, mixed cations (anions) permit a continuous variation in common anion (cation) solvents of broader range than possible in homologous organic solvents. Two final, and valuable, advantages of molten-salt solvents are the extensive range of the liquid state and the resultant low vapor pressures over a large segment of the liquid range. The former allows for a more precise evaluation of the heat of solution, i.e., the temperature dependence of the solubility, while the latter makes for a simple good approximation of the gas phase as uncontaminated solute gas in which gas-phase solute–solvent interactions can be neglected.

This survey is restricted to those molten-salt solvents which can be considered to be completely ionic. There exists an extensive body of data in melts having considerable covalent nature, such as the oxide mixtures, silicates, aluminates, etc., where solubility borders on chemical reactivity with solute gases, if in fact they are not specifically reactive, and these are considerably more complex in the evaluation of intermolecular interactions.

For those systems covered by this survey where temperature-dependent solubility data were published, the data have been reevaluated by means of a linear least-squares analysis of the van't Hoff relation log K_p versus $1/T$. The results of these calculations are given in the Appendix and include for each system: (1) the solvent composition, (2) the experimental temperature range, (3) the total number of specific temperatures and the total number of determinations, (4) the heat of solution, (5) the temperature at which the entropy of solution and the solubility were computed from the least-squares fit, (6) the entropy of solution, (7) the $T\Delta S$ product, (8) the solubility, and (9) the reference. Uncertainties listed for the heats and entropies of solution were computed at one standard deviation. The listing is divided between the nitrate and halide melts, with the order of solutes for each being presented: (a) rare gases; (b) nonpolar gases; (c) polar gases.

2. EXPERIMENTAL TECHNIQUES

Experimental methods employed for the determination of the solubilities of gases in molten salts can be broadly categorized into three classes. One technique is the absorption method where the solubility is determined by means of observing some property change in either the gas phase of the solute, generally a volumetric or manometric change, or the liquid phase of the solvent, usually a gravimetric change although some workers have utilized some reactive property such as voltammetry. A second general method is elution, where the solubility is determined by some means of analysis of the quantity of solute gas removed from a saturated solution. Usually the solute is stripped from solution by a nonequilibrium displacement with a second, or carrier, gas. Analysis of the resultant mixture of solute and carrier gases is then performed either with or without concentration of the solute gas and either by some physical method such as mass spectrometry or gas chromatography or by chemical methods which distinguish between the solute and carrier components. The third method is cryometry, which utilizes the depression of the freezing point of the solvent for the determination of the solubility. This method is considerably more limited in its application since only one temperature, i.e., the freezing point of the solution, can be determined. Moreover, the magnitude of gas solubility is sufficiently small in most cases that this method precludes systems having concentrations less than $\sim 10^{-4}$ mole fraction.

Battino and Clever[1] have reviewed the general field of gas solubilities in liquids and discuss the major experimental problems involved. One particular problem which presents a unique situation in molten-salt studies involves the degassing of solvents. The general technique of refluxing or boiling the liquid is not possible except under unusual circumstances for molten salts. Vacuum pumping at fixed temperature is completely unreliable. The most common technique for molten-salt solvents employs vacuum pumping over several freeze–thaw cycles of the melt. This problem presents one of the major drawbacks to the absorption method for determining solubilities and, on the other hand, represents an important advantage of the elution method.

An excellent analysis of the reliability of several experimental techniques was made by Bratland and co-workers.[2] Their study included an examination of the reliability of four techniques in the determination of the solubility of carbon dioxide in several pure alkali halide melts. These included the elution method developed at Oak Ridge National Laboratory by Newton and Hill[3a] and first described in the open literature by Grimes and co-workers[3b] using a (carrier) gas, and an alternate elution method in which about two-thirds of the saturated solution was transferred into a gas-tight chilling chamber from which the expelled solute was transported by means of a carrier gas for analysis. This latter technique depends on the negligible solubility of the gas in the solid, which is also the criterion of the freeze–

thaw method of solvent degassing. The remaining two techniques were absorption methods. One was the direct determination of the volume change of the solute gas at constant pressure caused by its dissolution into the melt. The other was a thermogravimetric technique employing a recording thermovacuum balance to measure the weight increase of the melt with absorption of the gas. The original apparatus of Grimes *et al.* is shown in

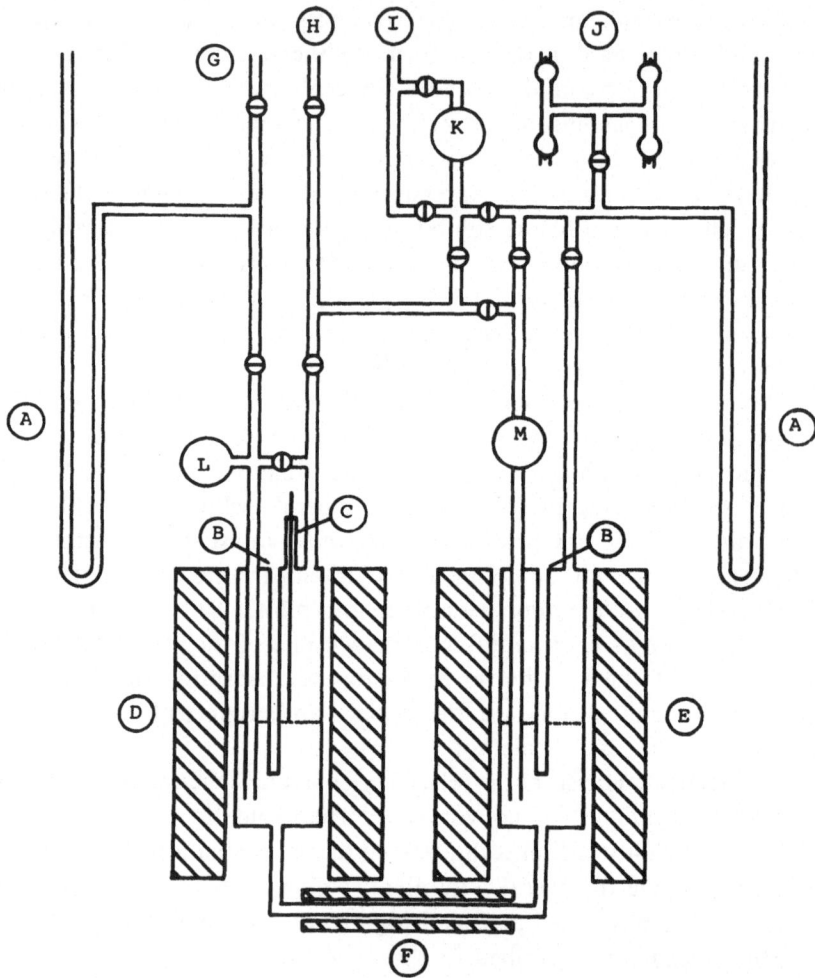

Fig. 1. Carrier gas elution apparatus (schematic) after Grimes: A, mercury manometers; B, thermocouple wells; C, level probe; D, saturation vessel and furnace; E, elution vessel and furnace; F, frozen seal in transfer line and heater; G, solute gas inlet; H, vacuum outlet; I, carrier gas supply; J, sampling bulbs; K, gasometer; L, pressure regulator; M, gas pump.

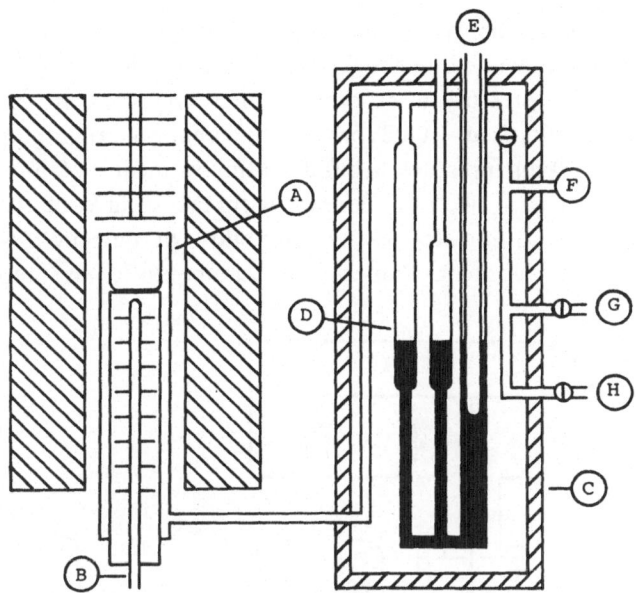

Fig. 2. Volumetric absorption apparatus (schematic) after Bratland: A, molten salt crucible and furnace; B, thermocouple well with baffles; C, solute gas thermostat; D, gas burette; E, mercury leveling piston; F, Pirani gauge; G, solute gas inlet; H, vacuum outlet.

Fig. 1. The apparatuses of Bratland and co-workers for the volumetric, gravimetric, and "chilling" techniques are shown in Figs. 2, 3, and 4, respectively.

Bratland et al.[2] were unable to evaluate the reliability of the stripping method due to reaction of the carbon dioxide with melt impurities. Although the volumetric method gave results in reasonable accord with the remaining two methods, the authors abandoned it because the gain in precision did not warrant the experimental problems which had to be resolved. Frame and co-workers[4] were unable to detect a freezing-point depression due to either O_2 or N_2 in melts of $NaNO_3$, KNO_3, or $CsNO_3$. Haug and Albright[5] likewise failed to determine the solubility of nitrogen dioxide in a fused mixture of $(Na–K)NO_3$.

Most work reported has employed variations on one or the other of the preceding methods with the exception of those utilizing the manometric absorption method where the pressure decrease of the solute gas phase is measured at constant volume. Desimoni, Paniccia, and Zambonin[6] have described a manometric absorption apparatus which directly measures the differential pressure change of the solute as a function of time at constant volume and thermostated temperature. Saturation is assumed when no further change in pressure is observed. Differential pressure is measured directly at a sensitivity of ± 0.02 torr for total changes ranging up to about 20 torr at a working pressure around 1 atm by means of a micrometer plunger which makes electrical contact with the surface of the mercury manometer. In addition to oxygen solubility in eutectic $(Na, K)NO_3$ reported by these workers, Panicca and Zambonin[7] have reported the

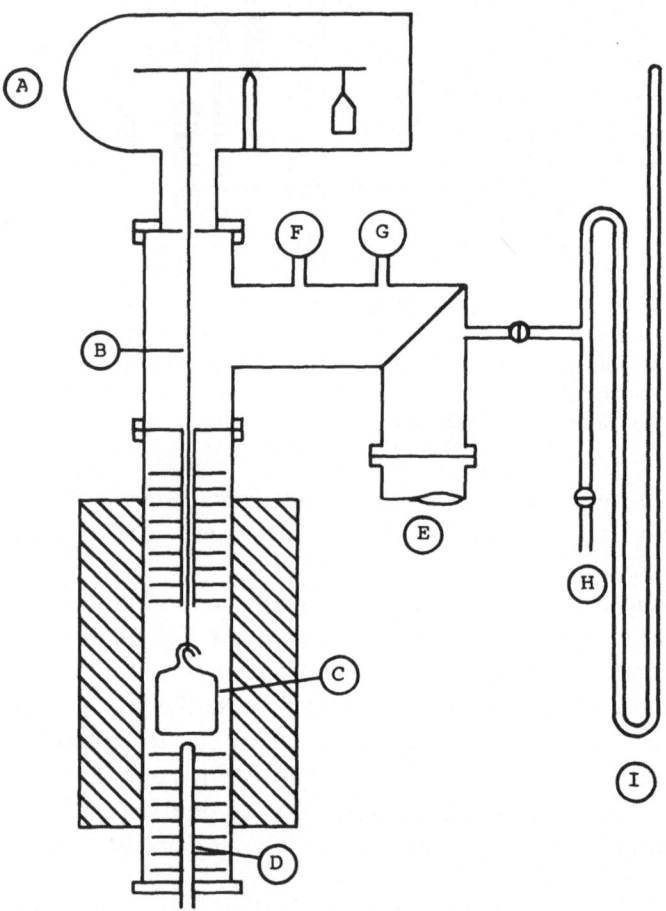

Fig. 3. Gravimetric absorption apparatus (schematic) after Bratland: A, recording vacuum microbalance; B, convection baffled Pt–Rh suspension wire; C, absorption vessel; D, thermocouple well with baffles; E, baffle valve vacuum outlet; F, Pirani gauge; G, Penning gauge; H, solute gas inlet; I, mercury manometer.

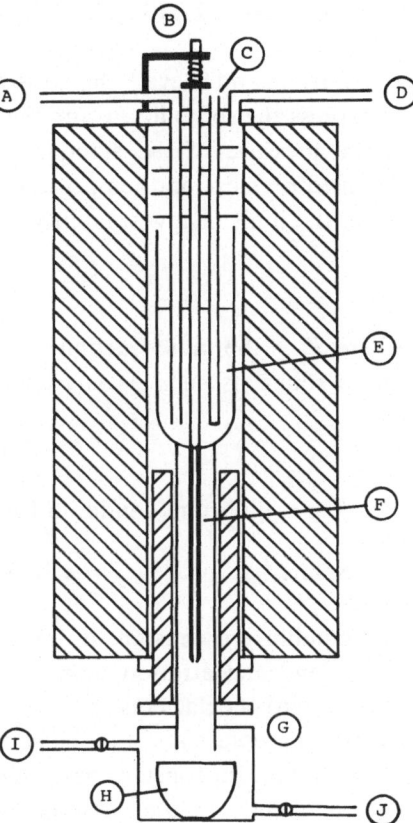

Fig. 4. Freeze elution apparatus (schematic) after Bratland: A, solute gas inlet; B, tap rod; C, thermocouple well; D, solute gas outlet; E, saturation vessel; F, capillary delivery tube; G, elution vessel; H, molten salt receiver crucible; I, carrier gas inlet; J, gas mixture outlet.

temperature-dependent solubilities of He, Ar, CH_4, and N_2 in the same melt using this apparatus. Copeland and co-workers[8] have employed an absorption technique at pressures up to about 500 atm. In their work, solubilities were calculated from the difference in initial and saturation pressure. Their results appear to be too high by at least an order of magnitude compared to the results of Cleaver and Mather[9] who also worked at comparable saturation pressures. Cleaver and Mather's apparatus, however, was a high-pressure variation of the chilling elution method.

One variation of the elution technique which depends on the quiescence of the saturated melt during vacuum pumping has been reported by Field and Green.[10] The advantage of this is the elimination of the second vessel and the experimental problems associated with transferring molten salts.

3. SOLUTION THERMODYNAMICS

Battino and Clever[1] have discussed in detail the various units employed to express the solubility. From the thermodynamic point of view, solubility is the measure of the standard Gibbs free energy under conditions of specified pressure and temperature. The pressure dependence of the solubility in the limit of low solubility is given by the Henry's law constant k_H:

$$P_g = k_H X_2 \tag{1}$$

where X_2 is the mole fraction of the solute gas in solution. In the limit of dilute solutions such as is found for nonreactive gases in molten salts,

$$X_2 = n_g/(n_s + n_g) \cong n_g/n_s \tag{2}$$

The equilibrium condition for solubility is represented by the reaction

$$\text{Solute (gas phase, } c_g = n_g/V_g) \rightleftarrows \text{Solute (solution, } C_d = n_g/V_s) \tag{3}$$

The equilibrium constant in dilute solution is $K_c = C_d/C_g$, where C_g and C_d are in concentration units of moles/cm³ under the approximation of unit activity coefficients, i.e., that the solute fugacity in the gas phase equals its pressure and that its activity in solution equals its concentration. This equilibrium constant has been called the distribution coefficient and is related directly to the Ostwald coefficient L, defined as the ratio of the volume of solute gas absorbed to the volume of solvent at specified temperature:

$$L = V_g/V_s = C_d/C_g = K_c \tag{4}$$

Solubilities have most commonly been reported as C_d. Where Henry's law is experimentally verified, the pressure dependence of the solubility has typically been incorporated into the modified form:

$$K_p = C_d/P_g \tag{5}$$

where P_g is the solute pressure and K_p is related to k_H on the approximation that the partial molal volume of the solvent is constant over the pressure range for which Henry's law is obeyed:

$$1/k_H = X_2/P_g = (n_g \bar{V}_s/V_s)/P_g \tag{6}$$

$$K_p = C_d/P_g = 1/k_H \bar{V}_s \tag{7}$$

Since the majority of solubilities have been found to obey Henry's law, the pressure-independent solubility, K_p, is the most commonly reported quantity.

In the approximation of ideal gas phase behavior, the distribution coefficient K_c is readily related to the Henry's law constant K_p:

$$K_c = C_d/C_g = C_d/(P_s/RT) = K_p RT_s \tag{8}$$

The physical relationship between K_c and K_p becomes apparent if the equilibrium reaction is considered as the two-step process in which the solute in the gas phase at pressure $P_s = C_g RT$ is first expanded to a pressure $P' = C_d RT$, followed by a dissolution step at constant concentration:

$$X \text{ (ideal gas, } P_s = C_g RT) \xrightarrow{I} X \text{ (ideal gas, } P' = C_d RT) \xrightarrow{II} X \text{ (solution, } C_d)$$

$$\tag{9}$$

Since we are considering the equilibrium condition, $\Delta G = \Delta G_I + \Delta G_{II} = 0$, $\Delta G° = -RT \ln K_p$; the separate enthalpy and entropy terms can be evaluated. From the van't Hoff relation, the heat of solution can be obtained as the slope $-Rd(\ln K_p)/d(1/T)$, where this enthalpy can be attributed to ΔH_{II} since, for an ideal gas expansion, $\Delta H_I = 0$. The entropy for the ideal gas expansion is well known and is given as

$$\Delta S_I = -R \ln(P'/P_s) = -R \ln K_c = -R \ln K_p R'T \tag{10}$$

where the gas constant in the logarithm is primed to emphasize that its value must be in appropriate units of K_p, i.e., 82.05 ml-atm/deg-mole, whereas the unprimed R will take the corresponding dimensions of the entropy.

Temperature-independent values of the solubility, even where the pressure dependence has been verified, are of substantially lesser value than studies over a range of temperatures and become tedious to catalog as well as interpret meaningfully in terms of physical-chemical solution theory. However, in as diverse a field of solvents as molten salts offer, considerable interest in mixed-salt solvents may dictate single-temperature studies where solvent components are systematically varied. The ideal experiment would, of course, examine solubility as a function of all three parameters: pressure, temperature, and solvent composition.

The simplest consideration of the thermodynamics of mixed solvents is based on the discussion by Brewer and Pitzer[11] and specifically applied to the case of gas solubility in mixed nonelectrolyte solvents by Hildebrand,

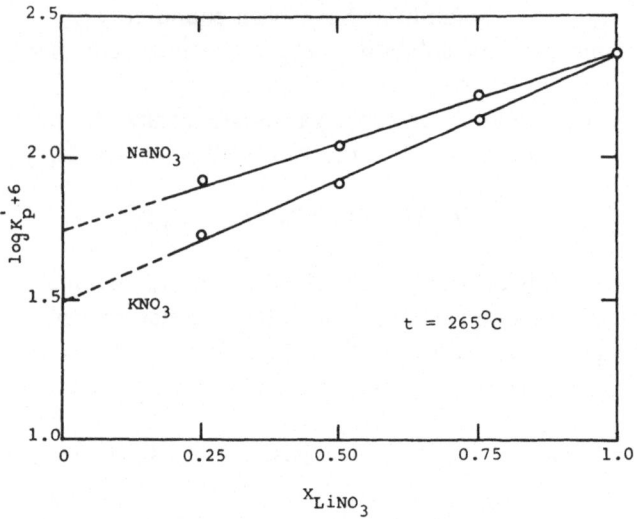

Fig. 5. Semilogarithmic plot of H_2O solubility vs $LiNO_2$, mole fraction for ideal mixed alkali nitrate solvents (units of K_p' are mole/mole/torr). (From Ref. 13.)

Prausnitz, and Scott.[12] The case for mixed ionic solvents having a common ion introduces no additional restrictions and may be treated identically for the limiting case. Identifying the pure and mixed solvents by the subscripts, the solubilities are related as

$$\log K_{AB} = X_A \log K_A + X_B \log K_B + w X_A X_B \qquad (11)$$

where w is the mixing parameter of the two-component solvent representing a first-order correction. For ideal mixtures $w = 0$.

Of the systems considered in this survey, Bertozzi's data[13] on the solubility of water in mixed lithium–sodium/potassium nitrates illustrate this relationship to be valid within experimental error. This system is particularly noteworthy since not only is the solute a highly polar molecule but the solvent is low-melting, which implies that the strength of the potential relative to the kinetic energy is more significant than at higher temperatures, and the ionic radii ratio is large, between approximately 1.5 and 2.0. These data are illustrated in Fig. 5.

Substitution of the enthalpy and entropy terms for the log K term in the equation for ideal mixed solvents and matching the temperature dependent and independent terms separately yields

$$\Delta H_{AB}^s = X_A \Delta H_A{}^s + X_B \Delta H_B{}^s \qquad (12)$$

and

$$\Delta S_{AB}^{\circ} = X_A \, \Delta S_A{}^{\circ} + X_B \, \Delta S_B{}^{\circ} \tag{13}$$

Examination of the separate heat and entropy terms will yield more stringent tests since we are no longer inspecting the small difference between two large quantities as is the case when the solubility is examined.

3.1. Solubility

In their review, Battino and Clever[1] point out that one of the more common gauges of gas-solubility accuracy is several independent determinations. An examination of Table I quickly reveals that very few systems have been studied independently by different techniques or in different laboratories. Although this condition is undesirable in terms of verifying solubility data, it is understandable in light of the paucity of systems which have been studied.

Seven solute–solvent systems have been reported by more than one group. Four of these are with alkali nitrate melts, and the remaining two are with alkali chloride solvents. Copeland and co-workers,[8] Cleaver and Mather,[9] and Field and Green[10] have reported solubilities for helium, argon, and nitrogen in sodium nitrate and argon in lithium nitrate. Recently, Copeland and Christie[14] have reported that the previous solubility results from their laboratory cannot be considered reliable. Therefore, their results will not be considered in this review.

The experimental results of Cleaver and Mather and Field and Green are shown in Fig. 6. The semilogarithmic plot of K_p versus the reciprocal temperature offers the advantage of displaying the data in linear form as well as providing a direct comparison of enthalpy effects. In order to facilitate comparison, the curves have been extended over the entire range of the abscissa although this extends below the melting point for sodium nitrate and well above the experimental range for lithium nitrate. The data reported by Cleaver and Mather are plotted as open circles in all four systems. The error bars were assigned by them as a consequence of several determinations at various pressures for each of the (usually) three temperatures they studied. Therefore, each of their data values represents an average of between four and six separate determinations. It should be noted that their highest temperature value for the $N_2/NaNO_3$ system is not shown in order to scale the figure. Field and Green's data were obtained by varying both temperature and pressure for each determination.

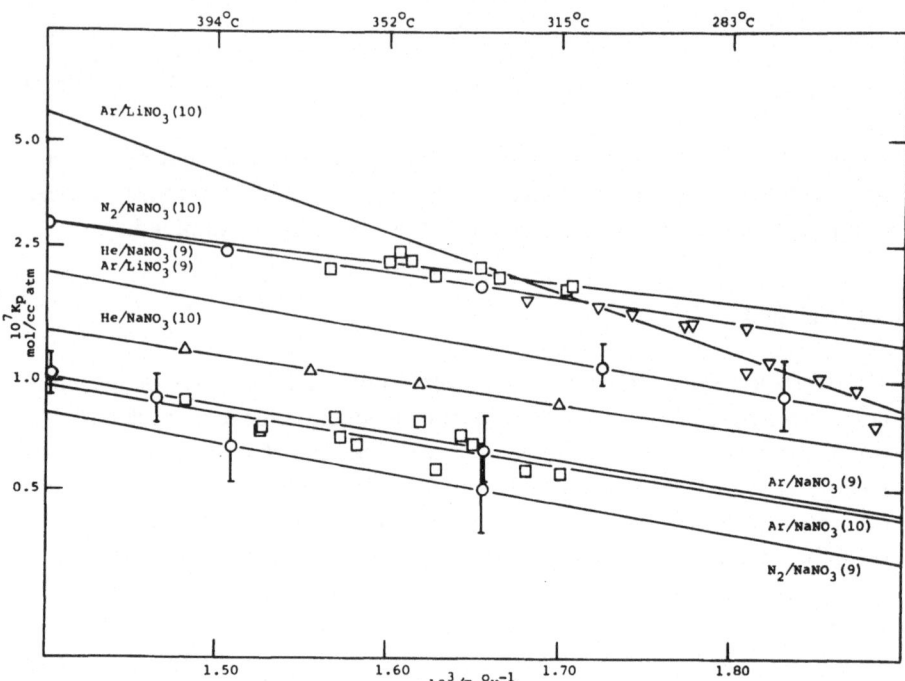

Fig. 6. Semilogarithmic plot of K_p vs $1/T$ for selected gases in alkali nitrate solvents.

A comparison of the results for the four systems calls for a separate discussion in each case. In the first place, the results for argon in sodium nitrate are in complete agreement within the experimental error of both groups. The difference obtained by the separate least-squares analyses lies within the 98% confidence interval. The excellent agreement for this system tends to confirm the reliability of both experimental techniques. For the Ar/LiNO$_3$ system, the results of the two groups are not as disparate as might at first glance be supposed. Although Field and Green's data appear to require a considerably larger heat of solution than in any of the other cases, it is apparent that below 283°C the solubility results of both groups are in agreement. Curiously, above this temperature the temperature dependences are in accord, but a systematic difference in the solubility values appears. The source of this difference is difficult to evaluate. At pressures around 1 atm corresponding to Field and Green's method, the decomposition of the nitrate, although not large enough to affect the freezing point, introduces sufficient oxygen that the eluted gas required pretreatment with heated copper mesh in order to remove the oxygen before chromatographic analysis. Moreover, the separation of the argon and oxygen peaks by gas chromatography is especially difficult. The strength of this argument, however, is weakened to a certain degree by the argon studies in sodium nitrate.

The remaining two systems of helium and nitrogen in sodium nitrate are even more perplexing. In both cases, the disagreement between the results of the two groups lies in the magnitude of the solubility (and therefore is reflected in the entropy term) rather than in the temperature dependence of the solubility; i.e., the enthalpy terms agree within experimental error. In the helium case, Cleaver and Mather report solubilities about two and one-half times larger than found by Field and Green, whereas, in the nitrogen studies, the latter group reports values about five times larger. It would appear that additional studies will be required in order to evaluate these results, particularly since there is no consistency which could be attributed to the experimental methods involved.

The remaining systems which have been reported independently are carbon dioxide in potassium chloride and hydrogen chloride in sodium chloride and potassium chloride. As was noted previously, Bratland and co-workers[2] reported the solubility of carbon dioxide in potassium chloride by the "chilling" method. Recently, Novozhilov and co-workers[15,16] published solubilities determined by a volumetric absorption technique on the CO_2–KCl system and also on the HCl–NaCl and HCl–KCl systems. The latter systems were previously reported by Lukmanova and Vil'nyanskii[17] based on an elution method. Bratland's and Novozhilov's reported data are shown in Fig. 7. The solubility curves calculated from the wt. % solubility at 840°C and the heats of solution reported by Lukmanova are shown as shortened plots. Bratland et al. found a reproducibility of $\pm 2\%$ and estimate the overall precision of their results at $\pm 5\%$, a value typically reported by other workers. Novozhilov et al. estimate their precision at $\pm 1\%$, a value significantly lower than typical. Examination of Fig. 7 indicates that the agreement between these workers on carbon dioxide's solubility in potassium chloride ranges between 15% at 800°C and 2% at 950°C. It would seem that both sets of results are most probably valid within $\pm 5\%$. The agreement is certainly good enough to warrant confidence in both experimental techniques to within the limits of this precision.

The results of Novozhilov and co-workers for the HCl–NaCl system are in very poor agreement with those of Lukmanova and Vil'nyanskii. With the already noted exception of the latter's curves, those curves drawn in Fig. 7 have been recalculated from the reported solubility values. This point is of significance with regard to the two sets of solubility data published by Novozhilov et al. on the HCl–NaCl system. In one of their pa-

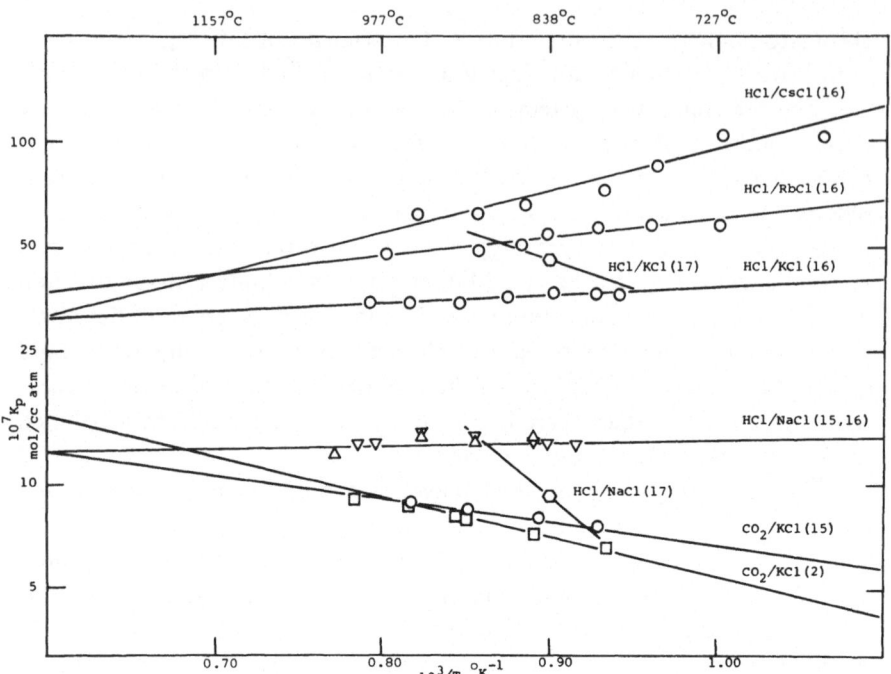

Fig. 7. Semilogarithmic plot of K_p vs $1/T$ for selected gases in alkali chloride solvents.

pers,[16] the authors neglect the two highest temperature values (of seven) reported in their calculation of the heat of solution because of a suspected change (decrease) in the solubility above 1213°K. They likewise suggest that the variation in the solubility of HCl in the other alkali chlorides displays a characteristic S-shape. There is also some confusion in that paper in that it does not include the values given in their first paper.* The present reevaluation of the solubility of HCl in NaCl includes both sets of data published by these workers. In this system, as well as the remaining alkali metal chloride solvents, it remains to be shown that non-linear behavior in the van't Hoff plots is not due to some experimental artifact, especially in dealing with such a corrosive solute. Caution would seem to be warranted since studies of HF in fluoride melts have all been found to be linear.[18]

   A comparison of the HCl solubility values of Novozhilov et al. and Lukmanova and Vil'nyanskii shows the latter's value to be approximately

---

* It might be noted that although Ref. 15 was published two months later than Ref. 16, it was received by the journal four months earlier.

30% smaller in NaCl and 30% larger in KCl. The temperature dependence of the solubility is even more inconsistent. Lukmanova's heats of solution are endothermic and similar to those reported for the nonpolar gases, whereas Novozhilov's heats are more in line with those found for other polar solutes (see Appendix).

Hydrogen solubilities have been reported to be particularly difficult to determine. Attempts to determine the solubility of H_2 in alkali metal hydroxides have been reported by Sullivan and co-workers.[19] They concluded that between 400° and 500°C the solubility was less than 60 mg/100 g hydroxide. More recently, Malinauskas and co-workers[20] discussed the problem in reference to the permeability of hydrogen in the container materials and described a modification of the Oak Ridge elution apparatus[3] where the salt contaminant vessels and transfer lines are doubly contained. Their preliminary results for hydrogen in 66–34 mole % LiF–BeF$_2$ at 600°C list $K_p = 4.34 \times 10^{-8}$ mole-cm^{-1}-atm^{-1} compared to a value for helium of 8.40×10^{-8}.

3.2. Heats of Solution

As indicated above, the experimental value of the heat of solution is obtained from the van't Hoff relation of the temperature dependence of the solubility. Evaluation of temperature-dependent solubility data is best performed by linear least-squares analysis in order to avoid introducing personal bias into the heat and entropy of solution values. All data presented below which were not so treated in the original publication have been reevaluated by least squares unless otherwise noted. The latter condition would be a consequence of either insufficient data or their unavailability in the reference. In those cases where uncertainties are reported they will be taken as one standard deviation unless noted otherwise, in which case they are average deviations.

A typical plot of the van't Hoff equation showing the results of Field and Green[10] for four gases in sodium nitrate is presented in Fig. 6. The heats of solution obtained from the slopes of these and additional studies made by Cleaver and Mather[9] and Field and Green are summarized in Table II. Of the four systems reported by both groups, agreement is found in all cases, although additional evidence discussed above indicates that Cleaver and Mather's value for argon in lithium nitrate is probably the more reasonable.

Bertozzi's solubility data for water in (Li-Na/K)NO$_3$[13] was discussed earlier in reference to ideal mixed solvents. The comparable test of the heats

TABLE II. Heats of Solution in MNO$_3$, ΔH^s, kcal/mole

Solute		M$^+$		
	Li	Na	K	Rb
He (F)^a		3.22 ± 0.18		
(C)^b		3.29 ± 0.08		
Ar (F)	8.00 ± 0.89	3.52 ± 0.81	2.53 ± 0.67	
(C)	3.34 ± 4.80	3.82 ± 1.71		4.86 ± 1.36
N$_2$ (F)	1.82 ± 0.18	2.74 ± 0.83	1.72 ± 0.78	
(C)		3.96 ± 2.20		
CO$_2$ (F)		−2.68 ± 0.12	−2.28 ± 0.44	

^{a} F: Field and Green.[10]
^{b} C: Cleaver and Mather.[9]

of solution as a linear function of solvent composition is presented in Fig. 8. The results are not nearly as concordant as in treating the solubility. One reason for this lies in the fact that five of the seven values were based on only two temperature points, with the remaining two values based on three temperature determinations. These values were obtained by least-squares fitting the van't Hoff relation of the published data. The solid curves were obtained from least-squares fits of the computed heats versus the lithium mole fraction. The broken lines connect Bertozzi's extrapolations to the pure melts. The 50–50 mole % (Li-K)NO$_3$ evaluated from Tripp and Braunstein's[21] study of vapor pressures of aqueous melts yields a heat of vaporization for water of 10.4 kcal/mole between 119–150°C, i.e., the negative heat of solution, compared to Bertozzi's −10.5 for the heat of solution between 230–265°C. Peleg[22] determined the solubility of water in pure LiNO$_3$ between 275 and 335°C (four temperatures) and in pure NaNO$_3$ between 310 and 342°C (three temperatures) by analysis of the voltammetric "water wave." He reports heats of solution of −9.35 and −8.15 kcal/mole for LiNO$_3$ and NaNO$_3$, respectively. These values are 4 kcal smaller than the determined value in LiNO$_3$, 2 kcal smaller than the determined value in LiNO$_3$, and 2 kcal smaller than the extrapolated value for NaNO$_3$. A similar method was used by Zambonin and co-workers[23] between 227 and 394°C in eutectic (Na, K)NO$_3$, presumably 50–50 mole %, to obtain a heat of solution equal to −8.4 kcal/mole. Although this value agrees within 0.6 kcal with the extrapolated value from Bertozzi's data, the extrapolation is too tenuous to enable any conclusion to be drawn.

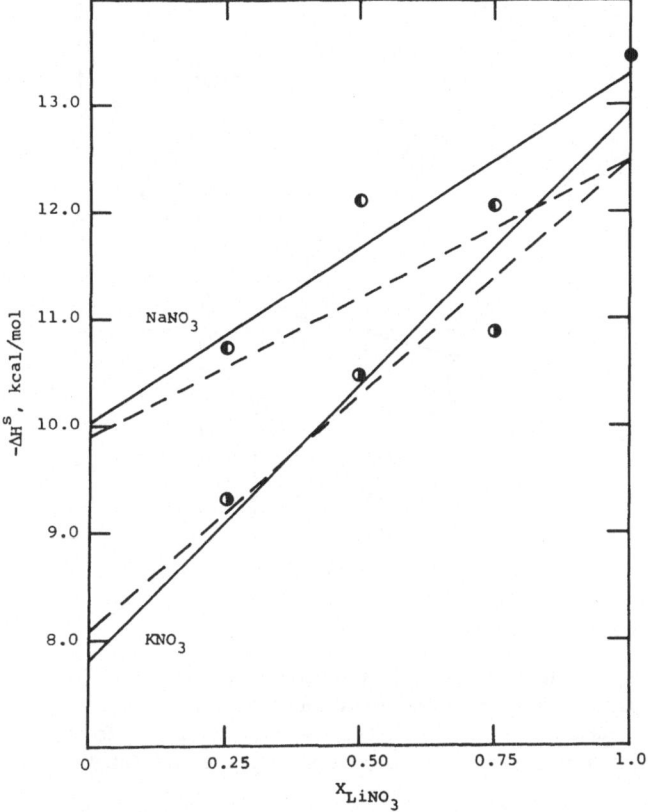

Fig. 8. Heats of solution of H_2O vs $LiNO_3$ mole fraction for ideal mixed alkali nitrate solvents; – – –, original slopes from Bertozzi; ——, least-squares slopes. (From Ref. 13.)

Duke and Doan[24] employed a manometric absorption method to determine the temperature dependence of water solubility in 30–23–37 mole % (Li, Na, K)NO_3 to obtain a heat of solution of -13 ± 4 kcal/mole compared to -10 kcal based on the ideal mixture extrapolation. The foregoing heats are presented in Fig. 9 as an orthogonal projection of the ternary solvent system. The enthalpy plane surface intersects the pure $LiNO_3$ coordinate at the mean of least-squares heats. It is apparent that there is no justification in ascribing any curvature due to the scatter although it is interesting to observe that the dependence of the magnitude of the exothermic heat of solution with cation radius is practically linear, decreasing at a rate of 7 kcal per angstrom. All values other than Bertozzi's are noted by reference number and listed in the text. The values from Bertozzi are

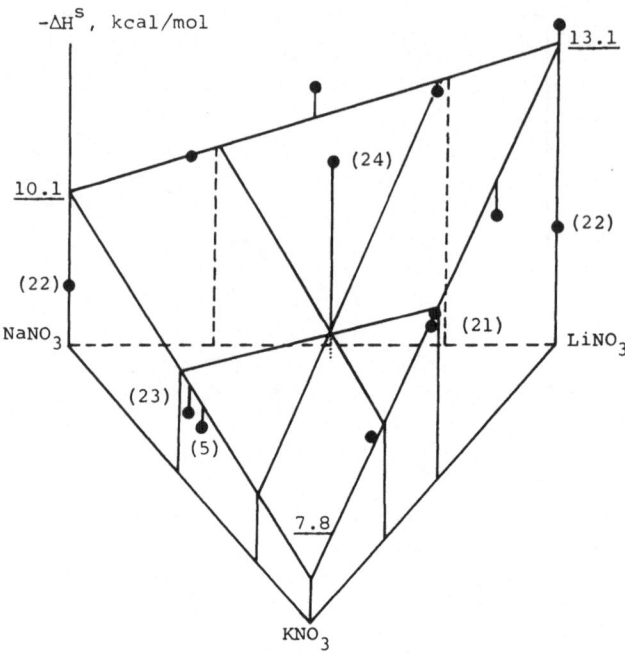

Fig. 9. Orthogonal projection of exothermic heats of solution of
H_2O for ideal mixed alkali nitrate solvents (base at 7.0 kcal/mole).
References are in parenthetes, unreferenced values are from Ref.
13, and extrapolated values for pure salts are underlined.

listed in Table III. Values which show no uncertainty are based on only
two solubilities. Parenthetical values are extrapolated. Haug and Albright[5]
determined the solubility of water in mixed solvents containing 54.3 mole %
$NaNO_3$, 0–5 mole % KCl, with a balance of KNO_3 by an elution method
between about 200° and 360°C. The heats of solution were unaffected
by the chloride ion at those concentrations and were reported to be −8.35
kcal/mole. Hull and Turnbull[25] have reported H_2O solubilities in 55.5%
KNO_3–44.5% $NaNO_2$ (nitrite) solvent with a heat of solution of −8.7
kcal/mole. This is effectively equal to that at the corresponding composition
in KNO_3–$NaNO_3$ of Haug and Albright.

Also included in Table III are the values of the heat of solution of
ammonia in alkali nitrate melts, due to Allulli,[26] from solubilities determined by a manometric absorption technique. Each solvent was investigated at four temperatures over a 60–90° range from 160°C in the
lower-concentration lithium ion solvents to around 320°C in the others.
Since these are the only data available in ammonia, an attempt to obtain

TABLE III. Polar Gas Heats of Solution in (Li,M)NO₃

Solute	Mole % M⁺	$-\Delta H^s$, kcal/mole	
		I[a]	II[b]
NH₃	0	15.91 ± 0.38	15.81 ± 0.34
	25 K	11.78 ± 0.13	11.90
	57 K	6.88 ± 0.24	6.84
	(100) K		(0.03 ± 0.26)
	18 Na + 55 K	5.53 ± 0.17	
	(100) Na		(6.94)
H₂O	0	13.45	13.38
	25 K	10.88	11.67

[a] Calculated from solubility data.
[b] Calculated from column I vs mole % Li.

estimates of the heats of solution in the pure components of the ternary system can be made using the ideal mixed solvents assumption and evaluating the plane surface through the heats of the four solvent systems. Figure 10 is an orthogonal projection plot of the (Li–Na–K)NO₃ ternary

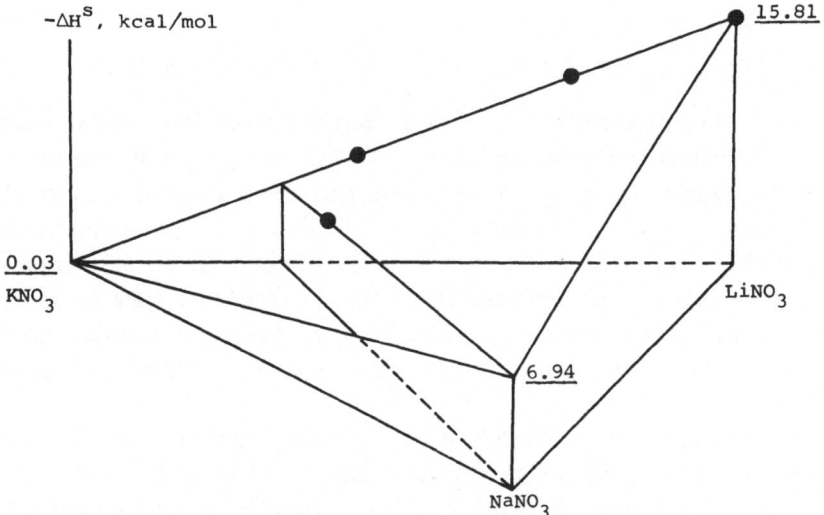

Fig. 10. Orthogonal projection of exothermic heats of solution of NH₃ for ideal mixed alkali nitrate solvents. (From Ref. 26.)

system. Extrapolated values for the heat of solution in pure $NaNO_3$ and KNO_3 obtained are 6.9 and 0.0 kcal/mole, respectively. The trend in magnitude of the exothermic heat of solution with cationic radii is consistent but considerably steeper than in the case of water.

Both water and ammonia solubilities have been reported for perchlorate solvent systems. Duke and Doan[24] reported a heat of solution for water in pure $LiClO_4$ of -9 ± 3.5 kcal/mole, and Allulli[26] reported a heat of solution for ammonia of -13.8 kcal/mole in 76–24 mole % (Li, K)ClO_4. Both studies were over a temperature range of approximately 240–300°C.

The temperature dependence of the solubility of oxygen in mixed alkali metal carbonates between 500–800°C has been reported by Schenke and co-workers.[27] The solubility was determined by an amperometric titration technique. Endothermic heats of solution considerably larger than found for nonpolar gases in nitrate melts were given at 14.9 and 17.7 kcal/mole for 53.3–46.7 mole % (Li, Na)$_2CO_3$ and 43.5–31.5–25.0 mole % (Li, Na, K)$_2CO_4$, respectively.

Of the alkali metal halide solvents, argon and carbon dioxide have been determined in the pure potassium salts of chloride, bromide, and iodide; in addition, carbon dioxide has been determined in sodium chloride. The heats of solution are given in the following list, in kcal/mole:

	NaCl	KCl	KBr	KI
Ar	—	9.45	6.28	6.46
CO_2	5.82	4.53	4.00	6.67

All solubilities were determined over at least the range 800–1000°C except CO_2 in KI which was determined around 700°C. The carbon dioxide studies were determined by Bratland and co-workers[2] as discussed under the Experimental section of this survey. Janz[28] quotes the argon studies by Woelk.[29] As can be seen, the enthalpies are all endothermic and larger in magnitude than those found for the nonpolar gases in nitrate melts. Insufficient numbers of systems are covered to enable any conclusions to be drawn either from the point of view of the solutes or from the effect of the anion.

Ryabukhin[30] has determined the temperature-dependent solubility of chlorine in the pure and equimolar mixtures of NaCl, KCl, and $MgCl_2$ by an elution technique. Kowalski and Harrington[31] have similarly studied the system Cl_2/KCl–$PbCl_2$. The lower limit of the temperature ranges for the solvents necessarily increase for the pure end members, but within the approximation of temperature independence of the enthalpy, comparison is feasible. The heat of solutions from both studies are given in Table IV. Ryabukhin's data were recalculated for the heats of solution by least squares. His reported values are given in parentheses. Kowalski and Harrington's data is listed in their paper as "sample results" which on least-squares evaluation agree with their reported values within one standard deviation in three of the four cases. Their published heats of solution are quoted.

For the three binary chloride systems reported by Ryabukhin the heats of solution are linear functions of the mole fraction within about one-half of a standard deviation. This is definitely not the case for the KCl–PbCl$_2$ system. Using the heat of solution of Cl$_2$ in pure KCl from Ryabukhin, we note that ΔH^s goes through a minimum which if extrapolated linearly on both sides of the minimum intersects at a composition of 0.66 mole fraction KCl with a value of -6 kcal/mole and intersects in pure PbCl$_2$ at a value of $+15$ kcal/mole. The minimum corresponds to a stable compound formation having the stoichiometry K$_2$PbCl$_4$ and is similar to the minimum found in the HF solubility in LiF–BeF$_2$ solvents at the composition Li$_2$BeF$_4$ discussed below.

Solubilities of polar gases in alkali metal chlorides have been reported for water in LiCl–KCl by Burkhard and Corbett,[32] for HCl in NaCl, KCl, RbCl, and CsCl by Novozhilov, Devyatkin, and Gribova[15,16] and in pure NaCl, KCl, and MgCl$_2$ and their mixtures by Lukmanova and

TABLE IV. Heats of Solution for Chlorine

Mole %				ΔH^s, kcal/mole
NaCl	KCl	MgCl$_2$	PbCl$_2$	
100				19.79 ± 1.00 (21.90)
	100			9.05 ± 0.56 (11.57)
		100		4.77 ± 1.26 (8.17)
50	50			13.59 ± 0.47 (16.10)
50		50		10.90 ± 1.01 (13.80)
	50	50		6.60 ± 0.41 (9.02)
	23		77	7.93
	40		52	-1.00
	60		40	-3.37
	70		30	-3.68

Vil'nyanskii,[17] and for carbon monoxide in equimolar (Na, K)Cl by Zezyanov and Il'ichev.[33] Hydrogen chloride solubilities have also been reported in the alkaline earth chlorides of calcium, strontium, and barium by Novozhilov, Gribova, and Devyatkin.[34] The water studies were made using a manometric absorption method in 50, 53, 60, and 68.6 mole % mixtures of LiCl in KCl at 490°C and in the 50 and 60 % compositions at 390°C. Henry's law behavior was observed up to 18 torr at 480°C and up to 14 torr at 390°C. Solubilities were found to show a strong dependence on the LiCl composition and were reported per mole LiCl. Heats of solution were reported for the two compositions studied at the two temperatures as -8 and -11 kcal/mole. These values are comparable to those discussed previously in the corresponding nitrate melts.

The HCl studies of Lukmanova and Vil'nyanskii were made by an elution method over a temperature range of 500–900°C. Heats of solution in the pure melts were reported as 19.1, 8.1, and 2.0 kcal/mole for NaCl, KCl, and $MgCl_2$ respectively, at 840°C. As discussed in the previous section, these results are more generally comparable in magnitude to the heats of solution (also endothermic) of Cl_2 in these solvents discussed previously. The heats of solution for HCl in both the alkali metal chlorides and the alkaline earth chlorides reported by Novozhilov and co-workers are calculated, in kcal/mole, as: NaCl (-0.1), KCl (-0.9), RbCl (-2.2), CsCl (-5.0), $CaCl_2$ $(+6.0)$, $SrCl_2$ $(+9.2)$, and $BaCl_2$ $(+10.6)$. The trend in comparing the Group I and II metal salts is interesting in that with increasing atomic number in the alkali metal chlorides the heats become increasingly exothermic while with the alkaline earth group the heats become increasingly endothermic. The same trend is observed below in considering the heats of solution of HF in molten fluoride mixtures of Groups I and II ions.

Gas solubilities in pure fluoride melts have not been reported. Extensive studies have been reported for the rare gases (He, Ne, Ar, and Xe) in mixed fluoride melts of LiF–NaF–KF eutectic (46.5–11.5–42.0 mole %, respectively),[35] in approximately equimolar $NaF-ZrF_4$ in some cases containing 4% UF_4,[3b] in 64–36 mole % $LiF-BeF_2$,[36] and in 57–43 mole % $NaF-BeF_2$.[37] The heats of solution reported for these systems are summarized in Table V. The temperature range for these systems was usually between 600–800°C. The heats are all endothermic and generally of the same magnitude as those found for nonpolar gases in other halide melts.

The solubility temperature dependence of carbon dioxide in a NaF–BeF_2[38] melt was reported to undergo a minimum around 600°C over a

TABLE V. Heats of Solution of Rare Gases in Fluoride Melts

Solute	Fluoride melt	ΔH^s, kcal/mole
He	Li–Na–K	8.0
	Na–Zr	6.2
	Li–Be	5.2
	Na–Be	4.7
Ne	Li–Na–K	8.9
	Na–Zr	7.8
	Li–Be	5.9
Ar	Li–Na–K	12.4
	Na–Zr	8.0
	Li–Be	8.6
	Na–Be	7.4
Xe	Li–Na–K	—
	Na–Zr	11.0
	Li–Be	12.1
	Na–Be	11.4

range between 400 and 800°C. Boron trifluoride was reported to have an exothermic heat of solution of -15.1 kcal/mole in a mixed Li–Be fluoride melt (65–28 mole %) containing smaller quantities of Zr, Th, and U (5–1–1 %).[39]

The solubility of hydrogen fluoride has been extensively investigated between 600–800°C as a function of solvent composition in $LiF-BeF_2$[40] and $NaF-BeF_2$ and $NaF-ZrF_4$[41] melts. Composition of the alkali metal component ranged between 54–89 mole %, 49–75 mole %, and 45–91 mole %, respectively. Heats of solution in all cases were found to be exothermic. Field and Shaffer[18] recalculated the $LiF-BeF_2$ data by least squares and found the composition dependence of the heat of solution to be linear on either side of a minimum of the 66 mole % LiF concentration in the Li–Be system. This behavior is similar to the Cl_2 behavior in the K–Pb chloride system mentioned previously and is illustrated in Fig. 11. Results for deuterium fluoride in the Li_2BeF_4 melts yielded a heat of solution of -6.43 ± 0.15 kcal/mole compared to -5.98 ± 0.19 kcal/mole for hydrogen fluoride. Since these uncertainties are calculated at one standard deviation based on small-sample statistics (Student's t), where the sample sizes (N) were 8 and 9, respectively, some additional measure of the sig-

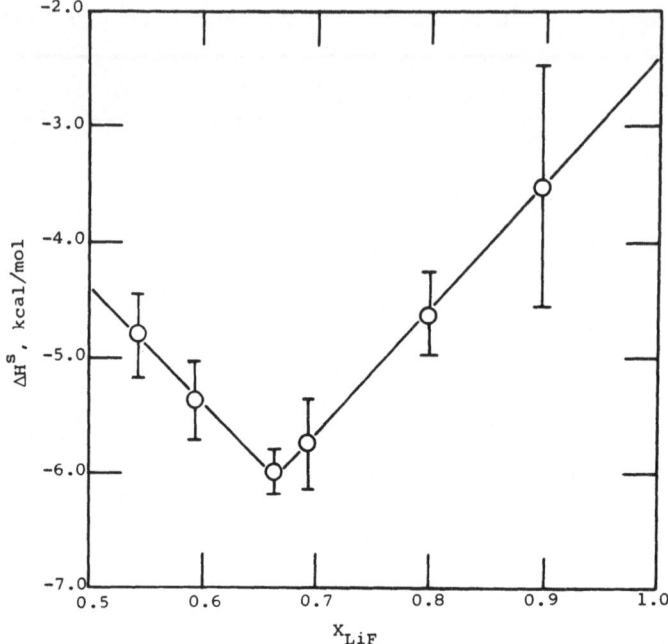

Fig. 11. Heats of solution of HF vs LiF mole fraction in BeF_2 mixed solvent.

nificance of their difference is desirable. The t-test confidence interval at which the error of the heat of solution for one system would include the mean of the other can be ascertained from the ratio of the error to the difference in the means, i.e., $0.15/0.45 = 3.0$ for DF and $0.19/0.45 = 2.5$ for HF. In either case, the probability that the heat of solution would be this large or larger lies between 5% and 2% for $N - 2$ degrees of freedom. Further discussion of the isotope effect found for these systems is included in the following section on the entropy of solution.

A somewhat different variation of the heat of solution is observed in the $NaF-BeF_2$ system as a function of solvent composition. A less well-defined point of inflection is observed having about a threefold decrease in the slope of ΔH^s vs X_{NaF} in the region of 0.6–0.7 mole fraction of NaF. Up to the composition Na_2MF_6 in both the Be and Zr systems with increasing Na^+, the heats of solution decrease at approximately the same rate as in the Li–Be system, but about 0.75 kcal more exothermic per mole. All three systems appear to have the same extrapolated value (hypothetical due to the immense increase in viscosity) in pure BeF_2, corresponding to $+2.5$ kcal/mole. As seen in Fig. 11, the extrapolated heat of solution in pure LiF appears to be -2.4 kcal/mole. A more tenuous extrapolation to pure NaF from the Na–Be system indicates heat of the order of -18 kcal/mole, if, in fact, the values of 70 and 75% do lie past the inflection point and do not represent a deeper minimum. Unfortunately only one composition at 80.5% in the Na–Zr was studied which lies above 65% NaF. Its value for ΔH^s differs only within experimental error from that at 65%.

3.3. Entropy of Solution

By considering the equilibrium process as the two-step mechanism where the solute gas phase is first expanded as an ideal gas followed by the dissolution step at constant concentration, the entropy of solution can likewise be conveniently divided into the corresponding terms. Except in the extreme cases of nonideality of the solute gas phase due to association, high saturation pressure, etc., the expansion term is not of any fundamental interest although, of course, it is significant in affecting the magnitude of the standard free energy and, consequently, the magnitude of the solubility. The second entropy term relating to the constant-concentration dissolution step is of potentially greater interest in the interpretation of solubility.

Although there is considerable confusion of terms in the literature, the entropy of solution associated with the overall equilibrium at constant pressure as reflected in the equilibrium constant K_p has been generally identified as ΔS_p, while that associated with the second step at constant concentration and related to the distribution coefficient K_c has been labeled ΔS_c. In some papers these two quantities have been referred to as ΔS_d and ΔS_2, respectively, where ΔS_1 is the ideal gas expansion term equal to the difference $\Delta S_d - \Delta S_2$.

The evaluation of the entropy terms from experimental solubility data can be accomplished in either of two equivalent ways. In either event it is necessary to have solubility data as a function of temperature in order to first calculate the heat of solution. Normally, the approximation that the heat of solution is independent of temperature is justified, and ΔS_p is simply taken as $\Delta H^s/T$. As noted above, ΔS_c is the difference given as $\Delta H^s/T + R \ln(K_p R'T)$. Alternately, where the data are subjected to linear least-squares analysis of $\ln K_p$ vs $(1/T)$, the slope is related to the heat of solution as $(-\Delta H^s/R)$, and the calculated intercept is equal to $\Delta S_c/R - \ln R'T$. The latter method has the advantage of a straightforward statistical evaluation of the precision of the thermodynamic quantities.

Solution entropies ΔS_c have been reevaluated by least squares from published solubilities and are listed in the Appendix. The values quoted are taken at $1.10T_m$, where T_m is the melting point of the solvent for pure melts or the melting point of the lowest-melting component for mixed melts, except for the fluoride melts where $1000°K$ was used.

One approach to examining the entropy of solution is the empirical correlation developed during the 1930's by Lannung,[42] Evans and Polanyi,[43] Bell,[44] and Barclay and Butler,[45] in which a linear relationship was proposed between ΔH and $T \Delta S$ of the form

$$T \Delta S = \alpha \Delta H + \beta$$

Similar examination of the results of gas solubilities in molten salts yield some striking similarities and some interesting differences from the earlier results. Green[46] observed that his results for nonpolar gases in alkali nitrates and those published on nonpolar gases in alkali halides showed a similar linear relationship but with a much larger α and a significantly more negative β of about 1.0 and -7.5, respectively, compared to Bell's corresponding values of 0.4 and -0.7. Further examination of additional sets of solubility results including (1) NH_3 and H_2O in alkali nitrates, (2) Cl_2 in KCl–$PbCl_2$, (3) HF in mixed metal fluorides, and (4) the rare gases in mixed metal fluorides yields in the first two cases parallel curves of α ca. 1.0 but with β of $+9.0$ and -3.0. The data for the latter two cases are more tightly clustered, as in fact are the majority of nonpolar gases in alkali nitrate melts. If a parallel curve is drawn through the HF data, β has a value of about -0.5. When the fourth case is examined, no conceivable parallel can be drawn. What is striking, however, is that the various clusters mentioned and the data for the rare gases in the fluoride solvents lie along a straight line that parallels the results of both Bell and Barclay and Butler having an α of 0.4 but with a β about -5.0. The net result appears as shown in Fig. 12. This result can be rationalized in terms of a base dependence of the $T \Delta S$ term on ΔH which follows the same relation observed for non-ionic solutions at normal temperatures with an additional effect dependent on the solute–solvent system in such a manner that displacement from the base line follows the condition $\Delta H° = T \Delta S_c°$, i.e., with a slope of unity. For this second condition, it follows that the corresponding contribution to the standard free energy is zero, where $\Delta G° = -RT \ln K_c$ would depend only on the base relation.

Besides the precedent set by the earlier studies noted above, the empirical correlation examined here has the additional significance that it spans a much broader temperature range than the earlier studies. It is at least as fruitful, even though the precision of the data is not as well sub-

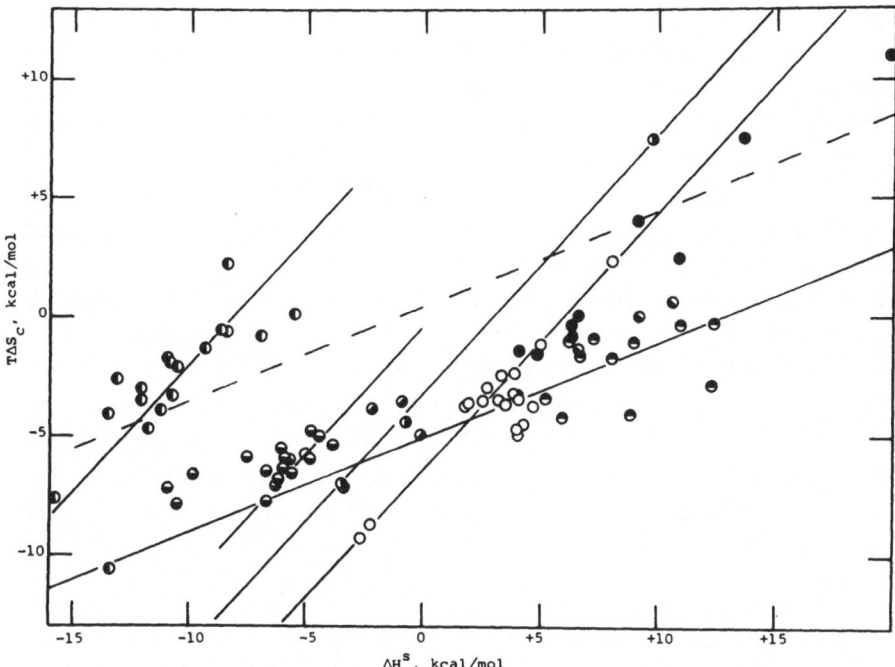

Fig. 12. Parametric relation of the standard free energy of solution, $T \Delta S_c$ vs ΔH^s: ○, He, Ar, N_2, O_2, CH_4, CO_2/MNO_3; ◐, NH_3, H_2O/MNO_3; ◑, Cl_2/KCl–$PbCl_2$; ◕, HCl/MCl; ◖, $HCl/M'Cl_2$; ●, Cl_2, CO_2, CO/MX; ◓, He, Ne, Ar, $Xe/MM'F$; ◒, $HF/MM'F$. (M = alkali metal, M' = alkaline earth metal, X = halide.)

stantiated, and warrants further investigation as more and better solubility data become available.

A second potential correlation in the contribution to ΔS_c for gaseous solutes which are not monatomic stems from the loss of internal degrees of freedom associated with the interaction of the solute in the potential field of the solvent. Except where very strong interactions between solvent ions and the solute molecule are involved, it would be anticipated that at the lower temperatures corresponding to smaller kinetic energies one would observe larger contributions to the entropy of solution due to the loss, particularly, of the rotational motion of the solute molecule in solution relative to the rotational contribution to its entropy in the gas phase.

In a study of the solubilities of the hydrogen fluoride isotopes HF and DF in a common solvent, Field and Shaffer[41] suggest that the difference in the entropies of solution can be attributed to the difference in the gas-phase rotational entropies of the isotopes, since the differences in the vibrational and translational contributions are an order of magnitude smaller. It is also noted that the magnitude of the rotational contribution for either gas is comparable to the magnitude of the entropy of solution, i.e., ca. 9 and 7 for HF and DF, respectively, while the translational contribution is much larger, 40 e.u., and that for vibration is negligible, 0.01 e.u.

Another interesting comparison is the high- and low-temperature results for carbon dioxide. The entropies of solution in alkali nitrates and chlorides given in the Appendix lists around -14 e.u. for the former at a temperature around $300°C$ and about -1 e.u. for the chloride solvent around $800°C$. Field and Green[10] pointed out that it is reasonable to assume that CO_2 has lost a significant portion of its rotational motion in solution based on calculated gas-phase contribution of 15.7 e.u. at $637°K$.

Further correlations are difficult to evaluate due to the large relative uncertainties in the values. Very little discussion on solution entropies has been forthcoming in the literature other than the two concepts discussed here.

4. CALCULATION METHODS

The development of suitable models to account for the solubility behavior of gases in molten salts has been based almost exclusively on a hard-sphere concept of the solute. Two separate approaches can be recognized, both of which have been applied to systems of interest in this review. Each model has as its basis the calculation of the work required to create a cavity in the solvent just large enough to accommodate a solute molecule. The two models differ in the means of evaluating the energy of cavity formation.

An alternative mechanism is based on the "free-volume" model of the liquid state where the solute is assumed to occupy those cavities already existing in the liquid which originated from the volume expansion on fusion of the solvent. The cavities of the free-volume model exist in a distribution of sizes, and the solubility of the gas is dependent on the number of cavities which are large enough to accommodate the particular solute molecule. The free-volume model has found little successful quantitative application although it has been used qualitatively to explain solution behavior in a few instances.

The significant difference between the cavity-formation models is in their treatment of the solvent. In the one case, the solvent is treated as a continuous medium, while in the other, it is treated as a collection of hard spheres. The former might thus be distinguished as a macroscopic or thermo-dynamic model, and the latter as a microscopic or statistical model. Two assumptions are common to both models. First, since the solutions under consideration are very dilute, they can be quite adequately treated as ideal solutions. Secondly, the approximation is made that the solute gas phase behaves ideally, which is reasonable at ordinary pressure and elevated temperatures.

The macroscopic model was developed by Uhlig[47] and applied by Blander and co-workers[35] to molten salt–gas solutions. The microscopic model is based on the rigid-sphere theory of Reiss and co-workers,[48] applied by Pierotti[49] to gas solubility in nonelectrolyte solvents, and ex-tended by Lee and Johnson[50] to molten-salt solvents.

4.1. Microscopic Model

The statistical mechanical theory of fluids developed by Reiss *et al.* is an equilibrium theory based on the properties of a distribution function which measures the density of rigid-sphere molecules in contact with a rigid-sphere solute of arbitrary size. From both geometrical considerations and the virial theorem, a number of exact relations which describe the functional form of the distribution function were derived.

An equation fundamental to the treatment of gas solubilities is derived for the calculation of the reversible work required for the production of a spherical cavity of radius r. The equation is approximate inasmuch as it is an expansion retaining terms up to cubic order. The partial molal Gibbs free energy of cavity formation is given by

$$\Delta \bar{G}_c = K_0 + K_1 a_{12} + K_2 a_{12}^2 + K_3 a_{12}^3$$

where the K's are functions of temperature T, pressure P, density ϱ, and hard-sphere diameter of the solvent a_1, a_{12} is the radius of a sphere which excludes the centers of solvent molecules, and $r = (a_1 + a_2)/2$, where a_2 is the cavity diameter.

In terms of the quantity, $y = \pi a_1^3 \varrho/6$, the K's were evaluated as

$$K_0 = RT\{-\ln(1 - y) + \tfrac{9}{2}[y/(1 - y)]^2\} - \pi P a_1^3/6$$

$$K_1 = -(RT/a_1)\{[6y/(1 - y) + 18[y/(1 - y)]^2\} + \pi P a_1^2$$

$$K_2 = (RT/a_1^2)\{[12y/(1 - y)] + 18[y/(1 - y)]^2\} - 2\pi P a_1$$

$$K_3 = \tfrac{4}{3}\pi P$$

Based on this contribution to the total free energy, but neglecting all pressure-dependent terms and including a contribution due to discharging the cavity, Lee and Johnson present an expression for the distribution coefficient:

$$\ln K_c = [+ \ln x - \tfrac{9}{2}(y/x)^2] + (r/a_1)[6(y/x) + 18(y/x)^2]$$
$$- (r/a_1)^2[12(y/x) + 18(y/x)^2] + 1$$

where $x = 1 - y$, and all other terms are as previously defined. The interaction term was derived from the dispersive portion of the Lennard-Jones 6–12 potential integrated between a_2 and infinity and is given by

$$-16\pi\varepsilon_2 a_2{}^3 P/3kT + 4\pi a_2{}^3 P/3 - kT$$

where ε_2 and a_2 are the potential parameters of the gas at uniform density, $\varrho_2 = P/kT$. The first two terms, however, are eliminated as negligible due to their pressure dependence, and only the $-kT$ contribution is retained.

The difficulty of the theory when applied to molten-salt solvents is the presence of two different hard spheres, i.e., cation and anion, as constituents of the solvent. Lee and Johnson introduced the results of Stillinger[51] and Mayer[52,53] from their development of a corresponding-states theory of molten salts in order to avoid this difficulty.

According to this theory as put forth by Reiss and Mayer the thermodynamic properties of a molten salt are dependent only on the sum of the radii of ions of opposite sign rather than the individual ionic diameters. The applicability of Lee and Johnson's equation is thus limited by the assumptions: (1) that the molten salt is characterized by the density and collision diameter, $\delta = a_1$, and (2) that the solution process depends on the packing of a hard-sphere solute into a hard-sphere solvent.

Based on the relation between the surface tension of the solvent and the collision diameter given by Reiss, Frisch, Helfand, and Lebowitz,[48] we have

$$\gamma = \frac{kT}{4\pi a_1{}^2}\left[\frac{12y}{1-y} + \frac{18y^2}{(1-y)^2}\right] - \frac{Pa_1}{2}$$

where, in an examination of this equation, Mayer[52] pointed out that the pressure-dependent term is negligible. Since the y term is dependent on the molar volume of the liquid (see above), both surface tension and density values are required for the evaluation of the collision diameter a_1. Using values for the temperature dependence of these quantities, the author has extended the computations of Mayer and evaluated the linear temperature dependence of the collision diameters for various alkali metal halides and nitrates using a reiterative computer calculation. Typical results for the 17 salts evaluated yielded collision diameters which reproduced the surface tension within 0.06% after around 25 reiterations at a given temperature. Values of the collision diameter were determined at 25° intervals beginning at the melting point and extending over a 225° range. The 10 values of the collision diameter for each salt were linear in their temperature dependence in all cases, and the least-squares values of the slope and intercept for each were evaluated. These results along with the input values taken from Janz[25] are presented in Table VI.

In a recent paper, Neff and McQuarrie[55] have applied Leonard, Henderson, and Barker's results on perturbation theory applied to a liquid mixture to the calculation of gas solubilities in liquids. They show that

TABLE VI. Temperature Dependence of the Collision Diameters of Molten Salts Calculated from Density and Surface-Tension Data

Salt	M.W.	M.P. (°C)	$\varrho = a_\varrho - b_\varrho T$ (g-cm^{-3})		$\gamma = a_\gamma - b_\gamma T$ (dyn-cm^{-1})		$\delta = a_\delta - b_\delta T$ (Å)	
			a_ϱ	$b_\varrho \times 10^3$	a_γ	b_γ	a_δ	$b_\delta \times 10^3$
LiF	25.94	845	2.3768	0.4902	373.2	0.1093	2.305	0.4365
NaF	41.99	995	2.655	0.560	289.6	0.082	2.606	0.4779
KF	58.10	856	2.6464	0.6515	226.1	0.075	2.945	0.5773
LiCl	42.39	610	1.8842	0.4328	197.9	0.0696	2.946	0.6872
NaCl	58.44	808	2.1393	0.5430	216.21	0.0930	3.454	0.9261
KCl	74.56	772	2.1359	0.5831	175.14	0.0730	3.627	0.8722
RbCl	120.92	717	3.1210	0.8832	179.5	0.0827	3.852	0.9713
NaBr	102.90	747	3.1748	0.8169	161.77	0.0610	3.407	0.8006
KBr	119.01	734	2.9583	0.8253	161.87	0.0720	3.862	0.9920
RbBr	165.38	680	3.7390	1.0718	156.2	0.058	3.721	0.6238
NaI	149.89	662	3.6274	0.9491	171.99	0.090	4.144	1.3523
KI	166.01	685	3.3594	0.9557	162.37	0.087	4.492	1.4440
RbI	212.37	640	3.9499	1.1435	143.4	0.06839	4.248	1.0576
LiNO$_3$	68.95	254	2.068	0.564	143.39	0.053	3.338	0.7561
NaNO$_3$	85.01	310	2.320	0.715	138.95	0.039	3.382	0.5267
KNO$_3$	101.10	337	2.315	0.729	148.48	0.0640	3.706	0.6801
RbNO$_3$	147.49	316	3.049	0.972	149.64	0.070	3.881	0.7274

Pierotti's equations can be obtained as a special case of their equations. In particular, Neff and McQuarrie maintain that Pierotti's equations underestimate the magnitude of the cavity potential charging because the low density limit of the radial distribution function is used. The pressure-dependent terms which were neglected by Pierotti (see above) apparently relate to a term which accounts for the effect on solvent–solvent interactions on adding solute due to the dependence of the radial distribution function of the solvent on the solute. Since Lee and Johnson's work likewise ignores the pressure-dependent terms, similar discrepancies, especially in the pressure and temperature dependence, i.e., solute partial molar volume and heat of solution, can be anticipated.

The results of the solubility calculations based on Lee and Johnson's equation will be examined following a discussion of the Uhlig model, so that the two methods may be compared.

4.2. Macroscopic Model

Using the Maxwell–Boltzmann distribution, the equilibrium distribution of solute molecules between the two states corresponding to the gas phase and the solution phase can be calculated as

$$c'/c = \exp(-\Delta u/kT)$$

where the ratio c'/c is either the number density or concentration ratio of gas to solution phases since the solute mass figures in both the numerator and denominator. This quantity has been shown previously to be the Ostwald coefficient of solubility γ or the distribution coefficient K_c. The energy change Δu appearing in the exponential term was identified by Uhlig as the change in free energy, not total energy, and related to the thermodynamic expression for the free-energy change of transfer of a perfect solute from one concentration to another. This free-energy change is obviously a state function independent of path and can therefore be traced by any convenient reversible path between states. The path adopted by Uhlig and subsequently employed in the later statistical mechanical models divides the state changes into two steps. The first step involves the creation of the molecular-sized cavity followed by a "charging up" of the cavity, i.e., insertion of the solute molecule, and the attendant interaction of the solute with the solvent. Recognizing the approximation, Uhlig identified the free energy of cavity formation with surface free-energy change in the solvent associated with the area A of the solute cavity. Since the surface free energy per unit area of the solvent is the macroscopic surface tension γ, the free energy of cavity formation becomes

$$\Delta G_{\mathrm{cav}} = A\gamma$$

The most important approximation in this development is the assumption that the macroscopic surface tension is a valid measure of the free energy of formation of cavities of molecular dimensions.

The second, or charging, step accounts for the solute–solvent interaction E. On inclusion in the total free energy and relating it to the equilibrium constant, we obtain the relation

$$\ln K_c = (-NA\gamma + E)/RT$$

Workers at Oak Ridge National Laboratory[35] employed this relationship based only on the cavity term for the comparison of experimental solubilities of rare gases in (LiNaK)F and (LiBe)F melts. The majority of the systems compared yielded predicted values smaller than the observed solubilities, but in all cases within an order of magnitude.

4.3. Comparison of the Methods

Although Lee and Johnson[50] have previously compared the two methods for the calculation of the solubility of gases in molten salts, there are several reasons for repeating the analysis. In the first place, it appears that the temperature-dependent equation for the surface tension used in calculating solubilities by the macroscopic (Uhlig) method incorrectly incorporated a positive temperature coefficient instead of the negative value given by Bloom *et al.*[56] For example, in Lee and Johnson's Table I they give the surface tension of KCl at $1073°K$ as 213.6 dyn/cm based on the sum of terms $155.2 + 58.4$, whereas the proper value according to both Bloom *et al.*[56] and Janz[28] is the difference of these terms, or 96.8 dyn/cm.

Second, it is not clear that the choice of diameters made by Lee and Johnson for the solute molecules is consistent in comparing the two methods. Calculating back from the K_p values listed by Lee and Johnson (their Table III), it appears that they chose the hard-sphere diameters, e.g., $Ar(= 2.92\ \text{Å})$ and $Cl_2(= 3.70\ \text{Å})$ for the Uhlig calculations, whereas it appears they employed "collision" diameters (their Table II), e.g., $Ar(= 3.418\ \text{Å})$ and $Cl_2(= 3.897\ \text{Å})$ for the scaled-particle calculations. Third, they estimated the collision diameter of $NaNO_3$ from the crystal radius of Na^+ and the thermochemical radius of NO_3^-. We have calculated the collision radii of all solvents by the method described by Mayer.[53] Although this latter calculation is questionable since the nitrate ion is not strictly spherical, the comparison made here will distinguish the nitrates from the halides so as not to becloud the comparison. As a point of interest, Lee and Johnson's value for sodium nitrate's collision diameter was 2.84 Å and assumed independent of temperature, compared to the value based on Mayer's method of 3.046 Å at 640°K and which decreases by 0.526×10^{-3} Å/deg. A fourth consideration is that Lee and Johnson used Stillinger's values for the collision diameter of the salt which were calculated from isothermal compressibility data. Although there are only minor differences between these values and those calculated from surface-tension data,* it was felt that a strictly fair comparison ought to be based on identical input data where possible. Finally, since additional solubility data are now available and all of the older data have been recalculated by the same criterion, a new comparison of the two methods is warranted.

The results of the calculations are presented in Table VII for Ar, Cl_2, and CO_2 in alkali halide solvents, and in Table VIII for He, Ar, N_2, and CO_2 in alkali nitrate solvents. Each system is evaluated at one temperature which is approximately 10% above the melting point of the solvent. In Table VII various choices for the solute diameter were taken in order to observe the effect of this variation. These values were either those listed by Lee and Johnson (their Table II, values listed with four digits) or those previously used in the literature of gas solubilities in molten salts. Table VIII contains only duplicate diameter for helium in sodium nitrate, as had been previously discussed by Field and Green[10] and Desimoni et al.[6] This table, however, contains the experimental values of both Cleaver and Mather[9] and Field and Green.[10] It should be noted that in a first-order comparison such as this, only those systems having solvents which are pure salts have been examined. It is also obvious that although both models require only properties of the pure components, the calculations are not without an adjustable parameter in the appropriate choice of gas size, i.e., diameter for the scaled-particle calculation and area for the Uhlig calculation.

Any broad generalization of the utility of either model is subject to ambiguity. The tables contain columns listing the relative solubilities, comparing the calculated solubility by each method to the experimental

---

* The salt collision diameters calculated from surface-tension data are about 1% smaller at the melting point than those calculated from the isothermal compressibility data. However, the former also decreases about 1% faster over a 100° interval than the latter.

TABLE VII. Calculated Solubilities in Halide Melts

Solvent	T, °K	Solute	a_2, Å	$K_p \times 10^7$, mole/cm³/atm			K_p(calc.)/K_p(expt.)		$K_p(U)/K_p$(S.P.)
				S.P.	Uhlig	Expt.	S.P.	Uhlig	
NaCl	1173	Cl_2	3.897	2.46	4.42	3.44[a]	0.72	1.29	1.8
		CO_2	4.115	1.59	3.07	6.17[b]	0.26	0.50	1.9
			5.12	0.16	0.45	6.17[b]	0.03	0.07	2.7
KCl	1173	Ar	3.418	9.79	13.65	6.23[c]	1.57	2.19	1.4
			3.84	5.03	8.02	6.23[c]	0.81	1.29	1.6
		Cl_2	3.897	4.58	7.43	12.87[a]	0.36	0.58	1.6
		CO_2	4.115	3.15	5.48	7.81[b]	0.40	0.70	1.7
			5.12	0.46	1.09	7.81[b]	0.06	0.14	2.4
KBr	1073	Ar	3.418	9.05	13.95	7.50[c]	1.21	1.86	1.5
			3.84	4.53	8.05	7.50[c]	0.61	1.07	1.8
		CO_2	4.115	2.79	5.43	8.84[b]	0.32	0.62	2.0
			5.12	0.38	1.03	8.84[b]	0.04	0.12	2.7
KI	973	Ar	3.418	8.47	14.97	7.90[c]	1.07	1.90	1.8
			3.84	4.15	8.58	7.90[c]	0.53	1.09	2.1
		CO_2	4.115	2.52	5.76	19.2[b]	0.13	0.30	2.3
			5.12	0.32	1.07	19.2[b]	0.02	0.06	3.3

[a] Reference 30.
[b] Reference 2.
[c] Reference 29.

TABLE VIII. Calculated Solubilities in Nitrate Melts

Solvent	T, °K	Solute	a_2, Å	$K_p \times 10^7$, mole/cm³/atm			K_p(calc.)/K_p(expt.)		$K_p(U)/K_p$(S.P.)
				S.P.	Uhlig	Expt.	S.P.	Uhlig	
LiNO₃	580	Ar	3.84	0.04	0.31	1.10[a]	0.04	0.28	7.6
		N₂	4.10	0.04	0.31	1.63[b]	0.03	0.19	8.9
				0.01	0.12	1.88[b]	0.01	0.07	
NaNO₃	640	He	2.44	6.05	17.04	2.18[a]	2.78	7.83	2.8
				6.05	17.04	1.07[b]	5.68	16.0	
			3.58	0.18	1.05	2.18[a]	0.08	0.49	5.8
				0.18	1.05	1.07[b]	0.17	0.99	
		Ar	3.84	0.07	0.48	0.76[a]	0.09	0.64	6.8
				0.07	0.48	0.72[b]	0.10	0.67	
		N₂	4.10	0.03	0.21	0.58[a]	0.05	0.36	8.0
				0.03	0.21	2.25[b]	0.01	0.09	
		CO₂	5.12	<0.01	0.01	1.04[b]	<0.01	0.01	>5.0
KNO₃	670	Ar	3.84	0.16	0.92	1.93[b]	0.08	0.48	5.9
		N₂	4.10	0.06	0.44	2.92[b]	0.02	0.15	6.9
		CO₂	5.12	<0.01	0.02	1.38[b]	<0.01	0.01	15.0
RbNO₃	640	Ar	3.84	0.11	0.78	1.64[a]	0.07	0.48	6.8

[a] Reference 9.
[b] Reference 10.

value, and also comparing the two methods directly. With respect to this latter ratio, the choice of solute diameter has least effect. In order to get some rough estimate, the means of this ratio $[K_p(\text{Uhlig})/K_p(\text{S.P.})]$ and their estimated standard deviations were calculated for each table. No great significance should be attached to the standard deviations (actually the best estimates of the standard error were computed) except that they define the range that includes roughly two out of every three of the values. For the alkali halides this ratio is 2.0 ± 0.5, while for the nitrates it is found to be 6.2 ± 1.9, indicating that in both solvent types the Uhlig method predicts solubilities twice to six times as large as the scaled-particle calculation.

A similar estimate of each of the calculation methods with the experimental solubilities, i.e., the ratio $K_p(\text{calc.})/K_p(\text{obs.})$, yields the following values:

1. Uhlig method in halides: 0.96 ± 0.70 (81%)
2. Scaled-particle method in halides: 0.51 ± 0.47 (92%)
3. Uhlig method in nitrates: 0.35 ± 0.09 (26%)
4. Scaled-particle method in nitrates: 0.06 ± 0.04 (67%)

These ratios indicate that both methods yield significantly better estimates at the higher temperatures of the alkali halide melts. This seems reasonable from the point of view that at these higher temperatures, and corresponding greater kinetic energies of the particles, both solute and solvent particles behave more like hard spheres since the potential-interaction energies are constant and independent of temperature. It is, of course, consistent with the preceding ratios of the methods that the Uhlig calculation estimates solubilities closer to those found experimentally.

It is likewise important to note that overestimation of the solubility contributes more to the failure of a model than underestimation, since one would anticipate that any potential interaction between solute and solvent ions would lead to a larger decrease in the total energy of the solution than the corresponding decrease in entropy and hence an increase in solubility. Thus, hard-sphere solubilities (by either method) should be less than those observed. In this respect, the "softer" gases, i.e., those having larger polarizabilities or permanent quadrupole moments, or both, should manifest the least agreement between model and nature. Conversely, the hardest gases should yield the best results. This is the case observed in the tables. Considering either melt system, i.e., nitrate or halide, helium has the lowest polarizability, with argon second in the list. In both instances these are either overestimated, i.e., with ratios to experimental values greater than unity, or the best estimated by either model. Carbon dioxide is the softest solute, having the largest polarizability and the largest permanent quadrupole moment. It is the most poorly estimated in most of the cases and is more underestimated in the lower-temperature solvents than in the higher-melting halides.

In comparing the relative merits of the Uhlig model and the scaled-particle model, it is necessary to differentiate between the practical and theoretical advantages and disadvantages of each. From the theoretical point of view, the Uhlig model suffers from the imposition of the macroscopic surface tension on the corresponding microscopic term, whereas the scaled-particle theory is rigorous in its analysis of the radial distribution function and the derivation of the reversible work for cavity formation. On this particular point, the discussion by Reiss and Mayer[54] is essential. Their basic argument for the development of a theory of the surface tension of molten salts and subsequently for a law of corresponding states for fused salts rests on the validity of the proposition that the expression for the reversible work $w(r)$, which must be expended in the production of a spherical cavity of radius r, is dependent on a surface tension term for which the real surface tension of the molten salt is a good approximation. In fact, this is the basis by which the collision diameter is calculated, i.e., from the same equations relying on the real surface tension. Therefore, both models are subject to the same limitation of substituting the real (macroscopic) surface tension for a microscopic one. Since the scaled-particle theory is a more general theory, the success of the surface-tension approximation in its other applications[52-54] confirms its applicability in both solubility models considered here. A final comparison of the two models that might be considered is the more practical consideration of the ease of calculation. With the availability of even minicomputers, however, this point is of little significance.

In conclusion, since the calculation of solubility is in fact a calculation of the free energy, a real distinction in the superiority of one model over the other may come with their extensions to the separate calculation of the enthalpy and entropy terms associated with the solution process. As noted above, both models are consistent in their underestimation of the solubilities of the soft gases and higher estimation of the harder gases. The apparent fact that the Uhlig model predicts solubilities nearer to the experimental values is not an adequate measure of its success until the differences of the interaction energies, i.e., ion-induced dipole, etc., can be evaluated. Nor will the overestimation negate one or the other model until the question of solute molecular size is satisfactorily resolved.

APPENDIX

Solute	Solvent (reference)	Composition (mole %)	Temp. range	N^a	$\Delta H^{s\,b}$ (kcal/mole)	T, °K^c	$\Delta S_c^{\,b}$ (cal/deg/mole)	$T\Delta S_c$ (kcal/mole)	$K_p \times 10^{7\,d}$ (mole/cm^3/atm)
He	(Li, Na, K)F(35)	46-12-42	600-800	4(16)	6.73 ± 0.36	1000	-1.59 ± 0.38	-1.59	1.85
	LiF-BeF$_3$(36)	64-36	500-800	4(12)	5.18 ± 0.16	1000	-3.39 ± 0.18	-3.39	1.63
	NaF-BeF$_2$(37)	57-43	500-800	4(NA)	4.72 ± 0.03	1000	-3.07 ± 0.03	-3.07	2.42
	NaF-ZrF$_4$(3)	53-47	600-800	3(20)	6.15 ± 0.70	1000	-1.01 ± 0.73	-1.01	3.32
Ne	(Li, Na, K)F(35)	46-12-42	600-800	3(9)	8.78 ± 0.22	1000	-1.12 ± 0.22	-1.12	0.84
	LiF-BeF$_2$(36)	64-36	500-800	4(13)	5.93 ± 0.11	1000	-4.23 ± 0.12	-4.23	0.73
	NaF-ZrF$_4$(3)	53-47	600-800	3(8)	7.32 ± 0.59	1000	-0.89 ± 0.61	-0.89	1.96
Ar	(Li, Na, K)F(35)	46-12-42	600-800	3(9)	12.35 ± 0.41	1000	-0.21 ± 0.43	-0.21	0.22
	LiF-BeF$_2$(36)	64-36	500-800	4(37)	8.78 ± 0.21	1000	-4.04 ± 0.23	-4.04	0.19
	NaF-BeF$_2$(37)	57-43	500-800	4(13)	7.41 ± 0.27	1000	-4.31 ± 0.30	-4.31	0.34
	NaF-ZrF$_4$(3)	53-47	600-800	3(12)	8.04 ± 0.05	1000	-1.70 ± 0.05	-1.70	0.91
Xe	LiF-BeF$_2$(36)	64-36	600-800	4(12)	12.43 ± 0.40	1000	-2.88 ± 0.42	-2.88	0.058
	NaF-BeF$_2$(37)	57-43	500-800	4(NA)	11.44 ± 2.07	1000	-2.79 ± 2.29	-2.79	0.094
	NaF-ZrF$_4$(3)	53-47	600-800	3(5)	10.97 ± 0.43	1000	-0.27 ± 0.44	-0.27	0.43
HF	LiF-BeF$_2$(40)	54-46	600-800	3(NA)	-4.80 ± 0.29	1000	-6.00 ± 0.30	-6.00	66.8
	(40)	59-41	600-800	3(NA)	-5.56 ± 0.09	1000	-6.59 ± 0.10	-6.59	72.5
	(18)	66-34	500-700	9(9)	-6.29 ± 0.17	1000	-7.11 ± 0.20	-7.11	80.5
	(40)	69-31	600-800	3(NA)	-6.23 ± 0.38	1000	-6.88 ± 0.39	-6.88	87.8
	(40)	80-20	600-800	3(NA)	-5.73 ± 0.09	1000	-5.98 ± 0.09	-5.98	108.0
	(40)	89-11	600-800	3(NA)	-4.83 ± 0.17	1000	-4.83 ± 0.18	-4.83	122.0

APPENDIX (*continued*)

Solute	Solvent (reference)	Composition (mole %)	Temp. range	N^a	ΔH^{sb} (kcal/mole)	T, °K^c	ΔS_c^b (cal/deg/mole)	$T\Delta S_c$ (kcal/mole)	$K_p \times 10^7{}^d$ (mole/cm³/atm)
HF	NaF–BeF₂(40)	49–51	600–800	3(NA)	−5.89 ± 0.14	1000	−5.89 ± 0.15	−5.89	122.0
	(40)	58–42	600–800	3(NA)	−6.10 ± 0.36	1000	−5.45 ± 0.37	−5.45	169.0
	(40)	66–34	600–800	3(NA)	−7.58 ± 0.01	1000	−5.92 ± 0.01	−5.92	282.0
	(40)	70–30	600–800	3(NA)	−10.62 ± 0.38	1000	−7.87 ± 0.39	−7.87	488.0
	(40)	72–25	600–800	3(NA)	−10.97 ± 0.10	1000	−7.23 ± 0.10	−7.23	802.0
	NaF–ZrF₄(41)	45–55	600–800	3(NA)	−3.92 ± 0.55	1000	−5.35 ± 0.57	−5.35	59.5
	(41)	53–47	550–800	6(37)	−4.37 ± 0.22	1000	−5.04 ± 0.24	−5.04	87.3
	(41)	60–40	600–800	3(NA)	−5.95 ± 0.48	1000	−6.93 ± 0.49	−6.93	97.8
	(41)	65–35	600^e–800	3(NA)	−6.67 ± 0.13	1000	−6.50 ± 0.14	−6.50	133.0
	(41)	81–19	600^e–800	3(NA)	−9.87 ± 0.02	1000	−6.63 ± 0.02	−6.63	624.0
DF	LiF–BeF₂(18)	66–34	500–700	9(8)	−6.65 ± 0.07	1000	−7.84 ± 0.08	−7.84	67.0
Cl₂	NaCl(30)		850–1030	5	19.79 ± 1.00	1181	+10.11 ± 0.83	+10.95	3.65
	KCl(30)		850–1050	6	9.05 ± 0.56	1154	+3.53 ± 0.46	+4.08	12.06
	(Na, K)Cl(30)	50–50	750–1030	10	13.59 ± 0.47	1154	+6.58 ± 0.41	+7.06	7.73
	MgCl₂(30)		780–950	4	4.77 ± 1.26	1079	−1.38 ± 1.11	−1.47	6.11
	NaCl–MgCl₂(30)	50–50	570–1020	6	10.90 ± 1.01	1079	+2.39 ± 0.99	+2.47	2.33
	KCl–MgCl₂(30)	50–50	560–1050	13	6.60 ± 0.41	1079	+0.06 ± 0.40	+0.06	5.37
	KCl–PbCl₂(31)	23–77	450–640	3(NA)	9.73 ± 0.13	850	+8.76 ± 0.15	+7.46	37.30
	(31)	48–52	440–690	3(NA)	−0.80 ± 0.04	850	−5.19 ± 0.05	−4.42	16.87
	(31)	60–40	510–670	3(NA)	−3.40 ± 0.11	850	−8.34 ± 0.13	−7.10	16.07
	(31)	70–30	590–700	3(NA)	−3.45 ± 0.35	850	−8.18 ± 0.38	−6.96	18.04
CO₂	NaCl(2)		830–1000	5(27)	6.24 ± 0.28	1181	−0.28 ± 0.24	−0.33	6.29
	KCl(2)		800–1000	6(29)	4.53 ± 0.08	1154	−1.32 ± 0.07	−1.52	7.56
	(15)		800–950	4	2.95 ± 0.38	1154	−2.52 ± 0.33	−2.91	8.23
	KBr(2)		770–930	4(24)	4.02 ± 0.12	1103	−1.27 ± 0.11	−1.40	9.32

Gas	Salt		Temp. range	N					
CO	(Na, K)Cl(33)	50-50	700-900	8	6.34 ± 0.17	1154	-0.64 ± 0.16	-0.74	4.81
HCl	NaCl(15, 16)		820-1020	11	-0.11 ± 0.38	1181	-4.17 ± 0.32	-4.93	13.3
	KCl(16)		790-990	7	-0.92 ± 0.13	1154	-3.01 ± 0.12	-3.45	34.8
	RbCl(16)		720-970	7	-2.17 ± 0.37	1089	-3.48 ± 0.34	-3.79	52.9
	CsCl(16)		670-950	7	-5.01 ± 0.31	1011	-5.62 ± 0.29	-5.69	86.5
	CaCl₃(34)		820-1000	4	6.60 ± 0.68	1150	-1.11 ± 0.58	-1.28	3.46
	SrCl₂(34)		890-1020	4	9.23 ± 0.28	1261	+0.09 ± 0.23	+0.11	2.55
	BaCl₂(34)		980-1080	4	10.62 ± 0.89	1359	+0.49 ± 0.68	+0.67	2.24
He	NaNO₃(9)		330-440	6(14)	3.19 ± 0.07	640	-3.91 ± 0.11	-2.40	2.17
	(10)		345-400	4	3.20 ± 0.09	640	-5.31 ± 0.14	-3.36	1.07
	(Na, K)NO₃(7)	50-50	235-330	4	3.95 ± 0.14	640	-5.09 ± 0.25	-3.26	0.66
Ar	LiNO₃(9)		270-310	4(8)	3.76 ± 4.80	580	-3.96 ± 8.53	-2.30	1.10
	(10)		260-320	12	7.79 ± 1.09	580	+3.78 ± 1.95	+2.42	1.64
	NaNO₃(9)		330-440	6(24)	3.83 ± 1.70	640	-5.03 ± 2.57	-3.20	0.75
	(10)		315-400	12	3.52 ± 0.82	640	-5.58 ± 1.31	-3.55	0.72
	KNO₃(46)		335-405	11	2.56 ± 0.65	670	-5.22 ± 1.05	-3.51	1.93
	(Na, K)NO₃(7)		235-330	4	4.67 ± 0.20	640	-5.77 ± 0.37	-3.67	0.27
	RbNO₃(9)		330-440	4(8)	4.86 ± 1.36	648	-1.84 ± 2.09	-1.14	1.85
N₂	LiNO₃(46)		260-315	11	1.82 ± 0.22	580	-6.23 ± 0.39	-3.62	1.88
	NaNO₃(9)		330-450	6(12)	3.96 ± 2.20	640	-5.34 ± 3.34	-3.39	0.58
	(10)		310-365	9	2.55 ± 0.74	640	-4.81 ± 1.21	-2.89	2.29
	KNO₃(46)		345-425	10	1.72 ± 0.78	670	-5.64 ± 1.20	-3.74	2.93
	(Na, K)NO₃(7)	50-50	235-330	4	4.07 ± 0.23	640	-7.62 ± 0.41	-4.88	0.17
O₂	(Na, K)NO₃(6, 7)	50-50	235-330	9	4.22 ± 0.14	640	-7.04 ± 0.26	-4.50	0.20
CO₂	NaNO₃(10)	315-365	315-365	5	-2.68 ± 0.27	640	-14.55 ± 0.44	-9.27	1.04
	KNO₃(46)		340-395	5(18)	-2.39 ± 2.06	670	-13.13 ± 3.23	-8.76	1.47
CH₄	(Na, K)NO₃	50-50	235-330	4	3.98 ± 0.14	640	-7.57 ± 0.25	-4.84	0.19

APPENDIX (*continued*)

Solute	Solvent (reference)	Composition (mole %)	Temp. range	N^a	ΔH^b (kcal/mole)	T, °K^c	$\Delta S_c{}^b$ (cal/deg/mole)	$T\Delta S_c$ (kcal/mole)	$K_p \times 10^7$ ^{d} (mole/cm³/atm)
NH₃	LiNO₃(26)	43–57	260–320	4	-15.91 ± 0.38	580	-13.08 ± 0.68	-7.59	3.78
	(Li, K)NO₃(26)	75–25	160–250	4	-6.88 ± 0.24	580	-1.38 ± 0.51	-0.80	0.54
	(26)		240–330	4	-11.78 ± 0.13	580	-8.32 ± 0.24	-4.83	1.15
	(Li, Na, K)NO₃(26)	27–18–55	160–250	4	-5.53 ± 0.17	580	+0.40 ± 0.37	+0.23	0.41
	(Li, K)ClO₄(26)	76–24	230–310	4	-13.11 ± 0.20	580	-4.48 ± 0.38	-2.60	25.2
H₂O	LiNO₃(22)		270–340	4	-11.03 ± 0.20	580	-2.91 ± 0.35	-1.68	9.17
	(13)		265–280	2	-13.45	580	-7.03	-4.07	9.41
	NaNO₃(22)		310–340	3	-8.39 ± 0.61	640	-0.90 ± 1.01	-0.58	1.16
	(Li, Na)NO₃(13)	75–25	230–280	3	-12.07 ± 0.39	580	-5.18 ± 0.73	-3.01	7.19
	(13)	50–50	230–280	3	-12.11 ± 0.22	580	-6.03 ± 0.42	-3.49	4.87
	(13)	25–75	265–280	2	-10.73	580	-4.00	-2.32	4.11
	(Li, K)NO₃(13)	75–25	230–265	2	-10.88	580	-3.31	-1.92	6.58
	(13)	50–50	230–265	2	-10.47	580	-3.56	-2.07	4.05
	(13)	25–75	230–265	2	-9.32	580	-2.26	-1.31	2.88
	(Na, K)NO₃(26)	50–50	230–290	11	-8.44 ± 0.09	580	+4.04 ± 0.18	+2.34	1.46
	(Li, Na, K)NO₃(24)	30–23–47	140–210	4	-13.43 ± 1.61	580	-18.50 ± 3.61	-10.64	0.029
	KNO₃–NaNO₂(25)	55–45	140–280	9	-8.71 ± 0.09	580	-1.10 ± 0.20	-0.64	3.06
	LiClO₄(24)		240–290	3	-11.21 ± 3.06	560	-7.00 ± 5.72	-3.92	2.01

^{a} N is the number of solubilities at separate temperatures listed in the reference and used in the least-squares analysis. The number in parentheses is the total number of pressure-dependent solubility measurements at all temperatures listed in the reference. (NA indicates that the total number was not available.)

^{b} Uncertainties are taken at one standard deviation.

^{c} Except for fluoride melts which are evaluated at 1000°K, all others are evaluated at $1.10T_m$, where T_m is the melting point of the pure solvent or the melting point of the lowest-melting pure component in mixed solvents.

^{d} Pressure unit of 1 atm except for H₂O and NH₃ solutes, which are in torr.

^{e} Extrapolated below melting point.

REFERENCES

1. R. Battino and H. L. Clever, *Chem. Rev.* **66**:395 (1966).
2. D. Bratland, K. Grjotheim, C. Krohn, and K. Motzfeld, *Acta Chem. Scand.* **20**:1811 (1966); *J. Metals* **1967**(October):13.
3. a. R. F. Newton and D. G. Hill, *U.S.A.E.C. Rep. ORNL-1771* (1954), p. 70; b. W. R. Grimes, N. V. Smith, and G. M. Watson, *J. Phys. Chem.* **62**:862 (1958).
4. J. P. Frame, E. Rhodes, and A. R. Ubbelohde, *Trans. Faraday Soc.* **57**:1075 (1961).
5. W. Haug and L. F. Albright, *Ind. Eng. Chem. Process Design Develop.* **4**:2023 (1965).
6. E. Desimoni, F. Paniccia, and P. G. Zambonin, *J. Electroanal. Chem. Interfacial Electrochem.* **38**:373 (1972).
7. F. Paniccia and P. G. Zambonin, *J. Chem. Soc., Faraday Trans. Part I* **68**:2083 (1972).
8. J. L. Copeland and W. C. Zybko, *J. Am. Chem. Soc.* **86**:4734 (1964); *J. Phys. Chem.* **69**:3631 (1965); *J. Phys. Chem.* **70**:181 (1966); J. L. Copeland and L. Siebles, *J. Phys. Chem.* **70**:1811 (1966); *J. Phys. Chem.* **72**:603 (1968); J. L. Copeland and S. Radak, *J. Phys. Chem.* **70**:3356 (1966); *J. Phys. Chem.* **71**:4360 (1967).
9. B. Cleaver and D. E. Mather, *Trans. Faraday Soc.* **66**:2469 (1970).
10. P. E. Field and W. J. Green, *J. Phys. Chem.* **75**:821 (1971).
11. G. N. Lewis and M. Randall, *Thermodynamics*, 2nd ed. (revised by K. S. Pitzer and L. Brewer), McGraw-Hill, New York (1961), pp. 282–290.
12. J. H. Hildebrand, J. M. Prausnitz, and R. L. Scott, *Regular and Related Solutions*, Van Nostrand Reinhold, New York (1970), pp. 130–134.
13. G. Bertozzi, *Z. Naturforsch.* **22A**:1748 (1967).
14. J. L. Copeland and J. R. Christie, *J. Phys. Chem.* **75**:103 (1971).
15. A. L. Novozhilov, V. N. Devyatkin, and E. I. Gribova, *Zh. Fiz. Khim.* **46**:2433 (1972).
16. A. L. Novozhilov, V. N. Devyatkin, and E. I. Gribova, *Zh. Fiz. Khim.* **46**:1856 (1972).
17. T. L. Lukmanova and Ya. E. Vil'nyanskii, *Izv. Vysshikh. Uchebn. Zavedenii, Khim. i Khim. Tekhnol.* **7**:510 (1964).
18. P. E. Field and J. H. Shaffer, *J. Phys. Chem.* **71**:3218 (1967).
19. E. A. Sullivan, S. Johnson, and M. D. Barns, *J. Am. Chem. Soc.* **77**:2023 (1955).
20. A. P. Malinauskas, D. M. Richardson, J. E. Savolainern, and J. H. Shaffer, *Ind. Eng. Chem. Fundamentals* **11**:584 (1972).
21. T. B. Tripp and J. Braunstein, *J. Phys. Chem.* **73**:1984 (1969).
22. M. Peleg, *J. Phys. Chem.* **71**:4553 (1967).
23. P. G. Zambonin, V. L. Cardetta, and G. Signorile, *J. Electroanal. Chem.* **28**:237 (1970).
24. F. R. Duke and A. S. Doan, Jr., *Iowa State Coll. J. Sci.* **32**:451 (1958).
25. H. S. Hull and A. G. Turnbull, *J. Phys. Chem.* **74**:1783 (1970).
26. S. Allulli, *J. Phys. Chem.* **73**:1084 (1969).
27. M. Schenke, G. H. J. Broers, and J. A. A. Ketelaar, *J. Electrochem. Soc.* **113**:404 (1966).
28. G. J. Janz, *Molten Salts Handbook*, Academic Press, New York (1967).
29. H. V. Woelk, *Nukleonik* **2**:278 (1960).
30. Yu. M. Ryabukhin, *Russ. J. Inorg. Chem.* **7**:565 (1962).
31. M. Kowalski and G. W. Harrington, *Inorg. Nucl. Chem. Letters* **3**:121 (1967).
32. W. J. Burkhard and J. D. Corbett, *J. Am. Chem. Soc.* **79**:6361 (1957).
33. S. P. Zezyanov and V. A. Il'ichev. *Zh. Neorg. Khim.* **17**:2541 (1972).

34. A. L. Novozhilov, E. I. Gribova and N. V. Devyatkin, *Zh. Neorg. Khim.* **17**:2570 (1972).
35. M. Blander, W. R. Grimes, N. V. Smith, and G. M. Watson, *J. Phys. Chem.* **63**:1164 (1959).
36. G. M. Watson, R. B. Evans, W. R. Grimes, and N. V. Smith, *J. Chem. Eng. Data* **7**:285 (1962).
37. W. T. Ward, R. B. Evans, G. M. Watson, and W. R. Grimes, *U.S.A.E.C. Rep. ORNL-2931* (1960), p. 29.
38. N. V. Smith, R. J. Shiel, R. B. Evans, and G. M. Watson, *U.S.A.E.C. Rep. ORNL-2931* (1960), p. 35.
39. J. H. Shaffer, *U.S.A.E.C. Rep. ORNL-3127* (1960), p. 12.
40. J. H. Shaffer and G. M. Watson, *U.S.A.E.C. Rep. ORNL-2931* (1960), p. 31.
41. J. H. Shaffer, W. R. Grimes, and G. M. Watson, *J. Phys. Chem.* **63**:1999 (1959).
42. A. Lannung, *J. Am. Chem. Soc.* **52**:68 (1930).
43. M. G. Evans and M. Polanyi, *Trans. Faraday Soc.* **32**:1333 (1936).
44. R. P. Bell, *Trans. Faraday Soc.* **33**:496 (1937).
45. I. M. Barclay and J. A. V. Butler, *Trans. Faraday Soc.* **34**:1445 (1938).
46. W. J. Green, Ph.D. Thesis, Virginia Polytechnic Institute (1969).
47. H. H. Uhlig, *J. Phys. Chem.* **41**:1215 (1937).
48. H. Reiss, R. L. Frisch, and J. L. Lebowitz, *J. Chem. Phys.* **31**:369 (1959); H. Reiss, H. L. Frisch, E. Helfand, and J. L. Lebowitz, *J. Chem. Phys.* **32**:119 (1960); E. Helfand, H. Reiss, H. L. Frisch, and J. L. Lebowitz, *J. Chem. Phys.* **33**:1379 (1960).
49. R. A. Pierotti, *J. Phys. Chem.* **67**:1840 (1963); *J. Phys. Chem.* **69**:281 (1965).
50. A. K. K. Lee and E. F. Johnson, *Ind. Eng. Chem. Fundamentals* **8**:726 (1969).
51. F. H. Stillinger, Jr., *J. Chem. Phys.* **35**:1581 (1961).
52. S. W. Mayer, *J. Chem. Phys.* **38**:1803 (1963).
53. S. W. Mayer, *J. Phys. Chem.* **67**:2160 (1963); *J. Chem. Phys.* **40**:2429 (1964).
54. H. Reiss and S. W. Mayer, *J. Chem. Phys.* **34**:2001 (1961); H. Reiss, S. W. Mayer, and J. L. Katz, *J. Chem. Phys.* **35**:820 (1961).
55. R. O. Neff and D. A. McQuarrie, *J. Phys. Chem.* **77**:413 (1973).
56. H. Bloom, F. G. Davis, and D. W. James, *Trans. Faraday Soc.* **56**:1179 (1960).

Chapter 3

ORGANIC REACTIONS IN MOLTEN TETRACHLOROALUMINATE SOLVENTS

H. Lloyd Jones

and

Robert A. Osteryoung

Department of Chemistry
Colorado State University
Fort Collins, Colorado

1. INTRODUCTION

The use of molten salts as a medium for organic reactions has been known since organic chemistry's beginning. The advantages such as short reaction times, ease of product recovery, possibility of high yields, and savings on solvent recovery together with the unusual physical properties (ionic liquids, high conductivity) shown by molten salts in comparison with other solvents opens new frontiers in the study of organic syntheses, separations, and mechanisms.

Only recently have organic reactions in fused salts been regarded as a class or subjected to a systematic investigation. Recent reviews by Sundermeyer[1] and Gordon[2] have covered the field in a general sense. Although these two reviews mentioned the use of fused aluminum chloride and aluminum-chloride containing melts, hereafter called $AlCl_3$ solvent systems, as a medium for organic reactions a comprehensive coverage was not attempted. It is the purpose of this chapter to present such a review.

The molten salt chemistry of the aluminum halides and their mixtures with other metal halides has been thoroughly reviewed by Boston,[3] and we will only mention the fact that the $AlCl_3$ solvent systems exhibit a wide range of acid–base properties. However, in all the reactions covered below other than the organic electrode reactions, only once were the acid–base properties of the $AlCl_3$ solvent systems considered.

While many different organic reactions have been carried out in $AlCl_3$ solvent systems, there are three types of reactions which have been used and studied extensively. These are (1) condensation–addition reactions, (2) dehydrogenation–addition reactions (normally called the Scholl reaction), and (3) rearrangement–isomerization reactions. The first two types are of considerable importance in the preparation of dyes, while the latter type includes the well-known Fries rearrangement as well as offering routes to certain heterocyclic compounds and to uniquely substituted aromatic materials. The bulk of this review is concerned with these three reactions types. Other types which will be covered are (1) halogen exchange, (2) dehydration, and (3) chlorination–reduction reactions. A section on organic electrode reactions is also included.

2. CONDENSATION–ADDITION REACTIONS

Condensation–addition reactions are formally analogous to the well-known Friedel–Crafts reactions. There appears to be only one physico-chemical study of the Friedel–Crafts type reaction in $AlCl_3$ solvent systems. Komagorov, Baeva, and Koptyug,[4] using infrared spectroscopy, studied the formation of acylium cations from aromatic carboxylic acids and acid chlorides. In a 60:40 mole % $AlCl_3$:NaCl melt in the temperature range 25–160°C the formation of the acylium ion $ArCO^+$, where Ar = phenyl, o-chlorophenyl, p-chlorophenyl, and o-methoxyphenyl, could be followed by its infrared absorption at 2240 cm^{-1}. Acylium ion concentration was found to increase with temperature and disappear reversibly on cooling, with appreciable concentrations of the ion occurring only at temperatures greater than 130°C.

Numerous examples of condensation–addition reactions taking place in $AlCl_3$ solvent systems to yield ketones are known. The acylating reactant can be derived either from carboxylic acids, acid chlorides, acid anhydrides, or lactones, while the other components can be either aromatic or heterocyclic compounds.

One of the simplest examples of a condensation–addition reaction is the reaction between phenol and benzoic acid in a 68:32 mole % $AlCl_3$:NaCl melt to give o- and p-hydroxybenzophenone,[5] as illustrated in Eq. (1):

$$(1)$$

Similar results are obtained with acid chlorides. However, they are also reported to undergo self-condensation to yield diketones, i.e., anthraquinone type products as shown for *m*-toluyl chloride[6] in Eq. (2):

$$(2)$$

Acid chlorides have also been shown to react with aromatic amines in $AlCl_3$ solvent systems.[7] When benzoyl chloride was reacted with di-phenylamine in $AlCl_3$ at 140°C for 1 hr, both N-benzoyldiphenylamine and 4,4'-dibenzoyldiphenylamine were obtained. Benzoylation must occur initially on nitrogen, since under similar conditions N-benzoyldiphenyl-amine rearranges to 4,4'-dibenzoyldiphenylamine and diphenylamine.

Aliphatic acids can also take part in condensation–addition reactions. Bruce *et al.*[5] have found that aryl-substituted aliphatic acids undergo intramolecular ring closure to yield cyclic ketones containing five to seven carbon atoms. For example, β-phenylproprionic acids gave 1-indanone and γ-phenylvaleric acid gave benzosuberone. Baddeley and Williamson[8] obtained identical results to those above even when the phenyl ring was deactivated toward electrophilic substitution by carboxyl or acyl substi-tuents. Thus, both γ-phenylbutyric[5] and γ-*p*-acetylphenylbutyric acid[8] gave α-tetralones as shown in Eq. (3):

$$(3)$$

It is to be noted that treating β-(2,5-dimethoxyphenyl)-propionic acid with 64:36 mole % $AlCl_3$:NaCl at 180°C for 5 min gave a dihydrocoumarin which was rearranged to 4,7-dihydroxy-1-indanone in the same solvent at longer times.[9] This sequence is illustrated in Eq. (4):

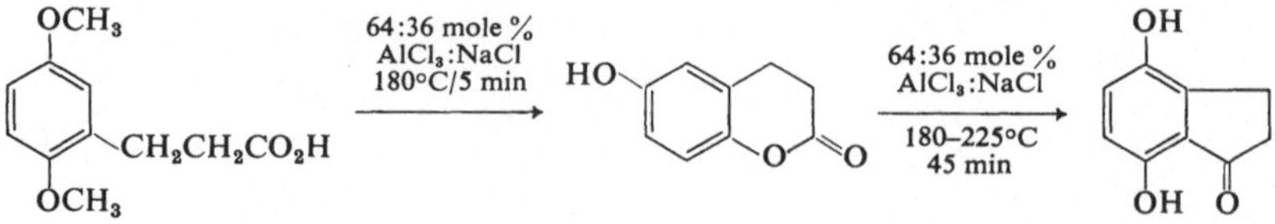

The 3-methylindanone derivative was obtained upon treating hydroquinone with crotonic acid in the same $AlCl_3$ solvent system.[5]

At this point, it is worthwhile to note that throughout this chapter we will have instances where aryl methoxy groups are cleaved to phenols in $AlCl_3$ solvent systems. This reaction is very common, with the final product probably arising from hydrolysis of an intermediate phenoxychloroaluminate during aqueous workup.

Hayes and Thomson[9] also obtained tetralones and a benzosuberone by condensation of phenols with unsaturated carboxylic acids in a $AlCl_3$ solvent system. The yields were very poor owing in part to the volatility of the acids. The same products were obtained by Fries rearrangement (see below) of aryl esters of the appropriate unsaturated acid. An example is illustrated in Eq. (5) for the formation of a tetralone:

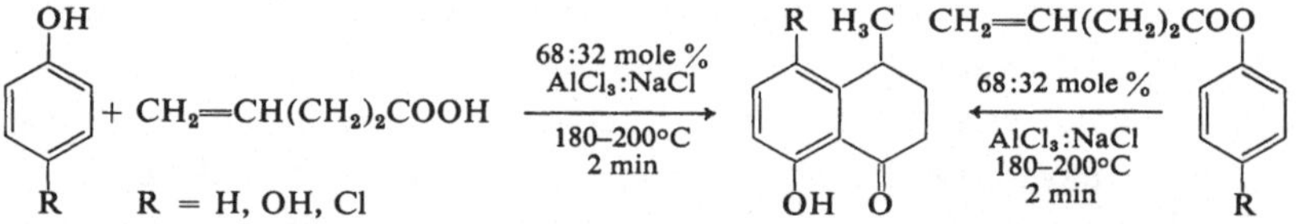

Reported condensation–addition reactions involving carboxylic acids or acid chlorides are listed in Table I. This list does not include a number of examples of various other types of acids, especially keto-acids which also undergo various reactions. These will be discussed later on.

The bulk of the reports on condensation–addition reactions in $AlCl_3$ solvent systems invariably involve the use of dicarboxylic acid anhydrides as a percursor for the acylating reagent. The products which one obtains are found to be very dependent on the reaction temperature. In the case of aromatic acid anhydrides, keto-acids are found at lower temperatures, while quinones are found at higher temperatures, no doubt by condensation of the keto-acid. In some cases condensation of the keto-acid does not give quinones but cyclic keto-acids (see below). We will first consider the reaction of phthalic anhydride with various aromatic and heterocyclic substrates.

Waldmann[10] has reported that when phthalic anhydride and catechol were heated at 130–140°C for 1 hr in a 69:31 mole % $AlCl_3:NaCl$ melt, the product was the keto-acid, 3,4-dihydrozybenzophenone-2'-carboxylic acid [Eq. (6)], and not the corresponding anthraquinones, hystazarin or alizarin:

$$\tag{6}$$

However, at temperatures of approximately 200°C, quinones are secured. In this way, hydroquinone with phthalic anhydride yields quinizarin[11] as shown in Eq. (7):

$$\tag{7}$$

Another example of the effect of temperature on the type of product formed is demonstrated by the reaction between phthalic anhydride and 4-amino-biphenyl[12] as shown in Eq. (8):

$$\tag{8}$$

Numerous examples of aromatic substrates including hydroxy-, alkoxy-, amino-, acetamido-, and chloro-substituted benzenes, polynuclear hydro-carbons, and heterocycles have been aroylated with phthalic anhydride in $AlCl_3$ solvent systems. These are listed in Table II. A number of these reactions deserve further comment.

As indicated in Table II, several reaction products are possible, especially in the condensation of polynuclear hydrocarbons and derivatives with phthalic anhydride. A thorough study of the products of the reaction

TABLE I. Acylation Reactions with Aromatic and Aliphatic Carboxylic Acids and Acid Chlorides

Acid or acid chloride	Compound acylated	Product(s)	Conditions[a]	Ref.
Benzoic	Phenol	Mixture of 2- and 4-hydroxybenzophenone	A	5
	Hydroquinone	2,5-Dihydroxybenzophenone	A	5
4-Bromobenzoic	Hydroquinone	2,5-Dihydroxy-4-bromobenzophenone	A	5
Crotonic	Hydroquinone	3-Methyl-4,7-dihydroxy-1-indanone	A	5
4-Pentenoic	Phenol	4-Methyl-8-hydroxy-1-tetralone	B	9
	Hydroquinone	4-Methyl-5,8-dihydroxy-1-tetralone	B	9
	4-Methylphenol	4,5-Dimethyl-8-hydroxy-1-tetralone	B	9
	4-Chlorophenol	4-Methyl-5-chloro-8-hydroxy-1-tetralone	B	9
5-Hexenoic	Hydroquinone	1',4'-Dihydroxy-7-methylbenzohept-1-en-4-one	B	9
β-Phenylpropionic		1-Indanone	A	5
β-3-Hydroxyphenylpropionic		Mixture of 5- and 7-hydroxy-1-indanone	A	5
β-2-Nitrophenylpropionic		4-Nitro-1-indanone	A	5
β-4-Acetylphenylpropionic		6-Acetyl-1-indanone	C	8

Acid / Chloride	Second reactant	Product	Conditions	Yield
β-4-Carboxylphenylpropionic		6-Carboxy-1-indanone	C	8
γ-Phenylbutyric		1-Tetralone	A	5
γ-4-Acetylphenylbutyric		6-Acetyl-1-tetralone	C	8
γ-4-Carboxyphenylbutyric		6-Carboxy-1-tetralone	C	8
γ-Phenylvaleric		Benzosuberone	A	5
Benzoyl chloride	1,4-Dichloronaphthalene	1,4-Dichloro-5-benzoylnaphthalene	D	99
	1,5-Dichloronaphthalene	1,5-Dichloro-4-benzoylnaphthalene	D	99
	Diphenylamine	Mixture of N-benzoyldiphenylamine and 4,4'-dibenzoyldiphenylamine	E	7
3-Methylbenzoyl chloride	1,4-Dichloronaphthalene	Mixture of 1,7- and 2,6-dimethylanthraquinone	F	6
2-Methylbenzoyl chloride	1,4-Dichloronapthalene	1,4-Dichloro-5-(2-methylbenzoyl)naphthalene	D	99
4-Methylbenzoyl chloride	1,4-Dichloronaphthalene	1,4-Dichloro-5-(4-methylbenzoyl)naphthalene	D	99
2-Chlorobenzoyl chloride	1,4-Dichloronaphthalene	1,4-Dichloro-5-(2-chlorobenzoyl)naphthalene	D	99
4-Chlorobenzoyl chloride	1,4-Dichloronaphthalene	1,4-Dichloro-5-(4-chlorobenzoyl)naphthalene	D	99
1-Naphthoyl chloride	1,4-Dichloronaphthalene	1,4-Dichloro-5-(1-naphthoyl)naphthalene	D	99
3,5-Dimethylbenzoyl chloride		1,3,5,7-Tetramethylanthraquinone	F	100

[a] Unless otherwise stated, the solvent was a 68:32 mole % $AlCl_3$:NaCl melt. A, 180–200°C/2 min; B, 64:36 mole % $AlCl_3$:NaCl melt/180°C/5 min; C, 180°C/1 hr; D, 200°C/1 hr; E, $AlCl_3$/140°C/1 hr; F, $AlCl_3$/130–40°C/16 hr.

TABLE II. Acylation Reactions of Phthalic Anhydride

Compound aroylated	Product	Conditions[a]	Ref.
Catechol	2-(3,4-Dihydroxybenzoyl)benzoic acid	A	10
3-Methylcatechol	3-Methylalizarin (1,2-dihydroxy-3-methyl-anthraquinone)	B	101
Dihydrobenzodioxin	Alizarin ethylene ether and hystazarin ethylene ether	A	102
Hydroquinone	2-(2,5-Dihydroxybenzoyl)benzoic acid	C	103
	Quinizarin (1,4-dihydroxyanthraquinone)	D	11
2-Methylhydroquinone	2-Methylquinizarin	E	104
2,3-Dimethylhydroquinone	2,3-Dimethylquinizarin	F	23
2-Methylphenol	Mixture of 1-hydroxy-2-methyl-, 2-hydroxy-3-methyl-, and 1-methyl-2-hydroxyanthraquinone	G	105
3-Methylphenol	Mixture of 1-hydroxy-3-methyl- and 1-methyl-3-hydroxyanthraquinone	G	105
4-Methylphenol	1-Hydroxy-4-methylanthraquinone	G	105
2-Methyl-4-chlorophenol	1-Hydroxy-2-methyl-4-chloroanthraquinone	G	105
3-Methyl-4-chlorophenol	1-Hydroxy-3-methyl-4-chloroanthraquinone	G	105
4-Aminophenol	1-Hydroxy-4-aminoanthraquinone	H	106
N-Acetyl-4-aminophenol	1-Hydroxy-4-aminoanthraquinone	H	106
N-Benzoyl-4-aminophenol	1-Hydroxy-4-aminoanthraquinone	H	106
N-Benzoyl-4-aminophenol	N-Benzoyl-1-hydroxy-4-aminoanthraquinone	I	107
Triacetyl-4-aminophenol	N-Benzoyl-1-hydroxy-4-aminoanthraquinone	H	106
Tribenzoyl-4-aminophenol	N-Benzoyl-1-hydroxy-4-aminoanthraquinone	H	106
1-Acetamido-3-methyl-4-hydroxybenzene	2-(2-Acetamido-4-methyl-5-hydroxybenzoyl)benzoic acid	J	12
Naphthalene	Mixture of 1,2- and 2,3-benzanthraquinone	K	108
1-Naphthol	Mixture of 1-hydroxy-2,3-benz- and 1-hydroxy-2,3-benzanthraquinone	G	14
1-Hydroxy-4-chloronaphthalene	1-Hydroxy-4-chloro-2,3-benzanthraquinone	L	23
1,5-Dihydroxynaphthalene	Mixture of 2-(1,5-dihydroxy-2-naphthoyl)- and 2-(4,8-dihydroxy-1-naphthoyl)benzoic acids and 6,14-dihydroxyhexacene-5,16:8,13-diquinone	M	13
2-Naphthol	1,8-Phthaloyl-2-hydroxynaphthalene	N	15

Reactant	Product		
2-Hydroxynaphthalene-3-carboxylic acid	1,8-Phthaloyl-2-hydroxynaphthalene-3-carboxylic acid	N	15
2,6-Dihydroxynaphthalene	1,8-Phthaloyl-2,6-dihydroxynaphthalene	N	15
2,7-Dihydroxynaphthalene	1,8-Phthaloyl-2,7-dihydroxynaphthalene	N	15
1-Methyl-2-hydroxynaphthalene	1-Methyl-2-hydroxynaphthalene-6-phthaloylic acid	O	109
2-Naphthylamine	2-(Aminonaphthoyl)benzoic acid and/or its N-acyl derivative	J	18
1-Acetamidonaphthalene	2-(1-Acetamido-2-naphthoyl)benzoic acid	A	110
Biphenyl	4,4'-Bis-(2-carboxybenzoyl)biphenyl	P	20
4,4'-Dihydroxybiphenyl	Mixture of 4,4'-dihydroxy-2-(2-carboxybenzoyl)biphenyl, 2,3-phthaloyl-4,4'-dihydroxy-2'-(2-carboxybenzoyl)biphenyl, and 4,4'-dihydroxy-1,1'-dianthraquinonyl	Q	111
2,2',4,4'-Tetrahydroxybiphenyl	2,2',4,4'-Tetrahydroxy-1,1'-dianthraquinonyl	Q	111
4-Aminobiphenyl	2-(4-Amino-4-phenylbenzoyl)benzoic acid	J	12
	2-(4-Aminophenyl)anthraquinone	R	12
4-Acetamidobiphenyl	2-(4-Acetamido-4-phenylbenzoyl)benzoic acid	J	12
	2-(4-Acetamidophenyl)anthraquinone	R	12
4-Acetamido-3-methylbiphenyl	2-[4-(4-Acetamido-3-methylphenyl)benzoyl]benzoic acid	J	12
	2-(4-Acetamido-3-methylphenyl)anthraquinone	R	12
Pyrene	Mixture of mono- and diphthaloylated products	S	112
Pyrene-3-phthaloylic acid	Diphthaloylpyrene	T	113
2,6-Dimethylthianthrene	2,6-Dimethyl-3,7-(2-carboxybenzoyl)thianthrene	U	114
Phenothiazine	2,3:6,7-Diphthaloylphenothiazine	U	114
2-Aminocarbazole	2-Amino-6,7-(or 7,8)-phthaloylcarbazole	R	12
N-Ethyl-3-aminocarbazole	N-Ethyl-3-amino-6,7-(or 7,8)-phthaloylcarbazole	R	12
2-Amino-diphenylene oxide	2-Amino-6,7-(or 7,8)-phthaloyldiphenylene	R	12
3-Aminophenanthrene	3-Amino-6,7-(or 7,8)-phthaloylphenanthrene	R	12
3-Aminopyrene	3-Amino-6,7-(or 7,8)-pyrene	R	12

[a] Unless otherwise stated, the solvent was a 69:31 mole % $AlCl_3$:NaCl melt. A, 130–140°C/1 hr; B, 180–200°C/5–10 min; C, 120–125°C/1 hr; D, 200°C/10 min; E, 160–170°C/10 min; F, 200–210°C/1 hr; G, 165°C/1.5 hr; H, 64:36 mole % $AlCl_3$:NaCl/200–210°C/45 min; I, 130°C/15 min; J, 57:43 mole % $AlCl_3$:NaCl/110°C/10 min; K, $AlCl_3$/130°C; L, 210°C/2 hr; M, 200°C/30 min; N, $AlCl_3$/210–220°C/10 min; O, $AlCl_3$/100°C; P, $AlCl_3$/100°C/9.5 hr; Q, $AlCl_3$/130–135°C/6 hr; R, 57:43 mole % $AlCl_3$:NaCl/150–155°C/1.75 hr; S, 57:43 mole % $AlCl_3$:NaCl/170°C/3 hr; T, 64:36 mole % $AlCl_3$:NaCl/180°C/2 hr; U, $AlCl_3$/100–104°C/7 hr.

between 1,5-dihydroxynaphthalene and phthalic anhydride has been carried out by Satchell and Stacey.[13] These authors found that the products ranged from the mono keto-acid, o-(1,5-dihydroxy-2-naphthoyl)benzoic acid, to the diquinone, 6,14-dihydroxyhexacene-5,16:8,13-diquinone.

One example which is of interest and has received considerable attention is the condensation of substituted naphthalenes with phthalic anhydride. While α-substituted naphthalenes give normal substituted and ring closed products,[14] β-hydroxynaphthalenes yield 7-membered peri-condensed phthaloyl products, as shown for β-naphthol[15] in Eq. (9):

$$\tag{9}$$

It has been conclusively shown that one obtains the peri-condensed product and not the lactone type compound as was believed by some.[16] Unfortunately, this misconception still exists today.[17] Cyclization of 2-(α-naphthoyl)benzoic acid under identical conditions gave 1,8-phthaloylnaphthalene and not the 1,2-derivative.[15] In the case of β-naphthylamine, the exact structure of the phthaloylated product was not determined,[18] and β-acetamidonaphthalene gave both 2-(6- and 7-acetamido-1-naphthoyl) benzoic acid.[19] A further study of the condensation of β-naphthylamine derivatives would be of interest.

In contrast to the behavior (N-acylation, intractable products) of primary amines under normal Friedel–Crafts conditions, Kränzlein[12] has shown that primary amines derived from several compounds related to biphenyl (including 3,4 bridges such as —O—, —S—, —NH—, —C≡C—) are readily phthaloylated in AlCl$_3$ solvent systems; see, e.g., the above case of 4-aminobiphenyl. It is interesting to note that phthaloylation takes place in the ring without the amino group, with the first substitution step occurring at the *para*-position to the biphenyl union. This is borne out by the fact that p-benzidine and m-tolidine, which have the 4'-position blocked, show no reaction under the above conditions. However, both 4,4'-dihydroxybiphenyl and 3,3',4,4'-tetrahydroxybiphenyl in AlCl$_3$ at 130°C yield 2,3,2',3'-diphthaloyl derivatives.[20] An explanation for the lack of reactivity of the benzidine derivatives is not readily available.

The importance of anthraquinone dyes accounts for the interest shown in the reaction of phthalic anhydride with various aromatic substrates especially in AlCl$_3$ solvent systems where ring closure of the intermediate keto-acid could be carried out very readily. In a similar manner, substituted phthalic anhydrides have been reacted with various substrates to yield substituted anthraquinones. Reported examples are listed in Table III. As was the case above, the structures of the products obtained were often incorrectly assigned. The reactions of chlorinated phthalic anhydrides are of importance since the reactivity of the phthaloylating agent goes up with increasing halogen substitution and thus deactivated substrates can be acylated. Also, the chlorines of the resulting anthraquinones can readily be replaced with other substituents yielding other substituted anthraquinones.

Anhydrides of benzenepolycarboxylic acids, polynuclear dicarboxylic acids, and heterocyclic dicarboxylic acids have also been used as acylating agents in AlCl$_3$ solvent systems. Table IV lists the reported examples. While almost all the tabulated examples yielded normal products, the intramolecular cyclization of 1-phenylnaphthalene-2,3-dicarboxylic anhydride gave primarily 7-oxobenz[de]anthracene-2-carboxylic acid and only a trace of the expected 7-oxobenzo[c]fluorene-6-carboxylic acid,[21] as illustrated in Eq. (10):

$$\text{69:31 mole \% AlCl}_3\text{:NaCl} \quad 140\text{--}150°C/2.5 \text{ hr} \tag{10}$$

However, the same treatment of several 1,4-diarylnaphthalene-2,3-dicarboxylic anhydrides gave rise to benzidenofluorenes,[22] as shown in Eq. (11):

$$\text{66:34 mole \% AlCl}_3\text{:NaCl} \quad 140°C/2 \text{ hr} \tag{11}$$

A complete explanation of these results has not been presented. However, it is known that 7-oxobenzo[c]fluorene can be isomerized to 7-oxobenz-[de]anthracene under similar conditions.[21]

TABLE III. Acylation Reactions with Substituted Phthalic Anhydrides

Substituent in phthalic anhydride	Compound aroylated	Product	Conditions[a]	Ref.
3-Methyl	Catechol	Found only two of the possible three compounds, 5,6-, 6,7-, and 7,8-dihydroxy-1-methylanthraquinone	A	115
	Resorcinol	1-Methyl-5,7-dihydroxyanthraquinone	A	115
	Hydroquinone	5-Methylquinizarin	A	115
	Hydroquinone	3-(or 6)-Methyl-2-(2,5-dihydroxybenzoyl)benzoic acid	B	115
	2-Methylhydroquinone	2,5-(or 2,8)-Dimethylquinizarin	A	115
	2-Hydroxyhydroquinone	1-Methyl-5,6,8-(or 5,7,8)-trihydroxyanthraquinone	A	115
4-Methyl	Catechol	2-Methyl-5,6(or 7,8)-dihydroxyanthraquinone	C	116
	Resorcinol	2-Methyl-6,8-dihydroxyanthraquinone	C	116
	Hydroquinone	6-Methylquinizarin	D	116
	2-Hydroxyhydroquinone	2-Methyl-5,6,8-(or 5,7,8)-trihydroxyanthraquinone	C	116
3-Chloro	Catechol	3-(or 6)-Chloro-2-(3,4-dihydroxybenzoyl)benzoic acid	E	10
	Hydroquinone	5-Chloroquinizarin	F	117
	2,2'-Dimethoxydiphenylurea	Urea derivative of 6-chloro-2-(3-amino-4-hydroxybenzoyl)benzoic acid	G	107

4-Chloro	Catechol	4-(or 5)-Chloro-2-(3,4-dihydroxybenzoyl)benzoic acid	E	10
	Hydroquinone	6-Chloroquinizarin	F	14
3-Bromo	Hydroquinone	5-Bromoquinizarin	F	117
4-Bromo	Hydroquinone	6-Bromoquinizarin	F	14
3,4-Dichloro	Hydroquinone	1,2-Dichloro-5,8-dihydroanthraquinone	F	117
3,5-Dichloro	Hydroquinone	5,7-Dichloroquinizarin	F	117
3,6-Dichloro	Catechol	3,6-Dichloro-2-(3,4-Dihydroxybenzoyl)benzoic acid	E	10
	Hydroquinone	5,8-Dichloroquinizarin	F	17
4,5-Dichloro	Hydroquinone	6,7-Dichloroquinizarin	F	14
3,4,5,6-Tetrachloro	Hydroquinone	3,4,5,6-Tetrachloro-2-(2,5-dihydroxybenzoyl)benzoic acid	H	118
		5,6,7,8-Tetrachloroquinizarin	I	118
	1,4-Dichlorobenzene	3,4,5,6-Tetrachloro-2-(2,5-dichlorobenzoyl)benzoic acid	J	119
4-Acetyl	Hydroquinone	6-Acetylquinizarin	K	120
	3-Methylhydroquinone	6-(or 7)-Acetyl-2-methylquinizarin	K	120
	1,4-Dihydroxynaphthalene	2,3-Benzo-6-acetylquinizarin	K	120

a Unless otherwise stated, the solvent was a 69:31 mole % $AlCl_3$:NaCl melt. A, 190°C/1 hr; B, 180°C/20 min; C, 160°C/45 min; D, 190°C/3 hr; E, 170°C/1 hr; F, 200–220°C/45 min; G, 120–130°C/20 min; H, 155°C/1.5 hr; I, 155°C/1.5 hr, then 210–215°C/1 hr; J, $AlCl_3$/95°C/5 hr; K, 180–190°C/45 min.

TABLE IV. Acylation Reactions with Anhydrides of Benzene Polycarboxylic, Naphthalenedicarboxylic, and Heterocyclic Dicarboxylic Acids

Anhydride	Compound aroylated	Product(s)	Conditions[a]	Ref.
5-Hydroxytrimellitic	Benzene	Either 2-benzoyl-5-hydroxyterephthalic acid or 4-benzoyl-6-hydroxyisophthalic acid or both	A	121
	Acenaphthalene	Either 2-(5-acenaphthoyl)-5-hydroxyterephthalic acid or 4-(5-acenaphthoyl)-6-hydroxyisophthalic acid or both	B	121
	1,4-Dichlorobenzene	Either 2-(2,5-dichlorobenzoyl)-5-hydroxyterephthalic acid or 4-(2,5-dichlorobenzoyl)-6-hydroxyisophthalic acid or both	C	121
1,2-Naphthalic	Hydroquinone	5,8-Dihydroxy-1,2-benzanthraquinone	D	122
	3-Methylhydroquinone	6-(or 7)-Methyl-5,8-dihydroxy-1,2-benzanthraquinone	D	122
	3-Hydroxyhydroquinone	5,6-(or 7)-8-trihydroxy-1,2-benzanthraquinone	D	122
	4-Chlorophenol	5-(or 8)-chloro-8-(or 5)-hydroxy-1,2-benzanthraquinone	D	122
	4-Methylphenol	5-(or 8)-methyl-8-(or 5)-hydroxy-1,2-benzanthraquinone	D	122
	1-Naphthol	5-(or 8)-hydroxy-1,2,6,7-dibenzanthraquinone	D	122
	1,4-Dihydroxynaphthalene	5,8-Dihydroxy-1,2,6,7-dibenzanthraquinone	D	122
2,3-Naphthalic	Hydroquinone	1,4-Dihydroxy-6,7-benzanthraquinone	D	14
	4-Chlorophenol	1-Hydroxy-4-chloro-6,7-benzanthraquinone	D	14
	3-Hydroxyhydroquinone	1,2,4-Trihydroxy-6,7-benzanthraquinone	D	14

		Product		Ref.
	3-Methylhydroquinone	2-Methyl-1,4-dihydroxy-6,7-benzanthraquinone	D	14
	3-Chlorohydroquinone	2-Chloro-1,4-dihydroxy-6,7-benzanthraquinone	D	14
	1,4-Dihydroxynaphthalene	1,4-Dihydroxy-2,3,6,7-dibenzanthraquinone	D	14
	1-Naphthol	1-hydroxy-2,3,6,7-dibenzanthraquinone	D	14
Mellitic	Hydroquinone	Hexahydroxyanthratriquinone	D	123
Benzophenone-3,3'-4,4'-tetracarboxylic acid	Hydroquinone	5,5',8,8'-Tetrahydroxy-2,2'-dianthraquinonyl ketone	D	124
Quinolinic (2,3-pyridinedicarboxylic acid)	Hydroquinone	3-(2,5-Dihydroxybenzoyl)pyridene-2-carboxylic acid	F	11
	Hydroquinone	1-Aza-5,8-dihydroxyanthraquinone	G	11
	1,4-Dihydroxynaphthalene	1-Aza-5,8-dihydroxy-6,7-benzanthraquinone	G	11
Cinchomeronic (3,4-pyridinedicarboxylic acid)	Hydroquinone	2-Aza-5,8-dihydroxyanthraquinone	G	125
	1,4-Dihydroxynaphthalene	2-Aza-5,8-dihydroxy-6,7-benzanthraquinone	G	125
Thionaphthene-2,3-dicarboxylic	Hydroquinone	1,4-Dihydroxybenzothiophanthrenequinone	H	126
	3-Hydroxyhydroquinone	1,2,4- or 1,3,4-Trihydroxybenzothiophanthrenequinone	I	126
	1,4-Dihydroxynaphthalene	1,4-Dihydroxy-2,3-dibenzothiophanthrenequinone	H	126
5-Chloro-7-methylthionaphthene-2,3-dicarboxylic acid	Hydroquinone	6-Chloro-8-methyl-1,4-dihydroxybenzothiophanthrenequinone	H	126

^{a} Unless otherwise stated, the solvent was a 69:31 mole % AlCl$_3$:NaCl melt. A, 80–100°C/30 min; B, 100–170°C/30 min; C, 140°C/2 hr; D, 200–10°C/3 hr; E, 180–190°C/3 hr; F, 120–130°C/1 hr; G, 180–200°C/10 min; H, 190°C/6 hr; I, 230°C/6 hr.

The condensations of aliphatic dicarboxylic acid anhydrides with aromatic substrates in $AlCl_3$ solvent systems have been extensively studied. Numerous examples are known of the reaction between maleic anhydride and its derivatives and 1,4-dihydroxyaromatics to yield structurally important naphthazarins and quinizarins. For example, condensation of α,β-dichloromaleic anhydride with hydroquinone yields 2,3-dichloronaphthazarin[23] [Eq. (12)]:

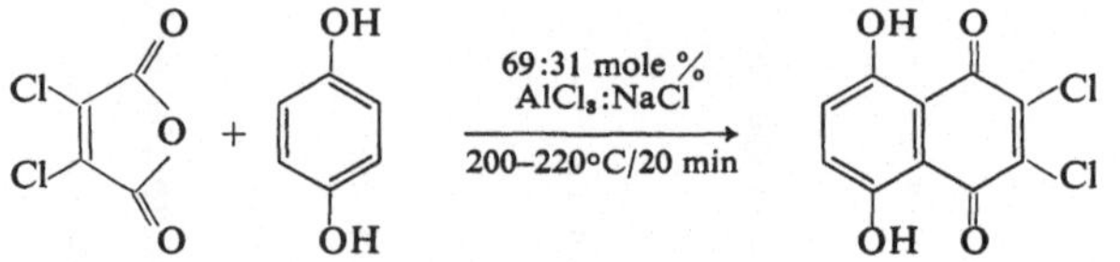

$$\text{(12)}$$

and α,β-dimethylmaleic anhydride reacts with 1,4-dihyroxynaphthalene to yield 2,3-dimethylquinizarin[23] [Eq. (13)]:

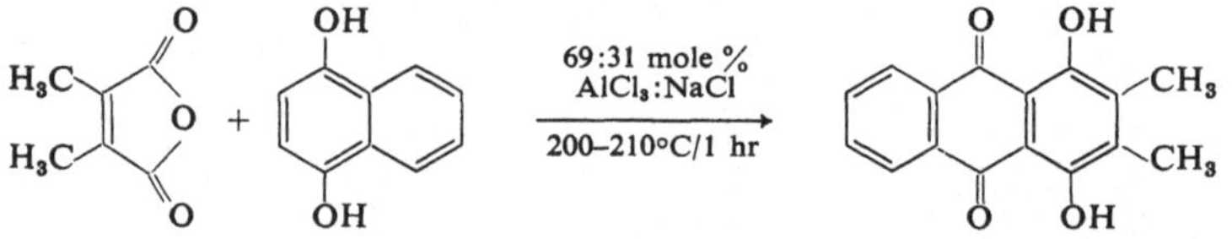

$$\text{(13)}$$

A complete list of the condensation reactions of aliphatic dicarboxylic acid anhydrides is presented in Table V.

As we have seen above for other condensation reactions, assignment of the structures of the products has not always been unequivocal. In the case where certain naphthazarins are formed, the quinoid and benzenoid states of the rings are interchangeable. For example, in the condensation of 2-alkylhydroquinones with maleic anhydride in a 70:30 mole % $AlCl_3$:NaCl melt, the 2-alkylnaphthazarin is formed,[24] as shown in Eq. (14):

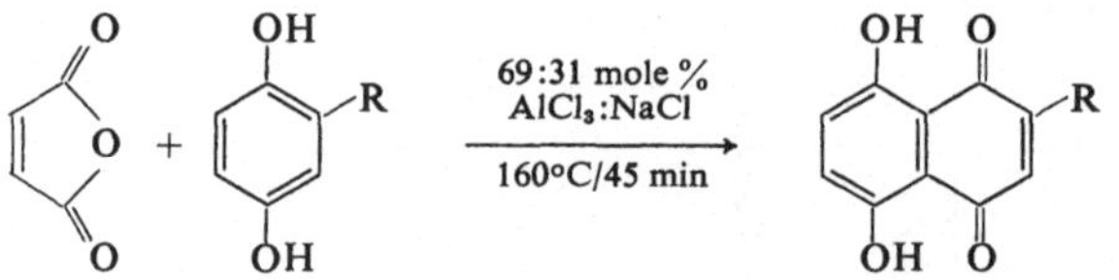

$$\text{(14)}$$

Similarly, in the condensation of 2-isohexylhydroquinonedimethyl ether with maleic anhydride, 1,1'-dimethyl-1,2,3,4-tetrahydroquinizarin is formed,[25] presumably by ring closure of the intermediate α-isohexylnaphthazarin [Eq. (15)]:

$$\text{(15)}$$

As was noted above, saponification of the ether linkages occurs. In the case of quinizarin formation, we see that tautomerism to the 9,10-anthroquinone always occurs. Even when 9,10-quinone formation is impossible, tautomerism to an anthrone derivative occurs, as shown in Eq. (16):

$$\text{(16)}$$

A number of aliphatic dicarboxylic acids have been reported to undergo condensation reactions with aromatic substrates in $AlCl_3$ solvent systems. The reported examples are listed in Table VI. The products are, in general, identical to those obtained when the corresponding acid anhydride is used. However, a number of examples do require some comment. First, in the series succinic, glutaric, and adipic acids, ring closure to the cyclic diketones was found in the case of the first two acids when they were condensed with hydroquinone, while adipic acid yielded the keto-acid, γ-(2,5-dihydroxybenzoyl)valeric acid[5] (6- and 7-membered ring formation vs 8-membered formation). Secondly, in comparison to the cyclic diketone which arises from glutaric acid and hydroquinone, condensation of glutaric anhydride and 4-chlorophenol yields the bis-condensation product, 1,3-di(2-hydroxy-5-chlorobenzoyl)propane.[26] As will be discussed below, it appears that the chelated configuration in the case of the p-chloro derivative is not strong enough to enhance ring closure. Finally, intramolecular cyclization of a diacid was observed when α,β-diphenylsuccinic acid was treated with a 68:32 mole % $AlCl_3$:NaCl melt,[5] as illustrated in Eq. (17):

$$\text{(17)}$$

Bruce, Sorrie, and Thomson[5] have also reported the condensation of lactones with several phenols under the above conditions. Upon reacting γ-butyrolactone with hydroquinone, one obtains 4,7-dihydroxy-3-methyl-

TABLE V. Acylation Reactions with Anhydrides of Aliphatic Dicarboxylic Acids

Anhydride	Compound acylated	Product(s)	Conditions[a]	Ref.
Maleic	Hydroquinone	Naphthazarin (5,8-dihydroxy-1,4-naphthoquinone	A	24
	2-R-hydroquinone; R = methyl, ethyl, propyl, butyl, isoamyl, isohexyl	2-R-naphthazarin	A	24
	2-R-hydroquinone dimethyl ether; R = ethyl, propyl, isobutyl, isoamyl	2-R-naphthazarin	B	25
	2-Isohexylhydroquinone dimethyl ether	1,1'-dimethyl-1,2,3,4-tetrahydroquinizarin	B	25
	2,3-Dimethylhydroquinone	2,3-Dimethylnaphthazarin	B	25
	2-Hydroxyhydroquinone	2-Hydroxynaphthazarin	C	127
	2-Bromohydroquinone	2-Bromonaphthazarin	D	23
	2-Methyl-3-methoxyhydroquinone	2-Methyl-3-hydroxynaphthazarin	A	24
	2,3-Dichlorohydroquinone	2,3-Dichloronaphthazarin	C	128
	1-Chloro-4-hydroxynaphthalene	1-Hydroxy-10-chloro-4,9-anthraquinone	E	129
	5,6,7,8-Tetrahydro-1,4-dihydroxy-naphthalene	5,6,7,8-Tetrahydroquinizarin	D	129
			E	130

Citraconic	Hydroquinone	2-Methylnaphthazarin	C	127
	2-Methylhydroquinone	Mixture of 2,6- and 2,7-dimethylnaphthazarin	B	25
	2,3-Dimethylhydroquinone	2,3,6-Trimethylnaphthazarin	B	25
	1,4-Dihydroxynaphthalene	2-Methylquinizarin	D	129
	1-Chloro-4-hydroxynaphthalene	Mixture of 2- or 3-methyl-1-hydroxy-10-chloro-4,9-anthraquinone	D	129
Dimethylmaleic	2,3-Dimethylhydroquinone	2,3,6,7-Tetramethylnaphthazarin	D	23
	2-Chlorohydroquinone	2,3-Dimethyl-6-chloronaphthazarin	D	23
	2-Bromohydroquinone	2,3-Dimethyl-6-bromonaphthazarin	D	23
	1,4-Dihydroxynaphthalene	2,3-Dimethylquinizarin	F	23
	1-Chloro-4-Hydroxynaphthalene	2,3-Dimethyl-Hydroxy-9-chloro-1,10-anthraquinone	D	23
Chloromaleic	2,3-Dimethylhydroquinone	2,3-Dimethyl-6-chloronaphthazarin	D	23
Bromomaleic	Hydroquinone	2-Bromonaphthazarin	D	23
	2,3-Dimethylhydroquinone	2,3-Dimethyl-6-bromonaphthazarin	D	23
Dichloromaleic	Hydroquinone	2,3-Dichloronaphthazarin	D	23
Dibromomaleic	Hydroquinone	2,3-Dibromonaphthazarin	D	23
	1,4-Dihydroxynaphthalene	2,3-Dibromoquinizarin	D	23
Glutaric	4-Chlorophenol	1,3-Di-(2-hydroxy-5-chloro-benzoyl)propane	G	9

[a] Unless otherwise stated, the solvent was a 69:31 mole % AlCl$_3$:NaCl melt. A, 52:48 mole % AlCl$_3$:NaCl/200°C/2 min; B, 200°C/5 min; C, 180–200°C/5 min; D, 200–220°C/20 min; E, 180°C/10 min; F, 200–210°C/1 hr; G, 64:36 mole % AlCl$_3$:NaCl/180°C/5 min.

TABLE VI. Acylation Reactions with Aliphatic and Aromatic Dicarboxylic Acids

Diacid	Compound acylated	Product(s)	Conditions[a]	Ref.
Succinic	Hydroquinone	1,2,3,4-Tetrahydro-5,8-dihydroxy-1,4-diketonaphthalene	A	5
Maleic	Hydroquinone	Naphthazarin	A	5
Glutaric	Hydroquinone	1',4'-Dihydroxy-1,2-benzocycloheptene-3,7-dione	B	131
	1,4-Dihydroxynaphthalene	1',4'-Dihydroxynaphtho(2',3'-1,2)-cycloheptene-3,7-dione	A	5
Glutaconic	Hydroquinone	4,7-Dihydroxy-3-oxo-1-indanylacetic acid	C	131
Adipic	Hydroquinone	γ-(2,5-dihydroxybenzoyl)-valeric acid	A	5
Homophthalic	Hydroquinone	1',4'-Dihydroxy-1,2-4,5-dibenzocycloheptadiene-3,7-dione	D	132

[a] Unless otherwise stated, the solvent was a 68:32 mole % AlCl$_3$:NaCl melt. A, 180–200°C/2 min; B, 57:43 mole % AlCl$_3$:NaCl/180–200°C/20–30 min; C, 66:34 mole % AlCl$_3$:NaCl/180–195°C/10 min; D, 180–195°C/10 min.

indan-1-one, the same product which one obtains from crotonic acid and hydroquinone [Eq. (18)]:

$$(18)$$

The product from the reaction of the lactone apparently arises by ring opening followed by elimination of hydroxide ion and a shift of the consequent double bond to form a vinyl ketone which, as we will see later, cyclizes to form the desired product. In a like manner, the product of the reaction between γ-valerolactone and hydroquinone under the same conditions is 4,7-dihydroxy-3-ethylindan-1-one. The reported examples are listed in Table VII.

The condensation of keto-acids in $AlCl_3$ solvent systems was briefly mentioned above. The importance of the intramolecular ring closure of o-aroylbenzoic acids obtained from phthalic anhydrides and aromatic substrates to yield anthraquinone derivatives is obvious. A thorough study of the ring closure of many o-benzoylbenzoic acid derivatives has been carried out by Baddeley, Holt, and Makar.[27] These workers found that certain derivatives afforded anthraquinones in which alkyl migration had occurred. For example, when o-(2,5-dimethylbenzoyl)benzoic acid is heated for 1 h at 175°C in a 75:25 mole % $AlCl_3$:NaCl melt, both 1,4-dimethyl- and 1,3-dimethylanthraquinone are obtained, while o-(2,4-dimethylbenzoyl)-benzoic acid under the same conditions gives only 1,3-dimethylanthraquinone. This sequence is outlined in Eq. (19):

$$(19)$$

Similarly, o-(2,4-dimethyl-6-hydroxybenzoyl)benzoic acid was cyclized to 1,2-dimethyl-4-hydroxyanthraquinone. The authors were able to show that migration of alkyl groups occurred prior to ring closure and, therefore, the products of the condensation reactions were determined by the relative

TABLE VII. Acylation Reactions with Lactones

Lactone	Compound acylated	Product(s)	Conditions[a]	Ref.
γ-Butyrolactone	Hydroquinone	4,7-Dihydroxy-3-methylindan-1-one	A	5
	Phenol	Mixture of 5- and 7-hydroxy-3-methylindan-1-one	A	26
	4-Methylphenol	Mixture of 4-hydroxy-3,7-dimethyl- and 7-hydroxy-3,4-dimethylindan-1-one	A	26
	4-Chlorophenol	4-Chloro-7-hydroxy-3-methylindan-1-one	A	26
γ-Valerolactone	Hydroquinone	3-Ethyl-4,7-dihydroxyindan-1-one	A	5

[a] A = 68:32 mole % $AlCl_3$:NaCl/180–200°C/2 min.

rates of isomerization and of ring closure of the o-aroylbenzoic acids. The results of Baddeley et al. together with the other reported examples of aroylaromatic acid condensation reactions are listed in Table VIII.

As we have discussed before, o-(1-naphthoyl)benzoic acid derivatives give exclusively the peri-condensed 7-membered ring product, while the 2-naphthoyl derivatives yield 2,3-diketo products. Peri-condensation to yield an 8-membered ring was also observed when 8-(1'-napthoyl)-1-naphthoic acid was fused in a 68:32 mole % AlCl$_3$:NaCl melt at 140°C,[28] as shown in Eq. (20):

$$\qquad (20)$$

However, it is curious that under the same reaction conditions 8-benzoyl-1-naphthoic acid showed no reaction.[28] Possible explanations for this are either that the reactions conditions were not severe enough (higher temperature needed) or that the phenyl ring is sterically hindered to electrophilic substitution.

The fusion of aroylaliphatic acids in AlCl$_3$ solvent systems has also been investigated, and the reported examples are listed in Table IX. The reactions of β-aroylacrylic acids have been studied thoroughly as these acids are believed to be intermediates in the condensation of maleic acid derivatives with aromatic substrates. Baddeley and co-workers[29] found that whereas β-2,5-dihydroxybenzoyl- and β-2-hydroxy-5-chlorobenzoyl-acrylic acid derivatives underwent intramolecular acylation, cyclization of β-2-hydroxy-5-methylbenzoyl and β-2-hydroxy-3,5-dimethylbenzoylacrylic acid occurred by intramolecular alkylation. Examples of these reactions are shown in Eqs. (21) and (22):

$$\qquad (21)$$

$$\qquad (22)$$

TABLE VIII. Intramolecular Acylation Reactions of 2-Aroylbenzoic Acids and Aroylnaphthoic Acids

Aroyl substituent in 2-aroylbenzoic acid or indicated aroylnaphthoic acid	Product(s)	Conditions[a]	Ref.
2,4-Dimethylbenzoyl	1,3-Dimethylanthraquinone	A	27
2,5-Dimethylbenzoyl	Mixture of 1,3- and 1,4-dimethylanthraquinone	B	27
2,5-Diethylbenzoyl	1,3-Diethylanthraquinone	C	27
2,4,5-Trimethylbenzoyl	Mixture of 2-(3,4,5-trimethylbenzoyl)benzoic acid and 1,2,3-trimethyl-anthraquinone	D	27
2,4,6-Trimethylbenzoyl	Mixture of 2-(2,4,5-trimethylbenzoyl)benzoic acid and 1,2,4-trimethyl-anthraquinone	E	27
	Mixture of 2-(3,4,5-trimethylbenzoyl)benzoic acid and 1,2,3-trimethyl-anthraquinone	F	27
2,3,5,6-Tetramethylbenzoyl	Mixture of 2-(2,3,4,5-tetramethylbenzoyl)benzoic acid and 1,2,3,4-tetramethylanthraquinone	G	27
	1,2,3,4-Tetramethylanthraquinone	H	27
2,4-Dimethyl-6-hydroxybenzoyl	Mixture of 2-(3,4-dimethyl-6-hydroxybenzoyl)benzoic acid and 1,2-dimethyl-4-hydroxyanthraquinone	I	27
	1,2-Dimethyl-4-hydroxyanthraquinone	J	27
3,4-Dimethyl-6-hydroxybenzoyl	1,2-Dimethyl-4-hydroxyanthraquinone	K	27

2,4-Diethyl-6-hydroxybenzoyl	Mixture of 2-(3,4-diethyl-6-hydroxybenzoyl)benzoic acid and 1,2-diethyl-4-hydroxyanthraquinone	L	27
	1,2-Diethyl-4-hydroxyanthraquinone	M	27
4-Phenylbenzoyl	2-Phenylanthraquinone	N	20
4-Acetamidobenzoyl	2-Acetamidoanthraquinone	O	19
2,5-Dimethyl-3-thenoyl	2,7-Dimethyl-β-thiophanthrenequinone	P	133
1-Naphthoyl	1,8-Phthaloylnaphthalene	Q	134
2-Naphthoyl	2,3-Benzanthraquinone	O	19
4-Chloro-1-naphthoyl	4-Chloro-1,8-phthaloylnaphthalene	O	19
5-Chloro-1-naphthoyl	5-Chloro-1,8-phthaloylnaphthalene	O	19
5-Bromo-1-naphthoyl	4-Bromo-1,8-phthaloylnaphthalene	O	19
3-Benzoyl-2-naphthoic acid	2,3-Benzanthraquinone	R	14
3-(4-Chlorobenzoyl)-2-naphthoic acid	2-Chloro-6,7-benzanthraquinone	R	14
1-(1-naphthoyl)-8-naphthoic acid	1,8-Naphthaloylnaphthalene	S	28

[a] Unless otherwise stated, the solvent was a 75:25 mole % $AlCl_3$:NaCl melt. A, 175°C/1 hr; B, 165° or 220°C/1.5 hr; C, 160°C/2 hr; D, 135°C/1 hr; E, 105°C/1 hr; F, 160°C/1 hr; G, 105°C/3 hr; H, 165°C/2 hr; I, 155°C/3 hr; J, 175°C/3 hr; K, 165°C/3 hr; L, 105°C/4 hr; M, 170°C/3 hr; N, $AlCl_3$/150–160°C; O, $AlCl_3$; P, 69:31 mole % $AlCl_3$:NaCl/140°C/5 min; Q, $AlCl_3$:NaCl; R, 69:31 mole % $AlCl_3$:NaCl/130°C/1 hr; S, 69:31 mole % $AlCl_3$:NaCl/130–140°C.

TABLE IX. Intramolecular Acylation Reactions of β-Aroylacrylic Acids

Aroyl substituent in β-aroylacrylic acid	Product(s)	Conditions[a]	Ref.
Benzoyl	3-Oxoindane-1-carboxylic acid	A	5
2,5-Dimethoxybenzoyl	Naphthazarin	B	29
2-Hydroxy-5-chlorobenzoyl	5-Hydroxy-8-chloronaphthaquinone	C	29
2-Hydroxy-5-methylbenzoyl	4-Hydroxy-7-methyl-3-oxoindane-1-carboxylic acid	D	29
2,5-Dimethylbenzoyl	4,7-Dimethyl-3-oxoindane-1-carboxylic acid	E	27
2,4-Dimethylbenzoyl	Mixture of 4,6- and 5,6-dimethyl-3-oxoindane-1-carboxylic acid	E	27
2-Hydroxy-4,6-dimethylbenzoyl	5,7-Dimethyl-4-oxochroman-2-carboxylic acid	F	29
2,4,6-Trimethylbenzoyl	4,6,7-Trimethyl-3-oxoindane-1-carboxylic acid	G	27
2,4,5-Trimethylbenzoyl	4,6,7-Trimethyl-3-oxoindane-1-carboxylic acid	G	27
2,3,5,6-Tetramethylbenzoyl	4,5,6,7-Tetramethyl-3-oxoindane-1-carboxylic acid	H	27
2,3,4,5-Tetramethylbenzoyl	4,5,6,7-Tetramethyl-3-oxoindane-1-carboxylic acid	H	27
1-Naphthoyl	3-Oxo-4,5-benzindane-1-carboxylic acid	I	30
2-Naphthoyl	3-Oxo-6,7-benzindane-1-carboxylic acid	I	30
s-Octahydro-9-anthroyl[b]	1,2,3,4,5,6,7,8-Octahydro-5'-oxo-9,10-cyclopentophenanthrene-3'-carboxylic acid	H	135
s-Octahydro-9-phenanthroyl[b]	1,2,3,4,5,6,7,8-Octahydro-5'-oxo-9,10-cyclopentophanthrene-3'-carboxylic acid	H	135

[a] Unless otherwise stated, the solvent was a 75:25 mole % AlCl₃:NaCl melt. A, 68:32 mole % AlCl₃:NaCl/180–200°C/2 min; B, 200°C/1 hr; C, 200°C/2 hr; D, 180°C/1 hr; E, 135°C/1 hr; F, 130°C/1 hr; G, 150°C/1 hr; H, 105°C/1 hr; I, 115°C/1 hr.
[b] The term s-octahydro is used when the central ring is aromatic.

The most reasonable explanation put forward by these authors was the suggestion that the trans-acid is converted by $AlCl_3$–HCl into the cis isomer which, only when conserved by formation of a chelate ring as depicted below for the intermediate acylium ion

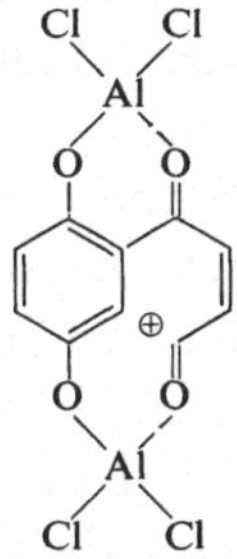

provides an intramolecular acylation route. It is claimed by these authors that intramolecular alkylation is not a general reaction of β-o-hydroxy-aroylacrylic acids, as β-2-hydroxy-4,5-dimethylbenzoylacrylic acid does not cyclize at 130°C and β-2-hydroxy-4,6-dimethylbenzoylacrylic acid affords 5,7-dimethyl-4-oxochroman-2-carboxylic acid. More work seems necessary to verify this claim.

The fusion of β-methylbenzoylacrylic and β-1- and β-2-naphthoylacrylic acids has also been studied in $AlCl_3$ solvent systems, and the results are reported in Table IX. When a methyl group was ortho to the carbonyl group, migration of the methyl group occurred prior to the intramolecular ring-closing alkylation. The same factors prevail here as in the condensation of the methylated o-benzoylbenzoic acids. In the case of the naphthoyl derivatives, what is both surprising and bothersome is that fusion of β-1-naphthoylacrylic acid in a 75:25 mole % $AlCl_3$:NaCl melt at 115°C for 1 h yields exclusively 3-keto-4:5-benzindane-1-carboxylic acid,[30] as illustrated in Eq. (23):

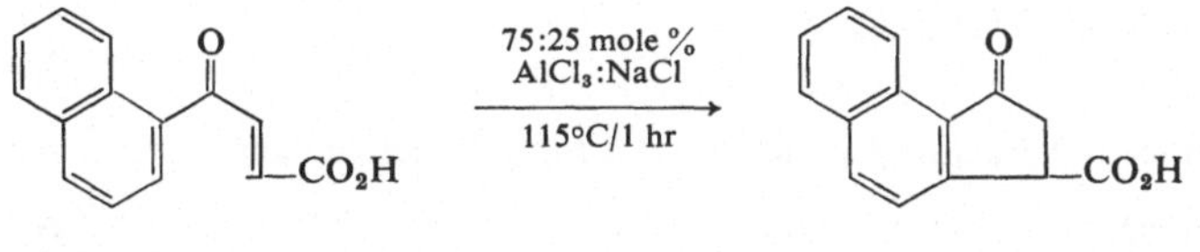

$$\tag{23}$$

This is remarkable when compared with other 1-naphthoyl aromatic and aliphatic keto-acids where, as we have seen above, peri-condensation is almost always favored. Further work is certainly needed to verify and explain these differences.

Condensation reactions in which Friedel–Crafts type of alkylation occurs also take place in $AlCl_3$ solvent systems. We have already seen that a number of β-aroylacrylic acids undergo ring closure via alkylation. Baddeley and Williamson[8] have thoroughly studied the reactions of a number of p-acetylphenylalkyl chlorides in a 82:18 mole % $AlCl_3$:NaCl melt at 100°C. Their results are listed in Table X. To account for the products having a single basic structure, 2-alkyl-5-acetylindane, the authors suggest that they are afforded by intramolecular alkylation via a primary carbonium ion which arises by isomerization of the chloroalkyl side-chains. Intramolecular alkylations were also found to occur via a secondary carbonium ion in the synthesis of 4-methyl-1-tetralone from 3-bromobutyl phenyl ketone and via a tertiary carbonium ion in the formation of 3,3-dimethyl-1-indanone from β-benzoyl-α,α-dimethylpropionic acid. The tertiary cation arises through decarbonylation of the keto-acid.

Ring closure via a secondary carbonium ion also probably occurs in the syntheses described above of peri-hydroxy-tetralones and -benzocycloheptenones obtained by the direct condensation of phenols with unsaturated aliphatic acids. The first step in the reaction is more than likely the formation of an intermediate olefinic ketone, followed by formation of a secondary carbonium ion which then undergoes ring closure. This carbonium ion must possess a fair degree of stability as no isomerization takes place, especially in the case where a 7-membered ring is formed.

A final example of a Friedel–Crafts hydrocarbon synthesis which has been carried out in an $AlCl_3$ solvent system is the reaction of benzene with

TABLE X. Intramolecular Alkylation Reactions of Aryl-Substituted Alkyl Halides

Compound cyclized	Product(s)	Conditions[a]	Ref.
1-(4-Acetylphenyl)-4-chlorobutane	2-Methyl-5-acetylindane	A	8
1-(4-Acetylphenyl)-3-chlorobutane		A	8
1-(4-Acetylphenyl)-2-methyl-3-chlorobutane		A	8
1-(4-Acetylphenyl)-3-chloropentane	2-Ethyl-5-acetylindane	A	8
1-(4-Acetylphenyl)-4-chloropentane		A	8
1-(4-Acetylphenyl)-5-chloropentane		A	8
ω-Bromobutyrophenone	3-Methyl-1-indanone	B	8
ω-Bromovalerophenone	4-Methyl-1-tetradone	B	8

[a] Unless otherwise stated, the solvent was a 82:18 mole % $AlCl_3$:NaCl melt. A, 100°C/1 hr; B, 77:23 mole % $AlCl_3$:NaCl/100°C/1 hr.

ethyl chloride or ethyl bromide at 150–170°C to yield ethylbenzene.[31] However, this reaction is hampered by formation of other tarry products.

3. DEHYDROGENATION–CONDENSATION REACTIONS

Aluminum chloride is known to catalyze nuclear dehydrogenative condensation of aromatic hydrocarbons with formation of higher ring compounds. The reaction may be intramolecular or intermolecular. Thus, 1,1'-binaphthyl yields perylene,[32] and 2 moles of chrysene yield 2,2'-bichrysenyl.[33] This process is commonly called the Scholl reaction, as it was this worker who was largely responsible for its application to the synthesis of a large number of compounds, especially to benzanthrone derivatives—an important class of dyes. The Scholl reaction has also been effectively used by Fieser in the synthesis of a number of powerful carcinogens. Although the Scholl reaction has been carried out in other solvents, in many cases the use of $AlCl_3$ solvent systems was found to remove hydrogen (the by-product of the reaction) more effectively.

In the original reports of the Scholl reaction, fused $AlCl_3$ itself was used as the solvent. For example, upon refluxing naphthalene with $AlCl_3$ at 180°C for 1 hr, perylene was obtained,[32] as shown in Eq. (24):

$$\text{(24)}$$

More recently, $AlCl_3$:NaCl melts have been used in the synthesis of polynuclear hydrocarbons. The preparation of 1.12;4.5;8.9-tribenzoperylene[34] is illustrated in Eq. (25):

$$\text{(25)}$$

Substituted binaphthyls have also been used as substrates in the Scholl reaction, as illustrated for 4,4'-dicyano-1,1'-binaphthyl[35] in Eq. (26):

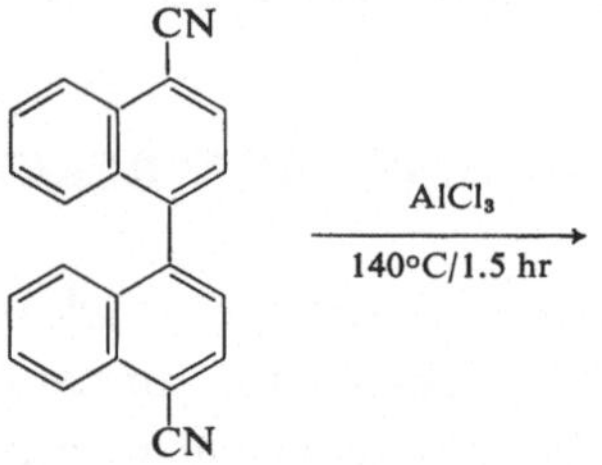

$$\text{(26)}$$

Balaban and Nenitzescu[36] have tabulated the many examples of poly-nuclear hydrocarbons and their derivatives synthesized via the Scholl reaction in $AlCl_3$ solvent systems as well as other solvent–catalyst mixtures.

The Scholl reaction has been used extensively in the synthesis of benzan-throne derivatives. Scholl and Seer[37] prepared 1,9-benzanthrone from phenyl 1-naphthyl ketone in 75% yield by heating the ketone with five parts of $AlCl_3$ at 150°C for 2.5 hr, as shown in Eq. (27):

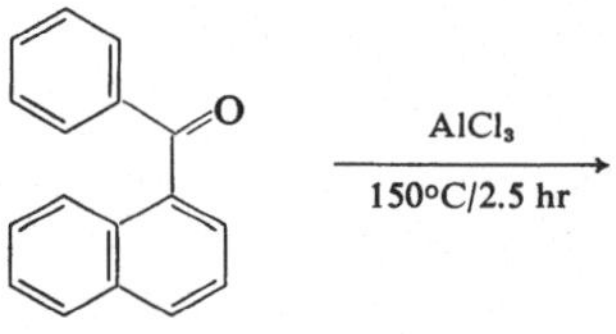

$$\text{(27)}$$

Similar procedures have been used for the intramolecular ring closure of a miscellany of phenyl 1-naphthyl ketones containing substrates such as methyl, chloro, carboxy, and hydroxy. Balaban and Nenitzescu[36] have tabulated the reported examples. Although it was not recognized initially, later workers[38] found that fusion of both 1-(2-methylbenzoyl)- and 1-(3-methylbenzoyl)naphthalene in $AlCl_3$ solvent systems gave the same product, 6-methyl-1,9-benzanthrone. It seems obvious, as we have seen above, that in the case of the 2-methyl derivative (methyl group ortho to a carbonyl group) methyl migration must have occurred prior to ring closure. Methyl migration was also observed during the condensation of 1-benzoyl-2-methylnaphthalene, with the subsequent formation of 3-methylbenzan-throne.[39] The failure of 1-(2-chlorobenzoyl)naphthalene to undergo the Scholl reaction[40] was attributed to steric interactions between the chlorine atom and the beta and peri hydrogens.

It is of further interest to consider the condensation in $AlCl_3$ solvent systems of the derivative and substitution products of 4-benzoyl-1,8-naphthalic acid or its anhydride. In all of the examples which have been reported, we find that the peri-functional group of the corresponding benzanthrone product remains intact. Thus, as illustrated in Eq. (28):

$$(28)$$

the condensation of 4-benzoyl-1,8-naphthal-N-phenylimide gives benzan-throne-peri-dicarboxylic acid-N-phenylimide[41] and not a product involving ring expansion of the imide ring as has been shown for N-phenylphthalimide[42] (see below).

Numerous other aromatic ketones containing more highly condensed aromatic nuclei undergo the Scholl reaction. For example, 3,9-dibenzoyl-perylene when fused in an AlCl$_3$:NaCl melt yields isoviolanthrone,[43] as illustrated in Eq. (29):

$$(29)$$

The reported examples have been tabulated.[36]

Furthermore, it is to be noted that the aroyl hydrocarbons used in these syntheses are ordinarily prepared by reaction of an aroyl chloride with the appropriate aromatic compound in the presence of AlCl$_3$. When the reaction is effected in an AlCl$_3$ solvent system, ring closure may occur during the preparation of the ketone. Dibenzopyrenequinone and its derivatives are thus obtained from α-aroylnaphthalenes and aroyl chlorides.[44] Polynuclear products from mixed diketones can also be formed by oxidative ring closure.[44] Thus, methylpyranthrone is formed by adding benzoylpyrene and *p*-toluyl chloride to an AlCl$_3$:NaCl melt at 110–120°C, raising the temperature quickly to 165°C, and passing in oxygen until the reaction is complete, the fused mass becoming blue. The reaction probably occurs according to Eq. (30):

$$\tag{30}$$

Although up to now we have only considered the Scholl reaction in terms of a dehydrogenative peri-cyclization process, other reactions are possible. Attempts to prepare fluorenone by heating benzophenone with AlCl$_3$ failed,[45] and it was believed that existence of polynuclear hydrogens was necessary. However, it has been shown that treating either azobenzene[35] or benzil[46] in an AlCl$_3$ solvent system yields their respective ortho-coupled products, 9,10-diazaphenanthrene and 9,10-phenanthrenequinone. Thus, we see that polynuclear hydrogens are not necessary. The possibility that five-membered ring formation was a limiting factor is dispelled by the observation that heating di-(-1-anthraquinonyl)-amine with AlCl$_3$ yields 1,2,7,8-diphthaloylcarbazole,[47] as illustrated in Eq. (31):

$$\tag{31}$$

Also, we have seen above that the Scholl reaction can be used for the synthesis of heterocyclic compounds. In their study (see below) of the dehydration of a number 1-arylamino-anthraquinones, Arient and Slavik[48] isolated small amounts of 1,2-phthaloylcarbazoles. The formation of a six-membered heterocyclic ring was observed when 1-amino-2,3-benzanthraquinone was heated with AlCl$_3$,[49] as illustrated in Eq. (32):

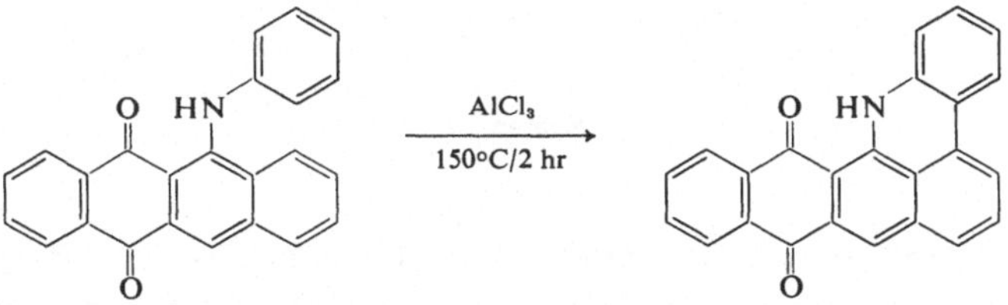

The by-product of the Scholl reaction is believed to be hydrogen gas. This belief is based on several observations of the apparent reduction of compounds containing hydrogen acceptors. For example, condensation of 1-naphthylphthalide yields 1,2,3,4-dibenzopyrene,[50] as shown in Eq. (33):

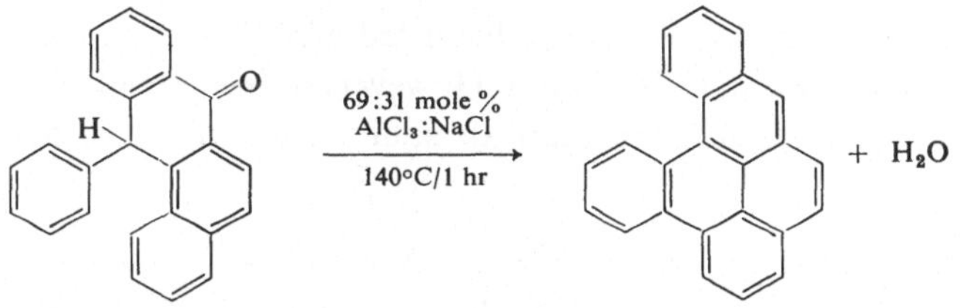

We cannot find any direct evidence for the formation of hydrogen; in fact, Cook and deWorms[51] state that during the condensation of 1,1'-dinaphthyl ketone, hydrogen chloride gas was given off. In the above example and others, the so-called reduced products could easily arise by dehydration of a ring-closed intermediate or via a redox process.

Although the mechanism of the Scholl reaction has not been studied in detail, a number of hypotheses (based almost entirely on product data with no physical organic studies having been made) on the mechanism have been presented. We need only discuss two of them here. The most detailed discussion has been given by Balaban and Nenitzescu[36] who reformulated the mechanism originally proposed by Baddeley.[52] These authors propose an ionic mechanism consisting of three steps—a protolytic or Lewis acid–base reaction, an electrophilic substitution reaction, and a dehydrogenation reaction—as outlined in scheme I. The most direct proof

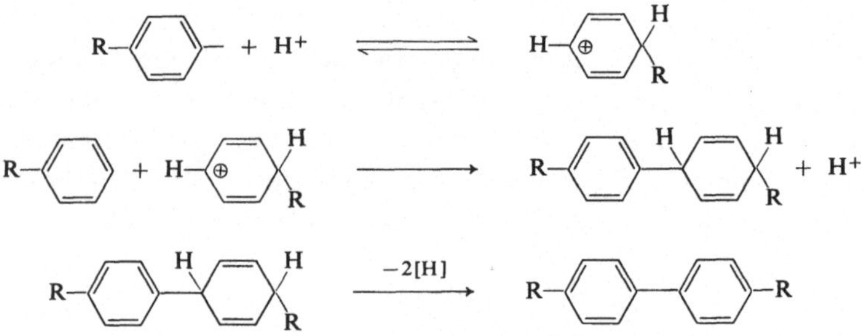

Scheme I

for this type of mechanism comes from the work of Baddeley[52] who found that hydrogen chloride was a necessary constituent of the reaction mixture. For example, 1-benzoylnaphthalene is converted in 70% yield into benzanthrone by fusion in an AlCl$_3$:NaCl melt at 125°C in the presence of hydrogen chloride; however, the reaction did not occur when the hydrogen chloride was expelled by passage of oxygen or nitrogen gas through the molten mixture. Examples are also known of the Scholl hydrocarbon synthesis occurring in either a purely protic solvent (liquid HF)[53] or in a mixture of an aprotic solvent and a Brönsted acid,[36] i.e., in the absence of AlCl$_3$ (or other Lewis acid catalysts). However, in the case of diaryl ketones Baddeley[52] found that more than an equivalence of AlCl$_3$ per carbonyl

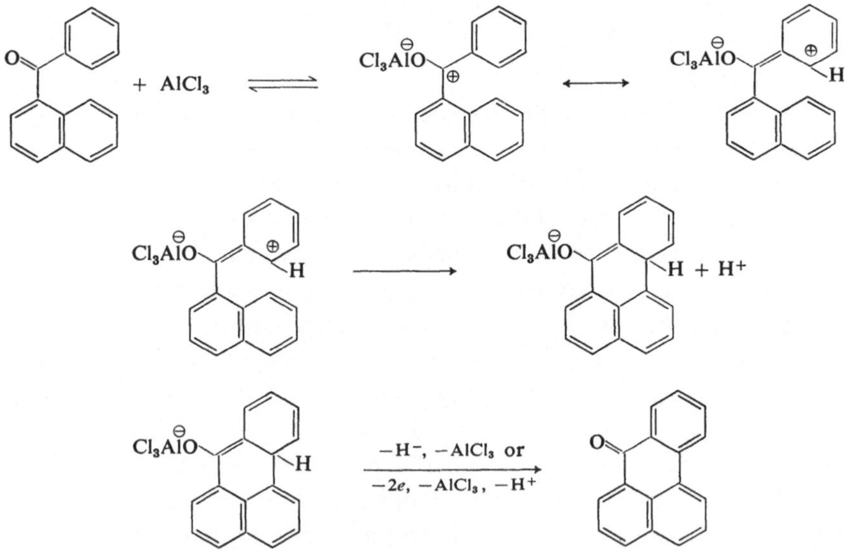

Scheme II

group was essential for the reaction to proceed. Thus, the mechanism for the formation of benzanthrone derivatives is perhaps better formulated as shown in scheme II. In either scheme the exact nature of the dehydrogenation step is not clear. Various redox or hydride-ion-abstraction reactions could be written which would give rise to the final product. The fact that the use of oxidative mixtures of fused salts or of an oxidizing agent with an $AlCl_3$ solvent system has been found effective in accelerating the condensation suggests that redox processes are important in this step. Oxidizing agents such as air or oxygen bubbled over or through the melt or reagents such as manganese dioxide or potassium nitrate added to the melt have been used. It should be kept in mind that Baddeley's result[52] regarding the necessity of hydrogen chloride also indicate that the Scholl reaction for diaryl ketones does not proceed in the presence of oxygen alone.

An alternative to the ionic mechanism, suggested by Rooney and Pink,[54] is that radical cations are the active intermediates in the Scholl ring-closure reactions of aromatic hydrocarbons. This was based on their observations that radical cations of certain aromatic hydrocarbons can be produced in inert solvents in the presence of $AlCl_3$. The existence of radical cations in $AlCl_3$ solvent systems has also been demonstrated electrochemically (see Sec. 6). The radical cation mechanism is outlined in scheme III. Whether dimer formation occurs via two radical cations or a radical cation plus parent molecule has not been established.

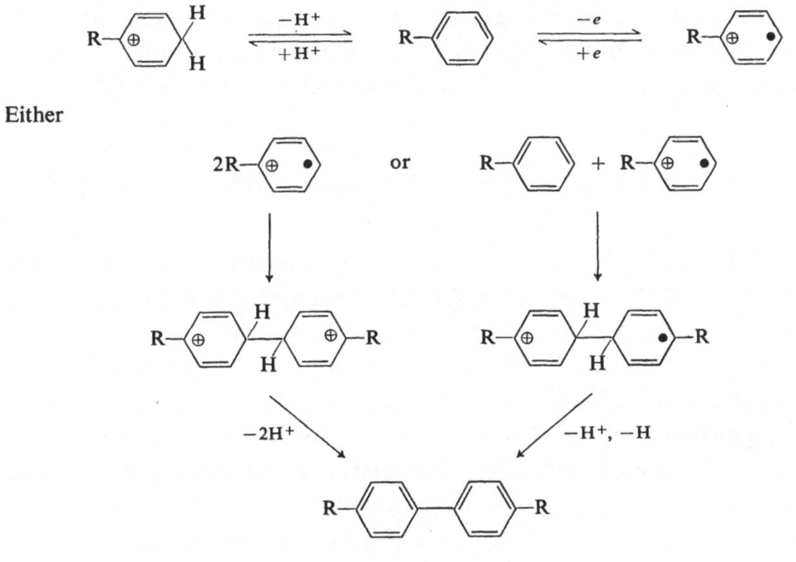

Scheme III

Although we have limited ourselves in the review to those reactions occurring in AlCl$_3$–alkaline metal chloride melts or a molten mixture of AlCl$_3$ and the reactant, the work of Clowes[55] carried out in anisole solution sheds considerable light on the mechanism of the Scholl reaction and will be discussed here. This author studied the binary coupling reactions of 1-ethoxynaphthalene and related compounds in anisole containing AlCl$_3$ and a variety of oxidants. He found that, depending upon the substrates and reaction conditions, his results were interpretable in terms of two complimentary mechanisms—one involving ionic intermediates. He also found that the pathway for a particular compound is predictable to a considerable extent through a consideration of the possible intermediates involved. That is, the least stable radical cations will participate most readily in the radical ion mechanism and least readily in the ionic process, the opposite being true for the most stable radical cations.

Although the above results offer insight into the mechanism of the Scholl reaction, what is missing especially in the case of the AlCl$_3$ solvent systems is a measure of their oxidizing or hydride ion abstracting power. That is, while a mixture of AlCl$_3$ and nitrobenzene appears to be both a strong oxidant and hydride ion abstractor, under the more conventional Scholl reaction conditions we have no measure of the oxidizing or hydride ion abstracting power for either hydrogen chloride dissolved in fused AlCl$_3$ (or binary melt) or of the free, protonated, or complexed (with AlCl$_3$) carbonyl or aromatic hydrocarbon substrate. Thus, at the present time it is impossible to differentiate between the two mechanisms or even the mode of dehydrogenation in the ionic mechanism. Certainly, further work on elucidating the mechanism of the Scholl reaction would be fruitful.

4. MOLECULAR REARRANGEMENTS AND ISOMERIZATIONS

Rearrangements and isomerizations of aromatic compounds are frequently facilitated by the use of AlCl$_3$ solvent systems. Aromatic esters, sulfonates, amides, ketones, hydroxyketones, sulfones, ketoacids, and sulfonic acids are known to rearrange and isomerize by both intra- and intermolecular pathways. Rearrangements leading to ring closure, contraction, or expansion have also been observed in AlCl$_3$ solvent systems.

Fries and co-workers[56] were the first ones to recognize that the rearrangements of phenolic esters to hydroxy aromatic ketones under the influence of Lewis acids represented a reaction of general applicability. The Fries reaction, as these rearrangements are now known, is readily carried out in $AlCl_3$ solvent systems. For example, upon heating phenyl acetate for 2 min at 180–200°C in a 68:32 mole % $AlCl_3$:NaCl melt, both o- and p-hydroxyacetophenone[5] were obtained, as shown in Eq. (34):

$$\tag{34}$$

An excellent up-to-date review with numerous tables of the Fries reaction has been presented by Gerecs,[57] and we will not attempt to duplicate it here. However, a number of omissions, particularly with regard to the reactions of diesters, do occur, and these are discussed below.

It should be mentioned here that Gerecs[58] and others[59,60,61,62] have concerned themselves with the role of hydrochloric acid and the effect of solvent composition on the ortho–para ratio of the products and hence the mechanism of the Fries reaction. However, no conclusive results could be drawn, and further work could certainly be performed.

Diesters consisting of either esters of dihydroxybenzenes or phenolic esters of dicarboxylic acids have been found to undergo a double Fries rearrangement in $AlCl_3$ solvent systems. For example, 1,4-dibenzoyloxy-benzene on treatment in an $AlCl_3$ solvent yields either 2-hydroxy-5-benzoyl-oxybenzophenone or 2,5-dihydroxybenzophenone depending on the number of moles of $AlCl_3$ present,[5,63] as illustrated in Eq. (35):

$$\tag{35}$$

When the 2,5-dihydroxy derivatives are formed, the fate of the other acyl group is unknown.

The double Fries rearrangement of di-p-tolyl esters of several di-carboxylic acids has been investigated by Thomas *et al.*[64] The series of aliphatic dicarboxylic acids from succinic to sebasic, and the aromatic acids, isophthalic and terephthalic acid, yielded the corresponding bis-

(-o-hydroxyketones), as shown in Eq. (36):

$$X = (CH_2)_n; \quad n = 3\text{-}8 \text{ or } m\text{- or } p\text{-phenylene}$$

Several diaryl esters of malonic acid were also shown by Chakravarti *et al.*[65] to undergo the double Fries reaction in molten $AlCl_3$ (140°, 3–4 hr). In each case 1,3-diketones were obtained in good yield.

Ziegler and co-workers[66] in a series of papers also studied the reactions of diaryl esters and amides of malonic acid and its derivatives in $AlCl_3$ solvent systems. In contrast to the Fries-rearranged 1,3-diketones found by Chakravarti, the present workers found products in which ring closure and loss (in most cases) of an aryl group had occurred to yield either coumarins or carbostyrils. The general reaction which was observed is illustrated in Eq. (37):

$$R = H, \text{ alkyl, phenyl}; \quad X = O, S, Se, NH; \quad Y = OH, NHAr$$

The reported examples are listed in Table XI. Several interesting features of the reaction are discussed below.

In a number of instances a direct comparison can be made between the normal Fries reaction and the ring-closure reaction for a given starting material. For example, in the case of di-(4-methylphenyl)malonate, we find that ring closure (in 96% yield) in comparison to 1,3-diketone formation occurs at a higher temperature (180° vs 140°C), at a shorter reaction time (20 min vs 4 hr), and with an $AlCl_3$-to-substrate ratio of about 2:1 (it is slightly higher in the ketone case). From these results and similar ones, we cannot at this time offer any rational explanation for these differences.

A number of exceptions to the general cyclization reaction [Eq. (37)] and other unusual features were also observed by Ziegler *et al.* However, as very little detail or description was given in any of the reports, we will limit our discussion to a brief description of the results.

Although we indicated in Eq. (37) that only a single product was formed, two other products, namely, derivatives of 3-acetyl-4-hydroxy-coumarin and of 4'-hydroxy-(1', 2'-pyrano-5',6':3,4-coumarin), have been observed. However, these additional products are found only when diesters in which the phenolic part contains a p-chloro or -bromo substituent undergo the ring-closure reaction. These results have not been fully explained.

Furthermore, while all of the diester derivatives including those derived from thia- and selenaphenol undergo ring closure with loss of phenoxy group, the loss of an anilide group during ring closure is dependent upon the substituent in the aniline ring. Thus, reactions of the unsubstituted and halogen-substituted anilides yielded derivatives of 4-anilinocarbostyrils, while the methyl-substituted anilides gave 4-hydroxycarbostyrils. Only in the case of the unsubstituted anilide were the reaction conditions somewhat different (ratio of $AlCl_3$ to substrate much less than 2). The apparent cleavage of an amide in the case of the methyl-substituted anilides is unusual, as we have seen above that amides are normally stable in $AlCl_3$ solvent systems (however, see later on below). This may suggest that the anilides are undergoing ring closure via two different mechanisms.

The reactions of a number of di-(2,6-disubstituted-phenyl)malonates and malonamides in $AlCl_3$ solvent systems were also investigated. In only one case, that of di-(2,6-dimethylphenyl)malonate, was cyclization observed. Depending on the reaction conditions, the product was either 5,8- or 6,8-dimethyl-4-hydroxycoumarin, as shown in Eq. (38):

$$(38)$$

Since it is quite likely that the 5,8 isomer can isomerize via methyl migration to the 6,8 isomer, then the initial product of the ring closure must be the 5,8 isomer. When one compares the ratios of $AlCl_3$ to substrate used, it appears that a ratio greater than 2 is needed for isomerization to take place.

Cyclization of several other 2,6-disubstituted esters did not occur when one or more halogen atoms were introduced into the phenyl rings either at the 2- or 4-position. For example, neither di-(2,6-dimethyl-4-bromophenyl)malonate nor di-(2,4,6-trichlorophenyl)malonate underwent ring closure. Two factors could be operating here. Firstly, the halogen atom could be deactivating the ring to electrophilic attack; secondly, sub-

TABLE XI. Reaction of Malonic Acid Diaryl Esters and Dianilides in AlCl₃ Solvent Systems

Malonic acid derivative	Product(s)	Conditions[a]	Ref.
Malonates			
Diphenyl	4-Hydroxycoumarin	A	66
Di-p-tolyl	4-Hydroxy-6-methylcoumarin	B	66
Di-(3,4-dimethylphenyl)	4-Hydroxy-6,7-dimethylcoumarin	D	66
Di-(2,4-dimethylphenyl)	4-Hydroxy-6,8-dimethylcoumarin	E	66
Di-(2,5-dimethylphenyl)	4-Hydroxy-5,8-dimethylcoumarin	F	66
Di-(2,6-dimethylphenyl)	4-Hydroxy-6,8-dimethylcoumarin	E	66
	4-Hydroxy-5,8-dimethylcoumarin	G	66
Di-(2,4,6-trimethylphenyl)	4-Hydroxy-5,6,8-trimethylcoumarin	H	66
Di-(2,4,5-trimethylphenyl)	4-Hydroxy-5,6,8-trimethylcoumarin	H	66
Di-(4-biphenyl)	4-Hydroxy-6-phenylcoumarin	I	66
Di-(1-naphthyl)	4-Hydroxy-7,8-benzocoumarin	J	66
Di-(2-naphthyl)	4-Hydroxy-5,6-benzocoumarin	K	66
α-Methyl-diphenyl	3-Methyl-4-hydroxycoumarin	L	66
α-Ethyl-diphenyl	3-Ethyl-4-hydroxycoumarin	L	66
α-Propyl-diphenyl	3-Propyl-4-hydroxycoumarin	L	66
α-(iso-Propyl)-diphenyl	4-Hydroxycoumarin	L	66
α-(n-Butyl)-diphenyl	3-(n-Butyl)-4-hydroxycoumarin	M	66
α-(n-Amyl)-diphenyl	3-(n-Amyl)-4-hydroxycoumarin	M	66

α-(n-Hexyl)-diphenyl	3-(n-Hexyl)-4-hydroxycoumarin	M	66
α-Phenyl-diphenyl	3-Phenyl-4-hydroxycoumarin	N	66
α-Benzyl-di-(2,6-dimethylphenyl)	4-Hydroxy-5,8-dimethylcoumarin	O	66
Di-(2-methyl-4-chlorophenyl)	4-Hydroxy-6-chloro-8-methylcoumarin	D	66
Di-(4-chlorophenyl)	Mixture of 4-hydroxy-6-chloro-, 3-acetyl-4-hydroxy-6-chlorocoumarin and 4'-hydroxy-6-chloro-(1',2'-pyrano-5',6':3,4-coumarin)	D	66
Di-(4-bromophenyl)	Mixture of 4-hydroxy-6-bromo-, 3-acetyl-4-hydroxy-6-bromocoumarin and 4'-hydroxy-6-bromo-(1',2'-pyrano-5',6':3,4-coumarin)	D	66
Di-(3,4-dichlorophenyl)	Mixture of 4-hydroxy-6,7-dichlorocoumarin and 4'-hydroxy-6,7-dichloro-(1',2'-pyrano-5',6':3,4-coumarin)	D	66
Di-(2,4-dichlorophenyl)	Mixture of 3-acetyl-4-hydroxy-6,8-dichlorocoumarin and 4'-hydroxy-6,8-dichloro-(1',2'-pyrano-5',6':3,4-coumarin)	D	66
Malonic acid-(thiophenol)-diester	1-Thia-4-hydroxycoumarin	C	66
Malonic acid-(selenophenol)-diester	1-Selena-4-hydroxycoumarin	P	66
Malonamides			
Diphenyl	4-Anilinocarbostyril	Q	66
Di-(3-chlorophenyl)	4-(3-Chloroaniline)-7-chlorocarbostyril	R	66
Di-(2,5-dichlorophenyl)	4-(2,5-dichloroanilino)-5,8-dichlorocarbostyril	S	66
Di-(2,3-dimethylphenyl)	4-Hydroxy-7,8-dimethylcarbostyril	T	66
Di-(2,5-dimethylphenyl)	Mixture of 4-hydroxy-5,8-dimethyl- and 4-hydroxy-6,8-dimethylcarbostyril	U	66

[a] Unless otherwise stated, the solvent was molten $AlCl_3$. A, 180–185°C/25 min; B, 180–185°C/20 min; C, 190°C/15 min; D, 180°C/15 min; E, 67:33 mole % $AlCl_3$:NaCl/180–185°C/6 min; F, 180–185°C/10 min; G, 235°C/5 min; H, 240°C/3 min; I, 210–25°C/15 min; J, 160°C/7 min; K, 180°C/7 min; L, 200°C/15 min; M, 195°C/15 min; N, 190°C/10 min; O, 230°C/3 min; P, 170°C/20 min; Q, 305°C/10 min; R, 275°C/15 min; S, 285°C/10 min; T, 64:36 mole % $AlCl_3$:NaCl/250°C/15 min; U, 59:41 mole % $AlCl_3$:NaCl/250°C/3 min.

stitution of an ortho halogen atom is an unlikely process. These results suggest that in the case of the 2,6-dimethyl derivative ring closure occurs via attack of an electrophilic reagent at an ortho carbon containing a methyl group which undergoes a 1,2 shift with loss of a proton to yield the 5,8-dimethyl-4-hydroxycoumarin. An explanation for the failure of di-(2,6-dimethylphenyl)malonamide to ring closure is not readily available.

Ring closure was also found to be dependent on the malonic acid substituents [R in Eq. (37)]. When R was either a phenyl or straight-chain alkyl group, cyclization was possible. However, ring closure did not occur when R was either a benzyl group or branched-chain alkyl group. Sufficient details are not given to allow us to comment further on this.

Finally, there are basically two possible mechanisms which can be written to account for most of the above results, and these are presented in Chart I. The first, Route I, involves initial cleavage of an ester or amide bond to give an acylium ion which undergoes intramolecular cyclization. Route II, on the other hand, involves an initial single Fries rearrangement followed by intramolecular cyclization to the product.

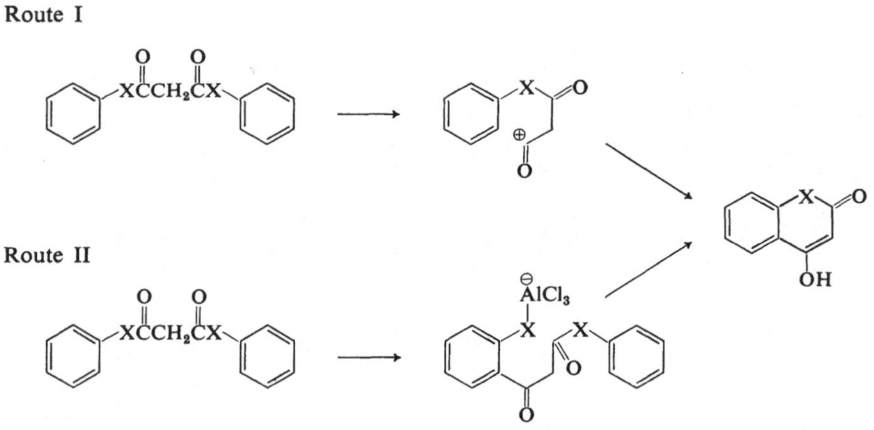

Route I

Route II

Chart I

Obviously, in certain cases, e.g., the anilides, neither path is entirely correct. Thus, at this time not enough data are available to make any binding decisions. We would only add that a study of the reactions of malonic acids esters and amides in $AlCl_3$ solvent systems could be a fruitful area of research.

A reaction which is formerly analogous to the Fries rearrangement is the isomerization of aryl sulfonates to hydroxyaryl sulfones. Upon fusing phenyl benzenesulfonate with $AlCl_3$ at 150°C for 1.5 hr, o- and p-hydroxydiphenyl sulfone are obtained,[67] as shown in Eq. (39):

$$\qquad (39)$$

A review with a tabulation of the reported examples of aryl sulfonate isomerization has been presented by Jensen and Goldman,[68] and this reaction will not be discussed further.

We have already seen that amides do not appear to rearrange or undergo cleavage in $AlCl_3$ solvent systems, although in principle this process should occur. Very recently Birchall and co-workers[7] have found that a number of N-aroyldiphenylamines undergo a Fries type rearrangement to give 4,4′-diaroyldiphenylamine and diphenylamine, as shown in Eq. (40):

$$\qquad (40)$$

The isolation of diaroyl and not monoaroyl diphenylamines, despite the presence of unsubstituted diphenylamine at the end of the reaction, indicates that the presence of an electron-withdrawing aroyl group in one ring of the diphenylamine molecule facilitates acylation in the other ring probably by limiting the extent to which the nitrogen atom will complex with either $AlCl_3$ or the acylium ion. Under similar conditions N-acetylcarbazole is converted into 3-acetylcarbazole.[69]

In his extensive study of the Fries reaction, Auwers[70] tabulated a number of instances in which the normal product of the reaction was accompanied by an isomer, derived by displacement by an alkyl group from its original position. Baddeley[72] later showed that the products depended on the number of moles of $AlCl_3$ used. In particular, this author studied the Fries rearrangement and subsequent isomerization of 3,5-dimethylphenyl acetate at 140–150°C. The results are shown in

Eq. (41):

$$(41)$$

Baddeley also obtained kinetic data for the isomerization in a 83:17 mole %
$AlCl_3$:NaCl melt at 140° and 150°C.

We have already mentioned above a number of instances where
migration of a methyl group ortho to a carbonyl group must have occurred
to account for the observed products. Although earlier workers recognized
these migrations, Baddeley and co-workers[71-75] were the first to undertake
and report on a thorough investigation of the isomerization reactions that
take place in alkylated phenols and aromatic ketones in $AlCl_3$ solvent
systems. These workers showed that for a variety of aromatic ketones
several different isomerization reactions could occur. Several representative
examples are shown in Eqs. (42) to (45):

$$(42)$$

$$(43)$$

$$(44)$$

$$(45)$$

From these and similar results, it was possible to make the following statements regarding the isomerization mechanisms:

1. In the case of ketones, one molecular proportion of $AlCl_3$ combines with the carbonyl group to form an oxonium complex, and an additional mole of $AlCl_3$ is needed to effect isomerization.

2. The presence of hydrogen halide is necessary.

3. The products are determined by the relative rates of (i) deacylation followed by reacylation, and (ii) intramolecular migration of the o-alkyl group to the adjacent meta-position.

4. As homologs of benzonitrile are not isomerized, it appears that steric factors determine the various isomerizations that take place. This is also seen in the case of methylated anthraquinones which also do not undergo isomerization (see above).

5. Alkyl groups migrate easily, the mobility being in the order propyl >ethyl >methyl.

Very recently some doubt has arisen regarding the validity of statement (3) above. This statement was based in part on Baddeley and Pendleton's[75] observations that in the presence of excess $AlCl_3$ at 100°C, acetyldurene (2,3,5,6-tetramethylacetophenone) was converted to acetylprehnitine (2,3,4,5-tetramethylacetophenone) (80%), aromatic hydrocarbon (10%), and diacetyldurene (10%). The latter two products were ascribed to fission of acetyldurene into durene and acetyl cation, followed by electrophilic attack on a second molecule of acetyldurene to produce diacetyldurene. Schlosberg and Woodbury[76] have repeated the isomerization of acetyldurene and could find no hydrocarbon or diacetyldurene product. Instead they found, besides acetylprehnitine (82%), 2,3,4,6-tetramethylacetophenone (5%), 2,3,4,5,6-pentamethylacetophenone (1%), and 2,4,5-trimethylacetophenone (10%). These authors concluded that a combination of Friedel–Crafts trans-alkylation and isomerization reactions (i.e., only intra- and intermolecular methyl migrations) could account for all the products formed.

The factors controlling isomerizations in $AlCl_3$ solvent systems are probably very similar to those involved in the rearrangements of alkylbenzenes in HF–BF_3 solvents.[77] Obviously, many more studies similar to those carried out in the HF–BF_3 system need to be done before we can fully understand what is happening in the $AlCl_3$ solvent systems.

Similar alkyl migrations have been shown to occur with some o-alkylbenzenesulfonic acids and o-alkyldiphenyl sulfones.[78] For example, in an 81:19 mole % AlCl$_3$:NaCl melt at 200°C for 10 hr, the following isomerizations shown in Eqs. (46) and (47) were observed:

$$(46)$$

$$(47)$$

Rearrangements in which ring closure occurs have been previously noted. For example, reactions of malonic acid dianilides and β-aroylacrylic acids in AlCl$_3$ solvent systems gave rise to carbostyrils and 3-ketoindane-1-carboxylic acids, respectively. Similarly, aryl vinyl ketones are found to cyclize in a 68:32 mole % AlCl$_3$:NaCl melt,[5] as shown in Eq. (48):

$$(48)$$

A number of other rearrangements involving ring contraction have been observed in AlCl$_3$ solvent systems. Loudon and Razdan[79] also found that chromanones and 3,4-dihydrocoumarins are rearranged to hydroxy-indanones. Some typical examples are shown in Eqs. (49) and (50):

$$(49)$$

$$(50)$$

Similarly, 2,3-dihydro-4-methyl-3-oxobenz-1-oxazine was rearranged to give two oxindole derivatives when fused with $AlCl_3$,[80] as illustrated in Eq. (51):

$$(51)$$

An example of a rearrangement involving ring expansion has also been observed. N-phenylphthalimide is converted to the lactam of 2-(2'-amino-benzoyl)benzoic acid,[81] as shown in Eq. (52):

$$(52)$$

Ring expansion of a different nature was observed when 2,2'-diphthal-imidobiphenyl was fused in an $AlCl_3$ solvent,[82] as shown in Eq. (53):

$$(53)$$

Further heating at higher temperature yielded the ring-closed flavan-threne.

About 25 patents have been granted for methods of isomerizing *n*-paraffins to branched-chain compounds. Aluminum chloride/alkali metal melts containing an excess of $AlCl_3$ have been used.[83] The reaction is generally carried out below 200°C. The longer the carbon chain of the *n*-paraffin, the lower is the temperature chosen, since otherwise undesirable cracking occurs. It has been found useful to add small quantities of hydrogen chloride;[84] for example, the isomerization of *n*-butane to isobutane then proceeds at 141°C with a 50% conversion. This process is of interest for increasing the octane rating of petroleum fractions.

5. MISCELLANEOUS REACTIONS

Although the following reactions are included under the general heading of miscellaneous reactions, they are of importance. We have classified them as dehydration, exchange, and reduction–chlorination reactions where possible.

5.1. Dehydration Reactions

One of the simplest dehydration reactions which has been observed is the formation of benzonitrile in 84% yield from benzamide,[85] [Eq. (54)]:

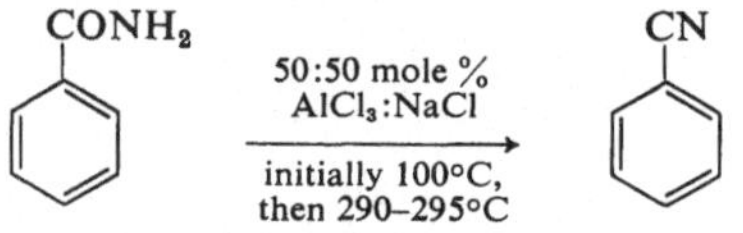

$$\underset{\substack{\text{initially 100°C,}\\\text{then 290–295°C}}}{\overset{\substack{\text{50:50 mole \%}\\\text{AlCl}_3\text{:NaCl}}}{\longrightarrow}}$$

(54)

The reaction is carried out by initially heating the amide at approximately 100°C with preformed sodium tetrachloroaluminate, then raising the temperature to 290–295°C whereupon the product usually distills out. Similar treatment of ammonium benzoate, using two molecular equivalents of $AlCl_3$, gives a 50% yield of benzonitrile. The reaction has also been applied to chloro and nitro derivatives of benzamide and to 1- or 2-naphthamide.

Recently an interesting dehydration reaction involving ring closure has been observed by Arient and Slavik.[48] When 1-arylamino-anthraquinones are heated in a 64:36 mole % AlCl$_3$:NaCl melt at 150–160°C for 15 min, ceramidonines are obtained [Eq. (55)], and in some cases small amounts of 1,2-phthaloylcarbazole derivatives:

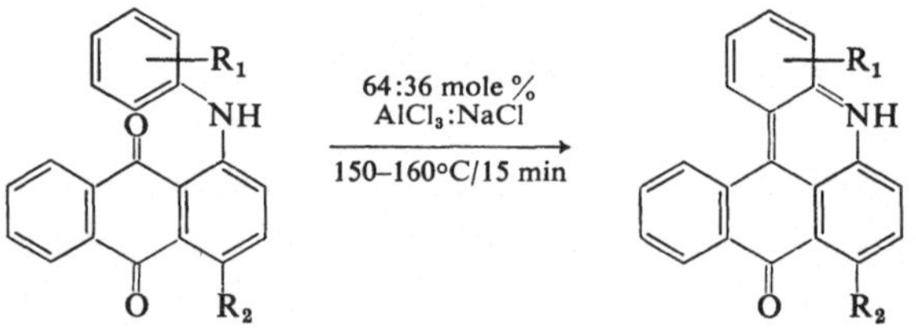

$$\underset{\substack{150\text{–}160°C/15 \text{ min}}}{\overset{\substack{\text{64:36 mole \%}\\\text{AlCl}_3\text{:NaCl}}}{\longrightarrow}}$$

(55)

Similar dehydration reactions were observed by Scholl and co-workers when either 4,8-diphenoxyanthraquinone-1,5-dicarboxylic acid or 4,8-diphenylmercaptoanthraquinone-1,5-dicarboxylic acid was fused in an $AlCl_3$ solvent to yield either hetero-cerobioxene- ōr hetero-cerobithiene-dicarboxylic acid.

5.2. Exchange Reactions

Several interesting exchange reactions have been observed in $AlCl_3$ solvent systems. A very useful example is the preparation of chlorosilanes. When a 1:1 mixture of silane and hydrogen chloride is led into either an $AlCl_3$:NaCl or $AlCl_3$:LiCl melt at 120°C, 80% of monochlorosilane is obtained together with other chlorosilanes,[86] [Eq. (56)]:

$$SiH_4 + HCl \xrightarrow[120°C]{AlCl_3:NaCl} H_3SiCl + H_2 \qquad (56)$$

To avoid evolution of hydrogen from the valuable Si–H compounds, reaction (57) under otherwise identical conditions is preferable:

$$SiH_4 + SiCl_4 \xrightarrow[120°C]{AlCl_3:NaCl} 2H_2SiCl_2 \qquad (57)$$

Alkylated silanes, as shown in Eq. (58), can also be prepared in this way using the same type melts:

$$(CH_3)_2SiH_2 + (CH_3)_2SiCl \longrightarrow 2(CH_3)_2SiHCl \qquad (58)$$

This provides a simple and extremely favorable method of preparing dimethylchlorosilane, which is used for the synthesis of α,ω-hydridopolysiloxanes.[87]

Recently, a new method was developed for the preparation of methyl metal compounds from methyl chloride and the chlorides of boron, aluminum, silicon, tin, phosphorus, antimony, and mercury in $AlCl_3$ solvent systems.[88] Metals are suspended in the melts to act as halogen acceptors. The reaction is illustrated for tetramethylsilane in Eq. (59):

$$3SiCl_4 + 12CH_3Cl + 8Al \xrightarrow[180°C]{AlCl_3:NaCl} 3(CH_3)_4Si + 8AlCl_3 \qquad (59)$$

The difference between this reaction and the usual Wurtz synthesis is that the acceptor metal is not an alkali metal, and that the metal halide formed is soluble in the melt. Thus, a free metal surface is always available until the metal has been completely consumed. Moreover, if the system is chosen correctly, more solvent (in the above case, $AlCl_3$) is formed in the course of the reaction. Another important advantage is that the acceptor metal halide can be reconverted into the acceptor metal (Al) and the halogen (Cl_2) by electrolysis of the melt. The recovered chloride can be used either for preparation of fresh silicon tetrachloride or for the oxychlorination of methane.

Another interesting exchange reaction which has been observed is the selective substitution of α-fluorine atoms by chlorine in perfluoro ethers. When, for example, perfluoro-2-methyl-tetrahydrofuran is heated with $AlCl_3$ at 180°C for 14 hr, α,α,α'-trichloroperfluoro-2-methyltetrahydrofuran[89] is formed, as shown in Eq. (60):

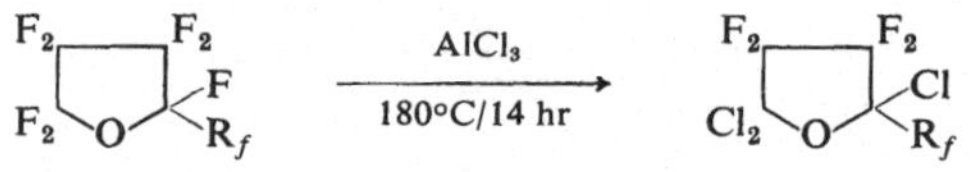

$$\qquad (60)$$

where R_f is a perfluoro alkyl. Several other perfluorotetrahydrofurans and tetrahydropyrans bearing a single perfluoroalkyl substituent in the α-position also undergo the above reaction. However, in the absence of an α-perfluoroalkyl substituent, perfluoro ethers undergo cleavages, as illustrated in Eqs. (61) and (62):

$$R_fCF_2OCF_2R_f \quad \xrightarrow[175°C/16\ hr]{AlCl_3} \quad (R_fCCl_2OCCl_2R_f) \quad \longrightarrow \quad R_fCOCl + R_fCCl_3 \qquad (61)$$

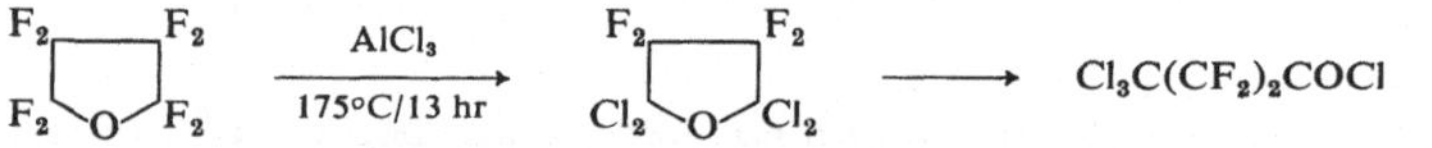

$$\longrightarrow \quad Cl_3C(CF_2)_2COCl \qquad (62)$$

It is interesting to note that when α-alkyl-substituted perfluoro ethers were heated at 230°C for 15 hr in a 50:50 mole % $AlCl_3$:NaCl melt, no reaction was observed. The authors attribute this observation to the speculated lower Lewis acidity of the binary system.

The replacement of fluoride by chloride has also been observed with perfluoro aromatic compounds. For example, when perfluorobenzene is fused with $AlCl_3$ at 140°C, perchlorobenzene results[90] [Eq. (63)]:

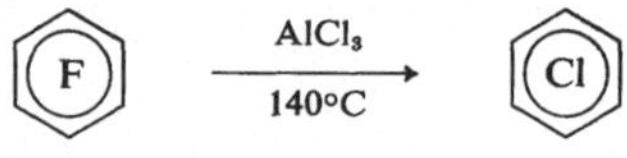

$$\qquad (63)$$

5.3. Reduction and Chlorination Reactions

At least two groups of workers have observed an interesting reaction of aromatic nitro compounds when they are brought in contact with $AlCl_3$,NaCl melts. In the cyclization reaction of 1-phenylaminoanthraquinones reported above,[48] Arient and Slavik found the following reaction scheme [Eq. (64)] when 1-(p-nitrophenyl)aminoanthraquinone was reacted in a 64:36 mole % AlCl$_3$:NaCl melt at 180°C for 15 min:

$$\tag{64}$$

We see that simultaneous reduction of the nitro compound and nuclear chlorination has occurred, the necessary hydrogen atoms being supplied by the starting material itself. A similar reaction occurs with the 3-nitrophenyl isomer.

A similar reaction has been observed by Russian workers.[91] When either m-dinitrobenzene or 2,4-dinitrochlorobenzene is heated in a 78:28 mole % AlCl$_3$:NaCl melt at 125–150°C and treated with a reducing agent such as stannous chloride and hydrogen chloride for 3–4 hr, 2,4,6-trichloro-1,3-diaminobenzene resulted [Eq. (65)]:

$$\tag{65}$$

When m-dinitrobenzene and 1,5-dibenzoylnaphthalene are treated in the same manner, one obtains I and 3,4,8,9-dibenzopyrene-5,10-quinone. In this, the hydrogen necessary for the reduction of the nitro compound no doubt is supplied by the naphthalene derivative when it simultaneously undergoes the Scholl reaction. One final point to be noted is that the above nitro compounds themselves are recovered unchanged on simple heating in an AlCl$_3$:NaCl melt. Other nitro compounds have also been found to be unreactive in AlCl$_3$ solvent systems.

Nuclear chlorination of unsubstituted aromatic compounds has also been found to occur in $AlCl_3:NaCl$ melts. When a mixture of benzene and chlorine are passed into a 61:39 mole % $AlCl_3:NaCl$ melt at 200°C, mono-, di-, tri-, and tetrachloro benzenes are obtained.[92]

Aliphatic chlorinations have also been carried out in $AlCl_3$ solvent systems. For example, 1,1,2-trichloroethane is prepared in 64% yield by passing chlorine and ethylene chloride, in a ratio of between 0.55 to 0.75 parts by weight of chlorine per part of ethylene chloride, through a 50:50 mole % $AlCl_3:KCl$ melt at 350°C.[93]

6. ORGANIC ELECTRODE REACTIONS

Although a considerable number of papers have appeared on the electrochemistry of inorganic species in $AlCl_3$-containing solvent systems,[94] only recently have organic electrode reactions in $AlCl_3$ solvent systems been studied. Fleischmann and Pletcher[95] have carried out a cyclic voltammetric study on a number of polynuclear hydrocarbons in a melt consisting of 50:36:14 mole % $AlCl_3:NaCl:KCl$ at 150°C using a pyrolytic graphite electrode. They found that the large polynuclear hydrocarbons show a series of electron transfers while the simple hydrocarbons show a single wave. At sweep rates of 0.1 V/sec, pyrene, anthracene, and diphenylanthracene gave peaks on the reverse sweep showing the initial products of the first electron transfer to have some stability, although it was clear from the peak size that they were not completely stable. Furthermore, the reverse peaks occurred at a potential close to that expected if the oxidation was electrochemically reversible. In no case could evidence be found for the reduction of the aromatic hydrocarbons.

A second report on electro-organic chemistry in $AlCl_3$ solvent systems concerns the electrochemical polymerization at a platinum electrode of a variety of aromatic compounds.[96] The compounds to be polymerized are dissolved in a ternary complex of an aromatic hydrocarbon, a hydrohalogen acid, and an aluminum halide with a mole ratio 1:1:2, respectively, with electrolysis then being initiated. Among the most interesting products are poly(p-phenylene), which is formed from benzene, and p-sexiphenyl, formed from biphenyl. The relation between the current density, electrode potential, and efficiency of formation of these products is very complex.

Recently we have reported[97] on a cyclic voltammetric study of the electrode reactions of several aromatic amines in $AlCl_3$ solvent systems. It was observed that the products of the electrode reactions, the amine radical cations from triphenylamine, diphenylamine, and N,N'-dimethylaniline, were markedly more stable in a 50:50 mole % $AlCl_3$:NaCl melt at 175°C than they are in other aprotic solvents at room temperature. Furthermore, the stabilization of the radical cations was found to be dependent on the acidity (or basicity), i.e., $AlCl_3$ content of the melt. In the case of dimethylaniline, it was postulated that the lack of a basic species of sufficient strength to extract a proton prohibited follow-up reactions of the initially formed radical cation. Similar results were obtained by Fung et al.[98] in their study of the anodic electrochemistry of two multisulfur heterocycles.

ACKNOWLEDGMENT

This work was supported in part by the Air Force Office of Scientific Research and by the American Chemical Society–Petroleum Research Fund.

7. REFERENCES

1. W. Sundermeyer, *Angew. Chem. Intern. Ed. Engl.* **4**:222 (1965).
2. J. E. Gordon, *Tech. Methods Org. Organometal. Chem.* **1**:51 (1969).
3. C. R. Boston, in: *Advances in Molten Salt Chemistry* (J. Braunstein, Gleb Mamantov, and G. P. Smith, eds.), Vol. 1, pp. 126–163, Plenum Press, New York (1971).
4. A. M. Komagorov, I. K. Baeva, and V. A. Koptyug, *Izv. Sibirsk. Otd. Akad. Nauk SSSR, Ser. Khim. Nauk,* **1966**:147.
5. D. B. Bruce, A. J. S. Sorrie, and R. H. Thomson, *J. Chem. Soc.* **1953**:2403.
6. C. Seer, *Monatsh* **32**:143 (1911).
7. J. M. Birchall, M. T. Clark, and D. H. Thorpe, *J. Chem. Soc. Perkin I* **1973**:442.
8. G. Baddeley and R. Williamson, *J. Chem. Soc.* **1956**:4647.
9. N. F. Hayes and R. H. Thomson, *J. Chem. Soc.* **1956**:1585.
10. H. Waldmann, *J. Prakt. Chem.* **150**:99 (1938).
11. H. Raudnitz and G. Laube, *Ber.* **62**:509 (1929).
12. P. Kränzlein, *Chem. Ber.* **71**:2328 (1938).
13. M. P. Satchell and B. E. Stacey, *J. Chem. Soc. (C)* **1971**:469.
14. H. Waldman and H. Mathiowetz, *Chem. Ber.* **64**:1713 (1931).
15. A. Rieche, H. Sauthoff, and O. Miller, *Chem. Ber.* **65**:1371 (1932).
16. L. F. Fieser, *J. Am. Chem. Soc.* **53**:3546 (1931).
17. S. Sethna, in: *Friedel–Crafts and Related Reactions* (G. A. Olah, ed.), Vol. 3, Part 2, p. 975, Interscience Publishers, New York (1964).
18. German Pat. 660,220; *Chem. Abstr.* **32**:6257 (1938).
19. T. Tsunoda, *Chiba Daigaku Kogakubu Kenkyu Hokoku* **7**:19 (1959); through *Chem. Abstr.* **54**:9861 (1960).
20. R. Scholl and W. Neovius, *Chem. Ber.* **44**:1075 (1911).
21. F. G. Baddar, *J. Chem. Soc.* **1941**:310.

22. W. I. Awad, F. G. Baddar, M. A. Omara, and S. M. A. Omran, *J. Chem. Soc. (C)* **1971**:3721.
23. H. Waldmann and E. Ulsperger, *Chem. Ber.* **83**:188 (1950).
24. T. Kurado and M. Wada, *Sci. Papers Inst. Phys. Chem. Res. (Tokyo)* **34**:1740 (1938); *Chem. Abstr.* **33**:2511 (1939).
25. H. Brockmann and K. Muller, *Ann. Chem.* **540**:51 (1939).
26. N. F. Hayes and R. H. Thomson, *J. Chem. Soc.* **1956**:1585.
27. G. Baddeley, G. Holt, and S. M. Makar, *J. Chem. Soc.* **1952**:2415, 3289.
28. W. Knapp, *Monatsh.* **67**:332 (1936).
29. G. Baddeley, S. M. Makar, and M. G. Ivenson, *J. Chem. Soc.* **1953**:3969.
30. G. Baddeley, G. Holt, S. M. Makar, and M. G. Ivinson, *J. Chem. Soc.* **1952**:3605.
31. W. Sundermeyer and O. Glemser, *Angew. Chem.* **70**:629 (1958).
32. R. Scholl, C. Seer, and R. Witzenbock, *Chem. Ber.* **43**:2202 (1910).
33. French Pat. 795,447; *Chem. Abstr.* **30**:5595 (1936).
34. H. Reinlinger and A. Overstraeter, *Chem. Ber.* **91**:2121 (1958).
35. R. Weitzenbock and C. Seer, *Chem. Ber.* **46**:1994 (1913).
36. A. T. Balaban and C. D. Nenitzescu, in: *Friedel–Crafts and Related Reactions* (G. A. Olah, ed.), Vol. 2, Part 2, pp. 979–1047, Interscience Publishers, New York (1964).
37. R. Scholl and C. Seer, *Monatsh* **33**:1 (1912).
38. L. F. Fieser and E. L. Martin, *J. Am. Chem. Soc.* **58**:1443 (1936).
39. F. Mayer, E. Flechtenstein, and H. Gunther, *Chem. Ber.* **63**:1464 (1930).
40. R. Scholl and C. Seer, *Chem. Ber.* **55**:109 (1922).
41. U.S. Pat. 1,892,241; *Chem. Abstr.* **27**:1895 (1933).
42. British Pat. 305,593; *Brit. Chem. Abstr. B* **1930**:603.
43. A. Zinke and K. Funke, *Chem. Ber.* **58**:2222 (1925).
44. R. Scholl, K. Meyer, and J. Donat, *Chem. Ber.* **70**:2180 (1937).
45. R. Scholl and C. Seer, *Ann. Chem.* **394**:111 (1921).
46. R. Scholl and G. Schwarzer, *Chem. Ber.* **55**:324 (1922).
47. A. K. Wick, *Helv. Chim. Acta* **54**:769 (1971).
48. J. Arient and V. Slavik, *Collection Czech. Chem. Commun.* **34**:3579 (1969).
49. H. Waldmann and K. G. Hendenburg, *J. Prakt. Chem.* **156**:157 (1940).
50. E. Clar and D. G. Stewart, *J. Chem. Soc.* **1951**:687.
51. J. W. Cook and C. G. M. deWorms, *J. Chem. Soc.* **1939**:268.
52. G. Baddeley, *J. Chem. Soc.* **1950**:994.
53. U.S. Pat. 2,258,394; *Chem. Abstr.* **36**:492 (1942).
54. J. J. Rooney and R. C. Pink, *Proc. Chem. Soc.* **1961**:142.
55. G. A. Clowes, *J. Chem. Soc.* **1968**:2519.
56. K. Fries and G. Finck, *Chem. Ber.* **41**:4271 (1908); K. Fries and W. Pfaffendorf, *Chem. Ber.* **43**:212 (1910).
57. A. Gerecs, in *Friedel–Crafts and Related Reaction* (G. A. Olah, ed.), Vol. 3, Part 1, pp. 499–533, Interscience Publishers, New York (1964).
58. A. Gerecs and M. Windholz, *Acta Chim. Acad. Sci. Hung.* **8**:295 (1955).
59. G. C. Amin and N. M. Shah, *J. Univ. Bombay* **17A**:5 (1948).
60. R. Baltzly, W. S. Ide, and A. P. Phillips, *J. Am. Chem. Soc.* **77**:2522 (1955).
61. A. F. Marey, F. G. Badder, and W. I. Awad, *Nature* **172**:1186 (1953).
62. A. W. Ralston, E. W. Segebrecht, and M. R. McCorkle, *J. Org. Chem.* **7**:522 (1942).

63. R. D. Desai and C. K. Mavani, *Proc. Indian Acad. Sci.* **29A**:269 (1949).
64. F. D. Thomas II, M. Shamma, and W. C. Fernelius, *J. Am. Chem. Soc.* **80**:5864 (1958).
65. D. Chakravarti, A. Chakravarti, and A. Sarkar, *J. Indian Chem. Soc.* **48**:1017 (1971).
66. E. Ziegler and H. Junck, *Monatsh.* **86**:29 (1955); E. Ziegler and H. Junck, *Monatsh.* **87**:503 (1956); E. Ziegler and H. Maier, *Monatsh.* **89**:143 (1958); E. Ziegler and H. Maier, *Monatsh.* **89**:551 (1958); E. Ziegler and E. Nolken, *Monatsh.* **89**:737 (1958); E. Ziegler and K. Gelfert, *Monatsh.* **90**:858 (1959).
67. A. A. Aleykutty and V. J. Baliah, *J. Indian Chem. Soc.* **31**:513 (1954).
68. F. R. Jensen and G. Goldman, in: *Friedel–Crafts and Related Reactions* (G. A. Olah, ed.), Vol. 3, Part 2, pp. 1319–1338, Interscience Publishers, New York (1964).
69. S. G. P. Plant and S. B. C. Williams, *J. Chem. Soc.* **1934**:1142.
70. K. v. Auwers and E. Risse, *Chem. Ber.* **64**:2216 (1931); K. v. Auwers and E. Janssen, *Ann. Chem.* **483**:44 (1930).
71. G. Baddeley, *J. Chem. Soc.* **1943**:273.
72. G. Baddeley, *J. Chem. Soc.* **1943**:527.
73. G. Baddeley, *J. Chem. Soc.* **1944**:232.
74. G. Baddeley, G. Holt, and W. Pickles, *J. Chem. Soc.* **1952**:4162.
75. G. Baddeley and A. G. Pendleton, *J. Chem. Soc.* **1952**:807.
76. R. H. Schlosberg and R. P. Woodbury, *J. Org. Chem.* **37**:2627 (1972).
77. D. A. McCaulay, in: *Friedel–Crafts and Related Reactions* (G. A. Olah, ed.), Vol. 2, Part 2, pp. 1049–1073, Interscience Publishers, New York (1964).
78. G. Holt and B. Pagdin, *J. Chem. Soc.* **1961**:4514.
79. J. D. Loudon and R. K. Razdan, *J. Chem. Soc.* **1954**:4299.
80. J. W. Cook, J. D. Loudon, and P. McCloskey, *J. Chem. Soc.* **1952**:3904.
81. British Pat. 305,393; *Brit. Chem. Abstr. B* **1930**:603.
82. V. Krepelka and R. Stefee, *Collection Czech. Chem. Commun.* **34**:3576 (1969).
83. U.S. Pat. 2,342,073; *Chem. Abstr.* **38**:4617 (1944).
84. U.S. Pat. 2,439,301; *Chem. Abstr.* **42**:4340 (1948).
85. J. F. Norris and J. Klemka, *J. Am. Chem. Soc.* **62**:1432 (1940).
86. W. Sundermeyer and O. Glemser, *Angew. Chem.* **70**:628 (1958).
87. W. Sundermeyer, Lecture, IUPAC Congress, London, September (1963).
88. W. Sundermeyer and W. Verbeek, *Angew. Chem. Intern. Ed. Engl.* **5**:1 (1966).
89. G. V. D. Tiers, *J. Am. Chem. Soc.* **77**:4837 (1965).
90. L. S. Kobrina, C. G. Furin, and G. G. Yakobson, *Izv. Sibirsk Otd. Akad. Nauk SSSR, Ser. Khim. Nauk* **1968**:98.
91. N. C. Dokunikhri and M. M. Sergeeva, *Dokl. Akad. Nauk SSSR* **88**:987 (1953).
92. O. Glemsen and K. Cleine-Weischede, *Ann. Chem.* **659**:17 (1962).
93. U.S. Pat. 2,140,549; *Chem. Abstr.* **33**:2540 (1939).
94. K. W. Fung and G. Mamantov, in: *Advances in Molten Salt Chemistry* (J. Braunstein, G. Mamantov, and G. P. Smith, eds.), Vol. 2, pp. 218–224, Plenum Press, New York (1973).
95. M. Fleischmann and D. Pletcher, *J. Electroanal. Chem.* **25**:449 (1970).
96. N. E. Wisdom, Abstracts, 135th Meeting, The Electrochemical Society, Abstr. No. 138, New York, May (1969).
97. H. L. Jones, L. B. Boxall, and R. A. Osteryoung, *J. Electroanal. Chem.* **38**:476 (1972).
98. K. W. Fung, J. Q. Chambers, and G. Mamantov, *J. Electroanal. Chem.* **47**:81 (1973).

99. German Pat. 495,332; *Chem. Abstr.* **24**:3248 (1930).

100. C. Seer, *Monatsh.* **33**:33 (1912).

101. J. C. Lovie and R. H. Thomson, *J. Chem. Soc.* **1959**:4139.

102. H. Raudmitz and W. Bohm, *J. Prakt. Chem.* **123**:284 (1929).

103. K. Zahn and P. Ochwat, *Ann. Chem.* **462**:72 (1928).

104. V. M. Chari, S. Needakanton, and T. R. Sashadri, *Indian J. Chem.* **4**:330 (1966).

105. H. Waldman and P. Sellner, *J. Prakt. Chem.* **150**:145 (1938).

106. V. P. Aggarwala, R. Gopal, and S. P. Garg, *J. Org. Chem.* **37**:1247 (1972).

107. German Pat. 538,457; *Chem. Abstr.* **26**:1619 (1932).

108. A. Eitel and R. Fialla, *Monatsh.* **79**:112 (1948).

109. R. Scholl and W. Neuberger, *Monatsh.* **33**:507 (1902).

110. W. Lageman, E. Lauria, and E. Fachinelli, *Farmaco (Pavia) Ed. Sci.* **11**:274 (1956); *Chem. Abstr.* **50**:13940 (1956).

111. R. Scholl and C. Seer, *Chem. Ber.* **44**:1091 (1911).

112. A. Zinke, H. Troger, and E. Ziegler, *Chem. Ber.* **73**:1042 (1940); A. Zinke, G. Gorbach, and D. Schimka, *Monatsh.* **48**:593 (1927).

113. R. Scholl, K. Meyer, and J. Dorat, *Chem. Ber.* **70**:2180 (1937).

114. R. Scholl and C. Seer, *Chem. Ber.* **44**:1233 (1911).

115. F. Mayer and O. Stark, *Chem. Ber.* **64**:2003 (1931).

116. F. Mayer and H. Gunther, *Chem. Ber.* **63**:1455 (1930).

117. H. Waldmann, *J. Prakt. Chem.* **130**:92 (1931).

118. H. Waldmann, *J. Prakt. Chem.* **147**:331 (1937).

119. N. S. Dokunikhin, Z. Z. Moiseeva, and V. A. Mayatnikova, *Zh. Organ. Khim.* **2**:516 (1966).

120. F. Mayer, O. Stark, and K. Schon, *Chem. Ber.* **65**:1333 (1932).

121. German Pat. 692,708; *Chem. Abstr.* **35**:4605 (1938).

122. H. Waldmann, *J. Prakt. Chem.* **131**:71 (1931).

123. H. Meyer and H. Raudnitz, *Chem. Ber.* **63**:2010 (1930).

124. F. Mayer and O. Hoffmann, *Chem. Ber.* **65**:1338 (1932).

125. H. Raudnitz and G. Laube, *Chem. Ber.* **62**:938 (1929).

126. F. Mayer, A. Mombour, W. Lassmann, W. Werner, P. Landmann, and E. Schneider. *Ann. Chem.* **488**:259 (1931).

127. A. K. McBeth, J. R. Price, and F. L. Winzor, *J. Chem. Soc.* **1935**:325.

128. F. L. Winzor, *J. Chem. Soc.* **1935**:336.

129. H. Waldmann and H. Poppe, *Ann. Chem.* **527**:190 (1937).

130. H. Raudmitz, L. Redlich, and F. Fiedler, *Chem. Ber.* **64**:1835 (1931).

131. A. J. S. Sorrie and R. H. Thomson, *J. Chem. Soc.* **1955**:2233.

132. A. J. S. Sorrie and R. H. Thomson, *J. Chem. Soc.* **1955**:2244.

133. W. Steinkopf, T. Barlag, and H.-J. v. Petersdorff, *Ann. Chem.* **540**:17 (1939).

134. German Pat. 512,229; *Chem. Abstr.* **25**:1100 (1931).

135. G. Baddeley and R. Williamson, *J. Chem. Soc.* **1953**:2120.

Chapter 4

EXPERIMENTAL TECHNIQUES
IN MOLTEN FLUORIDE CHEMISTRY*

Carlos E. Bamberger

Chemistry Division
Oak Ridge National Laboratory
Oak Ridge, Tennessee 37830

1. INTRODUCTION

Although research on molten halides is a current area of interest in applied and fundamental chemistry, a glance at the chemical literature reveals that molten fluorides are probably among the least studied halide systems. An important exception is the fluoride mixtures used and proposed for use in nuclear reactors as liquid fuels and coolants.[1] Other areas where considerable effort has been devoted to research and application of molten fluorides are the aluminum, beryllium, niobium, tantalum, and other metal industries, metallizing, and even production of fluxes for brazing metals.

Probably the main reason for the apparently limited interest in molten fluorides shown by the research community is, besides their known toxicity, the undeserved and indiscriminate label of highly corrosive media given them. The extensive research performed in order to build and successfully run the molten salt reactor experiment as well as to design other similar

* Research sponsored by the U.S. Atomic Energy Commission under contract with the Union Carbide Corporation.

nuclear reactors which have the capability of producing more fuel than is consumed has proven the fallacy of the intractability of molten fluorides.[2] Since most of these techniques have been described in the scientific literature under various specialized headings, we thought that it would be helpful to collect and review the many aspects of molten fluoride research with which we have been associated over the years. Since a large fraction of our experience is based on mixtures of LiF and BeF_2, they will be used mainly as examples with the understanding that the techniques reviewed here are not limited to these components; rather, they are applicable to most molten fluoride mixtures. BeF_2, one of the melt components used, exhibits many interesting physicochemical properties. It forms fluoroberyllates with several metal fluorides, and these coordination compounds have some very interesting properties. In the molten state, $LiF–BeF_2$ mixtures containing low concentrations of BeF_2 exhibit basic properties as defined by Lewis for aqueous solutions; correspondingly, high concentrations of BeF_2 in the mixture impart acidic properties.

In size and structure the divalent fluoroberyllate ion resembles both the divalent sulfate and the tetravalent silicate ions; most of the fluoroberyllates and sulfates are isomorphous and those with a common ion exhibit the same aqueous solubility behavior with temperature. One of the better-known aspects of fluoroberyllates is their close analogy in phase equilibria behavior, but at significantly lower temperatures, to that of silicates. Therefore, fluoroberyllates have been used also as low temperature "models" for silicates.

In writing this review, we have tried to describe the practical aspects of working with fluorides while assuming that the reader has no previous knowledge on the subject. Thus, we have included details and suggestions on techniques which, although some are common knowledge to the chemist, might not have been emphasized before in connection with the chemistry of molten fluorides.

The scope of this review includes several aspects of molten fluoride work such as the thermochemical basis for the purification of melts and the selection of containers, the design of equipment, miscellaneous techniques for determining viscosities and densities, absorption spectroscopy, electrochemistry, and, finally, the presentation of equilibrium data involving solutes in molten fluorides.

A very large fraction of the referenced work pertains to the open literature; however, some ingenious devices and techniques have been "borrowed" (with their knowledge and consent) from our colleagues at the Oak Ridge National Laboratory.

2. THERMOCHEMISTRY OF THE CONTAINMENT AND PURIFICATION OF MOLTEN FLUORIDES

These aspects of fluoride chemistry are probably the most important ones because they deal directly with the feasibility of performing a designed experiment, its cost, and the prediction of interfering reactions. Even in heterogeneous systems with molten fluorides, due to the relatively high temperatures involved, most reactions occur with rates that are reasonably fast for practical purposes. Because of this, thermochemical calculations become one of the most important tools available for studying molten halide systems, including the fluorides. These calculations are applied to homogeneous as well as to the more numerous heterogeneous equilibria occurring among the molten fluorides and solids (container metals, oxides, silicates, etc.) and/or gases. A significant amount of thermochemical data, mainly free energies of formation, is now available for many fluorides in the solid or dissolved state.[3] However, this kind of data is lacking for many "new" solid compounds, such as chalcogenides, borides, carbides, nitrides, etc., many of which have high thermal stability which makes them promising materials for the construction of containers and electrical insulators. Thus, these compounds comprise a new field of research, particularly with respect to reactions with molten fluorides. An updated collection of information on free energies of formation of solute fluorides in Li_2BeF_4, as well as on various other equilibria in molten fluorides, has been published by Baes.[3] In this review we will follow Baes' notation and use the same units and definition of standard states for the solvent and the solutes. The concentration scale is the mole fraction, e.g.,

$$X_{MF} = n_{MF}/(n_{MF} + n_{LiF} + n_{BeF_2})$$ (1)

where n represents number of moles. Gas pressures are measured in atmospheres, and, because of the high temperatures and the usually low pressures employed, the gases are assumed to behave ideally.

Throughout this chapter the reactions will be written using mainly molecular species, although the latter do not imply the actual state of the components in the melt.

The standard state for solutes is chosen as the hypothetical one mole fraction ideal solution in Li_2BeF_4 and, for the components of the solvent, the activities a_{LiF}, a_{BeF_2}, $a_{Be^{2+}}$, a_{Li^+}, and a_{F^-} are all taken as unity. Thus, the activity coefficients of solutes are unity at the reference composition of the solvent, Li_2BeF_4. A correlation of activity coefficients with solvent composition has been developed by Baes[4] using a polymer model[5] for molten BeF_2-containing mixtures together with available data on solubilities of metal fluorides and on distribution of solutes between molten fluorides and liquid bismuth. The activities of LiF and BeF_2 were redetermined in $LiF-BeF_2$ melts by Hitch and Baes using emf measurements.[6]

Some free energies of formation of fluorides and oxides are summarized in Table I as parameters a and b in the following expression:

$$\Delta G^f = a + b(T/10^3) \quad \text{(kcal/mole)}$$

in the temperature range 700–1000°K. A quick calculation reveals that LiF, BeF_2, PuF_3, ThF_4, UF_4, and ZrF_4, all extensively studied because of their use in the molten salt reactor concept, are significantly more stable than the fluorides of the structural or container metals copper, nickel, iron, and chromium. Thus, a pure melt composed of LiF, BeF_2, ThF_4,

TABLE I. Free Energies of Formation, $\Delta G^f = a + b(T/10^3)$, in kcal/mole, in the Range 700–1000°K[a]

Compound (crystalline or gaseous)			Compound (crystalline or gaseous)		
	a	b		a	b
AlF_3[b]	−359.73	51.82	SiF_4	−386.26	34.71
BeF_2[c]	−243.47	31.5	ThF_4[d]	−477.0	65.8
CF_4	−223.30	36.54	UF_4	−452.0	67.4
CaF_2[b]	−291.25	32.83	ZrF_4[b]	−453.7	75.7
CeF_3	−417.83	57.59	BeO	−146.02	24.94
CrF_2[d]	−182.0	34.2	Cr_2O_3	−270.79	61.64
FeF_2	−168.62	32.98	CuO[e]	− 37.73	18.3
HF	− 65.19	−1.01	FeO	− 62.71	15.15
KF[b]	−135.65	20.47	NiO	− 56.26	20.35
LiF	−146.50	23.11	SiO_2	−216.55	42.10
MgF_2[b]	−268.46	41.4	ThO_2	−292.40	44.55
NaF[b]	−137.46	21.11	UO_2	−258.0	40.0
NiF_2	−156.33	37.65	ZrO_2	−260.44	44.44
PuF_3	−370.0	57.5			

[a] Except where indicated, the values are from Ref. 3.
[b] JANAF Thermochemical Tables, 2nd ed. U.S. Dept. of Commerce, NSRDS-NBS-37 (1971).
[c] B. F. Hitch and C. F. Baes, Jr., *Inorg. Chem.* **8**:201 (1969).
[d] O. Kubaschewski, E. L. L. Evans, and C. B. Alcock, *Metallurgical Thermochemistry*, Pergamon Press, 4th ed. (1967).
[e] J. F. Elliott and M. Gleiser, *Thermochemistry for Steelmaking*, Vol. 1, Addison-Wesley (1960).

etc. can be contained with no corrosion in containers made of the above-mentioned metals or their alloys; e.g., the reaction

$$Ni^{\circ}_{(c)} + BeF_{2(d)} \rightleftarrows Ni^{2+} + 2F^- + Be^{\circ}_{(c)} \tag{2}$$

(where the subscripts c and d refer, respectively, to the crystalline and dissolved states) is so strongly displaced to the left that the concentration of Ni^{2+} is completely negligible. However, when uranium, which exists in solution as U^{4+} and U^{3+}, is present, an equilibrium with Cr^{2+} is possible; e.g., in the reaction

$$Cr^{\circ}_{(c)} + 2U^{4+} \rightleftarrows 2U^{3+} + Cr^{2+} \tag{3}$$

at $1000°K$, assuming $X_{UF_4}/X_{UF_3} = 10^2$ and $a_{Cr} \simeq 0.01$ in an alloy, the concentration of Cr^{2+} is $X_{Cr^{2+}} < 10^{-4}$. For this particular case, where U^{4+} is present, a better choice would be the use of graphite (as a liner) because the reaction

$$4UF_{4(d)} + C_{(c)} \rightleftarrows 4UF_{3(d)} + CF_{4(g)} \tag{4}$$

reaches equilibrium at $P_{CF_4} \approx 10^{-30}$ atm at $1000°K$ when $X_{UF_4}/X_{UF_3} = 100$.

Although the compatibility of molten fluorides with container materials is easy to ascertain by simple thermochemical calculations and many low cost metals and alloys can be used for the containment of purified solvent salts, life is never that simple, and somehow, somewhere, the melt components have to be purified. The purification of molten fluorides can be performed only in containers made of a limited number of metals, e.g., nickel, copper, platinum, gold, and graphite.

The main contaminant present in the fluoride components is water; it ranges from small amounts like those adsorbed on LiF or ThF_4 to significant concentrations present in hygroscopic fluorides such as BeF_2. When the components are mixed and melted, hydrolysis occurs and a significant concentration of O^{2-} (and, to a lesser extent of OH^-) appears in solution. The presence of this O^{2-} can cause saturation due to the low solubility of some oxides[7] (e.g., BeO, UO_2), i.e.,

$$BeF_{2(d)} + O^{2-} \rightleftarrows BeO_{(c)} + 2F^- \tag{5}$$

At this stage it should be remembered that the main objection to the presence of O^{2-} in a melt is not its oxidizing ability, which it lacks, but its ability to cause the undesirable precipitation of oxides. The OH^-, however, is an oxidizing species because its proton can be reduced to hydrogen.

Another important contaminant, especially in BeF_2, is sulfur present as either SO_4^{2-}, S^{2-}, or both. This element is particularly bothersome because it easily forms NiS, which in turn forms a low-melting eutectic with Ni.[8]

Other contaminants may be Cl^-, Fe^{2+}, and Fe^{3+}, graphite, compounds of silicon and of boron, etc. Although sparging with fluorine will remove all of the above-mentioned anions, fluorine is too strong an oxidant to be used with any metal in the presence of molten fluorides. Gaseous HF alone is also quite corrosive, but, when used mixed with hydrogen in a ratio of about 1:9, it removes the contaminants very successfully with little corrosion of the container.

The removal of oxide by HF sparging was studied by Mathews and Baes[9] and is shown by the following reactions:

$$O^{2-} + HF_{(g)} \rightarrow OH^- + F^- \tag{6}$$

$$OH^- + HF_{(g)} \rightarrow H_2O_{(g)} + F^- \tag{7}$$

The removal of sulfide can be accomplished similarly:

$$S^{2-} + 2HF_{(g)} \rightarrow H_2S_{(g)} + 2F^- \tag{8}$$

If the sulfur is present as SO_4^{2-}, it can be converted efficiently to S^{2-} by means of a reducing agent which is stronger than hydrogen, e.g., beryllium:

$$SO_4^{2-} + 4Be_{(c)}^\circ \rightarrow S^{2-} + 4BeO_{(c)} \tag{9}$$

The sulfide and oxide are then removed by treatment with HF, and the excess beryllium is converted to BeF_2.[10] The described hydrofluorination treatment also removes other contaminants; thus, Cl^- is volatilized as HCl, and compounds of silicon and boron as SiF_4 and BF_3, respectively. If graphite is present (occasionally it occurs finely divided in BeF_2), it is usually removed as an aerosol by gas sparging. The ratio $HF:H_2 \approx 1:9$ by volume is empirical and represents a compromise between the efficiency, the rate of removal of contaminants, and the extent of corrosion of the container. As can be seen from the equilibrium quotient of the reaction

$$Ni_{(c)}^\circ + 2HF_{(g)} \rightleftarrows NiF_{2(d)} + H_{2(g)} \tag{10}$$

$$Q_{10} = (P_{H_2}/P_{HF}^2)X_{NiF_2}$$

a decrease in the ratio P_{HF}/P_{H_2} decreases the concentration of NiF_2. Even when using pure HF, reaction (10) is arrested at the container surface above the melt by the formation of an impervious and adherent coating of nickel fluoride. However, where the surface is in contact with a melt, the metal fluoride dissolves continuously until saturation of the melt is reached. Based on practical reasons then, it is more expedient at this stage to tolerate some corrosion and follow the HF–H_2 sparging with sparging with pure H_2. This converts the products of corrosion (fluorides) into the metallic state, i.e.,

$$NiF_{2(d)} + H_{2(g)} \rightarrow Ni^\circ_{(c)} + 2HF_{(g)} \tag{11}$$

$$FeF_{2(d)} + H_{2(g)} \rightarrow Fe^\circ_{(c)} + 2HF_{(g)} \tag{12}$$

which are now as compatible with the melt as is the container itself. An exception is CrF_2, which is more difficult to reduce with hydrogen. In this case it is more expedient to use a stronger reductant,[10] e.g., beryllium, zirconium, etc., its choice depending upon the composition of the melt, e.g.,

$$CrF_{2(d)} + Be^\circ_{(c)} \rightarrow Cr^\circ_{(c)} + BeF_{2(d)} \tag{13}$$

The metals in finely divided form thus obtained may be separated, if the amount present is significant, by filtration of the salt through a porous metal filter. This filter may be located on a filter stick or between two containers, one with the melt and one empty for receiving the filtrate. Obviously, in either case the containers and/or filter must be at an adequate temperature to keep the salt molten. The rate of filtration can be enhanced by applying vacuum on one side of the filter and pressure (inert gas) on the other.

Based on the free energies of formation of metal fluorides, nickel is the preferred choice for containment of most molten fluoride systems, especially when their purification is performed *in situ*. An alternative choice, although not as easily available everywhere, is Hastelloy N.* Copper, although nobler than nickel, is recommended only for use as a liner because of its lack of rigidity at high temperatures. Graphite is also a good liner material for many fluoride systems, especially where oxides are present and known redox potentials are required. These can be obtained with CO_2–CO mixtures since the values of the equilibrium constant for the reaction

$$CO_{2(g)} + C_{(c)} \rightleftarrows 2CO_{(g)} \tag{14}$$

are well known.

---

* Hastelloy N is a nickel-based alloy containing (in w/o) 7 chromium, 5 iron, 17 molybdenum, 0.8 manganese, and other elements at concentrations below 0.5.

Finally, we would like to mention the use of silica, a material which has some very obvious advantages but also presents some limitations in its compatibility with molten fluorides. The main corrosion reaction that takes place between SiO_2 and the fluorides is

$$SiO_{2(c)} + 4F^- \rightleftarrows SiF_{4(g)} + 2O^{2-} \tag{15}$$

Although *a priori* one might not expect it, the SiF_4 pressures generated at equilibrium with molten Li_2BeF_4 are quite moderate,[11] of the order of 20 torr at 700°C. Thus, if one can tolerate small concentrations of O^{2-} in the melt, ($X_{O^{2-}} = 2 \times 10^{-5}$ under 1 atm SiF_4), silica permits experiments to be performed inexpensively. If a solute which has an oxide of very low solubility is present, its precipitation might be prevented by performing the experiment under a larger equilibrium pressure of SiF_4; this would shift reaction (15) to the left.

In our experience we have found the use of silica containers very helpful in performing visual tests of molten fluorides with other reactants. The main drawback that we have encountered has been the devitrification of silica into cristobalite, with the consequent loss of transparency, which takes place in 3 to 7 days at temperatures between 500–700°C.[12]

3. HANDLING FLUORIDES

3.1. Safety and Purity Aspects

The handling of fluorides requires a degree of precaution which stems from their recognized toxicity toward living organisms.[13a,b] Thus, it is recommended that the solid fluorides be handled in hoods provided with adequate ventilation in order to prevent breathing air containing fluoride dust. This is extremely important in the case of BeF_2, because beryllium is a very toxic element itself. Its effect is more damaging when it enters the mammal body via the respiratory tract in the form of dust. Whether beryllium is in its metallic form or in the form of compound is almost inconsequential to its effect. For some compounds, e.g., BeF_2, the damaging effect of HF resulting from its hydrolysis is added to that of Be^{2+}. The allowable maximum concentration of beryllium in air for an eight-hour workday is 2 μg/m³.[14] (For beryllium work, the hood exhaust system should be provided with absolute filters.) A relatively small fraction of people are very sensitive to direct contact by beryllium, which is manifested by the appearance of dermatitis. For these people, the use of protective clothing, including rubber gloves, is imperative. However, this practice is also recommended for handling fluorides in general. All of the above is applicable to HF whether anhydrous or in aqueous solutions. The experience of several colleagues on "burns" by HF has been very painful. Even when hoods with good ventilation are available, exhausted HF should be sparged through caustic solutions in order to protect the worker and the hood, including the filtering system.

With respect to the handling of fluorides from the point of view of maintaining the purity of the components, some general recommendations can be given. When possible, single crystals, large particles, or aggregates should be used in order to decrease the surface area exposed to moisture. This is particularly important with hygroscopic fluorides such as BeF_2, which should be stored, when possible, under dry inert gases or under vacuum. Although more expensive and to some extent more restrictive for working, inert-atmosphere glove boxes have obvious advantages for handling fluorides. If such a glove box is available and the components of the melt are of high purity, the resulting melt can, in many instances, be used without further purification. If work must be performed with exposure to air, the components should be weighed in closed containers and contact of the fluorides with air kept to a minimum before the melting and purification takes place.

The required weight of the batches of molten fluorides depends on the density of the molten salt, the type of experiment, and the configuration and size of the container. We have found that 300–800 g of molten fluorides is an upper limit adequate for most equilibration experiments with salts having densities of 2–3 g/cm^3, an exception being thermal gradient quenching (see Sec. 8.1) where $\sim$300 mg are sufficient. Since in most cases an accurate knowledge of the composition of the melt is required, it is generally recommended to weigh accurately the separate components rather than to resort to chemical analysis of the melt. Any subsequent change in composition due to sampling or addition of reagents can be accurately estimated by means of material-balance calculations, provided, obviously, that additions to or subtractions from the system are also accurately known.

The use of hydrogen at high temperatures for purifying the melt and for reduction experiments requires a special degree of caution of which the chemist is aware. Thus, we will describe briefly and recommend only the procedures that we have followed successfully for working with hydrogen in an apparatus exposed externally to air. The container, the gas lines leading to and from it, together with accessories (thermocouple well, valves, pressure gauges, etc.) should have as many connections welded or brazed as possible. The whole system should be tested for leaks, if possible using a helium leak detector or a sensitive pressure gauge or manometer, at pressures higher than those that will be used in the experiment. After the container has been loaded, the air should be removed at room temperature by alternately evacuating and filling with an inert gas (we prefer argon over helium because of its higher density and over nitrogen because of possible nitriding reactions). The system should be heated slowly under a stream of inert gas to remove surface moisture and to displace the air before the mixture of HF–H$_2$ is allowed to enter the container. For some long-term reduction experiments under static hydrogen pressure (at low partial pressures), we have successfully and safely used mixtures of Ar–4% H$_2$, the lower explosion limit for hydrogen in air. Oxygen should be avoided at all times, in the system not only because of the danger of explosion with hydrogen but also because at high temperatures undesirable reactions can occur even when it is mixed with an inert gas:

$$\text{Ni}^{\circ}_{(c)} + \tfrac{1}{2}\text{O}_{2(g)} \rightleftarrows \text{Ni}^{2+} + \text{O}^{2-} \tag{16}$$

The oxide ion in turn can precipitate the least soluble oxide, e.g.,

$$2\text{O}^{2-} + \text{Zr}^{4+} \rightleftarrows \text{ZrO}_{2(c)} \tag{17}$$

With respect to the handling of gaseous HF-containing mixtures, in addition to what has been said above, it is recommended that copper tubing be used (we prefer $\tfrac{1}{4}$ in. O.D.) and that the exit lines be heated with electric heating tape to $\geq 80°$C in order to avoid the condensation of H$_2$O–HF solutions resulting from the purification treatment and the subsequent dissolution of the protective coating of copper fluoride.

Since oxide is the main contaminant occurring in most fluorides, we will emphasize the procedures for its removal. From the extensively studied equilibria of HF with O^{2-} and OH$^-$ in LiF–BeF$_2$ melts by Mathews and Baes,[9] it can be concluded that the removal of O^{2-} from these melts by reaction with HF–H$_2$ mixtures is favored at lower temperatures (around 500°C). The reduction of metal fluorides by hydrogen may be estimated from available thermochemical data,[3] and, in general, this operation is favored by higher temperatures (600–700°C). In summary, a typical batch of 500 g of fluoride melt is purified by sparging for 24 hr with H$_2$–10% (vol.) HF followed by 8–12 hr of sparging with pure hydrogen, both at a flow rate of 100 std. cm^3/min.

The removal of oxygen (present in the form of oxide) from small batches ($\sim$10 g) of single or mixed metal fluorides can also be accomplished effectively by means of the following reactions with NH_4HF_2:

$$MO_{n/2(c)} + nNH_4HF_{2(l)} \rightarrow (NH_4)_nMF_{2n(l\,or\,c)} + \frac{n}{2} H_2O_{(g)} \qquad (18)$$

$$(NH_4)_nMF_{2n(l\,or\,c)} \rightarrow MF_{n(l\,or\,c)} + nNH_4F_{(g)} \qquad (19)$$

where $MO_{n/2}$ represents the oxide impurity. Usually the metal fluorides are mixed with excess NH_4HF_2 in a volume ratio of 1:2 or 1:3 in a nickel container which may be provided, if necessary, with a graphite liner. After evacuating and flushing with an inert gas, the mixture is slowly heated to ~160°C to melt the NH_4HF_2 (mp 124°C) and then sparged with an inert gas to accomplish thorough mixing and the removal of water. The mixture is then heated slowly to a higher temperature, 400–800°C, while sparging to volatilize the NH_4F resulting from the decomposition of the ammonium fluorometallates. If necessary, the procedure can be repeated.

The above purifications with HF (gaseous or as NH_4HF_2) are capable of producing fluorides of high purity. That a satisfactory degree of purity is obtained has been repeatedly demonstrated using spectral and electro-chemical methods of analyses. (Sects. 6 and 7).

If the situation arises in which the researcher needs to prepare his own fluoride or oxyfluoride, it is convenient to select the method of preparation on the basis of the available starting materials and facilities. Bougon et al.[15] have recently reviewed this subject comprehensively. Since many of the preparative methods described can be adapted to the measurement of equilibria, and thus provide thermochemical data, we recommend the above publication as a complement to this chapter.

3.2. Vapor Pressure of Molten Fluorides; the Distillation of BeF_2

Vapor-pressure measurements of metal fluorides provide important physical constants, such as boiling point, heat of vaporization, etc. With mixtures of molten fluorides, these measurements also provide quantitative information on the activities of the species in solution. Since most metal fluoride mixtures have relatively low vapor pressures, these have been measured using mainly the transpiration technique.[16–18]

BeF_2 is one of the few extensively used fluorides that has a significant vapor pressure, e.g., 89 mm at 1010°C.[18] This has led to the proposal of using distillation under reduced pressure for the separation of some solute fluorides from the solvent mixture $LiF–BeF_2$.[19] Another application is the purification, for research purposes, of commercially available BeF_2. Its most common impurities are oxide, sulfide, graphite, and small amounts of metals probably present as oxides or fluorides. The distillation of BeF_2 was first used successfully by Moynihan and Cantor[20] and has since become a frequently used technique in our laboratories. The apparatus consisted of a flanged vessel heated to 800°C with an air-cooled cold finger (at 500°C) attached to the top and a cup suspended from the cold finger for receiving the molten condensate. The whole apparatus was made of nickel, and a copper gasket was used for leak-tight operation under vacuum. Bronstein and Braunstein have developed a new design of the distillation apparatus,[21] which has some advantages over the earlier one. This still operates under a larger thermal gradient ($>400°$) which produces a condensate in the form of a fine powder. The powder can be treated with H_2–HF *in situ*, if desired, and then can be fused in order to reduce the surface area before further handling. The still consists of a flanged nickel container which can be slid vertically in a tubular furnace. The container can be visualized as divided into three sections. The upper section consists of a nickel cup used to contain the BeF_2 to be purified and is held at 800–850°C. The middle section at 400°C consists of a nickel funnel which collects all the fine powder condensate and delivers it into a receiving nickel cup located in the lower section of the still. The flanged container has a thermocouple inserted from each end and has provisions for evacuation and for flowing gases. To melt the BeF_2 powder, the whole container is simply moved upward in the furnace in order to place the lower section in the high-temperature region.

A typical analysis of a doubly distilled BeF_2 revealed the following impurities in ppm: O_2, 120; K, 7; Li, 15; Na, 70; Al, 70; Mg, 30; Cr, 10; Ti, 5; Fe <1; and all others below the limits of detection. The concentration of impurities in the "as received" BeF_2 varied widely from batch to batch and also among the different suppliers. An average analysis follows for comparison purposes: O_2, >1000; Sn, 3000; Na, 4000; Al, 100; Mg, 100; Cr, 40; Fe, 100; and all others in the range of 0–10 ppm.

The large affinity for water exhibited by glassy BeF_2 is reduced when the material is transformed into its crystalline form.[22] This is accomplished by holding the glass at approximately $\sim50°$ below its melting point (555°C) for a period of at least 18 hr.

4. ANALYSIS

Since the purity of each melt component varies widely, it is necessary to use some method or technique for ascertaining when the hydrofluorination has accomplished its purpose. Mathews and Baes[23] found that moni-

toring the evolution of HF when sparging with an inert gas, after stopping
the hydrofluorination, gives a direct indication of the purity of the melt
with respect to O^{2-} and OH^-. When the melt is pure, the number of moles
of HF evolved per unit time as a function of time decreases very rapidly,
being a function only of the gas volume in the system over the melt. The
solubility of HF in $LiF-BeF_2$ melts is relatively small[24] but should not be
neglected. When the melt is not pure, the concentration of OH^- is significant
and the following reaction takes place while sparging with an inert gas

$$OH^- + F^- \rightarrow HF_{(g)} + O^{2-} \tag{20}$$

This can be seen as a slowly decreasing evolution of HF as a function of
time, and the reaction can be used to determine the amount of OH^- in
the melt.

The course of the hydrogen reduction of metal fluorides is simply
followed by caustic titration of the evolving HF. This is best accomplished
by using a technique of continuous back titration in which the time for
neutralizing a known amount of NaOH is recorded. The end point can be
determined either with phenolphthalein or a pH meter. The use of KOH
permits a longer period of operation in a small-volume titration cell because
KF is more soluble in water than NaF.

The total oxygen content of a fluoride or fluoride mixture can be
determined by fusing a sample with $KBrF_4$.[25] The procedure is based on
measuring the oxygen evolved by the reaction

$$2MO + KBrF_4 \rightleftarrows KBr + 2MF_2 + O_2 \tag{21}$$

The reaction is performed in heavy-walled nickel containers, machined
from solid nickel stock. The top of the container consists of a threaded
brass cap with a Teflon gasket. A Hoke Inconel diaphragm valve* connects
the reactor to a gas manifold. A variable amount of nonhygroscopic sample
(50–250 mg, depending on its oxygen content) is rapidly introduced into a
reactor which already contains the $KBrF_4$ ($\sim$1 g), while the system is
sparged with helium to prevent the adsorption of atmospheric moisture. If
the sample is hygroscopic, it is loaded with special precautions in a dry-box
(see Sec. 5.5). The system is evacuated to a pressure of less than 10^{-3} torr
and is then heated to 450°C for 2 hr while the top of the reactor is kept at
approximately 120°C to prevent condensation of BrF_3. After cooling
to room temperature, the molecular oxygen is transferred with a Toepler

* Available from Hoke, Inc., Cresskill, New Jersey.

pump (through appropriate traps) to the measuring system which has been previously calibrated. The measuring system is based on differential manometry or on gas chromatography.[26]

The originators of the method applied it to a large number of fluoride mixtures and many other compounds with satisfactory results. A method for the preparation of the oxidizing flux, $KBrF_4$, is given.[25]

The analysis of samples of a fluoride melt is performed by means of wet chemical methods or by spectroscopic analyses. Since many fluorides are insoluble in water, the sample dissolution step deserves special comment. This step is usually performed by treatment with hot $HClO_4$ or H_2SO_4 which drives off the F^- as HF and converts the cations into more soluble compounds. The usual safety precautions should be observed. The dissolutions should be performed in an adequately ventilated hood because HF and other acid vapors are evolved. Work with beryllium-containing solutions does not require special handling and can be performed on laboratory benches. Care should be exercised, however, by cleaning any spilled or splashed material before it dries, thus preventing it from becoming airborne—its most dangerous state to the health of the individual. Contact with Be^{2+} solutions can be simply dealt with by adequate washing and rinsing. The dissolution of beryllium-containing solids, or other operations involving it, should always be performed in hoods.

When the sample is heterogeneous, e.g., a mixture of fluorides and refractory oxides, and interest lies only in the latter, the above treatment is too severe. Thus, we have used hot water in a Soxhlet extractor to dissolve Li_2BeF_4 and dissolution with "Verborcit"* for ThF_4-containing fluorides.

As will be discussed later in Sects. 6.1 and 7.3, the presently developing trend is to adapt, when possible, methods that can be used on line. The automation of analyses by the use of computers is presently receiving wide acceptance, and its application to molten fluoride analyses has already been demonstrated (see Sec. 7.3).

5. DESIGN OF EQUIPMENT

A review of the design of equipment for performing the large number of experiments with molten fluorides which is necessary for completely defining an equilibrium is a difficult task. There are probably as many

---

* "Verborcit" is a solution developed by members of the Analytical Chemistry Division of ORNL; it consists of 50 g of sodium versenate and 20 g of sodium citrate added to 200 ml saturated boric acid solution; final pH is ~9–10.

minor variations of a single design as there are researchers who use them. Thus, we will describe mainly the equipment that we have used and helped to design because we have learned its pros and cons "the hard way."

5.1. Containers and Accessories

Since research with molten fluorides may involve the study of numerous heterogeneous equilibria between the phases gas–liquid, gas–solid–liquid, and liquid–liquid, the size and shape of the containers required can vary considerably. However, most of the work has been performed in vertical containers of 5 to 15 cm O.D. fabricated from schedule* 40 pipe. This arrangement is advantageous because it provides a significant depth of liquid with a small surface area. However, the opposite, a small depth with a large surface area, may be required when using special techniques, such as electrochemical measurements[27] where a duplicate set of electrodes may be required and even small thermal gradients are undesirable. In the design of vertical containers we tend to distinguish two kinds: (a) welded and (b) flanged. Although this distinction may appear trivial, each kind has advantages and disadvantages of its own. In the first kind (a), the top is an integral part welded to the container, and all the service lines are welded to this top. In the second kind (b), the flanged top with its welded service lines is attached with bolts to a flanged container.

The main advantages of welded containers (Fig. 1) are the smaller likelihood of gas leaks and the possibility of heating the whole vessel, thus providing for a nearly negligible thermal gradient. Welded containers are more difficult to load initially with the melt components; the admissible size of the particles is limited by the internal diameter of the tubing (usually less than $\frac{1}{2}$ in. I.D.). Thus, smaller particles of hygroscopic materials are more likely to adsorb significant amounts of moisture when the container is loaded in air. When liners of other metals or graphite or large accessories are used, they have to be inserted before the bottom is welded on. It should also be mentioned that the cleaning and reuse of these containers is difficult and not always economically feasible, especially with fluorides which are insoluble in water.

Flanged containers (Fig. 2) overcome some of the above disadvantages, i.e., they can be loaded initially with large particles; liners and other accessories can be inserted without difficulty before bolting the top; and when disposable liners are used, the container can be reused with none or very

* Schedule indicates a ratio of thickness to diameter; see ASA Standards B36.10-1939.

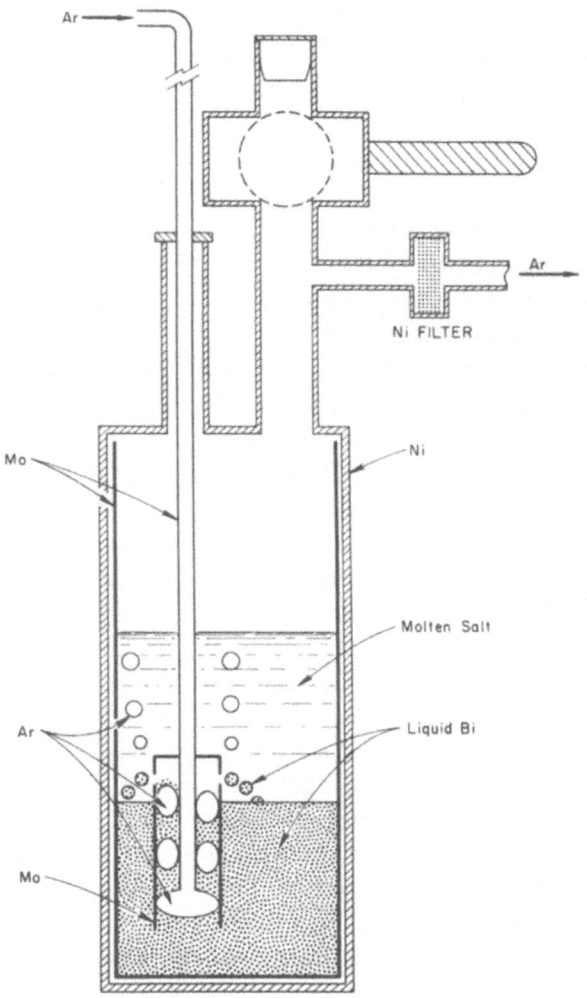

Fig. 1. Welded container with gas lifter for equilibrations of
two liquid phases.

little cleaning. On the other hand, flanged containers require a seal with
some material which is stable at relatively high temperatures or else needs
to be cooled to temperatures tolerable to the seal. The materials that are
resistant to high temperatures are limited to malleable metals of which the
most economical is copper. Copper can be used successfully for gaskets
in large-sized containers (>15 cm O.D.) where the flanges are provided
with "knife" edges. The malleability of the copper gaskets can be enhanced

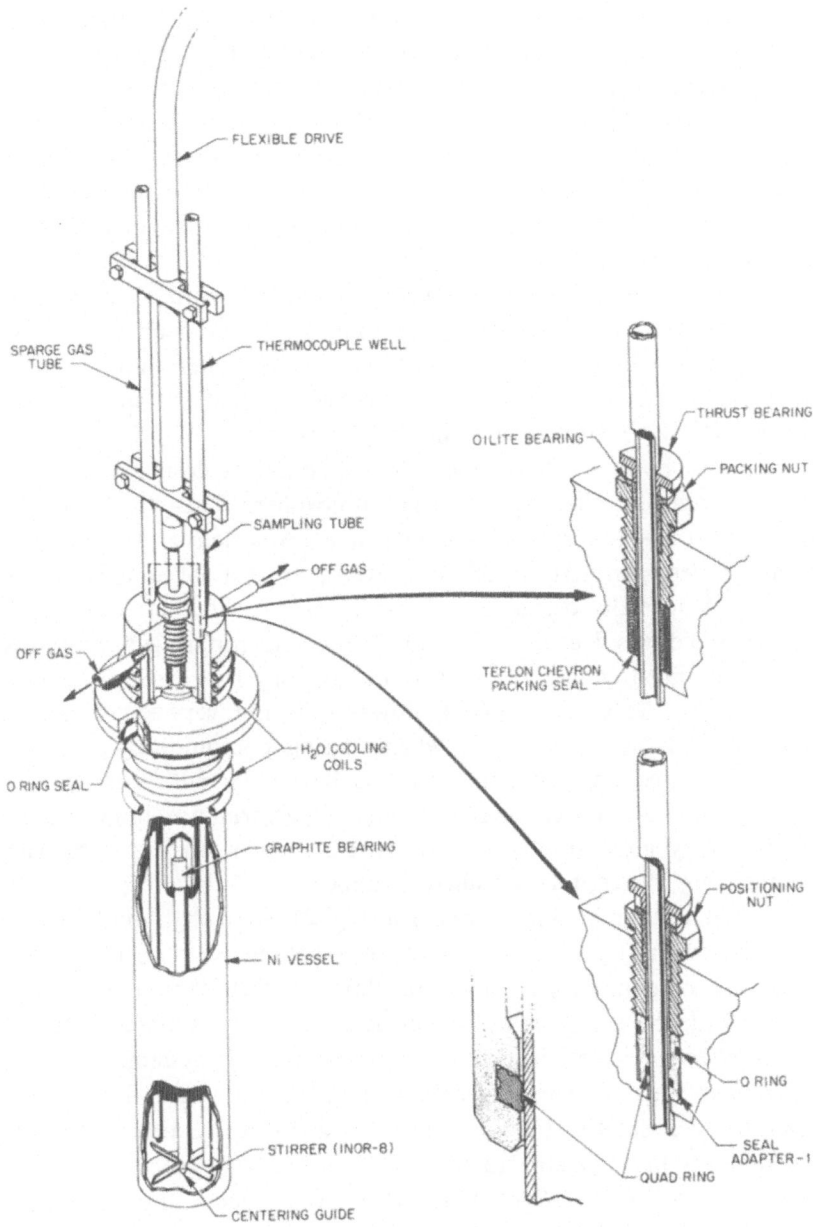

Fig. 2. Flanged container provided with mechanical stirrer. (Reprinted from Ref. 11, p. 565, by courtesy of the American Ceramic Society.)

by annealing. This consists of rapid heating in air to ~500°C and quenching in a reducing nonaqueous medium such as methyl or ethyl alcohol.

The necessity of cooling the flanged tops creates a thermal gradient along the vertical axis of the system. This thermal gradient in turn can result in significant mass transport of container metals through the melt. The thermal gradient, and consequently the mass transport, can be minimized to a negligible degree by agitating the liquid by stirring, gas sparging, or other means.

Figure 1 illustrates a nickel-welded container with a molybdenum liner inserted before welding on the bottom. The molybdenum "gas lifter" was used to provide mixing of liquid bismuth and molten fluorides since it was observed, using silica containers (see Sec. 2), that gas sparging alone was not effective in mixing both liquid phases.

Figure 2 shows a sophisticated flanged container designed for stirring molten fluorides in a hermetic inert gas atmosphere. This was required for suspending solids (density ≈ 10 g/cm^3) in a lower-density liquid fluoride mixture (3 g/cm^3) in a study of the distribution of a solute between both phases.[28] A specially designed seal for the stirrer shaft has been used in order to avoid gas leakage. Stacked Teflon chevron rings* compressed by a packing nut together with an "oilite" bearing have been occasionally used as a seal. This arrangement, however, is not very satisfactory for maintaining gas-tight conditions when stirring continuously at 500 rpm because it requires continuous tightening of the packing nut. With additional oil lubrication and lower stirring rates, its performance improved considerably. Very good results are obtained at higher stirring rates with a seal consisting of neoprene or silicone rubber "quad-rings"† positioned in a stainless steel adapter (see lower cut in Fig. 2). With slight oil lubrication they maintain gas-tight conditions under a helium pressure of 40 cm Hg for 20 days of continuous stirring at 500 rpm. One should be cautious, however, with the generous use of oil because it can produce undesirable reducing effects if it gets in contact with an oxidizing system.

Another way of providing adequate mixing of liquid fluorides with solids is by encapsulating them in containers sealed by welding, and locating the containers in a rocking furnace.[29] (Fig. 3). The advantages of this system are that it is leak-tight. If an adequate furnace is chosen (Marshall type), it can be made to operate practically with no significant thermal gradient. Its disadvantages are the impossibility of sampling the system

---

* Crane Packing Company, Morton Grove, Illinois.
† Minnesota Rubber Co., Minneapolis, Minnesota.

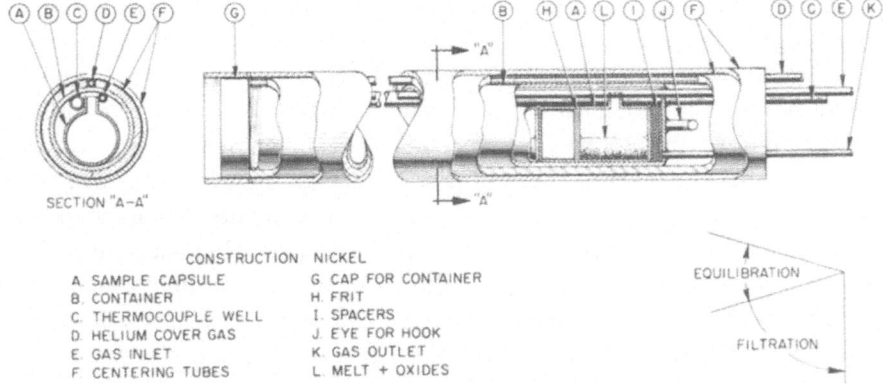

SECTION "A–A"

CONSTRUCTION NICKEL

A. SAMPLE CAPSULE
B. CONTAINER
C. THERMOCOUPLE WELL
D. HELIUM COVER GAS
E. GAS INLET
F. CENTERING TUBES

G. CAP FOR CONTAINER
H. FRIT
I. SPACERS
J. EYE FOR HOOK
K. GAS OUTLET
L. MELT + OXIDES

EQUILIBRATION

FILTRATION

Fig. 3. Arrangement for equilibrations using a rocking furnace. (Reprinted from Ref. 29, p. 1622, by courtesy of Microforms International Marketing Corporation.)

until the melt has been quenched. One must then rely on sedimentation for separating the phases, and this seldom provides a quantitative separation.

Reactions between gaseous and liquid phases are most frequently performed by sparging the gas once through the liquid (transpiration) or by recirculating the gases continuously with a pump while monitoring the progress of the reaction either with a suitable analytical method or by withdrawing aliquots of the gas and the liquid. This type of experiment can be performed using either welded or flanged containers. The use of a gas recirculating system is advantageous especially when working with relatively small amounts of melt (100–200 g) since it avoids the significant accumulative effect on the melt resulting from impurities in the gas, even in extremely low concentrations, which occurs with open systems.

It is always recommended to install a filter for particulate matter at the gas exit side of the container. This prevents gas-borne fluoride particles from contaminating the external environment. In the case where acidic gas flows are being monitored with a caustic titrator, the use of a filter also eliminates positive titration errors resulting from the hydrolysis of fluoride particles. In one sparging experiment where the evolution of HI was studied, it was found[30] that the HI was retained, probably by adsorption, on small particles, and these produced a fog when contacting the caustic solution resulting in the loss of a significant fraction of HI. This was avoided by the use of a suitable filter heated to ~80°C (see Sec. 5.2).

The welded and flanged containers described above are also used in glove boxes, often with small changes, such as facing valves and connections to the front of the glove box, in order to provide accessibility and easier handling.

High-temperature work inside glove boxes usually requires a special glove box or the modification of a standard one for locating the tubular furnace outside the confines of the box.[31] The advantages of this arrangement are that the furnace does not take up a large fraction of the available space and does not unduly heat the box and its contents. In our work the furnace was located under the floor of the glove box (Fig. 4a), which is a very satisfactory arrangement. In order to maintain leak-tight conditions, especially with respect to radioactive matter, a tubular extension is welded to the floor of the box (Fig. 4b), and the furnace is positioned in place from

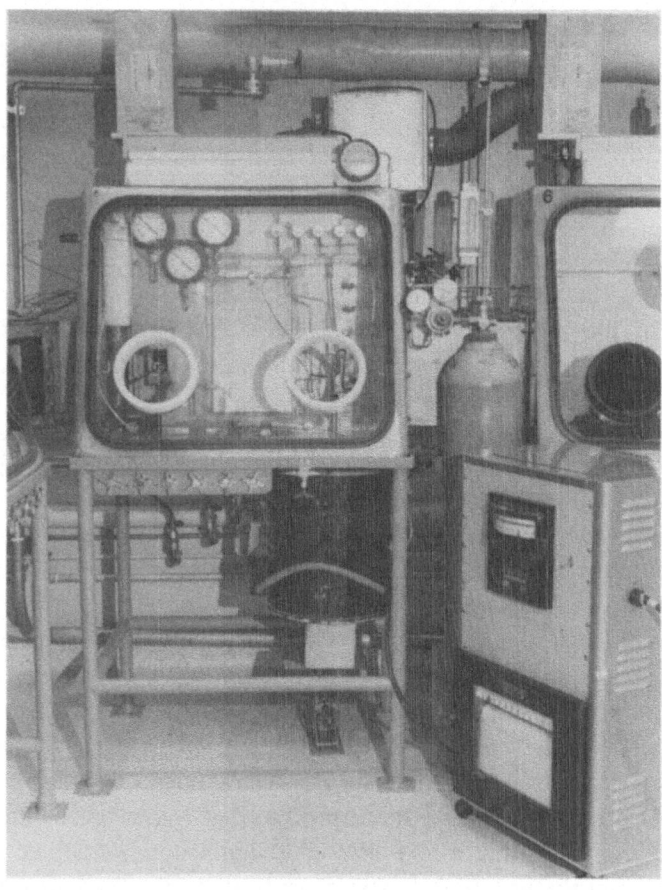

Fig. 4a. Glove box with attached heating compartment, prior to attaching the gloves.

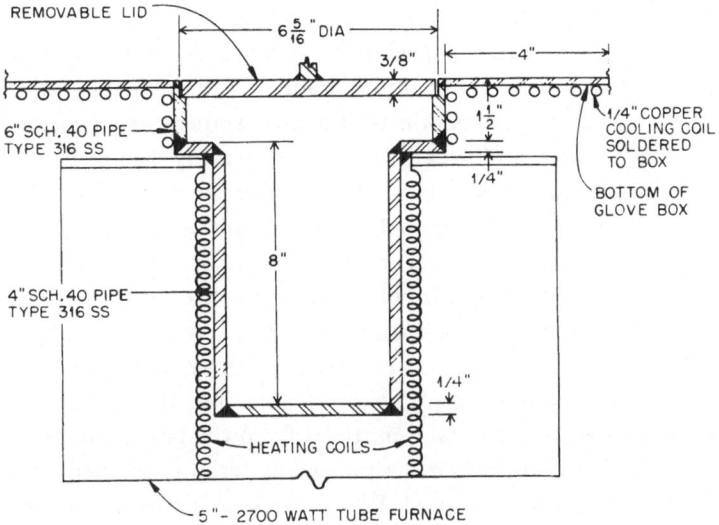

Fig. 4b. Heating compartment extension. The dimensions are typical and are shown as an example. (Reprinted from Ref. 31, p. 302, by courtesy of Interscience Publishers.)

the outside. This has the added advantage that the outside of the furnace has no possibility of being contaminated and is very accessible for maintenance or repairs. The area of the floor around the furnace is provided with water cooling by means of brazed copper tubing in order to maintain a low temperature inside the box. When the furnace is not used, the opening to the extension can be covered with a suitable lid. For a complete description of other glove box techniques, which fall outside the scope of this review, we refer the reader to the comprehensive publication by Barton.[31]

5.2. Connections and Filters

In general we have adopted the practice of connecting the accessories (thermocouple wells, sparging lines, electrodes, etc.) through nickel risers (stacks) welded to the top of the container. This adds significant flexibility to the design of the system, particularly since it permits moving these accessories in the vertical direction, allowing for their insertion or withdrawal even at high temperatures. The height of the risers is usually 15 cm, and their diameters depend on the accessories and/or on the area available on top of the container. We frequently use $\frac{1}{4}$, $\frac{3}{8}$, and $\frac{1}{2}$ in. O.D. tubing.

The connection is made with Swagelok,[*] Gyrolok,[†] or other fittings of similar design. To provide for sliding under leak-tight conditions, the metal ferrules are replaced by bushings made of Teflon. When noncorrosive gases are used and the connection is at a low temperature, Cajon[‡] fittings are also adequate. An additional advantage in using Teflon bushings lies in their low electrical conductivity which makes them good insulators for electrochemical work (see Sec. 7.11). The materials of construction of connectors and fittings most frequently used are stainless steel and brass (in the absence of HF) and Monel (in the presence of HF). In some instances, where Monel was not readily available, we successfully replaced it by gold-plated stainless steel, provided the fittings were at a sufficiently low temperature to avoid the diffusion of gold into the steel.

The filters used for removing particles from the gas stream are usually made with a nickel body, while the porous material is selected to be compatible with the gases to be filtered. Porous Teflon[§] is adequate for corrosive gases such as HF and for HI.[(30)] For the latter, the body should be gold plated. For mixtures of HF with H_2 or inert gases, all-nickel filters with nickel frit disks welded in place are quite adequate. HF-treated filter paper on a suitable support such as a nickel wire screen can also be used.

5.3. Thermocouple Wells

Since many readily available thermocouples are clad in stainless steel, it is recommended to shield them from the fluoride melt and from HF by means of nickel or copper tubing closed at one end. If liquid bismuth is also present, a molybdenum well is necessary because of the solubility of nickel and copper in bismuth. The size of the tubing obviously depends on the diameter of the thermocouple but should be as small as possible to reduce heat leakage.

5.4. Valves and Gas Flow Controls

Any good-quality needle or diaphragm valve can be used for noncorrosive gases. For gaseous mixtures containing HF, valves with Monel bodies and diaphragms are used.[‖] For throttling and control, a fine needle

---

[*] Trademark of Crawford Fitting Co., Solon, Ohio.
[†] Trademark of Hoke Inc., Cresskill, New Jersey.
[‡] Trademark of Cajon Co., Cleveland, Ohio.
[§] Pall Corporation, Cortland, New York.
[‖] Available from Hoke Inc., Cresskill, New Jersey.

valve is desirable. We have also successfully used a mass spectrometer valve* which uses a gold-plated steel diaphragm and a Monel or platinum capillary. When these are not readily available, proper use can be made of calibrated platinum orifices or capillaries. When possible, it is advisable to use a filter upstream of the capillaries to prevent their plugging.

When HF is absent, standard flow meters, Wet Test Meters, or Bubble-O-Meters can be used for monitoring gas flow rates. When HF is present in a gas mixture, as happens frequently at the exit of an experiment, it is removed previously by adsorption on $NaF^{(32)}$ or it is scrubbed with caustic solutions which can be back-titrated. The rate of the gas flow remaining after the HF removal can be measured with the equipment mentioned above.

When the necessity arises for measuring flow rates of pure HF, as in the case of preparing a gaseous mixture to be used in an experiment, it can be accomplished with the use of calibrated capillaries or differential pressure (DP) cells. The latter consist of a force-balance mechanism using encapsulated diaphragms which are interchangeable and operate at a wide range of temperatures. With Foxboro[†] models, the measurements can be transmitted either pneumatically or electronically. Very good results are also obtained using mass flow meters. The Hastings-Raydist[‡] models consist of an electronically heated tube and an arrangement of thermo-couples to measure the differential cooling caused by a gas flowing through the tube. Thermoelectric elements generate a dc voltage which is propor-tional to the rate of mass flow of the gas. The readout can be either linear or logarithmic, and the instruments are calibrated for air which happens to have the same calibration factor of unity as for HF.

5.5. Devices for Sampling Molten Fluorides and for Adding Reagents in Controlled Atmospheres

Because of the previously discussed reactivity of container materials and melts with atmospheric components, it is advisable, and sometimes imperative, to perform the operations of sampling the melt and of adding reagents in environments of inert gases. This requirement does not limit these operations to only inert-atmosphere glove boxes; with only a small

---

* Diaphragm-type adjustable leak valve (reference No. C-124492A) obtained from Oak Ridge Gaseous Diffusion Plant, Oak Ridge, Tennessee.
† Foxboro, Massachusetts.
‡ Hampton, Virginia.

degree of precaution and with simple devices, they can be performed in a conventional hood without exposing the contents to the atmosphere. To operate with total exclusion of air, a ball valve with Teflon gaskets is attached to one of the risers and a (holding) tube with a valved side arm is connected to the upper end of the ball valve, as shown in Fig. 5a. The top of the holding tube has a fitting with a Teflon bushing through which the sampler (filter stick) is inserted and maintained leak-tight. The sampler is connected by flexible tubing to a vacuum line, and the side arm is connected to an inert gas line. The assembly is operated as follows: with the ball valve closed and the filter stick in the holding tube, the inert gas is circulated generously through the system to displace all the air. This is usually estimated from the volume of the system and the flow rate employed. The vacuum line is closed, the flow of argon is stopped, the ball valve is opened, and the filter stick is slowly inserted into the melt. After allowing for thermal equilibrium, molten salt is sucked into the sampler by applying vacuum. The sampler is lifted into the holding tube, and the ball valve is closed; it can be removed into the atmosphere after cooling. If required, the addi-

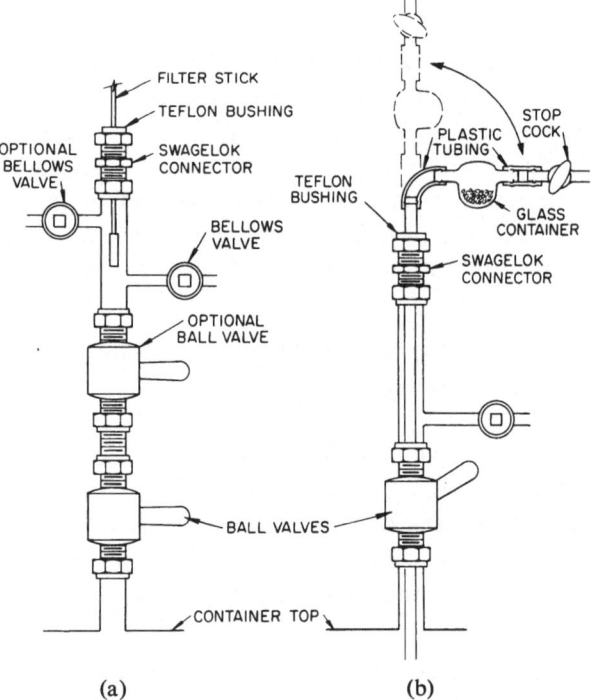

Fig. 5. Devices for sampling molten fluorides and for adding reagents at high temperature under controlled atmospheres.

tion of a second ball valve to the system (Fig. 5a) will permit the transfer of the sampler (in the holding tube under an inert atmosphere) to a suitable place. If rapid cooling of the sample is desired, it can be accomplished by applying a wet sponge to the holding tube, by wrapping a cooling coil around it, or by circulating an inert gas through the holding tube. The latter may require an additional valved side arm for the gas exit, although the same result is attained by loosening the Teflon bushing.

Since the amount of sample removed depends on the design of the filter stick, the porosity of the frit (see below), the viscosity of the melt, and the length of time under vacuum, no specific recommendations can be made; practice will be the best teacher. One word of caution, however, should be noted with respect to the pressure. Care should be exercised so that upon opening the ball valve to the system the pressure in the container does not increase suddenly. Otherwise the melt may back up through any open tube (sparging) immersed in the melt and cause plugging by freezing in a cold spot. This is easily avoided by previously equalizing the pressure between the inlet and the exit line, which is accomplished by opening a valve installed between those lines just for this purpose.

The addition of reagents can be handled similarly by means of a ball valve and a holding tube. If the reagent can be pelletized and only small amounts are required, the pellet is located in the holding tube, the system is purged with argon, and finally the pellet is dropped into the container. For the addition of reagents in powdered form or as microspheres, we have used successfully the arrangement showed in Fig. 5b. It consists of an open-ended section of metal tubing connected by flexible tubing to a glass container. This container is connected to a stopcock, a valve, or a mercury lock. The reagent is loaded into the glass container which is positioned at a 90° angle to the metal tube. After the air is displaced by argon entering through the valve on the side, the metal tube is inserted into the vessel, and the glass container is slowly tilted upwards in order for the reagent to flow. Since fine powders sometimes tend to cake inside small-diameter tubing, the addition should be slow and can be helped by applying a vibrator to the tubing. Microspheres do not present this problem; hence they should be used when available.

When the amount of added reagent must be known accurately, the device (metal tube and glass container) can be weighed before and after the addition. For the special case where these operations had to be performed in an air-filled glove box while the system was under hydrogen pressure, we designed and used the equipment shown in Fig. 6a, b. It consists of a shroud that can be placed around the Teflon bushing of the

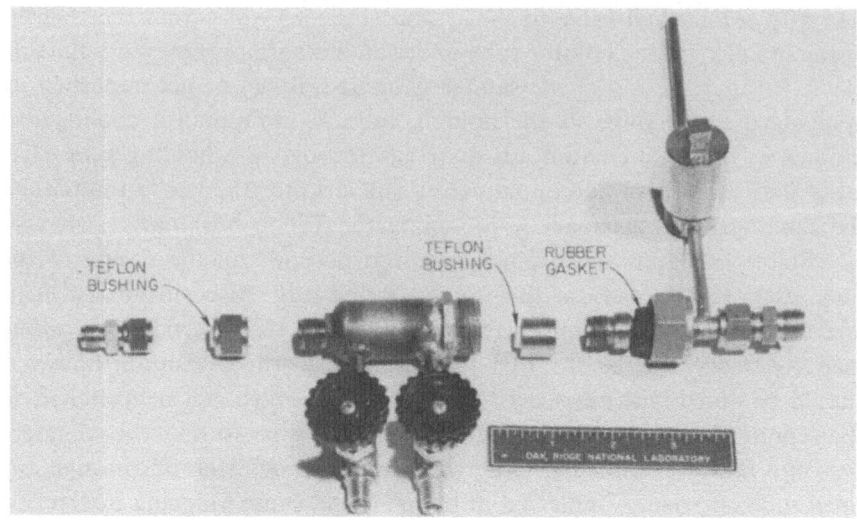

(a)

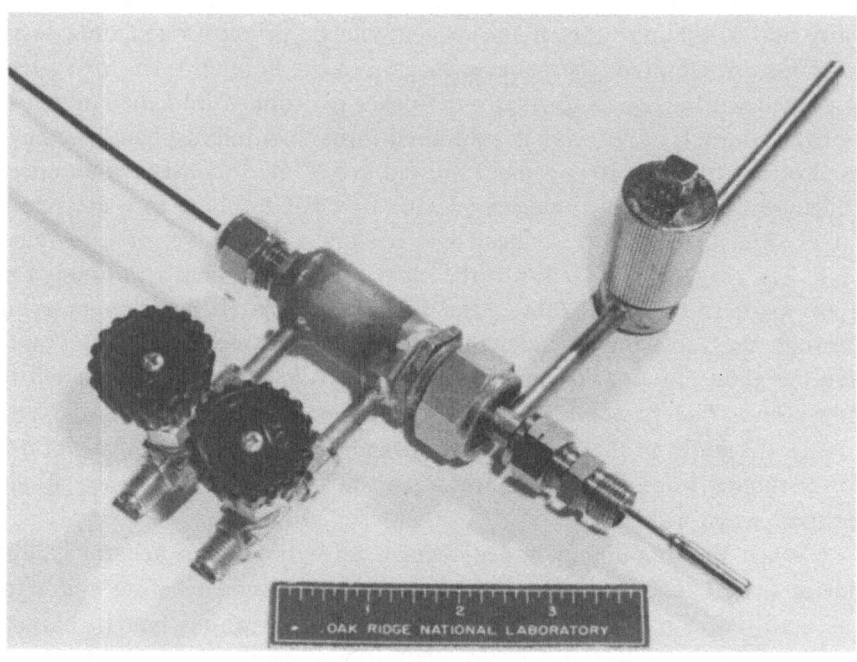

(b)

Fig. 6. Shroud for inert gas cover while sampling in a hydrogen atmosphere.

holding tube and provides a flow of argon. Thus, any hydrogen leak to the glove box is properly diluted, and only argon can leak into the vessel.

In the few instances where no filtration is required, a sample of the melt is obtained with an open-ended section of tubing. In the more numerous instances where a filtered portion of the molten fluorides is required, the tubing is provided with a plug of suitable filtering medium such as porous frits of sintered metal or felt metal. These frits are attached to the tubing by swaging and/or welding. The form and dimensions of the filter sticks vary according to the size of the sample desired. The porosity of the frit obviously depends on the size of the particles that are sought to be excluded from the sample. In our work, the designs more frequently used were $\frac{1}{4}$ in. O.D. tubing (up to 90 cm long) or 2–3 cm of $\frac{1}{4}$ in. O.D. tubing with the frit* at one end, welded to $\frac{1}{8}$ in. O.D. at the other end. The latter design is more suitable for frequently obtaining small and relatively constant amounts of sample. Almost any metal or alloy compatible with the liquids can be used, although we are partial to the use of copper because it is easier to crush, facilitating the removal of the frozen sample from the filter stick. Occasionally we have also used nickel, steel, and molybdenum filter sticks. In the few instances where contamination with oxide was no problem, fused silica filter sticks were used with advantage. Because molten fluorides do not wet silica, the sample can be readily recovered in the shape of a pellet rather than as a powder.

When contamination with oxide is a critical problem, any oxide coating on the filter sticks must be removed prior to their use. This is best accomplished by heating the copper and nickel samplers in flowing hydrogen which assures the removal of oxide from both outer and inner surfaces. (For copper filter sticks, hydrogen firing overnight at $\sim 600°C$ is suitable). Metals that show a tendency to form stable hydrides may be better cleaned by firing in vacuum. Stainless steel samplers cannot be thoroughly cleaned because the Cr_2O_3-containing coating is not reducible by hydrogen. Thus, only the exterior surface can be cleaned by abrasion, but there is no way to reach the interior or the bulk of the frit. To avoid reoxidation of the "clean" filter sticks it is recommended to store them in an inert atmosphere. It is also desirable to assess the filter sticks for cracks or large pores prior to their use. This can be easily determined by measuring the pressure differential (i.e., with a mercury manometer) at which a gas such as helium stops flowing through the frit immersed in a liquid, i.e., acetone. The

---

* Frits of 0.004 or 0.0015-in. mean pore size, of nickel and copper, have been generally used.

maximum pore size can thus be calculated from

$$\text{Maximum diameter of pore } (\mu) \approx 72/\Delta P \text{ (cm Hg)}$$

which is applicable to helium and acetone; for other media, adjustment should be made for the proper surface tension.

6. ABSORPTION SPECTROSCOPY OF MOLTEN FLUORIDE SOLUTIONS

As is the case with aqueous solutions, absorption spectroscopy of molten fluoride solutions can be used in a multipurpose fashion. It is a sensitive and suitable tool for the identification and quantitative determination of solute species. Thus, these properties of absorption spectroscopy are also useful for studying chemical equilibria *in situ* in the absorption cell.

When using high-melting fluoride solutions, it is necessary to rely more frequently on the self-absorbance of the solute species because few chromophore-containing organic reagents are stable at the high temperatures employed. Although this seemingly restricts the number of solute species that can be studied, it has its compensation because it allows this technique to be used on line for the continuous monitoring of selected solutes. Absorption spectroscopy is also widely used in studying the coordination geometry of solutes and the symmetry of the resulting species. We will not attempt to describe the details for estimating the latter, described elsewhere,[33] but rather concentrate on the experimental part.

6.1. Visible and Ultraviolet Spectroscopy

Visible and UV absorption spectroscopy of molten halides, specially of chlorides, has enjoyed a widespread use since the middle 1950s. The extension of this technique to molten fluorides took place in the late 1950s, mainly due to the work of Young and associates.[34]

Several models of spectrophotometers have been used for molten salt spectroscopy, e.g., Beckman DU,[35,36] Cary 11MS,[37] 14M,[38] and 14H,[39] Perkin-Elmer 12C,[40] and others. Some have required modifications to adapt them to high-temperature work, while others required only the removal of the conventional cell holders. The Cary model 14H has been specifically designed for high-temperature work. In this instrument, the light beam is chopped prior to passage through the sample. Afterward it is passed through a monochromator and then measured. Blackbody radiation from the sample is minimized as only the chopped signal is amplified on reaching the detector. Although some workers have used dual cell compartments in order to use a reference melt, presently the furnaces are designed for one container, and the spectra are measured against air as reference. For spectral measurements of highly radioactive melts, such as samples from the molten salt reactor experiment (17 mR at contact for a 200-mg sample), Young has designed a special system.[41] This consisted of an extended optical path which permitted the sample to be located in a hot cell while the spectrophotometer was located on the outside.

There are several furnaces described in the literature, but to our knowledge none is commercially available. They are all based mainly on the same design with slight variations. Some of the requirements for a furnace used in a spectrophotometer are obviously of geometrical nature; i.e., to fit adequately into the cell compartment and to provide easy access to the resistance heaters for maintenance work. Hermetic sealing of the cell compartment is a necessity for most of the work with molten fluorides because very few materials are available for constructing transparent cells which are compatible with the media (see below). Thus, most of the work is carried out in open cells while its compartment is leak-tight. The outer shell of the furnace is usually cooled by water to reduce thermal radiation. The resistance windings, the thermal insulators, and the temperature control can be chosen to meet the requirements of a specific type of work or to operate in a wide range of temperatures.

A typical furnace[42] design that we have used successfully at temperatures up to 800°C is shown in Fig. 7. It consists of a cell chamber (compartment), an inverted "tee" made of nickel pipe surrounded by massive metal in which are located the alumina-insulated platinum resistance heaters, and a thermocouple well. This arrangement is situated inside a double-walled nickel cylinder and insulated thermally from it by means of expanded alumina. The outer shell, which is double walled, is water cooled. Light passes through the horizontal section of the "tee" where the sample is located perpendicularly. The cell chamber is kept leak-tight by silica windows mounted on each end by means of "O" rings and threaded retainer rings. The top of the furnace has a provision for evacuating the cell chamber or filling it with gases. The cell chamber is hermetically closed at the top by means of a compartmented lid which has a sliding plate gate. Above the sliding plate is a holding chamber to contain the sample while the desired atmosphere is attained in the furnace. The lid and the gate are connected leak-tight by means of Viton "O" rings or Teflon gaskets, while

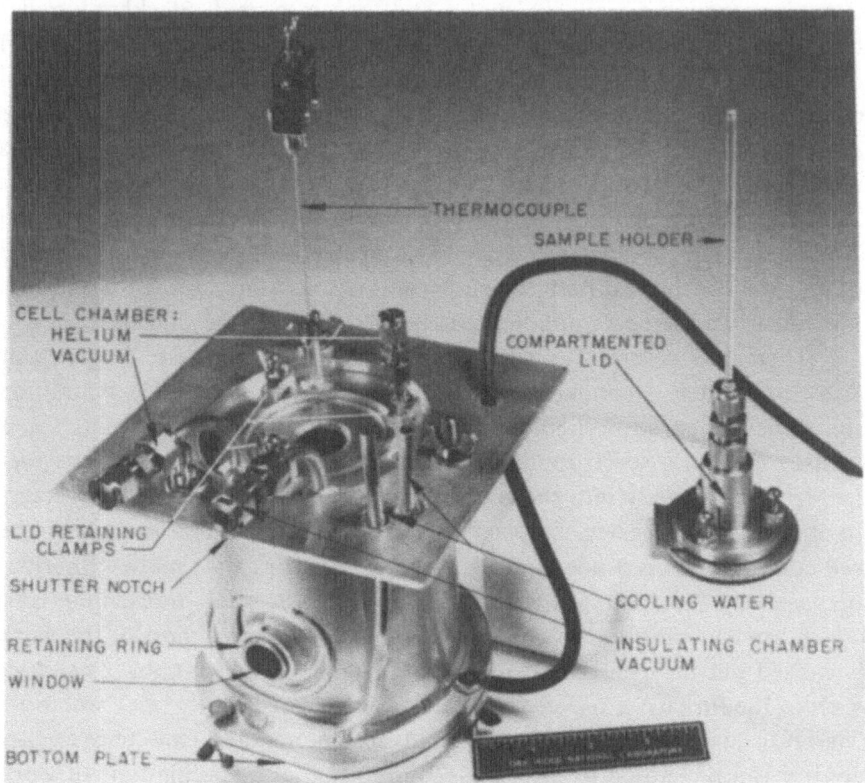

Fig. 7. High-temperature furnace for visible and UV spectroscopy. (Reprinted from Ref. 34, p. 192, by courtesy of Marcel Dekker, Inc.)

the sample holder and the lid are connected by means of a Teflon bushing. The compartmented lid can be very easily removed, and the furnace is then used in a conventional manner, i.e., exposed to air. The cell is usually attached to a rod-like holder by setscrews or other devices.

Very frequently, when using cells such as the windowless[43] or diamond-windowed cells[44] that are not hermetically sealed, the *modus operandi* is as follows. The solute and solvent are loaded into the cell which is then attached to its holder and placed into the compartmented lid. All these operations are preferably performed in a dry-box. The compartmented lid is attached to the furnace, which is evacuated and sparged with a suitable gas. The sliding gate is opened, the cell is lowered into the furnace chamber, and heating can proceed. Thus, the sample can be maintained in an air-free atmosphere while its spectrum is recorded.

The search for the ideal container material for spectral cells to be used with molten fluorides is complicated by the fact that the material must be transparent in addition to meeting the compatibility requirements described in Sec. 2. Many, if not most, of the spectral studies performed by Young were done in windowless containers, termed also captive liquid cells.[43] They eliminated the need for a transparent solid by the expedient procedure of simply not using one. Initially, wire loops which retained a film of molten fluoride were used.[45] They were later replaced by the containers[43] shown in Fig. 8; these had a transverse hole below the liquid level for the light beam and a set of smaller holes located above it for maintaining a constant level of liquid. The success in keeping the molten fluorides in such cells is due to the high surface tension of the solutions on most of the compatible construction materials employed—nickel, copper, platinum, graphite, molybdenum, etc. Thus, the melt forms a double convex lens of reasonably reproducible path length. As mentioned above, this cell design permits a wide choice of materials, and the worker can select one that is either practically inert toward solute and solvent or that purposely participates in a reaction; e.g., using an iron cell, the following reaction may be studied:

$$Fe^{\circ}_{(c)} + 2UF_{4(d)} \rightleftarrows 2UF_{3(d)} + FeF_{2(d)} \tag{22}$$

An interesting variation of the captive liquid cell described above is the use of containers made of porous metal proposed by Young.[46] These containers are fabricated with metal foils which contain small irregular holes generated by electrochemical etching. The salt would be retained in the container by virtue of its surface tension, and the small holes would provide for the transmission of light.

A variation of the simple captive liquid cell is that reported by Young et al. for the simultaneous voltammetric generation and spectrophotometric determination of soluble species with unusual or unstable oxidation states.[47] It consists (Fig. 9) of a working platinum microelectrode located in the optical path of the cell, and quasi-reference electrodes inserted below it by means of boron nitride plugs. This arrangement was used to generate and record the spectrum of the relatively unstable U^{3+} species in LiF–BeF_2–ZrF_4.

Although very useful, the captive liquid cells are not free of limitations, probably the most important being an uncertainty of $\pm 10\%$ in estimates of the path length. Homogenization of the liquid by gas sparging can be accomplished, but it is quite difficult. Since light transmission through

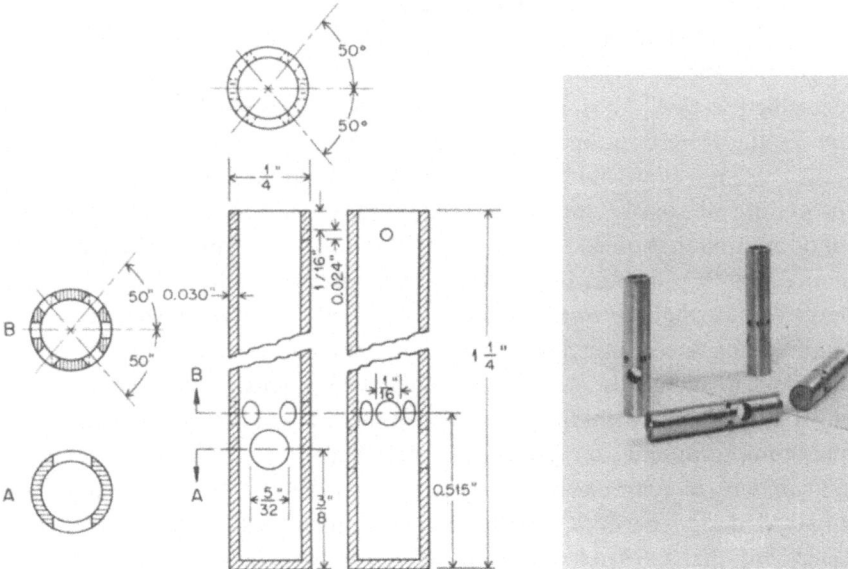

Fig. 8. Captive liquid cells for molten salts. (Reprinted from Ref. 34, p. 195, by courtesy of Marcel Dekker, Inc.)

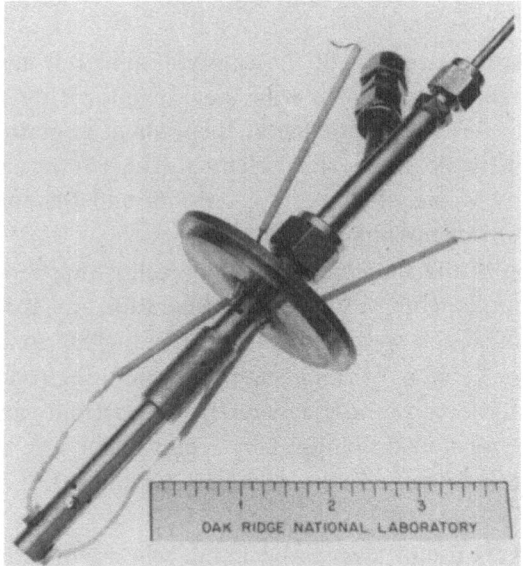

Fig. 9. Combination voltammetric–spectrophotometric (captive liquid) cell. (Reprinted from Ref. 47, p. 782, by courtesy of the American Chemical Society.)

the curved contour of the liquid is not rectilinear, the cells have to be carefully positioned at the focus of the light beam in order to decrease that effect.

The search for and testing of transparent windows compatible with molten fluorides has covered a wide range of materials, but none has been found that is of universal use. The candidate probably with the least limitations is diamond,[44] discussed below. The problem in using transparent windows is compounded by the necessity of mechanically attaching them to the cell body or of obtaining crystals large enough to be used as cells. Fortunately, due to the generally nonwetting behavior of most molten fluorides, the fastening of the window to the cell compartment seldom requires a gas-tight seal.[47]

MgO single crystal was successfully used to determine the NiF_2 spectrum in the eutectic mixture LiF–NaF–KF,[48] but it could not be used in LiF–BeF$_2$ mixtures because the more-insoluble BeO formed, coating the crystal.[48] An analogous result would be obtained using ZrF_4-, ThF_4-, or UF_4-containing melts, since ZrO_2, ThO_2, and UO_2 are generally less soluble than BeO. These results also apply to other transparent oxides, with the possible exception of SiO_2 for which the metathesis reaction can be hindered by applying an SiF_4 over-pressure, as discussed earlier. The devitrification of fused silica into cristobalite, apparently enhanced by SiF_4, becomes a serious problem after 2–3 days because of the accompanying loss of transparency.

Despite this disadvantage, good spectra of UF_4[12] and PuF_3[49] dissolved in molten Li_2BeF_4 contained in silica were obtained. The spectrum of PuF_3 is shown in Fig. 10. Solutions of PuF_3 and, similarly, of rare earth fluorides are particularly suitable when using oxide containers or windows since no precipitation of the solute oxide occurs. This is due to the higher thermodynamic stability of the fluorides, as shown in Table I. Obviously, when silica cells are used, they can be sealed with a torch under vacuum or reduced pressure and heated in the furnace in the absence of the protecting lid.

The use of very thin coatings of unreactive metals such as gold, silver, and palladium, on silica has been reported as a means of prolonging the life of the cell and of reducing oxide contamination of the fluoride melts.[50] We have not tried this approach, which is worth exploring further.

The use of other compounds, mainly the fluorides of rare earths and alkaline earths, as window or container materials has been reported[51] or suggested. The main disadvantage for this application is their significant solubility in molten fluorides.

The use of diamond windows was reported by Cocks *et al.* in 1957.[52] Although the cell design was complex and seemed to present some difficulties in handling, it demonstrated the feasibility of using diamond windows and confirmed its adequate optical transmission properties in the visible, ultra-violet, and near-infrared regions. Based on these findings, Toth *et al.*[44] developed and tested a new design of a diamond-windowed cell with a graphite body (Fig. 11) that overcame earlier inadequacies. It can be easily loaded and emptied, and the windows can be easily removed from the main body for cleaning. In this design the distance between windows is approx-imately 8 mm. Although the cell is not gas-tight, no leakage of molten fluorides occurs. The cell holder, like that for captive liquid cells, consists of a rod connected to a metal cylinder which fits over the cell and is held in place with setscrews or with a metal pin. The successful use of this cell was immediately demonstrated when Toth *et al.*[44] observed a predicted, but not yet seen, absorption band at 2039 nm of UF_4 dissolved in Li_2BeF_4.

It should be remembered, however, that diamond is slightly less stable than graphite, and therefore care should be exercised in selecting the systems to be studied.

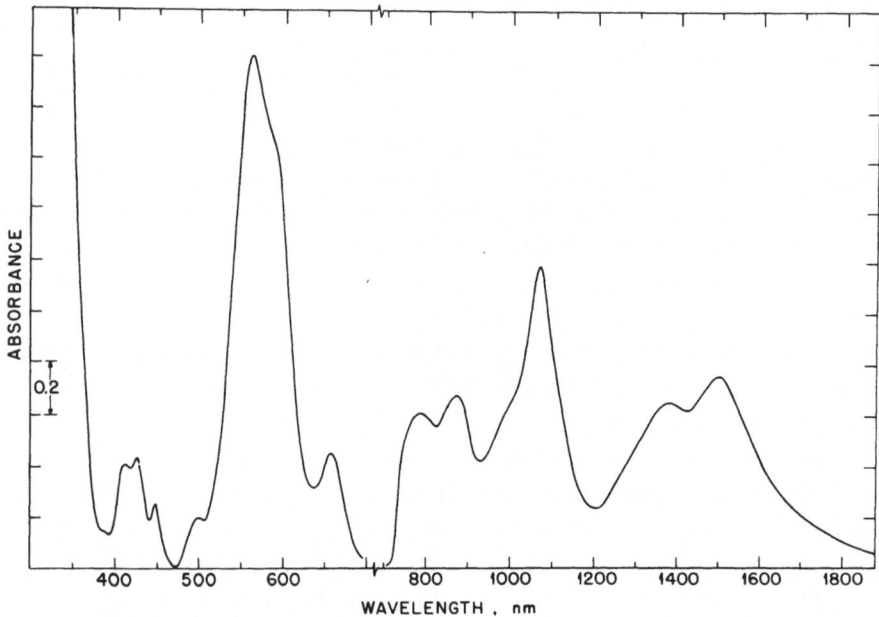

Fig. 10. Absorption spectrum of a 0.4 mole % PuF_3 solution in LiF–BeF_2–ThF_4 (72–16–12 mole %) at 575°C. (Reprinted from Ref. 49, p. 3594, by courtesy of Microforms International Marketing Corporation.)

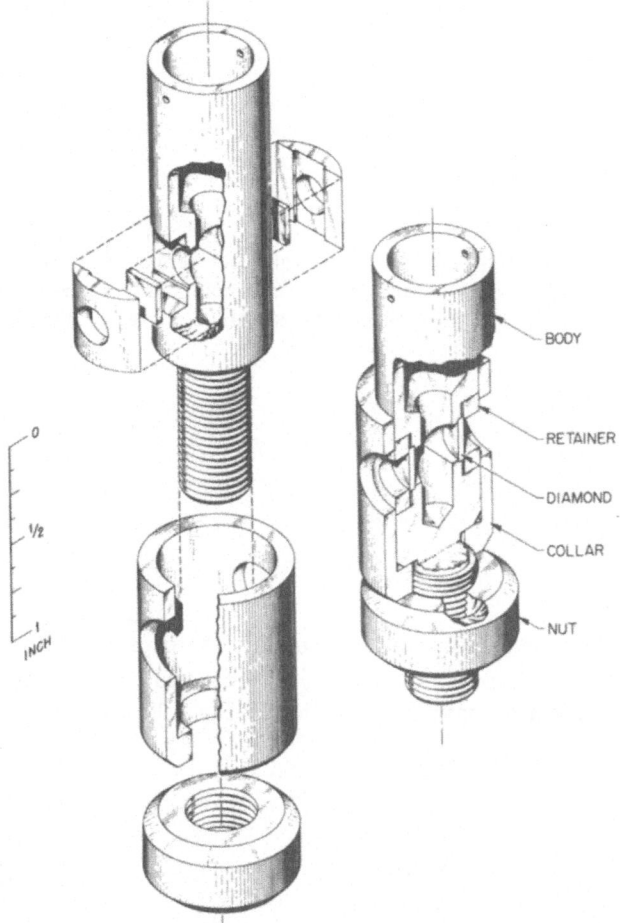

Fig. 11. Diamond-windowed cell for molten fluorides. The body and other parts are made of graphite. (Reprinted from Ref. 39, p. 683, by courtesy of the American Chemical Society.)

A novel approach to molten halide (including fluorides) spectrophotometry has been recently proposed by Young.[53] It consists of an immersible probe, shown in Fig. 12, which has a slot of appropriate width cut into it. The light beam, preferably from a laser, is subjected to multiple internal reflections and passes through the slot which is filled with liquid when the probe is immersed. It has been proposed to use LaF_3 to construct the probe; this would restrict its use to a discontinuous operation since, as mentioned above, LaF_3 exhibits a significant solubility in many molten fluoride mixtures.

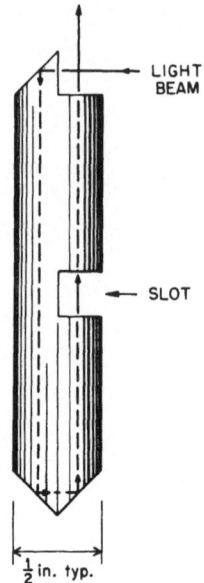

LIGHT
BEAM

SLOT

Fig. 12. Slotted optical probe proposed for spectral analysis of aqueous and nonaqueous solutions. For molten fluorides the probe would be fabricated of a rare earth fluoride.

$\frac{1}{2}$ in. typ.

With the above-described cells, mainly the captive liquid, quantitative characterizations (absorption peak positions, peak intensities and the assignment of spectra) have been made for the following species[2]:

Ni^{2+}	U^{3+}	Mn^{3+}	Pu^{3+}
Fe^{2+}	U^{4+}	Co^{2+}	Pr^{3+}
Cr^{2+}	U^{5+}	Mo^{3+}	Nd^{3+}
Cr^{3+}	UO_2^{2+}	CrO_4^{2-}	Sm^{3+}
Cu^{2+}	Mn^{2+}	Pa^{4+}	Er^{3+}
			Ho^{2+}

Semiquantitative characterizations (approximate peak intensities and possible assignment of spectra) have been made for the following species[2]:

Ti^{3+}	Eu^{2+}
V^{2+}	Sm^{2+}
V^{3+}	Cm^{3+}
	O_2^-

6.2. Infrared and Raman Spectroscopy

Excellent reviews on infrared and Raman spectroscopy of molten salts have been published recently[54,55,56]; however, they contain little information on molten fluorides. The recent availability of Fourier-transform spectrometers on the market has given, to those who can afford it, a new dimension in the application of infrared spectroscopy to molten halides, including fluorides.

We have chosen to describe only the latest equipment and techniques for infrared and Raman spectroscopy of molten fluorides developed at the Oak Ridge National Laboratory because it represents, probably, the largest single contribution on the subject.

To our knowledge, infrared spectra of molten fluorides, namely $NaBF_4$–NaF mixtures, were obtained for the first time using windowed containers. Bates et al.[57] obtained spectra of BF_3OH^- and BF_3OD^- in $NaBF_4$ using silica cells in a furnace similar to that described for visible

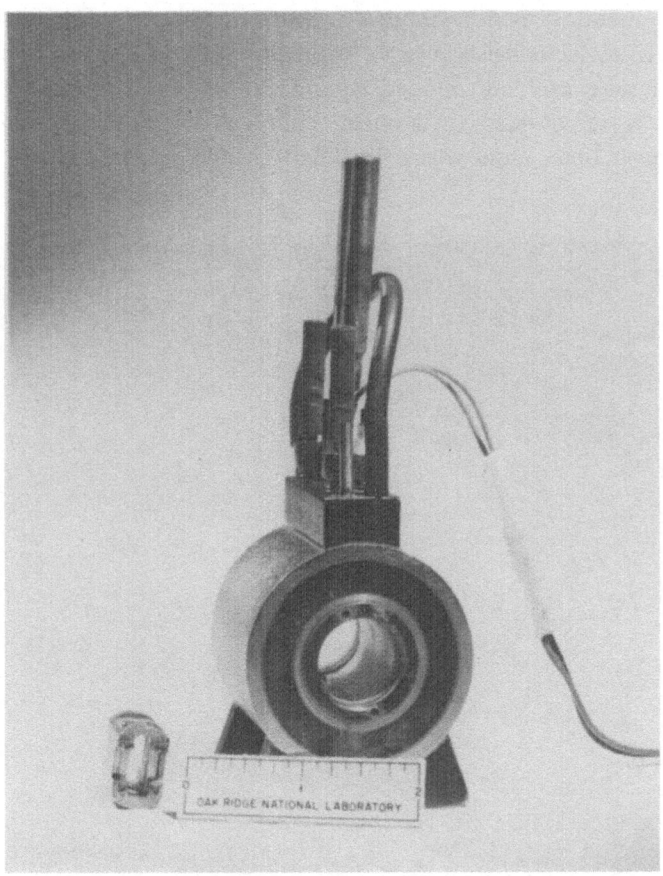

Fig. 13. Infrared furnace and LaF_3-windowed cell.

and ultraviolet spectroscopy. However, since silica exhibits absorption bands of OH⁻, the authors explored the possibility of using materials compatible with molten fluorides that would not exhibit the absorption of the OH⁻ group and thus allow the examination of a wider range of the spectrum. LaF_3 was chosen because it exhibits a low solubility in molten $NaBF_4$–NaF and because it transmits infrared between 2400 and 4000 cm⁻¹. Since the use of LaF_3 containers made from single crystals is not economically attractive, the material was used in the form of plates for windows in a nickel cell,[57] shown in Fig. 13. The windows are attached to the cell body by nickel retainers pressed by small screws. The use of thin gold gaskets is optional depending on the wetting behavior of the melt.

The cell is located in a furnace designed with small dimensions so that it can be used in several spectrometers (Perkin-Elmer, Digilab FTS-20, and even within a Perkin-Elmer 4X beam condenser). The furnace, also shown in Fig. 13, consists of a boron nitride tube, ¾ in. I.D., wound with resistance wire and surrounded by insulating boron nitride wool. The assembly is surrounded by a metal container which has provisions for cooling both open ends where leak-tight windows are located. The seal

Fig. 14. Assembled holder for infrared emission spectroscopy of molten salts.

Fig. 15. View of the holder attached to the heating element and of the cooled chamber.

between the AgCl or AgBr windows and the body is made with Viton "O" rings.

The furnace can be used under vacuum or with gas under moderate pressures introduced through fittings provided at the top. Good spectra have been obtained for OH⁻ and OD⁻-containing species in molten NaBF$_4$.[57] The only problem encountered with this equipment has been the occasional cracking of the windows, probably due to stresses developed by thermal cycling.

Although LaF$_3$ is more soluble in molten fluorides, such as BeF$_2$–LiF or LiF–NaF–KF mixtures, than in fluoroborates, it could be used to study these and similar melts in experiments of relatively short duration (1–3 days).

Bates has recently developed the equipment for obtaining emission spectra of molten salts.[58] This method offers several advantages over the classical ones (absorption, reflection, and attenuated total reflectance), but has been sparingly used to date mainly due to the limitations of existing equipment. Among the advantages of infrared emission spectra we may cite the absence of contact of the melt with windows and the fast collection of directly produced infrared spectra.

The method has been used successfully on molten nitrates and chlorates[59] and should also be applicable to molten fluorides.

The standard infrared source of a Digilab FTS-20 spectrometer was replaced by an emission sampling attachment consisting of two mirrors and an adjustable aperture. The attachment allowed radiation from an external source to be collected and directed to the interferometer. A plate with an "O" ring seal and window clamp was constructed to replace a blank-off plate on the vacuum box which contained the Michelson interferometer and the detector. Emitted radiation passed through an optical window into the evacuated chamber. This arrangement created about a 10-cm air space between the sample and the vacuum box. A KBr beam splitter and CsI window were employed in the spectral region above 500 cm^{-1}, while a 6-μ Mylar beam splitter and polyethylene window were used below 500 cm^{-1}. For molten salt work[58] a thin film of molten sample is held in a container, shown in Fig. 14, between the bottom of a graphite or metal cup and a metal screen (nickel or Hastelloy N) kept in place by a

Fig. 16. Holder, heating element, and chamber attached to vacuum box.

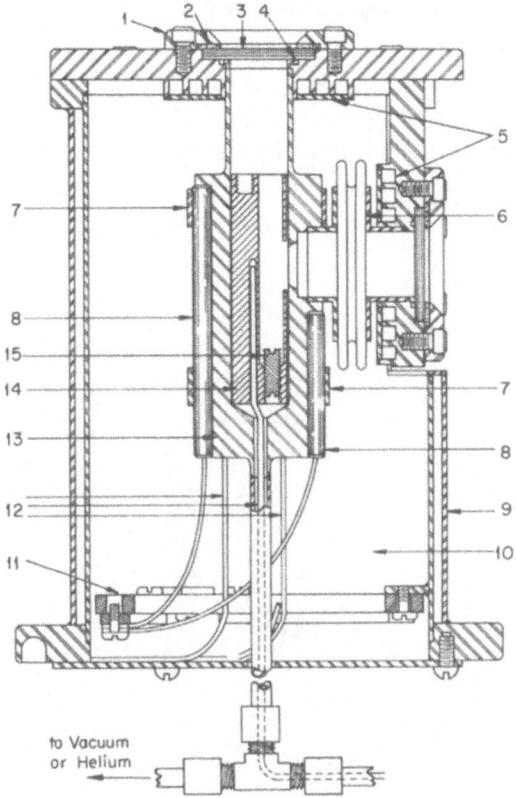

to Vacuum
or Helium

Fig. 17. Schematic cross-sectional view of furnace for laser Raman spectroscopy: (1) window flange; (2) Viton window pad; (3) quartz window; (4) "O" ring; (5) cooling water channels; (6) nickel bellows; (7) heater clamps; (8) cartridge heaters; (9) water jacket; (10) fibrous insulation; (11) heater connector assembly; (12) metal-sheathed thermocouples; (13) nickel furnace body; (14) cylindrical silver cell holder; (15) cell positioning screw. (Reprinted from Ref. 60, p. 83, by courtesy of the Society for Applied Spectroscopy.)

metal spring. The container is attached to a heating element (Fig. 15) which fits into an externally cooled chamber attached to the vacuum box, as shown in Fig. 16. The sample can be melted directly in the holder with its heat source in a horizontal position or can be melted in a separate environment. Since satisfactory results have been obtained with molten

NO_3^- and ClO_3^-, it is expected that similar results may be obtained also with molten fluorides when using this technique.

For laser Raman spectroscopy, Quist[60] designed and used a vacuum-tight furnace capable of operating up to 800°C (Figs. 17–18). Its salient features include the use of several types of cells (windowless, diamond-windowed, and conventional) and its ease of alignment with respect to the laser beam by means of the micrometer screws on a positioning table.

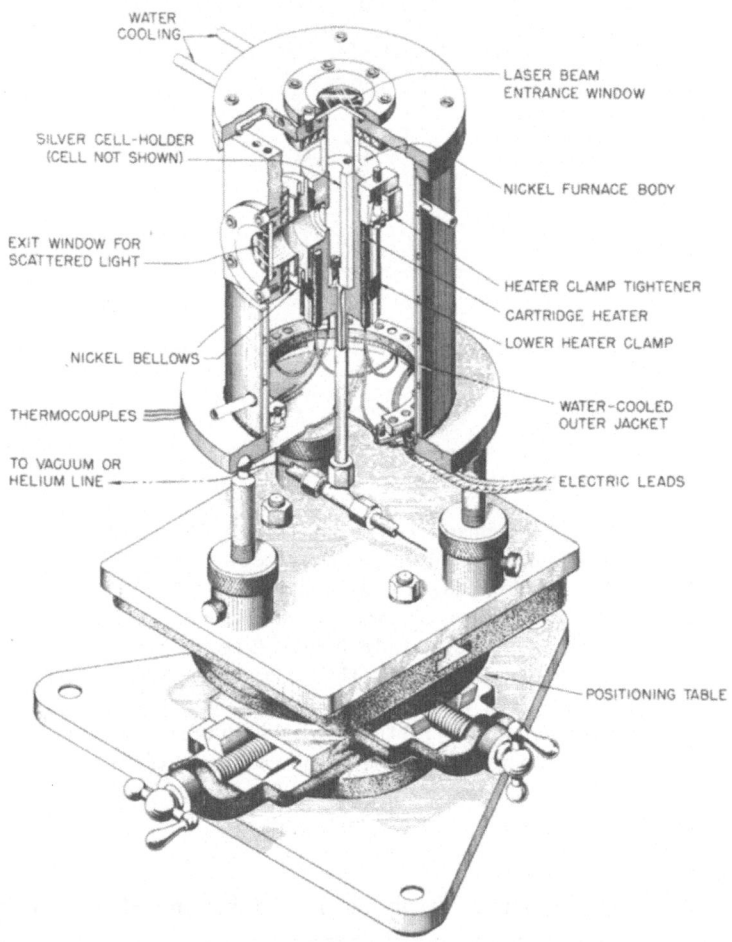

Fig. 18. Cutaway view of furnace for laser Raman spectroscopy of molten salts. (Reprinted from Ref. 60, p. 84, by courtesy of the Society for Applied Spectroscopy.)

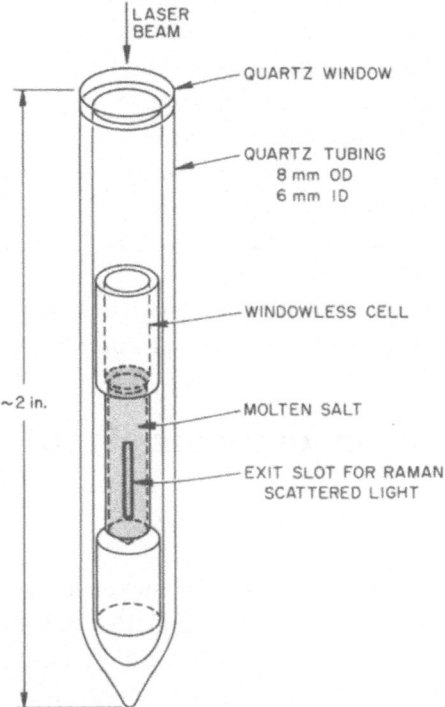

Fig. 19. Windowless cell for laser Raman spectroscopy of molten fluorides, sealed in a quartz tube.

Although the furnace design resembles that used in UV and visible spectroscopy in many aspects, the main difference is the direction of the light path; the laser beam enters the furnace through the top window and emerges through the side. Since the cells are located in metal blocks which are removable from the furnace, different cell designs can be used with different blocks. The complete furnace is approximately 40 cm high and weighs about 23 kg.

The fact that molten halides can be kept in a windowless cell by their surface tension was also used by Quist for constructing cells for Raman spectroscopy.[61] The schematic diagram of such a cell is shown in Fig. 19. The cell consists of a container of nickel, copper, or other material in which a vertical slot has been cut for the exiting laser beam. The cell is loaded with the solid sample and inserted into an evacuated silica tube and heated in order to melt the sample, fill the cell, and remove entrained gases. If the desirable amount of melt is present (the level is above the slot), the container is cooled and the cell is transferred to a silica tube to which an optically flat surface has been attached at one end. The other end is sealed under the desired gaseous atmosphere. The cell arrangement is introduced into the furnace block with the flat surface parallel to the upper window. The small beam divergence caused by the concave meniscus is not significant since most of the incident beam passes through the melt and is focused at the exit by an achromatic lens into the entrance slot of a double monochromator.

With the above-described furnace and cell, Quist[61] has reported on the Raman spectra of molten fluorides and fluoroborates.

7. ELECTROCHEMISTRY OF MOLTEN FLUORIDE SOLUTIONS

7.1. General Aspects and Electrical Insulators

The literature on electrochemistry of molten salts in general has been reviewed by Fung and Mamantov,[62] and that for molten fluorides by Mamantov.[63] Some interesting progress on the development of reference electrodes for molten fluorides has been made since the latter review. This will be described in detail below. The development of reference electrodes for use in molten fluorides for the measurement of the electromotive force of redox couples is one of the most important aspects of potentiometric measurements.

The care in handling the solutions and solutes described previously for other techniques should be exercised also with electrochemical techniques, and even more so, since some are very sensitive in detecting the presence of minute concentrations of contaminants. One of the problems encountered in the development of reference electrodes is that of finding suitable insulator materials which are compatible with the solutions at the high temperatures involved. The function of these insulators ranges from simply insulating the electrode lead from the container walls by its physical presence as spacers to its use as a container for the half-cell system itself. Since many electrochemical measurements are performed in containers provided with cooled risers, the insulation between the riser walls and the electrode leads is easily accomplished by using Teflon bushings. To prevent the deformation of the bushing at temperatures above 80°C, the top of the risers may be cooled with water. The choice of insulating container materials is not very wide, and the few available are not without limitations. Thus, boron nitride, beryllium oxide, and silica have been used recently, and others like silicon nitride hold promise but have not been fully investigated. One drawback of boron nitride, which has also been turned to an advantage, is its impregnation by molten fluorides; the other drawback is its apparent reducing behavior in an oxidizing environment.

7.2. Potentiometric Measurements

Boron nitride cups or containers have been used successfully by Jenkins et al.[84] to demonstrate the usefulness of the couple Ni/NiF_2 (constant concentration) as a reference couple. The action of the electrode was due to the slight impregnation of the boron nitride by the melt, the insulator thus acting as a diffusion barrier for ions. The long-range stability of the electrode (11 days at 500°C with $\pm$3–6 mV) was probably due to slow kinetics of the reaction

$$\tfrac{3}{2}NiF_{2(c)} + BN_{(c)} \rightleftarrows BF_{3(g)} + \tfrac{3}{2}Ni_{(c)} + \tfrac{1}{2}N_{2(g)} \qquad (23a)$$

and the relatively large excess of NiF_2, since the estimated free energy of the reaction is -34 kcal at 900°K. A schematic view of the electrode used by Jenkins et al. is shown in Fig. 20.

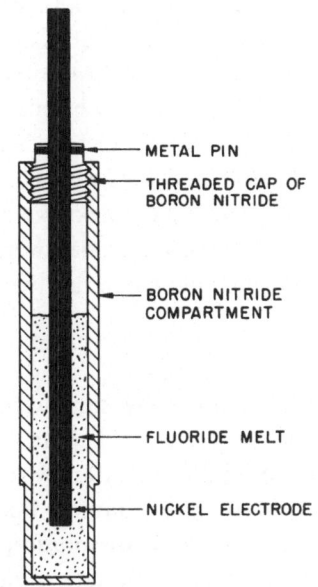

METAL PIN

THREADED CAP OF BORON NITRIDE

BORON NITRIDE COMPARTMENT

FLUORIDE MELT

NICKEL ELECTRODE

Fig. 20. BN reference electrode for emf measurements in molten fluorides. The nickel concentration of the fluoride melt can be fixed to a preselected value, preferably saturation. (Reprinted from Ref. 64, p. 386, by courtesy of Elsevier Sequoia SA.)

Winand[65] has developed reference electrodes for high-temperature (900–1100°C) work with a plug made of porous boron nitride using a sodium compound as binder. The half cells employed, $Ag°/AgCl$ (in NaCl) and $Ni°/NiF_2$ (in NaF) were tested in several designs. It was concluded that, at these high temperatures, the electrodes are better used in an intermittent fashion since prolonged contact with the melts employed, i.e., $NaF–ZrF_4$ (20–80 mole %), deteriorates the electrodes. This deterioration of the electrode containing NiF_2 might be explained by reaction (23a); however, for AgCl contained in BN, a reaction evolving BCl_3, such as

$$3AgCl_{(c)} + BN_{(c)} \rightleftarrows BCl_{3(g)} + 3Ag_{(c)} + \tfrac{1}{2}N_{2(g)} \tag{23b}$$

has a positive free energy of reaction. Thus, the mechanism for this deterioration remains unexplained.

Hitch and Baes[66] have made use of the insulating properties and the low solubility of BeO and SiO_2 to build a reference electrode of the third kind, shown in Fig. 21. It consists either of a tube of sintered BeO closed at one end by a plug of BeO sintered in place, or of a silica tube closed at one end by a silica frit of 5–10 μ porosity. Either tube contains Li_2BeF_4 saturated with BeO and NiO. A nickel rod to which sintered nickel has been welded is immersed in the heterogeneous mixture. Thus the concentration of Ni^{2+} is held constant ($X_{Ni^{2+}} = 3 \times 10^{-5}$ at 600°C) because of the low solubility product of NiO and the presence of the common ion O^{2-} which in turn is held constant by the presence of the saturating BeO. In order to support the oxide tubes and to reduce significantly the solubilization of the BeO or the SiO_2 in the bulk of the melt, the tubes were surrounded by a nickel tube in which small holes had been drilled. Although the walls of the BeO tube were too dense to permit salt penetration, it was found that the electrode worked successfully because the seal between the plug and the tube had microcracks which permitted contact with the outer (bulk) solution. These electrodes were tested in $LiF–BeF_2$ mixtures in which a Be° indicator electrode had been inserted. The net reaction was assumed to be

$$Be°_{(c)} + NiO_{(c)} \rightleftarrows BeO_{(c)} + Ni°_{(c)} \tag{24}$$

The results were similar for BeO or SiO_2, but the latter showed signs of corrosion, and the data obtained with it were associated with a larger uncertainty, ±5–10 mV versus ±0.5–2 mV in BeO. This electrode can be used in many molten fluoride mixtures, although one limitation to its use is the slow dissolution and diffusion of oxide to the bulk of the melt which may precipitate some solute ions such as Zr^{4+}, U^{4+}, Pa^{5+}, etc.

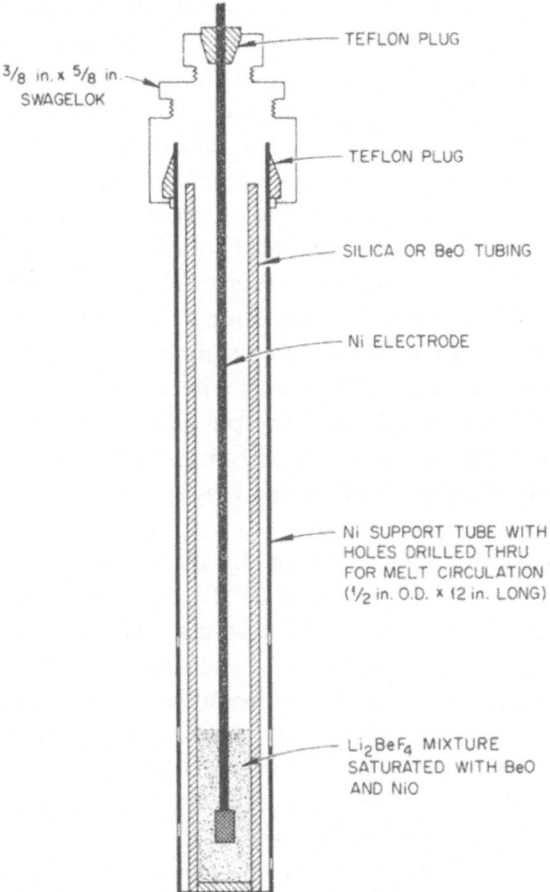

3/8 in. x 5/8 in.
SWAGELOK

TEFLON PLUG

TEFLON PLUG

SILICA OR BeO TUBING

Ni ELECTRODE

Ni SUPPORT TUBE WITH
HOLES DRILLED THRU
FOR MELT CIRCULATION
(1/2 in. O.D. × 12 in. LONG)

Li_2BeF_4 MIXTURE
SATURATED WITH BeO
AND NiO

Fig. 21. Reference electrode of the third kind for emf measurements in molten fluorides. (Reprinted from Ref. 66, p. 165, by courtesy of Pergamon Press.)

More recently, Bronstein and Manning[67] developed and used LaF_3 reference electrodes based on a different principle, that of anion diffusion through a crystalline structure. This is similar in concept to that used by Markin[68] in solid-state electrochemical cells with CaF_2. The LaF_3 was selected because of its known solid-state F^--conduction and because its solubility in molten fluorides[69] and fluoroborates was not excessively high (≤ 1 mole %).[70] Furthermore, this latter aspect was improved by using a metal frit (described below) that reduced significantly the dissolution rate of the LaF_3. The use of CaF_2 was not considered because its solubility in molten fluorides is estimated to be almost an order of magnitude larger.[71] The cell as originally developed by Bronstein and Manning consisted of a LaF_3 cup made by drilling a single crystal.* The cavity contained a small amount of Li_2BeF_4 saturated with a NiF_2 and a nickel wire inserted in the mixture. The crystal was inserted into a metal tube, preferably nickel or copper, with a frit welded at the end. Electrical insulation between the crystal and the metal walls was provided by a boron nitride tube. The top of the metal container was closed with a metal screw cap which was also insulated with BN. A concentric hole through the metal and the insulation of the cap allowed the nickel wire to extend above the system and provide for the electrical connection. The nickel wire was welded to a nickel disk which fitted on top of the crystal. The purpose of this disk is to act as a barrier for the salt in the cavity and, even more important, to allow the application of uniform pressure on the crystal against the frit to ensure a good contact.

Bronstein and Manning[67] tested this electrode in a cell formed with another electrode of known potential, $Be°/Li_2BeF_4$, and obtained potentials in very good agreement with calculated values. The electrode was also used to follow voltammetrically the half-wave potential of the U^{4+}/U^{3+} reduction wave (see below), and the measured potential agreed very well with the calculated value. The cell potential, however, exhibited a tendency to decrease with time in a reducing environment (presence of $Be°$). This was corrected by confining the beryllium electrode and a small amount of the solvent, Li_2BeF_4, in a porous BeO tube.

Clayton et al.[72a] used the same electrode configuration in molten mixtures of $NaBF_4$–NaF and filled the cavity with the same melt in order to eliminate any significant asymmetry potential. The concentration of Ni^{2+} in $NaBF_4$–NaF is apparently too low to generate a potential that can be measured with confidence. (This is caused by the low solubility of the $NaNiF_3$ formed in the melt[72b].) However, an iron electrode (fluoroborate saturated with FeF_2 and an immersed iron wire) gave good potentials which exhibited Nernstian behavior. The same electrode configuration was used in molten LiF–NaF–KF (46.5–11.5–42 mole %) and LiF–BeF_2–ZrF_4 (65.6–29.4–5.0 mole %); each melt was present in a LaF_3 electrode saturated with NiF_2.[72a] Again the results exhibited Nernstian behavior.

Ross and Bamberger[73] modified slightly the design of the electrode to facilitate filling the cavity, to ensure contact at any time between the melt

---

* Optovac Inc., North Brookfield, Massachusetts.

in the cavity and the metal wire, and to prevent any melt creeping up the interior walls from reaching and reacting with the BN insulator. This was accomplished by using a vertically sliding wire through a capillary and by preventing direct contact between the BN and the LaF_3 crystal by means of a thin nickel cup. The electrode thus modified is shown in Fig. 22.

Romberger et al.[74] used a $LiF–BeF_2$ concentration cell with transference to measure the liquidus of this system. It consisted of a nickel fritted tube, containing ~2 cm³ of one composition of melt, surrounded by 50–200 cm³ of melt of another composition contained in a molybdenum container. Short beryllium rods attached to nickel rods were used as electrodes in

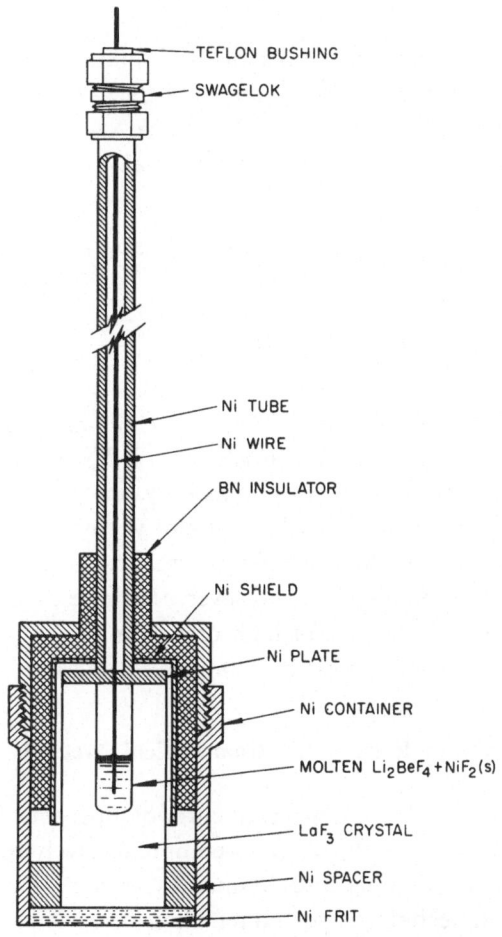

Fig. 22. LaF_3-membrane reference electrode for molten fluorides.

both compartments. The concentration cell can be represented as

$$
\text{Be} \left| \begin{array}{c|c} \text{LiF} & \text{LiF} \\ \text{BeF}_2 & \text{BeF}_2 \\ \text{I} & \text{II} \end{array} \right| \text{Be}
\tag{25}
$$

The emf E of cell (25) at constant temperature T can be written as

$$
2FE = \int_{\text{I}}^{\text{II}} (1 - t_{\text{Be}})\, d\mu_{\text{BeF}_2} - 2t_{\text{Li}}\, d\mu_{\text{LiF}}
$$

$$
= \int_{\text{I}}^{\text{II}} t_{\text{Li}} \left(\frac{1+X}{1-X} \right) d\mu_{\text{BeF}_2} = - \int_{\text{I}}^{\text{II}} t_{\text{Li}} \left(\frac{1+X}{X} \right) d\mu_{\text{LiF}}
\tag{26}
$$

where t_{Li} is the transference number of the lithium ion relative to that of F^-, X is the mole fraction of BeF_2, and μ is the chemical potential. If the composition of the reference half cell (I) is held constant while that of the bulk compartment (II) is varied, the isothermal change of emf with composition is given by

$$
2F \left(\frac{\partial E}{\partial X} \right)_T = t_{\text{Li}} \left(\frac{1+X}{1-X} \right) \left(\frac{\partial \mu_{\text{BeF}_2}}{\partial X} \right)_T
\tag{27a}
$$

$$
= -t_{\text{Li}} \left(\frac{1+X}{X} \right) \left(\frac{\partial \mu_{\text{LiF}}}{\partial X} \right)_T
\tag{27b}
$$

It can be seen from Eqs. (27a) and (27b) that if the melt composition in the bulk compartment (II) is being enriched in LiF, the emf will decrease continuously with composition as long as only one phase is present. When saturation of LiF is reached, the slope of emf versus composition becomes zero. For further discussions we refer the reader to the original publication.[74] It should be pointed out, however, that due to the sharp changes in slope this method is capable of high precision and can be used in other halide melts with similar success if suitable electrodes are available.

7.3. Voltammetric and Related Electroanalytical Methods

The application of voltammetry, chronoamperometry, and chrono-potentiometry to molten salts has been comprehensively reviewed by Fung and Mamantov.[75] Manning and Mamantov deserve much of the credit for applying these techniques to molten fluorides, which until 1962 were among the least studied media. As we have discussed previously, only recently have suitable reference electrodes been developed for molten fluorides. This is probably the reason that numerous applications of voltammetry to molten fluorides were developed using quasi-reference electrodes (QRE) of inert materials. The possibility of use of such electrodes results from the fact that only a negligible amount of the cell current flows through the electrode, and thus it does not become polarized and remains at a virtually constant potential although the origin of the potential may not be known. Although the half-wave potentials relative to the QRE are fairly reproducible when the same three-electrode system is used, these potentials are arbitrary and have no thermodynamic significance.

One of the major factors in the presently wider use of electroanalytical techniques is obviously the breakthroughs in the electronic industry which have produced more sensitive instruments with lower noise levels and faster scan rates. Another factor is the feasibility of using these methods for in-line analysis. One of the first applications of voltammetry to molten fluorides was a study of the behavior of Fe^{2+} in LiF–NaF–KF (46.5–11.5–42 mole %) by Manning.[76]

The behavior of pyrolytic graphite (PG) as an indicator electrode has been studied in various fluoride melts.[77,78] It was found that pyrolytic graphite electrodes can be used with satisfactory results. With no shielding, it is convenient to use small-diameter electrodes, i.e., $\frac{1}{8}$ in., to minimize the area in contact with the melt and thus decrease the creeping of melt by surface tension. The details for the construction of pyrolytic graphite electrodes, unshielded and shielded, can be found in Refs. (77) and (79), respectively. Typical dimensions of a shielded PG electrode are shown in Fig. 23.

Metals such as platinum, platinum–10% rhodium, molybdenum, and tungsten were also tested as indicator electrodes in U^{4+}-containing melts.[78] The voltammograms obtained with molybdenum and tungsten electrodes were less clearly defined. Electrodes of platinum and platinum–10% rhodium behaved as expected from theoretical considerations with scan rates up to

Fig. 23. Pyrolytic graphite indicator electrode for electroanalyses of molten fluorides. (Reprinted from Ref. 77, p. 104, by courtesy of Elsevier Sequoia SA.)

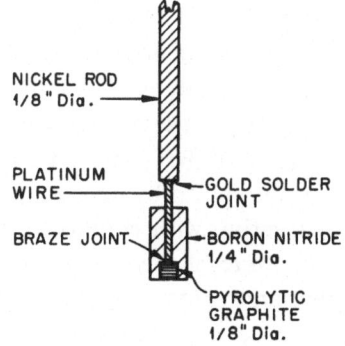

NICKEL ROD
1/8" Dia.

PLATINUM WIRE

BRAZE JOINT

GOLD SOLDER JOINT

BORON NITRIDE
1/4" Dia.

PYROLYTIC GRAPHITE
1/8" Dia.

0.3 V/sec; i.e., the peak current was proportional to the square root of the voltage scan rate. At higher scan rates the proportionality did not hold, indicating weak adsorption of U^{4+} on the electrode surface.[80]

The electroreduction of Bi^{3+} in molten $LiF-BeF_2-ZrF_4$ has recently been studied by Hammond and Manning[81] using voltammetry and chronopotentiometry. Since Bi^{3+} is more easily reduced than Fe^{2+}, Ni^{2+}, Cr^{3+}, and bismuth tends to alloy with metals, it required the testing of additional materials as indicator electrodes. The results of the study indicated that PG and iridium were both sufficiently inert and gave good reproducibility; i.e., we would expect, based on these results, that iridium will be increasingly used as an electrode in the future.

The electrochemical reduction of Ti^{4+} to Ti^{3+} in molten $LiF-NaF-KF$ (46.5–11.5–42.0 mole %) was studied by voltammetry, chronoamperometry, and chronopotentiometry by Clayton et al.[82] They found that further reduction of Ti^{3+} to Ti° involves the formation of an alloy when using a platinum electrode. The electrochemical oxidation of Ti^{3+} to Ti^{4+} in $LiF-BeF_2-ZrF_4$ was also studied by the authors using the same techniques. It was found that the electrode reaction was reversible.[82]

Interesting work has also been initiated on the voltammetry of molten fluoroborates.[83] The reduction characteristics of Cr^{3+} in $NaBF_4-NaF$ were studied using a platinum QRE and platinum and palladium working electrodes. Voltammetric studies of protonic species in $NaBF_4-NaF$ using evacuated palladium or silver–palladium electrodes have yielded preliminary information on the reduction of protons which were identified by the resultant hydrogen diffusing into the electrode.[83]

An $Fe^{2+}(NaBF_4)/Fe^\circ$ reference electrode, together with platinum and pyrolytic graphite electrodes, has been used to determine the electrochemical reduction of Ti^{4+} to Ti^{3+} in molten $NaBF_4$ by voltammetry, chronopotentiometry, and chronoamperometry.[84]

The demonstration that U^{4+}/U^{3+} ratios can be determined by voltammetry when the concentration of U^{4+} is much larger than that of U^{3+},[85] a real situation in a molten salt reactor, led to the application of voltammetry to in-line analysis of a molten fluoride mixture. A three-platinum-electrode system was inserted into a Hastelloy N thermal convection loop containing molten $LiF-BeF_2-ZrF_4-UF_4$ (65.4–29.1–5.0–0.5 mole %). The voltammeter was automated by interfacing with a PDP-8I computer.[86] Initial difficulties in locating with confidence the half-wave potential of the U^{4+} reduction wave were blamed on the presence of solids floating on the surface of the melt and contacting the electrodes. In order to eliminate this problem, the electrode assembly was surrounded by a $\frac{1}{2}$-in. nickel tube open at the bottom and extending below the surface of the melt. The tube could be emptied by flushing with helium. The compartment is kept filled with helium when no measurements are in progress.

After the system is started, the analyses are performed completely unattended through computer control. At the start of the analyses a signal from a logic circuit, added to the computer, starts a timer which causes the pressure in the electrode assembly to be released, and salt enters the electrode compartment. The computer then activates the voltammeter to perform five determinations of U^{3+} every hour and print out the results plus some diagnostic information on a Teletype after each analysis. At the end of the run, the average U^{3+} concentration and standard deviation are calculated and printed out. When the run is completed, the timer shuts off, causing the salt to be flushed from the electrode assembly. The system ran for 3600 hr, although not always continuously, giving satisfactory results.[87] The U^{3+}/U_{total} ratio varied from 0.02 to 0.30%, and a standard deviation of 2.1% was obtained for the results of hourly analyses made over a 72-hr period.[86]

The electrochemistry of U^{4+} in LiF–NaF–KF has been recently examined by Clayton et al.[88] using platinum indicator and counter electrodes and a LaF_3-membrane electrode saturated with NiF_2 as reference. The results obtained indicate that the reduction to U^{3+} occurs as a step complicated by disproportionation of U^{3+}; a second step involves the formation of uranium metal. Using the same solvent, the authors also studied the reduction of Th^{4+} and found that it is reversibly reduced to metal which then alloys with the nickel working electrode.

7.4. Electrical Conductivity Measurements of Molten Fluorides

Fluoride melts exhibit high ionic conductivity. Robbins and Braunstein[89] made conductivity measurements using silica cells, since it was estimated that the contribution of the small concentration of O^{2-} and $SiF_6^{=}$ generated by corrosion was negligible in the temperature and composition ranges used. The immersion conductance cell employed is shown in Fig. 24 and requires no explanation. To maintain a constant and controlled temperature, the cell was held in a molten nitrate bath contained in silica to permit observation of the system. The instrumentation used for measuring the electrical conductivity in molten fluorides has been comprehensively reviewed by Robbins.[90]

The cell constant was determined with an aqueous KCl solution at 25°C and with molten KNO_3 at temperatures up to 530°C, before and after

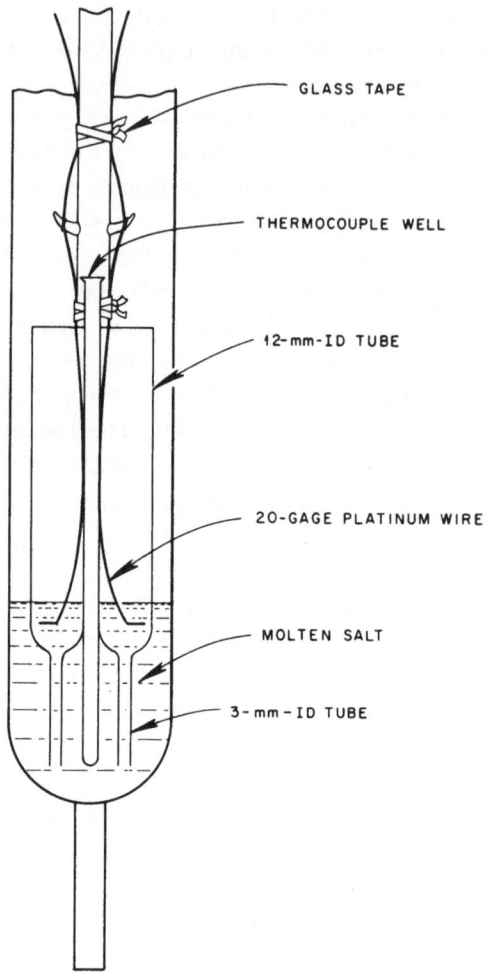

Fig. 24. Silica immersion conductance cell for
molten salts, including fluorides. (Reprinted from
Ref. 89, p. 456, by courtesy of Marcel Dekker,
Inc.)

the measurement of LiF–BeF$_2$. The concentration range studied varied
between 38 and 52 mole % BeF$_2$, and the temperature range was varied
between 400 and 530°C (the upper temperature being set by appreciable
decomposition of the nitrate bath).

After 4 days of contact with the molten fluorides, no attack of the
silica was evident below the surface of the melt. Some corrosion, however,
could be detected at the gas–liquid interface, but it had no effect on the

measurements since it was above the conduction path. It should be remembered, however, that the nature and concentration of the corrosion products formed from containing molten fluorides in silica depend on the composition of the melt and on the temperature. Thus, before using silica in conductivity cells for molten fluorides it is recommended to estimate, when possible, the expected concentrations of O^{2-}, $SiO_3^=$, $SiF_6^=$, etc. in order to predict their contribution to the uncertainty of the measurements.

8. MISCELLANEOUS MEASUREMENTS

Under this heading we have included methods for measurements which, although important, are less frequently reported in the literature. The methods described involve those used for determining equilibrium phase diagrams and measurements of viscosities, densities, and surface tension of molten fluorides.

8.1. Determination of Equilibrium Phase Diagrams of Fluoride Systems

The determination of equilibrium phase diagrams of fluoride systems is not essentially different from that of other systems; in fact, the same techniques are used, even with some advantage, due to the generally low vapor pressure of the fluorides. Thus, examination in hot stage microscopes, differential thermal analysis, and thermal-gradient quenching have been extensively used. These have been reviewed by Porter.[91] The main difference in the application of these methods to molten fluorides is only the choice of materials, namely nickel, used for their containment. The principal method used by Thoma and associates at ORNL for the determination of a large number of fluoride phase diagrams[92] is thermal-gradient quenching combined with petrographic examination and x-ray diffraction analysis of the samples. The thermal-gradient quenching method consists of loading the sample of finely ground fluorides, preferably in a dry-box, into a previously hydrogen-fired 0.10 in. O.D. (thin-walled 0.01 in.) nickel tube ~17 cm long closed at one end by welding. The open end is connected to a vacuum line through a suitable valve, and the tube is evacuated at 100°C to remove adsorbed moisture. The upper end is then tightly closed by crimping and is subsequently welded. If, after evacuation, the tube was filled with helium, the gas-tightness of the welds could be checked with a leak detector. The tube is then crimped into $\frac{1}{2}$-in. sections using a specially designed tool.[93] Several tubes (three to four) are held for different lengths of time in a

Marshall furnace provided either with 18 thermocouples every $\frac{1}{2}$ inch or with a thermocouple that travels the length of the tubes at a constant rate. The tubes can be separately quenched in oil at different intervals. During heating, a small stream of helium or argon is passed through the furnace to arrest the oxidation of the nickel surface. Examination of each section of the quenched tube then follows. A small amount of sample (10–20 mg) is sufficient for analysis with the petrographic microscope and by x-ray diffraction. Further details of the equipment used in this technique can be found in Ref. 93. Barton *et al.*[94] have used the thermal-gradient quenching technique to determine phase diagrams of PuF_3-containing mixtures. The furnace was mounted on top of a glove box. A metal container welded to the glove box isolated the samples from the outside, preventing contamination of the environment in case of leakage of the tubes. The quenching and subsequent handling of the samples was performed in glove boxes. One was specially modified to permit the eyepiece of the petrographic microscope to extend outside the glass plate by means of a rubber bellows.[31]

A high-temperature centrifuge (Fig. 25) developed by Friedman[95] has proven to be very useful for studying liquid–liquid immiscibilities in liquid fluoride mixtures as well as for separating solid phases equilibrated in molten fluorides. It consists of a stainless steel shaft that passes through the center of a tubular furnace. Provisions are made on the shaft for attaching as many as four samples. The furnace can be raised or lowered by means of counterweights for handling the samples. The rod is supported by ball bearings at both ends and by a graphite bushing in the center where it emerges from the top of the furnace. The rod is rotated by a $\frac{1}{12}$ hp motor by means of a belt and pulley arrangement. The sample holders (Fig. 25) are fabricated from nickel tubing (for fluoride melts) with a small-diameter side tube for loading the sample. On top of the container, a 310 stainless steel plate (with a hole) is welded to provide high-temperature strength at the coupling to the centrifuge rod. The tubes are filled and handled in a way similar to that described for the thermal-gradient quenching experiments.

Electromotive-force measurements for determining phase diagrams for systems other than fluorides have been used for more than 15 years,[91] but only recently have Romberger *et al.*[74] applied this technique to molten fluorides, namely, a redetermination of the phase diagram of the system $LiF–BeF_2$. The method as applied is extremely sensitive, in fact so much so that they found that the compound Li_2BeF_4 melts congruently at 459.1°C rather than incongruently at 458°C as believed earlier. Thus, this finding altered substantially the character of the $LiF–BeF_2$ phase diagram, since

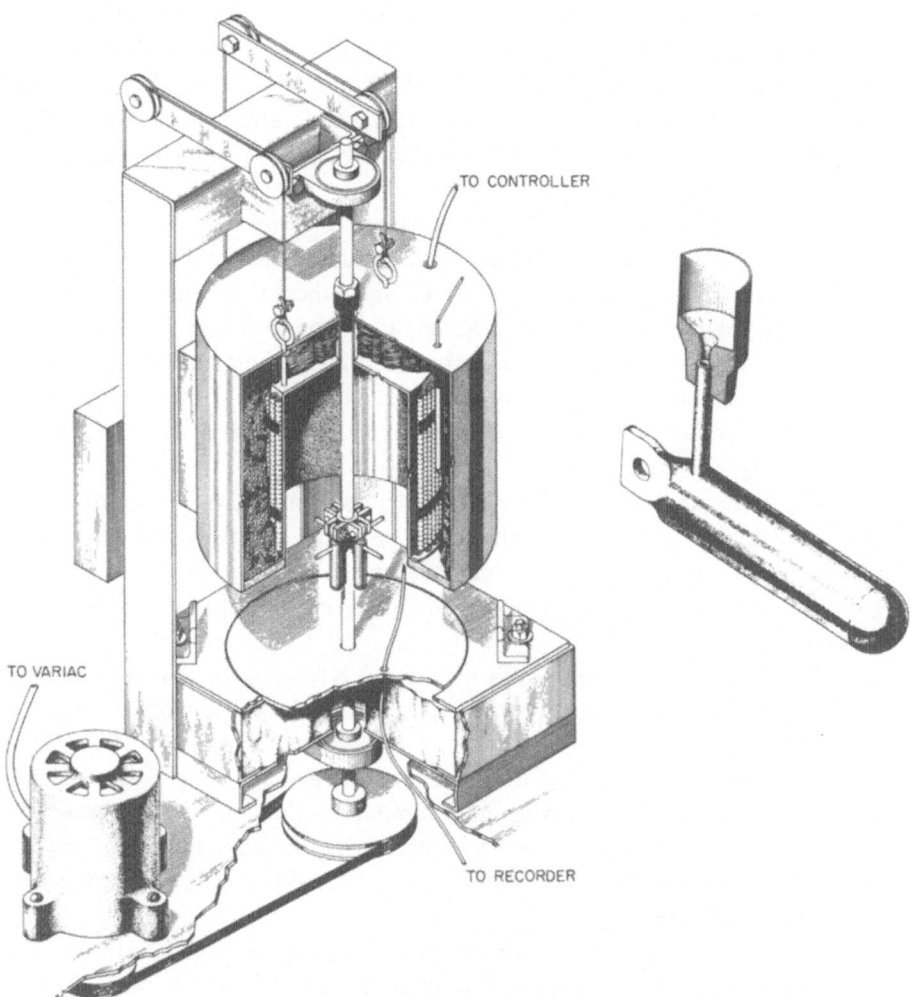

Fig. 25. High-temperature centrifuge and sample holder with loading funnel. (Reprinted from Ref. 95, pp. 454–455, by courtesy of The Institute of Physics.)

it indicated that the system has two eutectic invariant points rather than one peritectic and one eutectic.

8.2. Viscosity and Density Measurements of Molten Fluorides

The viscosities of mixtures of alkali fluorides or fluoroborates in the temperature range 500–700°C are quite low; by visual observation they can be approximated to that of water at room temperature. However, the presence of BeF_2 changes this behavior. The analogy between fluoroberyllates and silicates alluded to in the Introduction is very well reflected by the viscosities of fluoride mixtures containing variable amounts of BeF_2. The viscosity of pure BeF_2 at $\sim 600°C$ is extremely high,[20] $\sim 10^6$ P, and drops sharply with addition of alkali fluorides.

Moynihan and Cantor[20] have measured the viscosity of BeF_2 in the temperature range 574 to 980°C. The measurements were made using Brookfield Synchro-Lectric viscosimeters. Since these viscosimeters measure the torque required to rotate a cylinder, the apparatus was not leak-tight (a suitable seal requires additional torque to rotate a shaft) and the measurements were performed in a tall dry-box filled with helium. Two viscosimeters were used, a model LVT for the range 10 to 10^4 P, and a model HBF5X for the range 10^4 to 10^6 P.

The BeF_2 was contained in a nickel vessel provided with a thermocouple well, and the vessel was thermally insulated at both ends with lavite blocks. The rotating spindle was made of Inconel and was connected to the rotating shaft of the viscosimeter by means of a loose-fitting hook to eliminate eccentricity.

The viscosimeters were calibrated with NBS oils in a thermostated water bath. The procedure worked well, and the precision of the results obtained was estimated at $\pm 3\%$.

The viscosities of BeF_2–LiF melts were measured using a Brookfield LVT viscosimeter in an arrangement similar to that described above.[96] Only the sizes of the Inconel spindles were varied since the range of viscosities was significantly lower; at 600°C, 10^{-1} P for 40 mole % BeF_2 to 10^4 P for ~ 97 mole % BeF_2.

The densities of molten fluorides can be measured using several methods. The classical Archimedean method measures the weight changes of a known mass and volume when immersed in a fluid. The measurements of the volume of a known weight of melt can be accomplished with a suitable pycnometer. The Archimedean method was used to determine the densities of BeF_2–LiF mixtures,[96] liquid ThF_4,[97] and of LiF–ThF_4[98] mixtures. This method is particularly suited to melts which have low or negligible vapor pressures. It employs a platinum bob of bicone form, which can be immersed in the melt, suspended by a thin platinum wire from a balance. The melt is kept in a nickel container located in a tall dry-box that accommodates the furnace and the balance (Fig. 26). The volume of the bicone is determined by weighing it in air and in water. The suspension wire is thin to lessen the effect of surface tension. The bicone is immersed into the

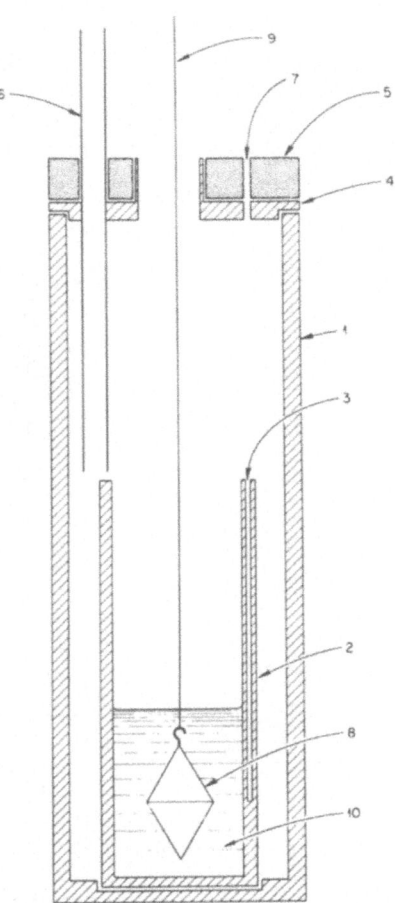

Fig. 26. Arrangement for density determinations by the Archimedian method. (1) Ni outer vessel: height 20 cm; O.D. 7.3 cm; (2) Ni crucible: height 12 cm; O.D. 4.8 cm; (3) thermocouple well; (4) nickel cover plate; (5) Lavite cover plate; (6) argon inlet; (7) thermocouple opening; (8) platinum density body; (9) platinum–10% rhodium suspension wire; (10) melt: depth in crucible approximately 5 cm. (Reprinted from Ref. 96, p. 2877, by courtesy of the American Institute of Physics.)

melt at least 0.5 cm below its surface. It is recommended to dip and remove the bob repeatedly to remove adherent bubbles before allowing for thermal equilibrium and weighing.

The dilatometric method is better suited for melts such as fluoroborates that have significant vapor pressure.[99] It consists[100] of a nickel container similar in shape to a volumetric flask where the melt level is determined electrically with a sharp-pointed metal probe, as shown in Fig. 27. Electrical insulation and leak-tight conditions are obtained by means of a Teflon bushing used with a Swagelok fitting. It is recommended that the neck of the container be constructed of high-quality tubing of constant internal diameter.

The vessel is calibrated with water at room temperature before and after the measurements. Measurements are made on known amounts of

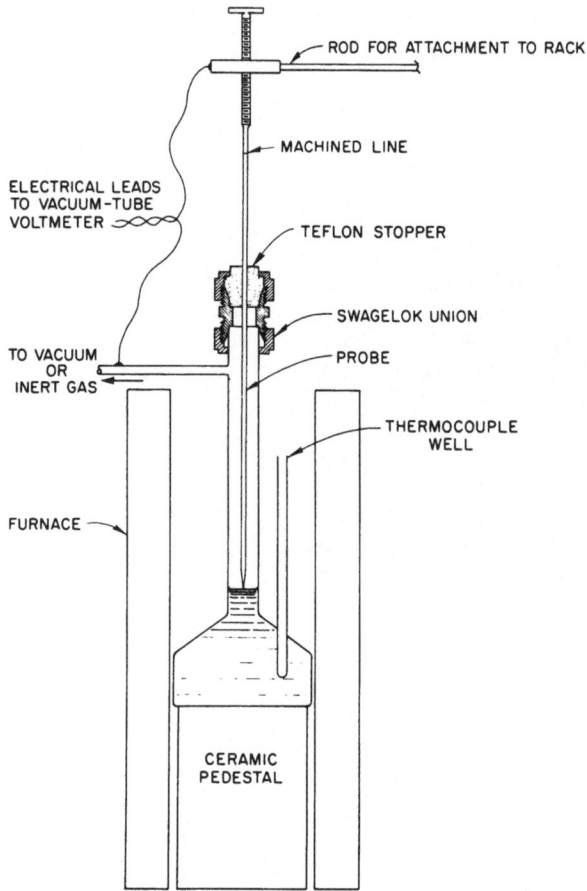

Fig. 27. Apparatus for measuring volumes (densities) of volatile liquids at elevated temperatures. The part of the probe that is above the Teflon stopper is longer, in scale, than indicated in the figure. (Reprinted from Ref. 100, p. 967, by courtesy of the American Institute of Physics.)

the components which are kept under argon pressures of 5–6 atm after melting. This prevents in-leakage of air and decreases the size of any gas bubbles in the melt, but has no significant effect on the volume of the liquid due to its generally low compressibility. After the measurements with melts of high vapor pressure, it is convenient to reweigh the sample to check for possible losses. It is also convenient to fill the container, leaving only a small gas space in order to lessen the dissociation of unstable samples and thus minimize changes of composition.

8.3. Determination of the Surface Tension of Molten Fluorides

The wetting behavior of molten fluorides on graphite, nickel, copper, and other metals was studied using the sessile drop method.[101] Measurements of the contact angles permitted the estimation of surface tension using established calculational procedures. The measurements were performed in a horizontal silica tube with a flat optical window at the end. The other end was connected to service lines through a tapered joint header. The silica tube was electrically heated with nichrome wiring. The graphite or metal plate specimens were located on a silica pedestal which was surrounded by titanium or uranium turnings to ensure the absence of traces of moisture or air. The solid fluoride mixture on the plate was loaded at room temperature either in air or in helium, according to the hygroscopicity of the fluorides. It was pushed into position with a rod moving through a fitting provided with an "O" ring. Evacuation of the silica chamber, filling with different gases, etc. was handled by means of a gas manifold in the

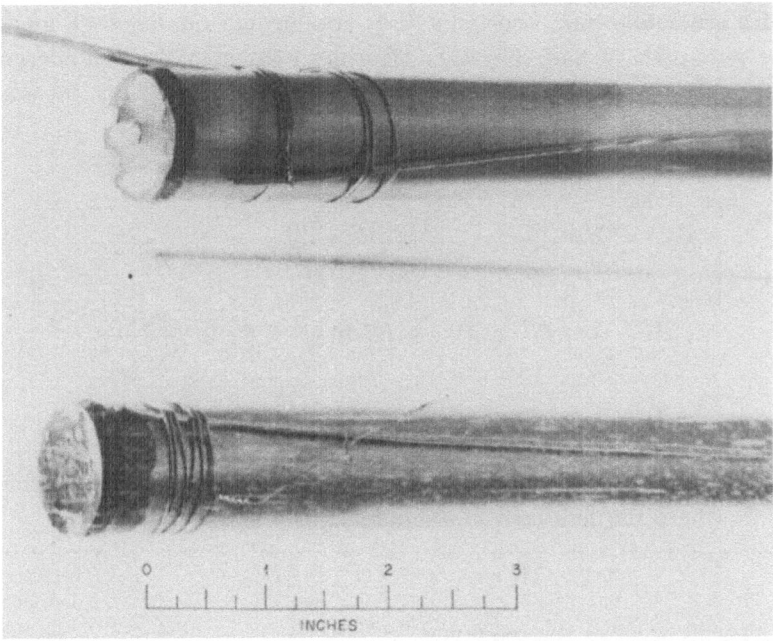

Fig. 28. Appearance of Li_2BeF_4 in contact with a nickel frit only (top) showing nonwetting behavior and with a nickel frit and a reductant (bottom) showing wetting behavior.

customary manner. The contact angle was measured from photographs of drops taken with a camera through the flat window. The sessile drop was also observed and partially measured with a cathetometer provided with a protractor. The contact angles of several BeF$_2$–LiF mixtures, LiF–NaF, LiF–NaF–KF, and some pure fluorides were measured in several gaseous environments. It was found that up to 400 ppm oxygen in moisture-free helium had no effect on the contact angles; however, as little as 10 ppm H$_2$O changed it drastically.

Young et al.[102] observed qualitatively, in the course of work with windowless cells and with filtration through nickel frits, that the presence of reductants also changed drastically the wetting behavior of fluoride melts. The presence of a reductant changed them from nonwetting to wetting, as shown in Fig. 28.

9. PRESENTATION OF DATA OF EQUILIBRIA IN MOLTEN HALIDES

The data most frequently obtained from equilibrations using molten halides are equilibrium constants K or equilibrium quotients Q, and electrode potentials of half cells, $E°$. They are related to the free energies of reaction ΔG_r by the well-known equations shown below. Thus, for a hypothetical reaction

$$\text{MF}_{n+1} + \tfrac{1}{2}\text{H}_{2(g)} \rightleftarrows \text{MF}_{n(d)} + \text{HF}_{(g)} \tag{28}$$

$$K = \frac{a_{\text{MF}_n} f_{\text{HF}}}{a_{\text{MF}_{n+1}} f_{\text{H}_2}^{1/2}} = \frac{X_{\text{MF}_n} \gamma_{\text{MF}_n} P_{\text{HF}}}{X_{\text{MF}_{n+1}} \gamma_{\text{MF}_{n+1}} P_{\text{H}_2}^{1/2}} = Q\,\frac{\gamma_{\text{MF}_n}}{\gamma_{\text{MF}_{n+1}}} \tag{29}$$

$$\Delta G_r = -RT \ln K = -RT \ln Q - RT \ln \frac{\gamma_{\text{MF}_n}}{\gamma_{\text{MF}_{n+1}}} \tag{30}$$

and

$$\Delta G° = -nFE° \tag{31}$$

As stated in Sec. 2, we have chosen to follow the notation established by Baes,[3] where the half-cell

$$\text{HF}_{(g)} + e^- \rightarrow \text{F}^- + \tfrac{1}{2}\text{H}_{2(g)} \tag{32}$$

is arbitrarily used as the reference, with $E° = 0$.

The dependence of K on temperature is conveniently represented by the expression

$$\log K = a + b(10^3/T) \tag{33}$$

but it should be borne in mind that the parameters a and b, generally obtained from plots of $\log K$ versus $1/T$, are strongly covariant and therefore can introduce large errors when used to calculate ΔH and/or ΔS.

Obviously, one way of presenting a collection of equilibrium data is by tabulation of either ΔG_r, $\log K$, or $\log Q$. However, a graphical presentation, although not as accurate, is often more convenient.

Ellingham type graphs[103] representing ΔG_r versus temperature for reactions such as

$$MF_{n(c)} \rightleftarrows M_{(c)} + \frac{n}{2} F_{2(g)} \tag{34}$$

$$MF_{n(c)} + \frac{n}{2} H_{2(g)} \rightleftarrows M_{(c)} + n HF_{(g)} \tag{35}$$

$$MO_{\frac{n}{2}(c)} \rightleftarrows M_{(c)} + \frac{n}{4} O_{2(g)} \tag{36}$$

$$MO_{n/2(c)} + \frac{n}{2} CO_{(g)} \rightleftarrows M_{(c)} + \frac{n}{2} CO_{2(g)} \tag{37}$$

can be constructed using as ordinates $P_{F_2}^{n/2}$, $P_{HF}/P_{H_2}^{1/2}$, $P_{O_2}^{n/4}$, and P_{CO_2}/P_{CO}, respectively. Although useful, their application to molten halide systems is limited because the free energies of reaction are only equivalent to the free energies of formation of the solid compounds, and because they cannot be made to show simultaneously all the occurring reactions.

The Pourbaix diagrams which have been extensively used for representing corrosion reactions in aqueous systems[104] seem well suited for representing reactions occurring in molten halide systems. The fruitful adaptation of Pourbaix diagrams to chloride systems by Harder et al.[105] prompted us to represent the various equilibria involving molten fluorides using the same method. It is in this context that we will describe briefly the construction and use of Pourbaix diagrams. These diagrams, which may be called "chemical phase diagrams," represent the various equilibria by boundary lines in a system of coordinates of redox potential versus oxide concentration. The areas limited by the various equilibria are proportional to the stability (formation free energy) of the species they represent.

Bockris[106] has recently criticized the use of Pourbaix diagrams for the teaching of corrosion because they do not take into account localized effects and neglect the importance of slow reaction rates. However, at the high temperatures involved in molten halide work, the above considerations become of much less importance. Before getting into a detailed description of the algebra involved, we would like to remind the reader of the obvious fact that Pourbaix diagrams reflect only the knowledge and accuracy available for the equilibria to be represented; thus they are useless for predicting the formation of species or compounds for which no information exists.

Because the reference half-cell depicted in Eq. (32) is a convenient way to define redox potential, we have chosen $\log(P_{HF}/P_{H_2}^{1/2})$ for the ordinate. The abscissa represents oxide concentration in terms of $\log X_{O^{2-}}$. However, the reader is not restricted to these choices, i.e., $\log P_{F_2}$, $\log(X_{U^{4+}}/X_{U^{3+}})$, or any other redox couple may be used. A hypothetical example is shown in Fig. 29, where a metal M°, also present in the oxidation states M^{y+} and M^{z+} $(z > y)$, is in equilibrium with its oxides $MO_{y/2}$ and $MO_{z/2}$ in molten fluorides at a given temperature. Each boundary line in Fig. 29 is identified by a circled roman numeral which refers to the equilibrium it represents.

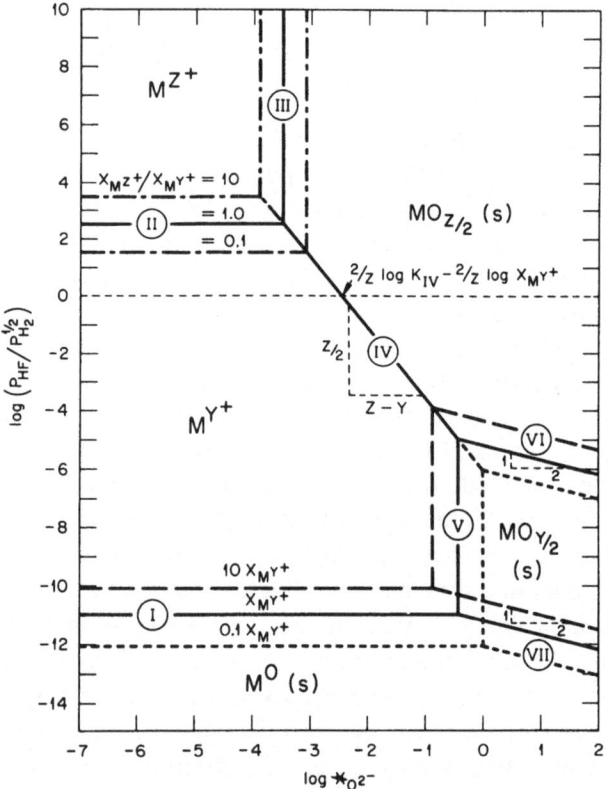

Fig. 29. Hypothetical Pourbaix diagram for several species of a solute M at various concentrations in equilibrium with a molten fluoride at a constant temperature.

I. Boundary $M°/M^{y+}$

It represents the equilibrium

$$M^{y+} + \frac{y}{2} H_{2(g)} + yF^- \rightleftarrows M^°_{(c)} + yHF_{(g)} \tag{I}$$

where

$$Q_I = \frac{P^y_{HF}}{P^{y/2}_{H_2}} \bigg/ X_{M^{y+}} \tag{Ia}$$

from (Ia)

$$\log\left(\frac{P_{HF}}{P^{1/2}_{H_2}}\right) = \frac{1}{y} \log Q_I + \frac{1}{y} \log X_{M^{y+}} \tag{Ib}$$

Equation (Ib) thus shows that the only independent variable is the concentration of the species M^{y+}.

II. Boundary M^{y+}/M^{z+}

$$M^{z+} + \left(\frac{z-y}{2}\right)H_{2(g)} + (z-y)F^- \rightleftarrows M^{y+} + (z-y)HF_{(g)} \tag{II}$$

$$Q_{II} = \frac{X_{M^{y+}}}{X_{M^{z+}}} \left(\frac{P_{HF}}{P^{1/2}_{H_2}}\right)^{(z-y)} \tag{IIa}$$

from (IIa)

$$\log\left(\frac{P_{HF}}{P^{1/2}_{H_2}}\right) = \frac{1}{z-y} \log Q_{II} + \frac{1}{z-y} \log\left(\frac{X_{M^{z+}}}{X_{M^{y+}}}\right) \tag{IIb}$$

III. Boundary $M^{z+}/MO_{z/2}$

$$MO_{z/2(c)} \rightleftarrows M^{z+} + \frac{z}{2} O^{2-} \tag{III}$$

$$Q_{III} = X_{M^{z+}} X^{z/2}_{O^{2-}} \tag{IIIa}$$

from (IIIa)

$$\log X_{O^{2-}} = \frac{2}{z} \log Q_{III} - \frac{2}{z} \log X_{M^{z+}} \tag{IIIb}$$

IV. Boundary $M^{y+}/MO_{z/2}$

$$MO_{z/2(c)} + \left(\frac{z-y}{2}\right)H_{2(g)} + (z-y)F^- \rightleftarrows M^{y+} + (z-y)HF_{(g)} + \frac{z}{2}O^{2-} \tag{IV}$$

$$Q_{IV} = Q_{III}Q_{II} = X_{M^{y+}}X_{O^{2-}}^{z/2}\left(\frac{P_{HF}}{P_{H_2}^{1/2}}\right)^{(z-y)} \tag{IVa}$$

from (IVa)

$$\log\left(\frac{P_{HF}}{P_{H_2}^{1/2}}\right) = \frac{1}{z-y}\log Q_{IV} - \frac{1}{z-y}\log X_{M^{y+}} - \frac{z}{2(z-y)}\log X_{O^{2-}} \tag{IVb}$$

It can be seen from (IVb) that the slope of the line is always given by $-z/2(z-y)$; thus it is not necessary to know explicitly the value of Q_{IV}. A line with the above-indicated slope can be drawn from the intersection of the lines representing (IIb) and (IIIb).

V. Boundary $M^{y+}/MO_{y/2}$

This case is similar to (III) because it represents the solubility product of another oxide.

VI. Boundary $MO_{y/2}/MO_{z/2}$

Here both oxides remain as pure phases.

$$MO_{z/2(c)} + \left(\frac{z-y}{2}\right)H_{2(g)} + (z-y)F^- \rightleftarrows MO_{y/2(c)} + (z-y)HF + \left(\frac{z-y}{2}\right)O^{2-} \tag{VI}$$

$$Q_{VI} = X_{O^{2-}}^{(z-y)/2}\left(\frac{P_{HF}}{P_{H_2}^{1/2}}\right)^{(z-y)} \tag{VIa}$$

from (VIa)

$$\log\left(\frac{P_{HF}}{P_{H_2}^{1/2}}\right) = \frac{1}{z-y}\log Q_{VI} - \tfrac{1}{2}\log X_{O^{2-}} \tag{VIb}$$

The slope of (VIb) is always $-\tfrac{1}{2}$.

VII. Boundary $M°/MO_{y/2}$

$$MO_{y/2(c)} + \frac{y}{2}H_{2(g)} + yF^- \rightleftarrows M°_{(c)} + yHF_{(g)} + \frac{y}{2}O^{2-} \tag{VII}$$

$$Q_{VII} = \left(\frac{P_{HF}}{P_{H_2}^{1/2}}\right)^y X_{O^{2-}}^{y/2} \tag{VIIa}$$

from (VIIa)

$$\log\!\left(\frac{P_{HF}}{P_{H_2}^{1/2}}\right) = \frac{1}{y}\log Q_{VII} - \tfrac{1}{2}\log X_{O^{2-}} \qquad \text{(VIIb)}$$

Since the slope of (VIIb) likewise always has a slope $-\tfrac{1}{2}$, neither Q_{VI} nor Q_{VII} need to be known explicitly, and it is only necessary to draw lines with slope $-\tfrac{1}{2}$ from the intersections of (Vb) and (Ib) and from (Vb) and (IIIb).

The additional lines shown in Fig. 29 indicate the effect of the concentration of the species M^{y+} at constant temperature. Alternatively, Pourbaix diagrams can be made for a constant concentration and varying temperatures. Such a case is shown in Fig. 30 for the equilibria of plutonium

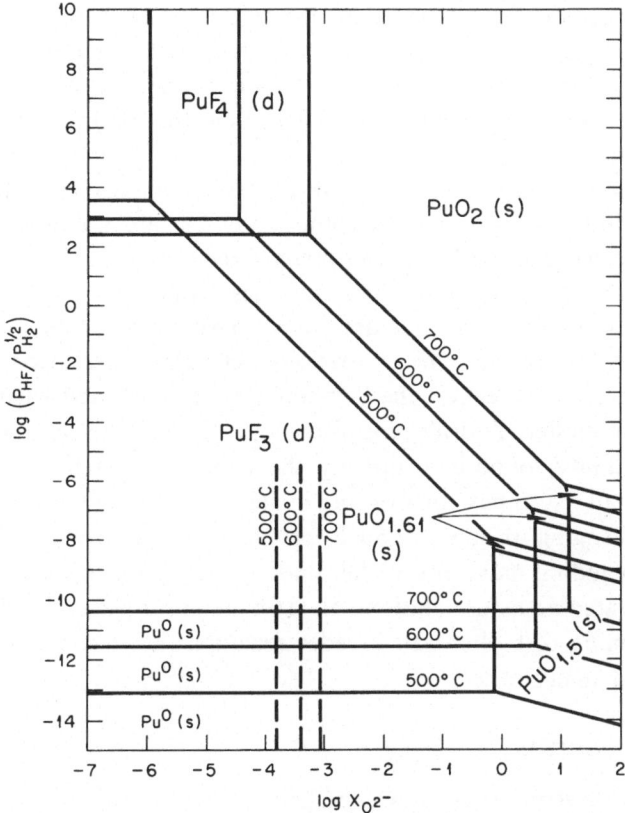

Fig. 30. Pourbaix diagram for plutonium (0.2 mole % PuF_3) in Li_2BeF_4 at various temperatures.

species in molten Li_2BeF_4. When two or more solid oxide phases are present and form solid solutions, their boundary lines become curved; however, they can be plotted using the above methodology.[107]

When several Pourbaix diagrams of different metals are drawn on transparent sheets, all using the same coordinates, they can be overlaid. This additional useful feature permits a visual estimate of which metal will be more readily oxidized or reduced, which oxide will precipitate first, etc.

10. CONCLUSIONS

Although various technological applications for molten fluoride systems have been in use for many years (e.g., in the aluminum industry), it is expected that through additional research performed by groups with different interests new applications for molten fluorides will be developed in the near future. It is hoped that this collection of the many and varied techniques used in molten fluoride chemistry will convince the chemist inexperienced with these systems that their properties and behavior can be measured with only a little more effort than that required with aqueous solutions.

The vast information collected to date demonstrates that molten fluorides, although corrosive to some materials, are tractable substances that can be handled with not too much difficulty.

The toxic aspects of molten fluorides, compounded when BeF_2 is present, deserve attention but are by no means worse than many other "established" chemicals. The experimental chemist experienced with aqueous solutions might feel, at the beginning, a bit frustrated when initiating research in molten fluoride chemistry, mainly because of the often-occurring inability to observe directly the course of a reaction. However, this is a small fee to pay for working in such an exciting field that provides many challenging problems. Such problems range from theoretical to applied, including ways for better understanding the physical chemistry of concentrated aqueous solutions, the storage of off-peak electrical energy, the functioning of molten salt breeder reactors, the electroslag refining of metals, and many others.

ACKNOWLEDGMENTS

It is a pleasure to acknowledge helpful discussions with J. P. Young, D. L. Manning, and J. B. Bates concerning their work. I also express my gratitude to Professors J. Braunstein and G. Mamantov, and to

L. M. Ferris and C. F. Baes, Jr., for their many helpful suggestions concerning the manuscript of this paper. Because the experience of an individual is largely composed of that of his colleagues, I thank my colleagues at the Oak Ridge National Laboratory for sharing with me their success and "unsuccess."

11. REFERENCES

1. *Nucl. Appl. Tech.* **8**(2) (1970).
2. MSR Staff, The Development Status of Molten Salt Breeder Reactors, USAEC Rept. ORNL-4812 (1972).
3. C. F. Baes, Jr., in: Symposium on Reprocessing of Nuclear Fuels, *Nuclear Metallurgy* (P. Chiotti, ed.), Vol. 15, p. 615, CONF-690801 (1969).
4. C. F. Baes, Jr., MSR Program Semiannual Progr. Rept., Feb. 28, 1970, USAEC Rept. ORNL-4548, p. 149.
5. C. F. Baes, Jr., *J. Solid State Chem.* **1**(2):159 (1970).
6. B. F. Hitch and C. F. Baes, Jr., *Inorg. Chem.* **8**:201 (1969).
7. B. F. Hitch and C. F. Baes, Jr., Reactor Chemistry Div. Annual Progr. Rept., Dec. 31, 1966, USAEC Rept. ORNL-4076 (1967), p. 19.
8. R. P. Elliot, *Constitution of Binary Alloys*, 1st Suppl., p. 668, McGraw-Hill, New York (1965).
9. A. L. Mathews and C. F. Baes, Jr., *Inorg. Chem.* **7**:373 (1968).
10. J. H. Shaffer, Preparation and Handling of Salt Mixtures for the Molten Salt Reactor Experiment," USAEC Rept. ORNL-4616 (1971), p. 11.
11. C. E. Bamberger and C. F. Baes, Jr., *J. Am. Ceram. Soc.* **55**(11):564 (1972).
12. C. E. Bamberger, C. F. Baes, Jr., and J. P. Young, *J. Inorg. Nucl. Chem.* **30**:1979 (1968).
13a. B. C. Saunders, in: *Advances in Fluorine Chemistry* (M. Stacey, J. C. Tatlow, and A. G. Sharpe, eds.), Vol. 2, p. 183, Butterworth (1961).
13b. World Health Organization, *Fluorides and Human Health*, Geneva (1970).
14. *Beryllium: Its Industrial Hygiene Aspects* (H. E. Stockinger, ed.), Academic Press, New York (1966).
15. R. Bougon, J. Ehretsmann, J. Portier, and A. Tressaud, in: *Preparative Methods in Solid State Chemistry* (P. Hagenmuller, ed.), Academic Press, New York (1972).
16. K. A. Sense, M. J. Snyder, and J. W. Clegg, *J. Phys. Chem.* **58**:223 (1954).
17. S. Cantor, R. F. Newton, W. R. Grimes, and F. F. Blankenship, *J. Phys. Chem.* **62**:96 (1958).
18. K. A. Sense and R. W. Stone, *J. Phys. Chem.* **62**:453 (1958).
19. J. R. Hightower, Jr., and L. E. McNeese, *J. Chem. Eng. Data* **17**(3):342 (1972).
20. C. T. Moynihan and S. Cantor, *J. Chem. Phys.* **48**(1):115 (1968).
21. C. E. Vallet, H. R. Bronstein, and J. Braunstein, *J. Electrochem. Soc.* **121**:1429 (1974).
22. C. E. Bamberger, unpublished results.
23. A. L. Mathews, Oxide Chemistry and Thermodynamics of LiF–BeF$_2$ by Equilibration with Gaseous Water–Hydrogen Fluoride Mixtures, Ph.D. Thesis, University of Mississippi (1965); also issued as USAEC Rept. ORNL-TM-1129 (1965).

24. P. E. Field and J. H. Shaffer, *J. Phys. Chem.* **71**:3218 (1967).
25. G. Goldberg, A. S. Meyer, and J. C. White, *Anal. Chem.* **32**:314 (1960).
26. D. Constanzo, Anal. Chem. Div. Annual Progr. Rept., Sept. 30, 1972, USAEC Rept. ORNL-4838 (1973), p. 48.
27. K. A. Romberger and J. Braunstein, *Inorg. Chem.* **9**:1273 (1970).
28. C. E. Bamberger and C. F. Baes, Jr., *J. Nucl. Mat.* **35**:177 (1970).
29. K. A. Romberger, C. F. Baes, Jr., and H. H. Stone, *J. Inorg. Nucl. Chem.* **29**:1619 (1967).
30. C. E. Bamberger and C. F. Baes, Jr., Reactor Chem. Div. Annual Progr. Rept., Dec. 31, 1966, USAEC Rept. ORNL-4076 (1967), p. 32.
31. C. J. Barton, in: *Technique of Inorganic Chemistry* (H. B. Jonassen and A. Weissberger, eds.), Vol. 3, p. 259, Interscience Publishers, New York (1963).
32. A. R. Miller, *J. Phys. Chem.* **71**:1144 (1967).
33. K. E. Johnson and J. R. Dickinson, in: *Advances in Molten Salt Chemistry* (J. Braunstein, G. Mamantov, and G. P. Smith, eds.), Vol. 2, p. 83, Plenum Press, New York (1973).
34. J. P. Young, in: *Molten Salts, Characterization and Analysis* (G. Mamantov, ed.), p. 189, Marcel Dekker, New York (1969).
35. B. R. Sundheim and J. Greenberg, *Rev. Sci. Instr.* **27**:703 (1956).
36. D. M. Gruen, S. Freid, P. Graf, and R. L. McBeth, A/Conf. 15/P/940, 2nd United Nations International Conference on the Peaceful Uses of Atomic Energy, Geneva. Switzerland (1958).
37. C. R. Boston and G. P. Smith, *J. Phys. Chem.* **62**:409 (1958).
38. J. P. Young and J. C. White, *Anal. Chem.* **31**:1892 (1960).
39. L. M. Toth, J. P. Young, and G. P. Smith, *Anal. Chem.* **41**:683 (1969).
40. N. W. Silcox and H. M. Haendler, *J. Phys. Chem.* **64**:303 (1960).
41. J. P. Young, Proceedings of the 10th Rare Earth Research Conference, CONF-730402-P2, Vol. II, p. 912 (1973).
42. J. P. Young, in: *Molten Salts, Characterization and Analysis* (G. Mamantov, ed.). p. 192, Marcel Dekker, New York (1969).
43. J. P. Young, *Anal. Chem.* **36**:390 (1964).
44. L. M. Toth, J. P. Young, and G. P. Smith, *Anal. Chem.* **41**:683 (1969).
45. J. P. Young and J. C. White, *Anal. Chem.* **32**:199 (1960).
46. J. P. Young, Molten Salt Reactor Program Semiannual Progr. Rept., Aug. 31, 1969, USAEC Rept. ORNL-4449, p. 161.
47. J. P. Young, G. Mamantov, and F. L. Whiting, *J. Phys. Chem.* **71**:782 (1967).
48. J. P. Young and J. C. White, *Anal. Chem.* **32**:199 (1960).
49. C. E. Bamberger, R. G. Ross, C. F. Baes, Jr., and J. P. Young, *J. Inorg. Nucl. Chem.* **33**:3591 (1971).
50. G. Joos and A. Klopfer, *Z. Physik.* **138**:251 (1954).
51. L. M. Toth, USAEC Rept. ORNL-TM-2047 (1967).
52. G. G. Cocks, J. B. Schroeder, and C. M. Schwartz, Progress Relating to ANP Applications, February–April 1957, Battelle Memorial Institute Rept. BMI-1185, p. 13.
53. MSR Staff, The Development Status of Molten Salt Breeder Reactors, USAEC Rept. ORNL-4812 (1972), p. 162.
54. J. P. Devlin, P. C. Li, and R. P. J. Cooney, in: *Molten Salts, Characterization and Analysis* (G. Mamantov, ed.), p. 209, Marcel Dekker, New York (1969).

55. V. Maroni and Elton J. Cairns, in: *Molten Salts, Characterization and Analysis* (G. Mamantov, ed.), p. 231, Marcel Dekker, New York (1969).
56. R. E. Hester, in: *Advances in Molten Salt Chemistry* (J. Braunstein, G. Mamantov, and G. P. Smith, eds.), Vol. 1, p. 1, Plenum Press, New York (1971).
57. J. P. Young, J. B. Bates, and G. E. Boyd, Molten Salt Reactor Program Semiannual Progr. Rept., Feb. 29, 1972, USAEC Rept. ORNL-4782, p. 59.
58. J. B. Bates, Furnace for Infrared Emission Spectra of Molten Salts and Solids, private communication.
59. J. B. Bates and G. E. Boyd, *Appl. Spectry.* **27**:204 (1973).
60. A. S. Quist, *Appl. Spectry.* **25**:82 (1971).
61. A. S. Quist, *Appl. Spectry.* **25**:80 (1971).
62. K. W. Fung and G. Mamantov, in: *Advances in Molten Salt Chemistry* (J. Braunstein, G. Mamantov, and G. P. Smith, eds.), Vol. 2, p. 199, Plenum Press, New York (1973).
63. G. Mamantov, in: *Molten Salts, Characterization and Analysis* (G. Mamantov, ed.), p. 529, Marcel Dekker, New York (1969).
64. H. W. Jenkins, G. Mamantov, and D. L. Manning, *J. Electroanal. Chem.* **19**:385 (1968).
65. R. Winand, *Electrochim. Acta* **17**:251 (1972).
66. B. F. Hitch and C. F. Baes, Jr., *J. Inorg. Nucl. Chem.* **34**:163 (1972).
67. H. R. Bronstein and D. L. Manning, *J. Electrochem. Soc.* **119**:125 (1972).
68. T. L. Markin, in: *Electromotive Force Measurements in High-Temperature Systems* (C. B. Alcock, ed.), p. 91, American Elsevier (1968).
69. C. J. Barton, M. A. Bredig, L. O. Gilpatrick, and J. A. Fredriksen, *Inorg. Chem.* **9**:307 (1970).
70. C. E. Bamberger and J. P. Young, private communication.
71. C. E. Bamberger and C. F. Baes, Jr., Molten Salt Reactor Program Semiannual Progr. Rept., Feb. 28, 1970, USAEC Rept. ORNL-4548 (1970), p. 140.
72a. F. R. Clayton and G. Mamantov, *High-Temperature Science* **5**:358 (1973).
72b. C. E. Bamberger, B. F. Hitch, and C. F. Baes, Jr., *J. Inorg. Nucl. Chem.* **36**:543 (1974).
73. C. E. Bamberger and R. G. Ross, private communication.
74. K. A. Romberger, J. Braunstein, and R. E. Thoma, *J. Phys. Chem.* **76**:1154 (1972).
75. K. W. Fung and G. Mamantov, Electroanalytical Chemistry of Molten Salts, in: *Comprehensive Analytical Chemistry* (C. L. Wilson and D. W. Wilson, eds.), American Elsevier, pp. 305–370 (1975).
76. D. L. Manning, *J. Electroanal. Chem.* **6**:227 (1963).
77. D. L. Manning and G. Mamantov, *J. Electroanal. Chem.* **7**:102 (1964).
78. G. Mamantov and D. L. Manning, *Anal. Chem.* **38**:1494 (1966).
79. D. L. Manning and G. Mamantov, *J. Electroanal. Chem.* **17**:137 (1968).
80. G. Mamantov and D. L. Manning, *J. Electroanal. Chem.* **18**:309 (1968).
81. J. S. Hammond and D. L. Manning, *High Temp. Sci.* **5**:50 (1973).
82. F. R. Clayton, G. Mamantov, and D. L. Manning, *J. Electrochem. Soc.* **120**:1193 (1973).
83. A. S. Meyer, in: Anal. Chem. Div. Annual Progr. Rept., Sept. 30, 1972, USAEC Rept. ORNL-4838, pp. 23, 25, 26.
84. F. R. Clayton, G. Mamantov, and D. L. Manning, *J. Electrochem. Soc.* **120**:1199 (1973).

85. H. W. Jenkins, G. Mamantov, D. L. Manning, and J. P. Young, *J. Electrochem. Soc.* **116**(12):1712 (1969).
86. J. M. Dale and A. S. Meyer, Molten Salt Reactor Program Semiannual Progr. Rept., Aug. 31, 1971, USAEC Rept. ORNL-4728, p. 69.
87. J. M. Dale and A. S. Meyer, Molten Salt Reactor Program Semiannual Progr. Rept., Feb. 29, 1972, USAEC Rept. ORNL-4782, p. 77.
88. F. R. Clayton, G. Mamantov, and D. L. Manning, *J. Electrochem. Soc.* **121**:86 (1974).
89. G. D. Robbins and J. Braunstein, in: *Molten Salts, Characterization and Analysis* (G. Mamantov, ed.), p. 443, Marcel Dekker, New York (1969).
90. G. D. Robbins, *J. Electrochem. Soc.* **116**(6):813 (1969).
91. R. F. Porter, in: *Annual Review of Physical Chemistry* (H. Eyring, ed.), Vol. 10, p. 219, George Banta Co., Stanford (1960).
92. R. E. Thoma, in: *Advances in Molten Salt Chemistry* (J. Braunstein, G. Mamantov, and G. P. Smith, eds.), Vol. 3, p. 275, Plenum Press, New York (1975).
93. H. A. Friedman, G. M. Hebert, and R. E. Thoma, USAEC Rept. ORNL-3373 (1962).
94. C. J. Barton, J. D. Redman, and R. A. Strehlow, *J. Inorg. Nucl. Chem.* **20**:45 (1961).
95. H. A. Friedman, *J. Sci. Instr.* **44**:454 (1967).
96. S. Cantor, W. T. Ward, and C. T. Moynihan, *J. Chem. Phys.* **50**(7):2874 (1969).
97. S. Cantor, D. G. Hill, and W. T. Ward, *Inorg. Nucl. Chem. Letters* **2**:15 (1966).
98. D. G. Hill, S. Cantor, and W. T. Ward, *J. Inorg. Nucl. Chem.* **29**:241 (1966).
99. S. Cantor, D. P. McDermott, and L. O. Gilpatrick, *J. Chem. Phys.* **52**(9):4600 (1970).
100. S. Cantor, *Rev. Sci. Instr.* **40**:967 (1969).
101. P. J. Kreiger, S. S. Kirslis, and F. F. Blankenship, Reactor Chem. Div. Annual Progr. Rept., Jan. 31, 1964, USAEC Rept. ORNL-3591 (1964), p. 38.
102. J. P. Young, K. A. Romberger, and J. Braunstein, Molten Salt Reactor Program Semiannual Progr. Rept., Feb. 28, 1969, USAEC Rept. ORNL-4396, p. 205.
103. H. J. T. Ellingham, *J. Soc. Chem. Ind. (London)*, **63**:125 (1944).
104. M. Pourbaix, *Atlas of Electrochemical Equilibria in Aqueous Solutions*, Pergamon Press (1966).
105. B. R. Harder, G. Long, and W. P. Stanaway, in: Symposium on Reprocessing of Nuclear Fuels, *Nuclear Metallurgy* (P. Chiotti, ed.), p. 405, CONF-690801 (1969)
106. J. O'M. Bockris, *Corrosion Sci.* **11**:889 (1971).
107. C. E. Bamberger, *Energia Nucl. (Madrid)* **17**:133 (1973).

Chapter 5

THE CHEMISTRY OF THIOCYANATE MELTS

D. H. Kerridge

Department of Chemistry
The University
Southampton SO9 5NH

---

1. INTRODUCTION

1.1. Versatility of Thiocyanates

Among the numerous potentially interesting classes of molten salts, thiocyanates have certain unique features and have in recent years been the subject of an increasing number of investigations, though not hitherto the subject of a review article.

The classification of the thiocyanate ion as a pseudohalide emphasizes that it is chemically similar to the halides in a number of ways, though fortunately these similarities are combined with materially lower melting points and this makes the application of experimental techniques easier and broadens the chemical possibilities. The polyatomic anion, besides lowering the melting point, also confers a much wider chemistry, and when it acts as a reductant, a surprising variety of oxidation products result. The extensive possibilities are further illustrated by the range of thio products formed when sulfur is exchanged for oxygen, and by the linkage isomers and bridged coordination complexes formed because of the almost equal ease of electron donation to metal ions from both sulfur and nitrogen.

1.2. Melting Points and Availability

The melting points of the pure alkali metal thiocyanates have been given as KSCN 177°C,[1] NaSCN 323°C,[2] 310°C,[3] 287°C,[4] and LiSCN 237°C,[5] <185°C.[6] The only eutectic reported (NaSCN–KSCN) has been the subject of three phase-diagram studies[2,7,8] with somewhat variable eutectic melting points and compositions. This has had the result that reactions have been studied in a variety of melts. The Italian workers[9–12] have used a 26.3 mole % NaSCN melt, which is the eutectic composition found both by Piantoni et al.[7] and by Oparina and Dombrovskaya,[8] though the melting point generally quoted (129°C) is that determined by the latter workers rather than the value of 134°C found by the former group. Crowell and his associates[13–15] have used a 25 wt. % NaSCN melt (i.e., 36 mole % NaSCN) for the study of organic reactions with a melting point of 133°C, while Panzer and Schaer[16] have quoted the eutectic as 30 mole % NaSCN with a melting point of 123.5°C, which is presumably a reference to the early phase diagram of Wrecesnewsky.[2] However, Jindal and Harrington[17] gave the same eutectic composition (30 mole % NaSCN) but with a melting point of 133°C. Obviously, it would help in making comparisons if in the future the same melt composition were used by all workers, i.e., that of the most recent determinations by Piantoni et al.[7] and Oparina and Dombrovskaya,[8] though it is fortunate that the range of variation is reasonably small and there is at present no indication that melt composition plays a very significant part in determining the course of a reaction. It must be said, however, that there are very few comparable measurements made in more than one melt by the same workers, and so this assertion may have to be modified in the future.

The low melting points of the thiocyanates may be compared with those of the corresponding alkali metal chlorides [KCl 776°C, NaCl 801°C, LiCl 613°C, NaCl–KCl eutectic (50 mole % NaCl) 670°C][18] in order to appreciate the considerable easing of the experimental problems conferred by the use of molten thiocyanates.

Notwithstanding the number of workers quoted above who have used a sodium thiocyanate–potassium thiocyanate melt, the great majority of investigators have used the potassium salt alone, partly because it is readily available in high purity (whereas the anhydrous sodium salt is difficult to obtain pure, and the lithium salt, crystallizing as the monohydrate, is not easy to dehydrate) and partly no doubt because the potassium salt has a low melting point not far above that of the (uncertain) eutectic. The generally low melting points have been correlated with the unusually small volume change on melting and attributed to the possible formation of association complexes in the melt[4] (c.f. the 1:3 compound inferred from the phase diagrams[2,7]), so that the introduction of a second cation would be expected to have a relatively small effect. The melts are customarily dried before use under vacuum at 150°C for several hours[12] or at 200°C for two days.[20]

1.3. Thiocyanate as Ligand

X-ray diffraction has shown the thiocyanate anion to be linear with a C–N distance close to 1.2 Å and C–S close to 1.6 Å. Infrared and Raman spectroscopy also indicate a linear ion ($C_{\infty v}$ point group) in the molten salts where the intraionic vibrations of potassium thiocyanate are ν_1 2068 cm^{-1} (C–N stretching), ν_2 478 cm^{-1} (N–C–S bending), and ν_3 745 cm^{-1} (C–S stretching)[6] and are similar to those in the disordered solid.[21]

The shift in position of these absorption bands after coordination to metal ions has often been used in determining the nature of the metal–ligand bond. N-bonded thiocyanate (isothiocyanato) complexes produce absorptions in the ranges ν_1 2040–2080, ν_2 465–480, and ν_3 780–860 cm^{-1}, whereas S-bonded (thiocyanato) complexes have absorptions in the ranges ν_1 2080–2120, ν_2 410–470, and ν_3 690–720 cm^{-1}.[22] However, since the ν_3 vibration is of low intensity, a 800–880 cm^{-1} band can be assigned either as the ν_3 vibration of an N-bonded complex or as the first overtone of the ν_2 vibration of an S-bonded complex.[23] Thus, the positions of all three fundamentals should be considered, or use made of the diagnostic difference in the integrated intensity of the ν_1 band.[24] Alternatively, ^{14}N NMR chemical shifts have been used to identify the bonding in metal–thiocyanate complexes.[25]

Nitrogen bonding from the ambidentate thiocyanate ligand has been correlated with "class a" or "hard" metal ions, and sulfur bonding with "class b"[26] or "soft"[27] metal ions, though the bonding is also influenced by the other ("symbiotic") ligands present[28] and possibly also by steric effects.[29] The spatial geometry of the two forms is rather different—that of the thiocyanato ligand, where the M–S bond involves π orbitals, being bent

$$
\begin{array}{c}
\text{M–S} \\
100° \diagdown \\
\diagdown \text{C} \\
\diagup \\
\text{N}
\end{array}
$$

while the isothiocyanato ligand, where the M–N bond involves only σ orbitals, is linear

$$M\underbrace{-N-}C-S}_{180^\circ}$$

An integrated explanation of class a and class b behavior in terms of Klopman's polyelectronic perturbation theory has recently been given by Norbury.[30]

The tendency to ambidentate nature is shown by the molecular orbital calculations of Di Sipio et al.,[31] where almost equal terminal charge distributions (S, 0.48, C, 0.01, N, 0.51) were calculated for the free ion. Thus the bond formed depends principally on the rearrangement of electrons which accompanies its formation.

1.4. Physical Properties

The calculation of Di Sipio et al.[31] agrees with the observation that the allowed electronic charge transfer band of the free thiocyanate ion is well into the ultraviolet, calculated to be at 65,100 cm^{-1}, and that when thiocyanate is bonded to a metal ion the initially forbidden lower-energy transition calculated to be at 36,800 cm^{-1} may become allowed. Such a low-energy high-intensity band is frequently found in thiocyanate melt spectra where it often limits the number of d–d transitions which can be observed.

The density of pure potassium thiocyanate has been reported as ϱ (g-ml^{-1}) $= 1.6005 \pm 0.007 - 0.00080(t - t_f)$, where $t_f = 173.5°C$;[32] and as 1.60 at 185°C[33] (c.f. Ref. 32: 1.596 at 185°C). The densities of 10, 20, and 30% solutions of sodium thiocyanate in potassium thiocyanate have been reported to follow the equation $\varrho = \varrho_0 - bt$, with $\varrho_0 = 1.728$, 1.711, and 1.719 g-ml^{-1} and $b = 6.86$, 5.95, and 6.23×10^{-4} g-ml^{-1}-°C, respectively.[17] The latter measurements suggest that the density is almost insensitive to sodium thiocyanate addition (calculated as 1.601, 1.601, 1.603 g-ml^{-1} at 185°C, respectively, the pure potassium melt at this temperature having a density of 1.60 g-ml^{-1}[32,33]).

The surface tension of potassium thiocyanate has been measured as 101.5 dyn-cm^{-1} with the unexpected high temperature coefficient of -1.36 dyn-cm^{-1}-°C for a fresh melt.[4] This was attributed to bond rearrangements, rather than to surface-active impurities, possibly of the type

$$3SCN^- \; \rightleftharpoons \;$$

Among the other properties of molten thiocyanate reported are the viscosity,[3] electrical conductivity[3] (NaSCN 0.67 and KSCN 0.16 ohm^{-1}-cm^{-1} at the melting points), compressibility (β_{corr} for KSCN $= 25.6 \pm 1$ bar^{-1} at 200°C),[34] heat of fusion and molar volume at the melting point,[3] refractive index, and molar refraction.[17]

1.5. Thermal Stability of Thiocyanates

It has been known for many years that on heating thiocyanate the initially clear colorless melt first turns green and then blue, the color continuously deepening with temperature (in pure potassium thiocyanate the change to green occurs at about 300°C and to blue at 340°C).[35] The blue color disappears upon the addition of potassium cyanide,[36] and addition of sodium hydroxide precipitates sulfur,[37] the postulated equilibrium being

$$x\text{KSCN} \rightleftharpoons x\text{KCN} + \text{S}_x$$

Infrared spectra have shown no absorption due to sulfur over the 500–700 cm^{-1} range, though longer wavelength absorptions may be so attributed.[37] Electronic spectra show an intense band at 17,900,[37] 17,250[32] or 16,500[36] cm^{-1} ($\varepsilon > 16{,}000$ liter-mole^{-1}-cm^{-1})[36] which has been attributed to the presence of sulfur polymers S$_2$[38] and S$_4$.[37] Certainly sulfur, potassium polysulfide, and sodium thiosulfate when dissolved in potassium thiocyanate give the same absorption band.[36] More recently, the blue color (band at 17,000 cm^{-1}) has been attributed to the radical ion S$_2^-$, and the absorption band at 25,000 cm^{-1} to a polysulfide ion, possibly S$_2^{2-}$.[39] These species were postulated to be formed by the reactions

$$2\text{KSCN} \xrightarrow{>250°C} \text{K}_2\text{S} + (\text{CN})_2$$

$$\text{K}_2\text{S} + \text{KSCN} \xrightarrow{>300°C} \text{KCN} + \text{K}_2\text{S}_2 \text{ (yellow)}$$

$$\text{K}_2\text{S}_2 + 2\text{KSCN} \underset{\;}{\overset{350°C}{\rightleftharpoons}} \text{K}_2\text{S}_4 + 2\text{KCN}$$

$$\text{K}_2\text{S}_4 \rightleftharpoons 2\text{K}^+ + 2\text{S}_2^- \text{ (blue)}$$

(Cyanogen was detected by qualitative tests.)

Similar species were found to be formed in molten chlorides and carbonates and in borate glasses.[39] According to another recent study on lithium chloride–potassium chloride involving electrochemical methods and Raman as well as visible-ultraviolet spectroscopy, the 16,800-cm^{-1} band is to be attributed to the species S_3^-, and the 31,400 and 25,600 cm^{-1} bands to S_2^-.[40] Though such studies have not so far been applied to molten thiocyanates, it is likely that similar results will be obtained with these melts.

Raising the temperature still further, to 400°C, is said to result in the evolution of nitrogen and "elemental carbon."[38] Thiocyanates have been heated under atmospheres of various gases with the reported formation of numerous products.[41] Hydrogen at low red heat is said to react with thiocyanates in accordance with the equation

$$4KSCN + 4H_2 \rightarrow 2KCN + K_2S + 3H_2S + 2HCN$$

while thiocyanates react with water at high temperature to form potassium sulfide and carbonate, carbon dioxide, hydrogen sulfide, and ammonia; with hydrogen sulfide to form potassium and hydrogen cyanides and sulfur; and with carbon dioxide to form potassium cyanate; with sulfur dioxide, thiocyanates react in accordance with the equations[41]

$$2KSCN + 2SO_2 \rightarrow K_2S + 2CO_2 + N_2 + 3S$$

and

$$K_2S + 2SO_2 \rightleftarrows K_2SO_4 + 2S$$

Oxygen was reported to form sulfate and polysulfide ions with thiocyanates at temperatures around 400°C.[42]

1.6. Electrochemical Stability of Molten Thiocyanate

Electrolysis of molten sodium thiocyanate–potassium thiocyanate has shown it to be reduced at -1.75 V and oxidized to parathiocyanogen at $+0.25$ V (vs. the Ag/AgCl electrode).[16] The latter compound forms a passivating film on the anode and has recently been the subject of extensive investigation using a rotating platinum disk electrode in 190°C in molten potassium thiocyanate. Repetitive and single-pulse linear sweep voltammetry has given detailed information on nucleus formation and growth. The cathodic process has been shown to involve both reduction to sulfide and cyanide ions at the metal–film interface and mechanical detachment of the film.[43,44]

2. CHEMISTRY OF THE MAIN GROUP ELEMENTS

2.1. Complexes with Alkaline Earth Cations

Magnesium, calcium, and barium thiocyanates have been shown to be very soluble (>7 M at 170°C in KSCN) and to be nitrogen-bonded to form tetraisothiocyanato complexes in the case of calcium and barium but lower complexes in the case of magnesium.[45]

2.2. Indium and Thallium Cations

Indium(III) sulfate has been found to be soluble in sodium thiocyanate–potassium thiocyanate eutectic at 150–190°C, and the standard electrode potential has been determined.[12] The potential of thallium(I) thiocyanate was determined under the same conditions.[9] The thallium(I) ion has also been studied polarographically at 137°C[9] and by electromigration in potassium thiocyanate at 210°C.[46] The ion could not, however, be amperometrically titrated with electrogenerated sulfide ions presumably due to the greater stability of the complex formed with cyanide ions, which are also generated in the course of the reduction[11]:

$$SCN^- + 2e \rightarrow S^{2-} + CN^-$$

2.3. Reactions with Organic Compounds

The alkali metal carboxylates are both soluble and stable in molten thiocyanates, and many phase diagrams have been determined. A critical compilation of such systems, ranging from formate to caproate, has just been published.[47]

The solubility of 15 very varied organic compounds in sodium thiocyanate–potassium thiocyanate has been reported at 150°C by Crowell and Hillery.[48] Pentaerythritol, hydroquinone, glycol, and methanol were shown to be very soluble; sucrose, p-nitroaniline, and glucose, fairly soluble; triphenylmethyl chloride, triphenylacetic acid, hexyl bromide, phenol, borneol, p-toluidine, and pyridine, relatively insoluble; and benzyl alcohol, soluble but subliming unchanged. These solubilities show similarity to those in the aqueous system. The hydroxy compounds are apparently ionized. Hillery[15] has also studied the effect of the sodium–potassium cation ratio and of magnesium sulfate and calcium chloride additions on the reaction of anilinium thiocyanate to form phenylthiourea and *sym*-

diphenylthiourea, together with small amounts of the corresponding ureas at 139°C.

The nucleophilic displacement reactions with esters proceed by an S_N2 mechanism,

$$\underset{\text{HO}}{\overset{\text{HO}}{\bigcirc}}\!\!-COOCH_3 + SCN^- \longrightarrow \underset{\text{HO}}{\overset{\text{HO}}{\bigcirc}}\!\!-COO^- + MeSCN + MeNCS$$

96% of methyl thiocyanate being obtained with the methyl ester, but nearly equal quantities of thiocyanate and isothiocyanate with the ethyl ester, a reaction which proceeds at only $\frac{1}{49}$th the extent of that of the methyl.[14] The 2,4-dihydroxybenzoate gives similar products but whith decarboxylation and formation of resorcinol and the unreactive phenoxide ion in equimolar quantities.[13]

Halide–pseudo-halide exchange was found to occur when the lower alkyl chlorides were passed through sodium thiocyanate–potassium thiocyanate melt at 200–300°C; alkyl isothiocyanate was the main product with up to 60% yield.[50]

Organic electrochemical reactions are just receiving attention. Thus anodic oxidation of potassium thiocyanate at 250 and 400°C with a stream of methanol vapor flowing through the cell has recently been found to give a 50% yield of methyl cyanide together with sulfur dioxide and hydrogen sulfide. The reaction was postulated to proceed by reaction of methanol with discharged thiocyanogen and subsequent thermal decomposition of methyl thiocyanate, rather than by a free-radical mechanism.[51] When electrolytic reduction proceeded in the presence of quinolinium halides, dark-red insoluble products were formed, seemingly involving a 1:1 compound with sulfide and probably also with cyanide.[52]

2.3.1. Acid–Base Indicators in Thiocyanate Melts

Like most other organic compounds, many acid–base indicators are stable in potassium thiocyanate at 200°C but also respond to the presence of acidic and basic solutes with the formation of colored species very similar to those found in aqueous systems. For example, phenolphthalein, initially insoluble, dissolves in the presence of sodium monoxide (or of sodium peroxide, strontium and barium hydroxides, and potassium bicarbonate) to give a purple solution [absorption bands at 18,050 cm^{-1} (st.) and 26,600 cm^{-1} (m.)] which disappears on addition of potassium disulfate, disulfite, hydrogen oxalate, dihydrogen phosphate, sodium hydrogen sulfate, or ammonium chloride. Similarly, the color changes and absorption bands of methyl red, alizarin, and *m*-cresol purple paralleled those in water, while six other indicators were soluble and gave reproducible color changes, and eight further indicators were found to be insoluble or unstable.[53]

2.4. Reactions with Tin and Lead Compounds

Tin(II) solutions are stable, clear yellow in color, and have been studied by polarography[54] and electromigration.[46] A cyanide complex is apparently formed, as large negative errors are obtained on amperometric titration. Tin(IV) in the form of potassium hexachlorostannate(IV) acts as an oxidant forming parathiocyanogen and tin(II).[55] Tin(IV) oxide is reduced to the sulfide at 300°C:

$$SnO_2 + 2KSCN \rightarrow SnS + K_2S + 2CO + N_2$$

and to the polysulfides at high temperatures.[56]

Lead(II) is stable at low temperatures, having been the subject of polarography at 137°C,[9] emf,[54,57,58] and electromigration studies. It gave a light green solution at 160°C, was very soluble ($>7\ M$), and mainly sulfur-bonded. Some evidence of nitrogen bonding may indicate a certain amount of polymerization.[45] The solutions have also been reported as colorless at 195°C,[54] and to form a cyanide[56] but not a chloro complex.[57] At 300°C lead(II) chloride reacts to form the sulfide,[59] as does the oxide[56]:

$$PbO + KSCN \rightarrow PbS + KOCN$$

2.5. Reactions with Nitrogen, Antimony, and Bismuth Compounds

Although alkali metal nitrate–potassium thiocyanate phase diagrams have been reported,[7] the nitrate ion has also been claimed to react at 195°C forming nitrogen oxides,[54] and at 190°C to liberate cyanogen with precipitation of sulfur.[61] The sole report on antimony is that "Sb^{3+} reacts to form sulfur and a gas with a cyanogen odor" i.e., it reacts as an oxidizing agent. The final form of the antimony is not stated.[54]

Bismuth(III) is apparently stable at 210°C in potassium thiocyanate and has been studied by electromigration.[46] The oxide forms the sulfide at 300°C:

$$Bi_2O_3 + 3KSCN \rightarrow Bi_2S_3 + 3KOCN$$

and potassium tetrathiobismuthate(III) at higher temperatures.[56]

2.6. Reactions with Oxygen and Sulfur Compounds

Calcium and barium oxides are reported as being insoluble in potassium thiocyanate at 195°C. Lithium hydroxide is said to be insoluble, but sodium and potassium hydroxides are said to be slightly soluble and to decompose the melt.[54]

The sulfate, thiosulfate, and sulfite ions are stable in potassium thiocyanate and are generally soluble with the exception of the alkaline earth sulfates.[54]

Little other sulfur chemistry has been studied, though it is known that parathiocyanogen is the product of chemical[55] as well as electrochemical[63,44] oxidation of the thiocyanate ion. Sulfur itself is almost insoluble[54] but becomes slightly soluble in the presence of sulfide ions. The formation of polysulfide ions and radical ions (see Sec. 1.5) is possibly the beginnings of a chemistry of considerable complexity.

2.7. Reactions with Halogen Compounds

The alkali metal halides are stable and soluble in thiocyanate melts.[16] Phase diagrams[8,62,63] and cryoscopic studies[64] have been reported. Iodine forms low-melting eutectics with potassium thiocyanate (54.3°C, 33 mole % KSCN) and ammonium thiocyanate (38.1°C, 75 mole % NH_4SCN and 40.3°C, 20% mole NH_4SCN) but reacts on heating to 120–140°C with the formation of parathiocyanogen and iodide ions.[65]

There is only one brief mention of the reaction of halogens in higher oxidation states, which is that iodates and bromates react with potassium thiocyanate at 195°C and that iodine, or bromine, vapor is evolved.[54]

3. CHEMISTRY OF THE TRANSITION METALS

Apart from a brief reference to the synthesis of zirconium and hafnium hexathiocyanate anions in molten potassium thiocyanate,[66] the chemistry of the transition metals begins with that of the vanadium group, where, however, vanadium itself is the only metal so far to receive attention.

3.1. Reactions with Compounds of Vanadium

Vanadium pentoxide has been noted as reacting in molten thiocyanate in the course of voltaic cell experiments.[16] It has since been found to react in potassium thiocyanate at 200°C to form a green oxovanadium(IV) solution (maximum concentration 0.005 M) with the remaining vanadium precipitating as black vanadium dioxide. The oxidation products of the thiocyanate are numerous and included sulfate, cyanide, nitrate and oxide, sulfur, traces of sulfide and cyanate, and in oxidations above 250°C, carbon dioxide. Ammonium metavanadate reacts similarly, though additionally forming thiosulfate among the oxidation products, no doubt largely because the melt is above the solute thermal-decomposition temperature, i.e.,

$$2NH_4VO_3 \xrightarrow{170°C} V_2O_5 + 2NH_3 + H_2O$$

Sodium orthovanadate dissolves at 200°C initially to form a colorless solution which nevertheless contains oxide ions (purple color with phenolphthalein). On heating at 200°C for 24 hr, the color changes to yellow due to exchange of sulfur to form the monothioorthovanadate ion, and after 7 days to red with formation of the trithioorthovanadate. Under vacuum the purple tetrathioorthovanadate is formed, but this reacts to the red trithio ion on admission of air. Sulfide, oxide, and cyanate ions are detected in the melt, probably due to reactions such as

$$2VO_4^{3-} \rightleftarrows V_2O_7^{4-} + O^{2-}$$

$$O^{2-} + SCN^- \rightleftarrows OCN^- + S^{2-}$$

$$V_2O_7^{4-} + S^{2-} \rightleftarrows VSO_3^{3-} + VO_4^{3-}, \text{ etc.}$$

However, there are undoubtedly further complexities to be unravelled since the addition of cyanate (1.8 M) to the initial solutions considerably increases the rate of formation of the thio complexes, and the rate in air is approximately twice that in vacuum.

Similar orange, red, and purple solutions can be obtained directly by adding vanadium pentoxide to thiocyanate melts containing sodium sulfide or sodium oxide, though, as would be expected, the times required are considerably shorter (1 hr, 10 hr, and 2 days, respectively).

In its reactions sodium metavanadate resembles sodium orthovanadate rather than ammonium metavanadate, initially giving a colorless solution containing oxide ions, then a yellow species (possibly a thiometavanadate), and finally trithioorthovanadate on addition of sodium monoxide or sulfide.

Despite the number of thiooxovanadium anions formed, a relatively small increase of temperature causes them to become unstable with respect to reduction by thiocyanate ions, and a green oxovanadium(IV) solution and black vanadium dioxide precipitate are formed at 250°C. Similar green solutions are formed by solution of oxovanadium(IV) sulfate and potassium tetrathiocyanatooxovanadate(IV).[67] The spectrum is found to consist of two bands [12,500 cm^{-1} ($\varepsilon = 69$) and 16,313 cm^{-1} ($\varepsilon = 43$ liter-mole^{-1}-cm^{-1}] as well as a charge-transfer edge at 20,000 cm^{-1}.[33]

Vanadium(III) solutions of vanadium trichloride or potassium hexathiocyanatovanadate(III) are golden brown in color and stable in vacuum but are oxidized to green oxovanadium(IV) solutions on admission of air.[67]

3.2. Reactions with Compounds of Chromium and Molybdenum

The oxoanions of chromium(VI) show a limited stability in potassium thiocyanate at 200°C, dissolving to give yellow solutions with similar charge-transfer absorptions [K$_2$CrO$_4$ 27,400 cm^{-1} ($\varepsilon = 3200$), K$_2$Cr$_2$O$_7$ 27,400 cm^{-1} ($\varepsilon = 2380$), and K$_2$Cr$_3$O$_{10}$ 27,300 cm^{-1} ($\varepsilon = 2100$ liter-mole^{-1}-cm^{-1}].[68] Reaction takes place over a period of time; the oxoanion does not give thiooxoanions as in the case of vanadium but is reduced to insoluble chromium(III) oxide which further slowly exchanges oxide for sulfide ions, ultimately forming insoluble black chromium(III) sulfide.[69] The time required for precipitation decreases with increased concentration of chromium oxoanion and with addition of cyanate and oxide ions. Addition of sodium sulfide causes a rapid precipitation of chromium(III) sulfide. The products of the oxidation of thiocyanate ion include sulfate, thiosulfate, cyanate, cyanide, sulfide, oxide, and nitrate ions, and carbon dioxide and nitrogen, the proportions varying with initial concentration of the oxoanion and the time elapsed. The sequence of reactions is again very complicated, but the first three anions account for 60% of the oxygen of dichromate after allowing for the precipitation of chromium(III) oxide.[68] In contrast, the insoluble calcium chromate was found not to react in the course of tests of voltaic cells.[16] Chromium(VI) oxide also reacts in potassium thiocyanate at 180°C, and in sodium thiocyanate–potassium thiocyanate at 145°C to give green chromium(III) solutions and sulfide, sulfate, cyanate, nitrate, and oxide ions.[67]

Chromium(III) compounds, as the trichloride and potassium hexaisothiocyanate chromate(III), are stable in pure thiocyanate melts, giving green solutions. The latter compound was shown by cryoscopy to contribute one foreign ion[64] and to be anionic by electromigration.[46] The infrared spectrum of a 1 M solution (stated to be deep violet[45] in color, in contrast to the green of more dilute melts, for example, of 0.05 M concentration[68]) shows coordination by nitrogen bonding.[45] The main electronic absorption band [16,500 cm^{-1} ($\varepsilon = 128$),[68] 16,740 cm^{-1} ($\varepsilon = 191$),[33] or 16,700 cm^{-1} ($\varepsilon = 15$ liter-mole^{-1}-cm^{-1}][70] is assigned to the $^4T_{2g} \leftarrow {}^4A_{2g}$ octahedral transition, the red shift of about 1000 cm^{-1} from the aqueous solution, being attributed to the second sphere environment of potassium ions around the complex lowering the ligand field of the thiocyanate ions.[33] The shoulder at 23,250 cm^{-1} cm^{-1} is assigned to the $^4T_{1g}(F) \leftarrow {}^4A_{2g}$ transition.

The dilute, green solutions are said to be stable for at least 14 days but to precipitate the oxide/sulfide within 2 hr when heated to 300°C,[68] though the more concentrated deep violet solutions decomposed in the course of the infrared measurements at 170°C.[45] Addition of chloride ions caused no change in the spectrum at 200°C and hence indicated no displacement of coordinated thiocyanate. Cyanate ions, however, caused a slight red shift [15,200 cm^{-1} ($\varepsilon = 94$) and 22,700 cm^{-1} ($\varepsilon = 88$ liter-mole^{-1}-cm^{-1})], attributed to the displacement of nitrogen by oxygen ligands and confirmed by a similar shift with carbonate, nitrate, and thiosulfate ions, and the ready precipitation of chromium(III) oxide.[68] Chromium(III) oxide was also rapidly precipitated on addition of sodium monoxide and chromium(III) sulfide, and slightly less rapidly on addition of sodium monosulfide.[68]

Addition of potassium cyanide gave a large blue shift, indicating displacement of thiocyanate but retention of octahedral coordination; but the red melts reverted to green on further addition of cyanate.[68]

Chromium(II) chloride was stable, giving a brown solution (strong absorption >20,000 cm^{-1}) in the absence of air,[68] but was oxidized to green chromium(III) in the presence of air[33,68] together with the formation of sulfide, cyanide, and cyanate ions.[68]

At higher temperatures Milbauer[56] reported somewhat similar reactions:

$$Cr_2O_3 + 3KSCN \rightarrow Cr_2S_3 + 3KOCN$$

$$2KOCN + KSCN \rightarrow K_2S + KCN + 2CO + N_2$$

$$Cr_2S_3 + K_2S \rightarrow K_2Cr_2S_4$$

and at red heat:

$$Cr_2S_3 + 4\tfrac{1}{2}O_2 \rightarrow Cr_2O_3 + 3SO_2$$

The only published reaction of a molybdenum compound, also by Milbauer,[56] is that a molybdenum oxide (not specified) formed molybdenum di- and trisulfide in molten potassium thiocyanate at 300°C.

3.3. Reactions with Compounds of Manganese and Rhenium

Manganese in the higher oxidation states appears to be insoluble in thiocyanate melts and, as might be expected of such oxidants, frequently reacts rapidly, first precipitating brown manganese(III) oxide and, finally, after exchange of oxide by sulfide ions, the insoluble green form of manganese(II) sulfide.[71] Among the oxidation products of the thiocyanate produced in the reactions with potassium permanganate, potassium manganate(VI), and sodium manganate(V) were sulfide, sulfate, cyanide, cyanate, oxide, and nitrate ions, oxygen, nitrogen, carbon dioxide, and nitrous oxide in gaseous form, and sulfur which dissolved to form a blue solution at 300°C. Thiosulfate was also produced in the extremely rapid reaction of sodium manganate(V), brown oxide being precipitated within 1 min at 200°C. Manganese dioxide reacted more slowly (3 hr at 200°C) to give the same solid products and the first five anions listed above. Manganese(III) compounds ($K_2MnF_5 \cdot H_2O$, $MnPO_4 \cdot H_2O$, and Mn_3O_4) were stable at 200°C but reacted to the green manganese(II) sulfide at 290°C.[71]

Manganese(II) chloride has been stated to be insoluble at 177°C[70] but also reported to be stable in polarographic studies at 195°C[72] and to give stable yellow solutions at 200°C.[71] On addition of sodium sulfide it precipitated the normally more stable pink-brown form of manganese(II) sulfide, which was converted to the green form in 15 mins. (In the aqueous system a source of sulfide ions is required for this conversion to take place.) At 290°C the pure manganese(II) solutions were unstable, precipitating the green sulfide in 30 min.[71] Polarography indicated manganese(II) ions were complexed by cyanide at 195°C.[72] Manganese(II) oxide was converted to the sulfide at 300°C.[56]

The only publication on lower oxidation states reports that manganese metal reacted rapidly at 195°C to give manganese(II) sulfide and cyanide ions.[72] The reactive relativity of the oxidation states, as judged by the time required to form manganese(II) sulfide at 200°C, appears to be

$$V > IV \approx 0 > VI > VII \gg III \approx II$$

No work has so far been reported for technetium compounds in molten thiocyanates, but several papers dealing with rhenium compounds have appeared.

Perrhenate has been solvent-extracted by dioctyl- and tetra-*iso*-pentyl-ammonium thiocyanate from sodium thiocyanate–potassium thiocyanate melts[73] and thus exhibits its less powerful oxidizing character as compared to permanganate. Rhenium(V), as $K_3[ReO_2(SCN)_4]$, proved stable and migrated anionically on application of a potential.[44]

Rhenium pentachloride also dissolves in potassium thiocyanate at 230°C. After heating for 1 hr the hexathiocyanato complex was extracted from the quenched melt with ethanol and precipitated by addition of cesium chloride.[74] Further examples of the use of thiocyanate melts as preparative media are provided by the addition of potassium cyanide and cyanate to the rhenium pentachloride solutions, the hexacyanide and hexacyanate compounds being eventually precipitated.[75] On heating the rhenium(V) solutions at 250°C for 3 hr reduction to rhenium(IV) takes place.

Potassium hexachlororhenate(IV) dissolves at 225°C to give a dark brown melt from which hexathiocyanatorhenate(IV) can eventually be obtained,[76] though at 200°C a series of thiocyanato-chloro complexes are extracted from the light brown melt solution, as well as the hexacyanorhenate(IV) complex from the olive green melt formed when potassium cyanide had been added.[77]

Rhenium(III) chloride reacted violently within 15 min, evolving gas and producing a black powder which proved difficult to analyze but which did not appear to contain rhenium–rhenium bonds.[75]

3.4. Reactions with Compounds of Iron

Iron(III) is unstable in thiocyanate melts. An early report stated that the iron(II) formed was insoluble,[70] though the insoluble product originally postulated as an "oxide"[48] has since been claimed to be sulfur, formed together with cyanogen[54,72] and also parathiocyanogen.[55] Recently, however, De Haas *et al.*[78] have shown, in the course of solvent-extraction measurements with tri-*n*-butyl phosphate (TBP) solutions, that dilute iron(III) solutions have rather greater stability (a solution of 10^{-4} m concentration requiring over 18 hr to decompose at 150°C). The complex contained 2 TBP per mole. On heating in potassium thiocyanate iron(III) sulfide forms the potassium diiron tetrasulfide at 600°C.[56]

Iron(II) is formed in solution from the reaction of ferric chloride or potassium hexathiocyanatoferrate(III) tetrahydrate[55] and by the solution of iron(II) sulfate heptahydrate. The greenish solution displays a charge transfer band at 34,300 cm^{-1} ($\varepsilon = 6340$[61]) and d–d absorption bands at 10,000 cm^{-1} ($\varepsilon = 16$) and 7900 cm^{-1} ($\varepsilon = 13$ liter-mole^{-1}-cm^{-1})[79] (attributed to a nitrogen-bonded octahedral complex) and, at concentrations greater than 0.1 M, reacts to form iron(II) sulfide, sulfur, and cyanogen.[72]

In solutions of the sodium salt of ethylenediamine tetraacetic acid (EDTA) in potassium thiocyanate, iron(II) sulfate heptahydrate forms a complex which, by spectroscopic measurements of the charge transfer absorption at 34,300 cm^{-1} ($\varepsilon = 635$ liter-mole^{-1}-cm^{-1}), was shown to have a 1:1 mole ratio[80] with an equilibrium constant of 1.3×10^5 at 180°C. A similar complex is formed in aqueous solution though with a much larger equilibrium constant (2.1×10^{15} at 20°C) attributed to the greater stability of the solvated metal ions in the melt.[61] A similar study using potassium cyanide solutions showed the appearance of a new absorption band at 26,300 cm^{-1} ($\varepsilon = 2580$ liter-mole^{-1}-cm^{-1}) and the formation of complexes of 1:3 and 1:6 mole ratios.[81] The latter complex is clearly hexacyanoferrate(II), and the former the insoluble $\{Fe[Fe(CN)_6]\}^{2-}$ which is oxidized in air from yellow to the familiar blue color.[61]

Iron metal apparently reacts rapidly at 195°C forming sulfide and cyanide ions.[72]

3.5. Reactions with Compounds of Cobalt

Cobalt(III) compounds, ($Na_3[Co(NO_2)_6]$ and $[Co(NH_3)_6]Cl_3$), act as oxidants precipitating orange parathiocyanogen and forming blue cobalt(II) solutions.[55] Cobalt(II) solutions are stable, blue in color, and have been the subject of polarographic study.[54] The standard electrode potential in sodium thiocyanate–potassium thiocyanate has been measured,[12] but the cell was found irreversible in potassium thiocyanate.[57] Electromigration experiments indicate an anionic complex,[46] and cryoscopic measurements of the chloride indicate the introduction of three foreign ions.[64] Infrared spectra of the dichloride solution, of more than 1 M concentration, indicate nitrogen bonding. The d–d absorption bonds at 17,400[70] or 17,100[33] cm^{-1} (sh.) 1600 cm^{-1} ($\varepsilon = 510$[33], 480[78], or 90[70]), and 7700 cm^{-1} ($\varepsilon = 64$ liter-mole^{-1}-cm^{-1})[33] are interpreted as those of a tetrahedral, nitrogen-bonded complex and assigned as $^4T_1(P) \leftarrow {}^4A_2$ (split by spin-orbit coupling) and $^4T_1(F) \leftarrow {}^4A_2$, respectively.

The charge-transfer absorption is at 31,000 cm^{-1} ($\varepsilon = 4870$ liter-mole^{-1}-cm^{-1}).[61] Cobalt(II) has been solvent-extracted with TBP in diphenyl (the complex having 3 TBP per mole) at 160°C[70] but was found to decompose to the insoluble sulfide when extraction was attempted with dioctylammonium thiocyanate dissolved in tetralin, though the decomposition was somewhat less rapid when the complexing agent was dissolved in 1-chloronaphthalene.[73] The sodium salt of EDTA has been shown by spectrophotometric measurement to form a 1:1 complex[80] ($K = 2.3 \times 10^5$

at 190°C, c.f. $K = 2 \times 10^{16}$ in water at 20°C).[59] No halide complexes were detected,[70] but with potassium cyanide a complex series of reactions occurred which are still the subject of study. Initially, it was noticed that cyanide ions caused the precipitation of cobalt(II) sulfide,[70] which could be dissolved in a large excess of cyanide, while a further initial product was the soluble hexacyanocobaltate(III) ion.[81] The cyanide/cobalt ratio was determined to be 3.5 when extrapolated to zero cobalt concentration,[72] and the reaction has now been expressed in the form of the following equations, the initial reaction being

$$3Co^{2+} + SCN^- + 11CN^- = 2[Co(CN)_6]^{3-} + CoS \downarrow$$

the cobalt(II) sulfide then dissolving in excess cyanide

$$2CoS + SCN^- + 11CN^- = 2[Co(CN)_6]^{3-} + 3S^{2-}$$

thus giving an overall equation[61]

$$6Co^{2+} + 3SCN^- + 33CN^- = 6[Co(CN)_6]^{3-} + 3S^{2-}$$

The only other fragmentary information on cobalt chemistry is the early report that cobalt(II) oxide forms the complex sulfide ($K_4Co_{11}S_{10}$) at 400°C.[56]

3.6. Reactions with Compounds of Nickel and Platinum

Nickel(II) solutions have generally been recorded as stable, though the colors reported have included yellow,[70] green,[61,70] yellow-green[54] and blue.[45] Polarographic[54] and emf measurements[57] have been made on these solutions, though Bailey and Steger[46] found that nickel(II) gave an insoluble product ("oxide") at 210°C. However, Hester and Krishnan[45] found that nickel(II) chloride was more than 1 M soluble in the course of their infrared measurements, which indicated bonding via nitrogen. Cryoscopic measurements on nickel(II) bromide in potassium thiocyanate indicated three foreign ions.[64]

The visible absorption bands attributable to d–d transitions have been the subject of three studies which show reasonable agreement on the positions, i.e., 14,300 cm^{-1} ($\varepsilon = 18^{[33]}$ or $2.4^{[78]}$) or 14,000 cm^{-1} ($\varepsilon = 4.6$),[70] and 8550 cm^{-1} ($\varepsilon = 13$)[33] or 8300 cm^{-1} ($\varepsilon > 24$ liter-mole^{-1}-cm^{-1}).[78] However, there is some disagreement on the assignments, De Haas et al.[78] preferring a deformed tetragonal or square planar geometry while Egghart[33] maintaining that the coordination was octahedral because of the large ligand field stabilization energy occurring with octahedral thiocyanate ligands and a d^8- ion. The charge-transfer band was found to occur at 32,000 cm^{-1} ($\varepsilon = 4870$ liter-mole^{-1}-cm^{-1}).[61]

The complex with tri-n-butyl phosphate was investigated by solvent extraction and shown to have 1 TBP per mole. Chloride and bromide ions apparently did not displace the solvated thiocyanate ligands,[70,78] but spectrophotometric measurements with sodium ethylenediamine tetra-acetate indicated a 1:1 complex[80] ($K = 5.9 \times 10^5$ at 190°C in potassium thiocyanate, again much smaller than the corresponding value in water at 20°C, $K = 7.9 \times 10^{18}$). Potassium cyanide solutions were initially reported to produce an undefined black precipitate[70] and later a tetracyano complex,[81] described as yellow-gray in color with a charge-transfer band at 34,400 cm^{-1} ($\varepsilon = 6180$ liter-mole^{-1}-cm^{-1}) and with a stability constant of 2.5×10^{16} at 190°C (c.f. $K = 3.2 \times 10^{31}$ in water at 25°C).[61]

Nickel(II) oxide is reported to react with potassium thiocyanate at 400°C to form a complex sulfide analogous to the cobalt compound,[56] and nickel metal to react rapidly at 195°C to give sulfide and cyanide ions.[72]

Platinum electrodes have been used in the study of the anodic oxidation of thiocyanate ions to form insoluble parathiocyanogen, and its subsequent cathodic stripping as sulfide and cyanide ions, with no mention of any direct reaction of the platinum.[43,44] However, a less detailed study reported the formation of a lime-green layer on platinum cathodes, in potassium thiocyanate at 200°C, and a golden-yellow melt which contained platinum in solution.[82] A platinum emf cell (Pt/Pt^{2+} in KSCN) has been found to be irreversible.[57] The sulfur layer formed on platinum by anodic electro-deposition has very recently been reported to be markedly photoelectric in visible light at 180°C in sodium thiocyanate–potassium thiocyanate melts containing sulfide ion.[83]

3.7. Reactions with Compounds of Copper, Silver, and Gold

It is generally agreed that copper(II) ions are unstable in molten thiocyanates and are reduced to soluble copper(I) together with the precipitation of orange parathiocyanogen[55] or of sulfur and cyanogen.[54,72]

Similarly, the early work of Milbauer[56] indicated that copper(II) oxide was converted to copper(I) sulfide in potassium thiocyanate at 300°C:

$$2CuO + 2KSCN = Cu_2S + K_2S + 2CO + N_2$$

and to the complex sulfide $K_2Cu_8S_6$ at higher temperatures.

Copper(I) solutions are stable,[9] yellow in color[55,61] with an absorption band at 34,200 cm^{-1} ($\varepsilon = 715$ liter-mole^{-1}-cm^{-1}), and form a 1:1 complex with EDTA ($K = 9.2 \times 10^5$).[61] With potassium cyanide, though complexes are certainly formed, their relative stability is the subject of conflicting claims. Thus Metzger,[54] after a polarographic study with dilute cyanide solutions, reported a di- and a tetracyano complex with stability constants of 3.2×10^5 and 5×10^9, respectively. However, Eluard and Trémillon[72] found by voltammetry that dicyanocuprate(I) was the stable form, as did Hennion *et al.*[61] from spectrophotometric measurements, the latter deriving a stability constant of 1.9×10^7 at 190°C (cf. $K = 10^{24}$ in water at 25°C); while Kordes *et al.*[64] reported cryoscopic measurements which indicated that potassium tetracyanocuprate(I) formed five foreign ions in potassium thiocyanate, and thus that neither complex was stable.

A recent extensive study of an organic thiocyanate melt, utilizing measurements of electrical conductivity, viscosity, density, and infrared and visible spectra, showed that copper(I) thiocyanate initially reacted with tetra-*n*-pentylammonium thiocyanate at 60–80°C to form a network polymer which first went into solution as chain polymers and later dissociated into dimeric complexes on dilution.[84] Unfortunately, there is no evidence at present as to how far this interesting series of reactions may be paralleled in alkali metal thiocyanate melts.

Silver(I) ions have been found to be stable in thiocyanate melts and to migrate as a cation on application of an electrical potential,[46] and have been the subject of polarographic and emf studies.[57,58] Silver(I) chloride, bromide, and iodide apparently dissociate to form 1.5 foreign ions per mole, according to the cryoscopic measurements of Kordes,[64] and are interpreted on the basis of the reactions

$$2AgX + 3SCN^- \rightarrow [Ag(SCN)X]^- + [Ag(SCN)_2]^- + X^-$$

though addition of chloride ions is reported not to effect the (charge transfer) spectrum of silver(I).[57]

As would be expected, cryoscopy indicates only one foreign ion is produced by silver(I) thiocyanate.[64] However, silver(I) cyanide supplies one foreign ion, and potassium dicyanoargentate(I) two foreign ions,[64] thus indicating the stability of the monocyano compound which has a reported stability constant of 1.7×10^3 in potassium thiocyanate and 1.4×10^3 in sodium thiocyanate–potassium thiocyanate.[58] However, Eluard and Trémillon[72] claim the dicyano complex to be the stable form, while Jensen[57] states that silver(I) sulfide is precipitated on addition of excess potassium cyanide. Silver(I) ions are precipitated when sulfide ions are formed by the electrolytic reduction of sodium thiocyanate–potassium thiocyanate at 155°C,[10] and solubility products have been calculated, though the influence of the cyanide generated along with the sulfide ions was not considered. Analogous reactions at 170°C could not be carried out "because of the formation of polysulfides by the sulfur arising from thermal decomposition of the solvent." However, it seems most improbable that sufficient sulfur would be produced at these temperatures (cf. Sec. 1.5).

Gold(III) ions apparently oxidize thiocyanate to sulfur, cyanogen, and gold(I) ions. The latter are stable except at high concentrations when they are precipitated as sulfide:

$$2Au^+ + 2SCN^- \rightarrow Au_2S + S + (CN)_2$$

Addition of potassium cyanide to the gold(I) solutions gives the dicyano complex, the relative stabilities of the group IB metal dicyanide complexes being $| Au(CN)_2 |^- > [Cu(CN)_2]^- > (Ag(CN)_2]^-$.[72]

According to a brief remark, both gold and silver cathodes became covered with a lime-green insoluble layer at 200°C, and metal ions were found in the golden-yellow melt of potassium thiocyanate.[82]

3.8. Reactions with Compounds of Zinc, Cadmium, and Mercury

Zinc(II) ions have been generally considered to be stable in thiocyanate melts at low temperatures, have been studied cryoscopically[62] and polarographically at 137°C[9] and 195°C,[54] and have been shown to electromigrate as an anionic complex at 210°C.[45] Zinc(II) thiocyanate is soluble to concentrations of greater than $7 M$ in potassium thiocyanate at 170°C, giving a brownish solution. Infrared spectra of these solutions indicated both nitrogen and sulfur bonding to zinc as well as free thiocyanate ions, observations interpreted as indicating polymer formation.[45]

Evidence of reaction of zinc(II) chloride has been found at 300°C, though not at 250°C, zinc(II) sulfide being precipitated.[59] Zinc sulfide has also been rather surprisingly reported to be formed at 195°C if the concentration of Zn(II) exceeded 0.1 M.[72] Zinc(II) oxide forms the sulfide at 300°C.[56]

$$ZnO + SCN^- \rightarrow ZnS + OCN^-$$

Zinc(II) ions are also precipitated by electrogenerated sulfide at 170°C in sodium thiocyanate–potassium thiocyanate, the cyanide ions produced at the same time by the process

$$SCN^- + 2e = S^{2-} + CN^-$$

apparently not interfering since the amperometric titrations were correct within 2.5%.[11] In contrast, the di-, tri-, and tetracyano complexes are stated to be stable in potassium thiocyanate at 195°C.[72]

Cadmium(II) ions are reported to behave in a very similar fashion to those of zinc(II). They have generally been found to be stable at low temperatures[86] and have been studied cryoscopically,[64] by emf,[12] and by polarography.[9,54] Cadmium(II) thiocyanate dissolves to give a colorless solution, of greater than 2 M concentration, which showed, by infrared, nitrogen and possibly sulfur bonding.[45] Cadmium(II) thiocyanate complexes have also been reported in electrical conductivity measurements in a molten organic thiocyanate.[87]

Reaction of cadmium(II) chloride has been found at 300°C, but not at 250°C, with formation of the sulfide[59] and at 195°C if present in concentrations greater than 0.1 M; an insoluble product has also been reported at 210°C.[46] Again as with zinc, distinctly different conclusions have been reported as to whether cyanide complexes are formed or not;[10,72] in each case, the results reported with cadmium are analogous to those of zinc except that the postulated complexes are the mono-, di-, and tricyanides. Similarly, cadmium oxide reacts to form the sulfide.[56] Cadmium metal also reacts rapidly at 195°C to form sulfide and cyanide ions.[72]

Mercury(II) appears to be stable in low-temperature thiocyanate melts. The chloride, bromide, and iodide have been shown by cryoscopy to produce two foreign ions per mole, possibly by formation of monohalogen complexes and free halide ions.[64] Mercury(II) thiocyanate is very soluble, dissolving initially to give a yellow-green color but turning pink on standing. The infrared absorption bands of the $>7 M$ solution indicate sulfur bonding to mercury and very little free thiocyanate ion, i.e., the formation of tetrathiocyanate complexes.[45]

Further evidence implying stability is the report that the electronic spectrum of mercury(II) chloride is not affected by additions of chloride ion,[57] though in the presence of excess cyanide, polarography indicates the formation of uncharged mercury(II) cyanide in potassium thiocyanate at 195°C.[72] However, mercury(II) is apparently amperometrically titrated with sulfide ions at 170°C in sodium thiocyanate–potassium thiocyanate without interference from cyanide ions.[11]

In contrast to the above reports indicating the stability of mercury(II) ions, Eluard and Trémillon[72] found that black mercury(II) sulfide precipitated at 195°C in potassium thiocyanate:

$$Hg^{2+} + 2SCN^- \rightarrow HgS + S + (CN)_2 \uparrow$$

when the concentration was $>0.1\ M$, while Metzger[54] reported that mercury(II) gave black solutions.

Mercury(I) also gave black solutions at 195°C in potassium thiocyanate,[54] indicating instability; in addition, disproportionation to mercury(II) (which was subsequently precipitated as the sulfide) and mercury metal has been reported.[72]

Mercury metal itself slowly reacts to give mercury(II) sulfide and cyanide ions.[72]

4. CHEMISTRY OF THE "F" BLOCK ELEMENTS

The chemistry of the lanthanide elements is at present little studied. The sole, mainly incidental, fragments of information available are that cerium(III) and neodymium(III) are stable for 10 hr in potassium thiocyanate at 190°C;[86] that the distribution coefficients of terbium(III), holmium(III), thulium(III), and lutetium(III) between sodium thiocyanate–potassium thiocyanate and dioctylammonium thiocyanate in tetralin at 150–190°C has been determined at 150–190°C;[88] and that cerium(IV) reacts in potassium thiocyanate at 195°C liberating sulfur and cyanogen.[54] Hence it may be deduced that the ions of oxidation state III are probably all stable, but that cerium(IV) displays the usual oxidizing tendencies.

Of the actinide elements, the only information reported is on uranium(VI). Panzer and Schaer[16] found that uranium(VI) oxide and hydroxyperuranic acid (i.e., $UO_4 \cdot 2H_2O$) reacted, while Milbauer[56] reported that uranyl acetate formed uranyl sulfide at 400°C. Yanagi[86] also found uranyl ions to react at 190°C in potassium thiocyanate to give a black precipitate of uranyl sulfide. No other products (e.g., parathiocyanogen or sulfur and cyanogen) or chemistry have been reported.

5. CONCLUSION

It will be apparent to readers that though the chemistry of molten thiocyanates is a particularly rich and varied one, by no means are all of its facets are thoroughly explored. However, it is hoped that the considerable gaps in this field, only too evident when reading this review, will be filled soon; and that the occasional completely contradictory assertions, sometimes only narrowly avoided by some small experimental difference in concentration, temperature, or melt composition (and thus demanding such laborious description), will stimulate workers in the field to remove the inconsistencies so that the intrinsically interesting topic of molten thiocyanate chemistry may be more clearly understood.

6. REFERENCES

1. E. Rhodes and A. R. Ubbelohde, *Trans. Faraday Soc.* **55**:1705 (1959).
2. J. B. Wrecesnewsky, *Z. Anorg. Allgem. Chem.* **74**:95 (1912).
3. D. W. Plester, S. E. Rogers, and A. R. Ubbelohde, *Proc. Roy. Soc. A* **235**:469 (1956).
4. J. P. Frame, E. Rhodes, and A. R. Ubbelohde, *Trans. Faraday Soc.* **55**:2039 (1959).
5. H. E. Williams, *Cyanogen Compounds*, Edward Arnold and Co. (1948).
6. C. B. Baddiel and G. J. Janz, *Trans. Faraday Soc.* **60**:2089 (1964).
7. G. Piantoni, M. Braghetti, and P. Franzosini, *Ric. Sci.* **38**:942 (1968).
8. A. F. Oparina and N. S. Dombrovskaya, *Zh. Neorgan. Khim.* **3**:413 (1958).
9. M. Francini, S. Martini, and H. Giess, *Electrochim. Metal.* **3**:355 (1968).
10. F. Pucciarelli, P. Cescon, and V. Bartocci, *Anal. Chim. Acta.* **57**:224 (1971).
11. A. Cescon, F. Pucciarelli, and M. Fiorani, *Talanta* **17**:647 (1970).
12. P. Cescon, R. Marassi, V. Bartocci, and M. Fiorani, *J. Electroanal. Chem.* **23**:255 (1969).
13. E. M. Wadsworth and T. I. Crowell, *Tetrahedron Letters* **13**:1085 (1970).
14. E. W. Thomas and T. I. Crowell, *J. Organ. Chem.* **37**:744 (1972).
15. P. S. Hillery, *Dissertation Abstr.* **30B**:4993 (1970).
16. R. E. Panzer and M. J. Schaer, *J. Electrochem. Soc.* **112**:1136 (1965).
17. H. R. Jindal and G. W. Harrington, *J. Phys. Chem.* **71**:1688 (1967).
18. G. Charlot and R. Trémillon, *Chemical Reactions in Solvents and Melts*, Pergamon, Oxford (1969).
19. A. R. Ubbelohde, *Melting and Crystal Structure*, Oxford University Press (1965).
20. D. H. Kerridge and M. Mosley, *J. Chem. Soc. A* **1967**:352.
21. Z. Iqbal, L. H. Sarma, and K. D. Moller, *J. Chem. Phys.* **57**:4728 (1972).
22. A. H. Norbury and A. I. Sinha, *Quart. Rev.* **1970**:69.
23. A. Sabatini and I. Bertini, *Inorg. Chem.* **4**:1665 (1965).
24. C. Pecile, *Inorgan. Chem.* **5**:210 (1966).
25. O. W. Howarth, R. E. Richards, and L. M. Venanzi, *J. Chem. Soc.* **1964**:3335.
26. S. Ahrland, J. Chatt, and N. R. Davies, *Quart. Rev.* **12**:265 (1958).
27. R. G. Pearson, *J. Am. Chem. Soc.* **85**:3533 (1963).
28. C. K. Jørgensen, *Inorgan. Chem.* **3**:1201 (1964).
29. F. Basolo, W. H. Baddley, and J. L. Burmeister, *Inorg. Chem.* **3**:1202, 1587 (1964).
30. A. H. Norbury, *J. Chem. Soc. A* **1971**:1089.
31. L. Di Sipio, L. Oleari, and G. De Michelis, *Coord. Chem. Rev.* **1**:7 (1966).
32. L. J. B. Husband, *J. Sci. Instr.* **35**:300 (1958).
33. H. C. Egghart, *J. Phys. Chem.* **73**:4014 (1969).
34. P. N. Spencer and B. Cleaver, private communication.

35. M. Mosley, Ph.D. Thesis, University of Southampton.
36. H. Lux and H. Anslinger, *Chem. Ber.* **94**:1161 (1961).
37. J. Greenberg, *J. Chem. Phys.* **39**:3158 (1963).
38. J. Greenberg, B. R. Sundheim, and D. M. Gruen, *J. Chem. Phys.* **29**:461 (1958).
39. W. Giggenbach, *Inorgan. Chem.* **10**:1308 (1971).
40. D. M. Gruen, R. L. McBeth, and A. J. Zielen, *J. Am. Chem. Soc.* **93**:6691 (1971).
41. J. Milbauer, *Z. Anorg. Allgem. Chem.* **49**:46 (1906).
42. E. Patterno and A. Mazzucchelli, *Atti Accad. Naz. Lincei, Rend. Classe Sci. Fis. Mat. Nat.* **16**:465 (1907).
43. A. J. Calandra, M. E. Martins, and A. J. Arvia, *Electrochim. Acta* **16**:2057 (1971).
44. A. J. Arvia, A. J. Calandra, and M. E. Martins, *Electrochim. Acta* **17**:741 (1972).
45. R. E. Hester and K. Krishnan, *J. Chem. Phys.* **48**:825 (1968).
46. R. A. Bailey and A. Steger, *J. Chromatog.* **11**:122 (1963).
47. P. Franzosini, P. Ferloni, and G. Spinolo, Molten Salts with Organic Anions, Università di Pavia, Italy, 1973.
48. T. I. Crowell and P. Hillery, *J. Organ. Chem.* **30**:1339 (1965).
49. E. M. Wadsworth, *Dissertation Abstr.* **31B**:5301 (1971).
50. D. I. Packham and F. A. Rockley, *Chem. Ind.* **1966**:899.
51. C. Iwakura, K. Sawada, and H. Tamura, *Electrochim. Acta,* **17**:2381 (1972).
52. P. Cescon, F. Pucciarelli, and V. Bartocci, *Chim. Ind. (Milan)* **54**:222 (1972).
53. B. J. Brough, D. H. Kerridge, and M. Mosley, *J. Chem. Soc. A* **1966**:1556.
54. G. Metzger, C.E.A.N. Rept. 2566, 1964 [in *Nucl. Sci. Abstr.* **19**:1333 (1965)].
55. D. H. Kerridge and M. Mosley, *Chem. Commun.* **1969**:429.
56. J. Milbauer, *Z. Anorg. Allgem. Chem.* **42**:433 (1904).
57. W. P. Jensen, *Dissertation Abstr.* **25**:819 (1966).
58. W. E. Bennett and W. P. Jensen, *J. Inorgan. Nucl. Chem.* **28**:1829 (1966).
59. T. Yanagi, K. Hattori, and M. Shinagawa, *Rev. Polarog. (Kyoto)* **14**:11 (1966).
60. R. Ricardi and P. Franzosini, *Gazz. Chim. Ital.* **92**:386, 395 (1962).
61. J. Hennion, J. Nicole, and G. Tridot, Analusis **1**:48 (1972).
62. P. Dingemans, *Rec. Trav. Chim.* **58**:559 (1939).
63. G. Piantoni, M. Braghetti, P. Franzosini, *Z. Naturforsch. A* **1968**:2075.
64. E. Kordes, W. Bergmann, and W. Vogel, *Z. Electrochem.* **55**:600 (1951).
65. V. S. Yakovlev, *Ukr. Khim. Zh.* **31**:113 (1965).
66. A. Nobile, M.S. Dissertion, Rensselaer Polytechnic Institute, Troy, New York, 1968.
67. D. H. Kerridge and M. Mosley, *J. Chem. Soc. A* **1969**:2211.
68. D. H. Kerridge and M. Mosley, *J. Chem. Soc. A* **1967**:1874.
69. D. H. Kerridge and M. Mosley, *Chem. Commun.* **1965**:505.
70. G. Harrington and B. R. Sundheim, *Ann. N.Y. Acad. Sci.* **79**:950 (1960).
71. D. H. Kerridge and M. Mosley, *J. Chem. Soc. A* **1967**:352.
72. A. Eluard and B. Trémillon, *J. Electroanal. Chem.* **13**:208 (1967).
73. J. David, as cited by Y. Marcus, in: *Advances in Molten Salt Chemistry,* Vol. 1 (J. Braunstein, G. Mamontov, and G. P. Smith, eds.), Plenum Press, New York (1971).
74. R. A. Bailey and S. L. Kozak, *Inorgan. Chem.* **6**:2155 (1967).
75. S. L. Kozak, Ph.D. Thesis, Rensselaer Polytechnic Institute, Troy, New York, 1968.
76. R. A. Bailey and S. L. Kozak, *Inorgan. Chem.* **6**:419 (1967).
77. M. I. El-Guindy, Ph.D. Thesis, Rensselaer Polytechnic Institute, Troy, New York, 1967.

78. K. S. De Haas, P. A. Brink, and P. Crowther, *J. Inorgan. Nucl. Chem.* **33**:4301 (1971).
79. K. B. Jacymirski, S. W. Wolkow, I. I. Mastowska, and N. I. Buriak, *Roczniki Chem.* **46**:1999 (1972).
80. G. Tridot, J. Nicole, J. Hennion, and R. Pellerioux, *Compt. Rend. Acad. Sci. Paris Ser. C* **1969**:1290.
81. J. Hennion, J. Nicole, and G. Tridot, *Compt. Rend. Acad. Sci. Paris Ser. C* **1968**:831.
82. K. F. Denning and K. E. Johnson, *Electrochim. Acta* **12**:1391 (1967).
83. F. Pucciarelli, P. Cescon, F. Diomedi-Camassei, and M. Heyrovsky, *J. Chem. Soc. Chem. Commun.* **1973**:154.
84. M. A. Kowalski and G. W. Harrington, *Anal. Chem.* **44**:479 (1972).
85. M. A. Kowalski, *Dissertation Abstr.* **31B**:5905 (1971).
86. T. Yanagi, *Bunseki Kagaku*, **18**:195 (1969).
87. P. Keller and G. W. Harrington, *Anal. Chem.* **41**:523 (1969).
88. J. David and M. Zangen, in: *Solvent Extraction Research* (A. S. Kertes and Y. Marcus, eds.), John Wiley (Interscience), New York (1969).

Chapter 6

PHASE DIAGRAMS OF BINARY AND TERNARY FLUORIDE SYSTEMS

Roy E. Thoma

Oak Ridge National Laboratory
Oak Ridge, Tennessee 37830

1. INTRODUCTION

Investigations of the equilibrium phase diagrams of metal fluoride systems have existed as a classical scientific effort since Gibbs, Roozeboom, *et al.* made their major contributions to science around the turn of the century. It remained for the workers in the Oak Ridge Aircraft Reactor Program and later in the Molten Salt Reactor Program to develop the means to acquire definitive data in massive amounts from a variety of interrelated techniques before the complex phase diagrams pertaining to the materials aspects of the Program could be defined with rapidity and precision. Because of their success in developing the experimental methods for investigation of the fluoride systems, the Oak Ridge group is responsible for a considerable fraction of the binary and ternary systems of the metal fluorides for which detailed phase diagrams appear in the literature.

In the earlier stages of the Molten Salt Reactor Program, research in fluoride systems prior to its fruition proceeded not only in the United States, but also in Western Europe and Russia. As a consequence, reports of the metal fluoride systems began to appear in the literature at an unprecedented rate over the last few decades. These investigations were conducted to define the phase relationships in numerous binary, ternary, and quaternary systems of the metal fluorides as a means of evaluating their potential as nuclear-fuel-enriching solutions, fuel solvents, coolants, and fuel-reprocessing media. Within this scope lay systems that could have application to single- or two-region breeder and converter reactors, to advanced-design reactors that employ liquid lead in direct contact with fuel salt, to extractive schemes for separation of rare earths from mixed salts, and to the assessment of accident conditions that might exist if fuel solutions and coolant media were mixed. Obviously, the scope of these goals is so broad that their pursuit was necessarily intermittent; their dynamic character, inherent in such a technology, caused some of the investigations to proceed through only limited stages of development.

With the culmination of the research and development efforts required to select the most suitable combinations of fluorides for application in the technology, it is fitting that the cumulative information which has been developed be recapitulated in summary form. To this end, all of the phase diagrams known to the author that can be regarded as condensed systems, and that delineate the liquidus, have been collated and included in the present review. Excluded from the collection are the recently reported systems in which ammonium fluoride, hydrazinium fluoride, water, or the highly volatile fluorides are components, for, in general, reports on these systems are fragmentary. Information is included, if known, pertaining to structural properties of the intermediate compounds that form in binary and ternary fluoride systems, and in some cases where such information comprises the only known information for a particular system. It will be noted that occasionally in this collection structural data, obtained subsequent to the termination of phase investigations with a particular system, indicate stoichiometry of some compounds to be different from that reported previously. Since such structural determinations originate from single-crystal studies, the compound stoichiometry designated by the results of these determinations is probably correct.

The following sections of this review are developed essentially as a catalog, similar to the format of the early report, *Phase Diagrams of Nuclear Reactor Materials*, ORNL-2548 (1959). They are arranged in group combinations, I–I, I–II, I–III, etc., and by increasing atomic numbers and oxidation states within groups. Many of the phase diagrams in the current review have been included in *Phase Diagrams for Ceramists*, published by the American Ceramic Society. A number of the diagrams, especially those originating in the Russian literature, were available for reproduction here through the courtesy of the American Ceramic Society. Much of the structural data that have been collected is summarized in the report, *Crystallographic Data for Some Metal Fluorides, Chlorides, and Oxides*, D. D. Brunton, H. Insley, and R. E. Thoma, ORNL-3761 (1965).

It has not been possible to assess the accuracy of the information pertaining to the phase transitions listed in the following sections, for this can only be done experimentally. Similarly, no effort has been given to appraisal of the various experimental methods which have been employed in the determination of the phase diagrams. Except in a few instances, the data are less precise than is desired in the sophisticated researches that are now commonplace with these systems. The accuracy of most of the transitions range from ± 3 to $\pm 10°$. The reader will generally find that the accuracies given are within $5°$.

Annotation of the invariant and singular equilibrium points in the phase diagrams of the fluoride systems follows an abbreviated style in which the temperature and composition of these invariant equilibrium points are designated. For example, $E = 500°$, 50% MF indicates that within the system a eutectic occurs at $500°C$ containing 50 mole $\%$ MF. All other designations follow the same pattern, which designates temperature in degrees centigrade and composition in mole percent, exclusively. Where crystal lattice dimensions are given, the units are in angstroms.

In the most current report of a system one may occasionally find a review of the previous investigations pertaining to the system, thus affording the reader the opportunity to trace the evolutionary stages of a system. Therefore, the practice has been adopted to include in the present collection only the most recent phase diagram of a system unless significantly disparate versions have been reported.

The reader is cautioned to note that it is commonplace in the Russian literature for authors to construct phase diagrams in terms of equivalent compositions, e.g., the composition of LiF–NaF–CaF_2 will be given diagrammatically in terms of Li_2F_2, Na_2F_2, and CaF_2. Phase diagrams for such systems are shown in their original form, for otherwise it would entail complete reconstruction (with possible attendant interpretational errors) of the phase diagram to conform to Western style.

2. BINARY SYSTEMS

The System LiF–NaF. See Fig. 1. Source: J. L. Holm, *Acta Chem. Scand.* **19**:638 (1965). $E = 649°$, 39% NaF. Results of earlier work by A. G. Bergman and E. P. Dergunov, *Compt. Rend. Acad. Sci. URSS* **31**:753 (1941), are in good agreement, showing the eutectic at 40 mole $\%$ NaF with m.p. $652°C$. Holm reported the maximum solid solubility of LiF in NaF to be 4 ± 0.5 mole $\%$, as compared with the value $8 \pm 1\%$ reported by J. M. Short and R. Roy, *J. Am. Ceram. Soc.* **47**:149 (1964).

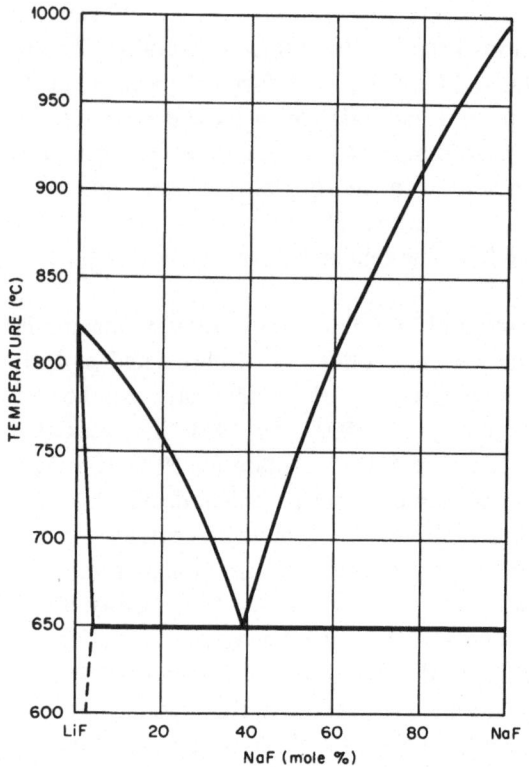

Fig. 1. The system LiF–NaF.

The System LiF–KF.   See Fig. 2. Source: A. G. Bergman and E. P. Dergunov, *Compt. Rend. Acad. Sci. URSS* **31**:753 (1941). $E = 492°$, 50% KF.

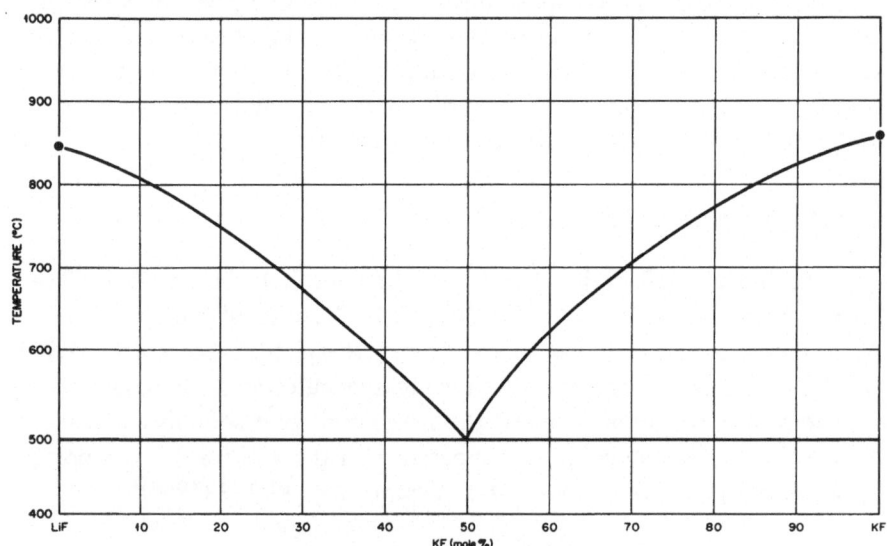

Fig. 2. The system LiF–KF.

The System LiF–RbF. See Fig. 3. Source: C. J. Barton, T. N. McVay, L. M. Bratcher, and W. R. Grimes, ORNL, unpublished work (1954). The eutectic is at 44 mole % LiF, m.p. 470°; the peritectic is at 47 mole % LiF, and at 475°. The data used for the construction of this phase diagram were obtained from cooling-curve experiments. The general

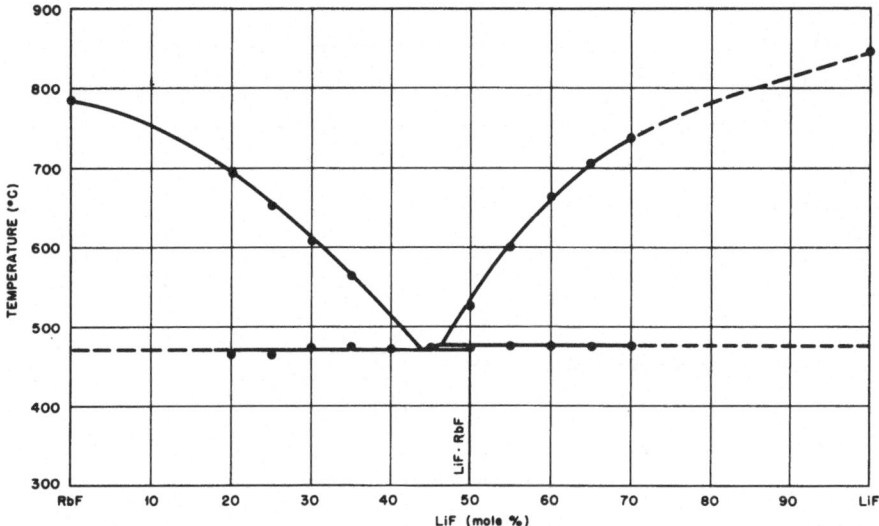

Fig. 3. The system LiF–RbF.

characteristics of the system were reported earlier by E. P. Dergunov, *Dckl. Akad. Nauk SSSR* **58**:1369 (1947). The structure of the intermediate compound, LiF · RbF, was established by Burns and Busing, *Inorg. Chem.* **4**:1510 (1965). Crystals of the compound are monoclinic, space group $C2/c$, $a = 5.83$, $b = 11.16$, $c = 7.86$, $\beta = 94°55'$.

The System LiF–CsF. See Fig. 4. Source: C. J. Barton, L. M. Bratcher, T. N. McVay, and W. R. Grimes, ORNL, unpublished work (1954), $E = 495 \pm 5°$, 55% CsF; $E = 475 \pm 5°$, 63% CsF. The data for construction of this phase diagram were obtained from cooling-curve analysis and thermal gradient quenching experiments (designated hereafter as TGQ). The structure of the intermediate compound, LiF · CsF, has been determined by J. H. Burns and W. R. Busing, *Inorg. Chem.* **4**:1510 (1965), S. G. $C2/c$, $a = 6.01$, $b = 11.64$, $c = 8.18$, $\beta = 90°45'$.

A phase diagram of the system was published by G. A. Bukhalova and D. V. Sementsova, *Zh. Neorg. Khim.* **10**:1880 (1965), showing LiF · CsF as a congruently melting compound, and with $E = 490°$, ∼48% CsF, and $E = 479°$, 60% CsF. More recent phase diagrams of ternary systems limited by the LiF · CsF binary system (Figs. 153, 164, 166, 179–181) indicate that LiF · CsF melts congruently at 494°. Although the phase diagrams of the ternary systems reported by Bukhalova and co-workers indicate conclusively that congruent melting occurs, they are based on thermal analysis, whereas inference of incongruency was based on petrographic

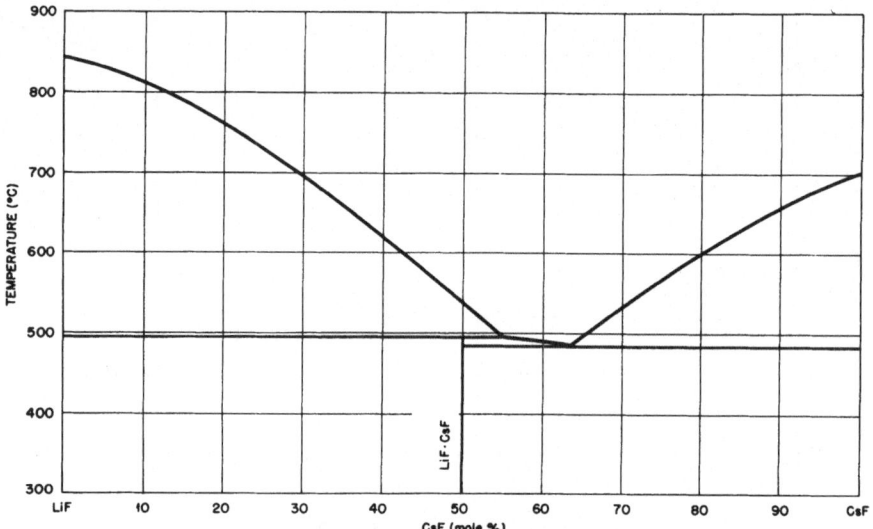

Fig. 4. The system LiF–CsF.

examination of quenched specimens as well as on thermal analysis of cooling-curve data. Evidence of the existence of two eutectics adjacent to Alkamede lines, as shown in Figs. 153, 164, and 166 leading from LiF · CsF, is strongly indicative of congruent melting of LiF · CsF. However, because of the frequency of experimental artifacts in fluoride systems, it is the reviewer's conclusion that the melting relations of LiF · CsF have not yet been established unequivocally.

The System NaF–KF. See Fig. 5. Source: J. L. Holm, *Acta Chem. Scand.* **19**:638 (1965). $E = 721°$, 60% KF. The phase diagram indicates the eutectic temperature to be 11° higher than that found by A. G. Bergman and E. P. Dergunov, *Compt. Rend. Acad. Sci. URSS* **31**:753 (1941), although of the same composition. In contrast to previous investigators, Holm found that KF can contain $5 \pm 0.5\%$ NaF at the eutectic temperature. No solid solubility of KF in NaF was found.

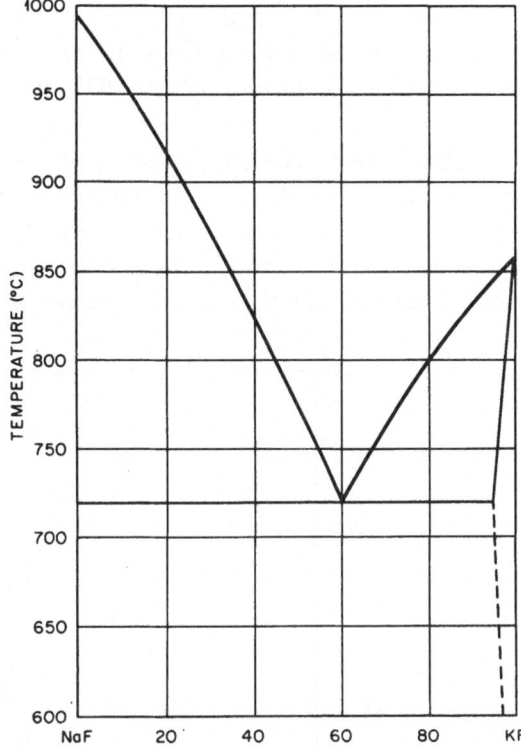

Fig. 5. The system NaF–KF.

The System NaF–RbF. See Fig. 6. Source: C. J. Barton, L. M. Bratcher and W. R. Grimes, ORNL, unpublished work (1951). Barton

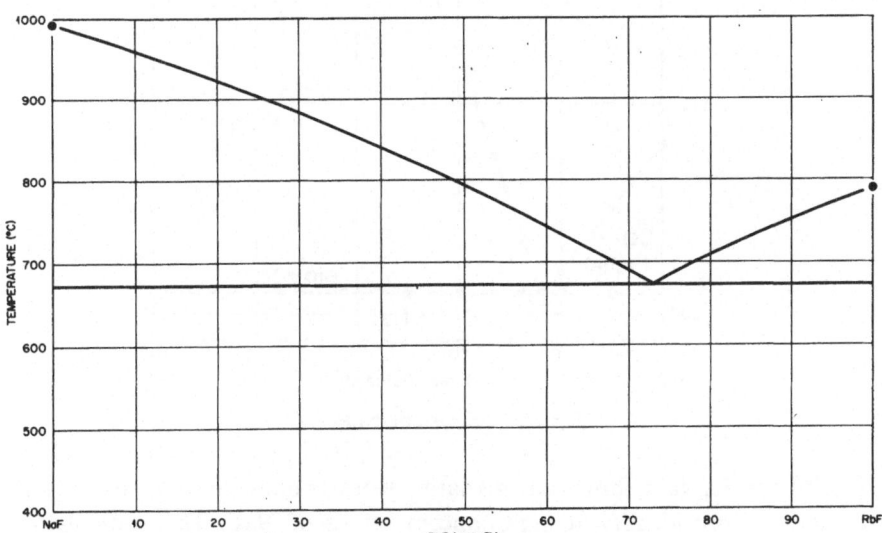

Fig. 6. The system NaF–RbF.

and co-workers found the eutectic at 73 mole % RbF, 675 ± 10°. In a
more recent examination, Holm [*Acta Chem. Scand.* **19**:639 (1965)] concluded that the eutectic exists at 667° and at 67.2 mole % RbF.

The System NaF–CsF. See Fig. 7. Source: D. L. Deadmore and J. S.
Machin, *J. Phys. Chem.* **64**:824 (1960). The eutectic composition was not
reported, but is at approximately 24 mole % NaF. A report by R. G. Samuseva and V. E. Plyushchev, *Russ. J. Inorg. Chem.* (*English Transl.*) **6**:1093
(1961), indicates that the eutectic occurs at 20 mole % NaF and at 630°.

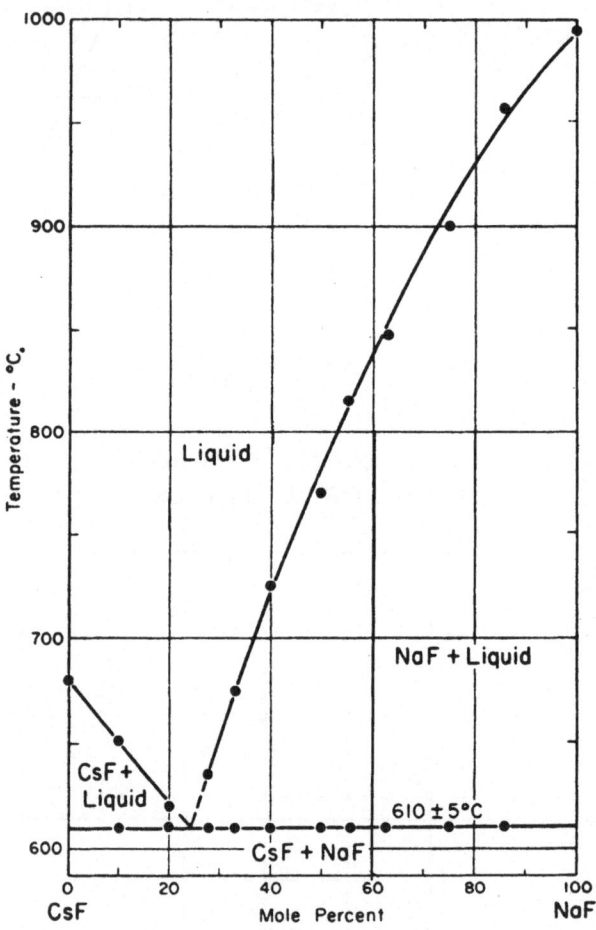

Fig. 7. The system NaF–CsF.

The phase diagram shown in this latter work exhibits anomalous negative
deviation from ideality at high concentrations of NaF. The anomaly was
not discussed by the authors, nor did they indicate cognizance of the 1960
diagram shown here.

The System KF–RbF. See Fig. 8. Source: C. J. Barton, L. M. Bratcher, and W. R. Grimes, ORNL, unpublished work (1951). The system was found to exhibit continuous solid solution of the components with no evidence of immiscibility in the solid state. The minimum temperature of the liquidus is $770 \pm 10°$ at 72% RbF.

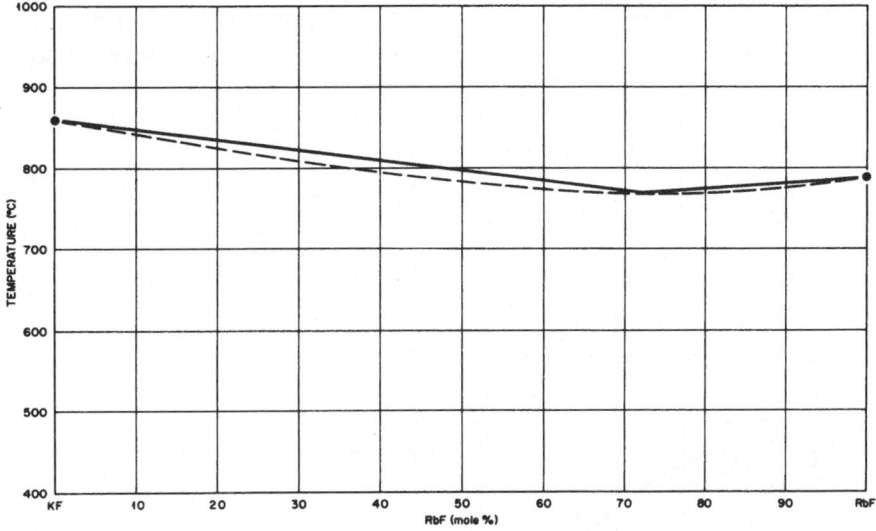

Fig. 8. The system KF–RbF.

The System KF–CsF. See Fig. 9. Source: R. G. Samuseva and V. E. Plyushchev, *Russ. J. Inorg. Chem. (English Transl.)* **10**:689 (1965). The

Fig. 9. The system KF–CsF.

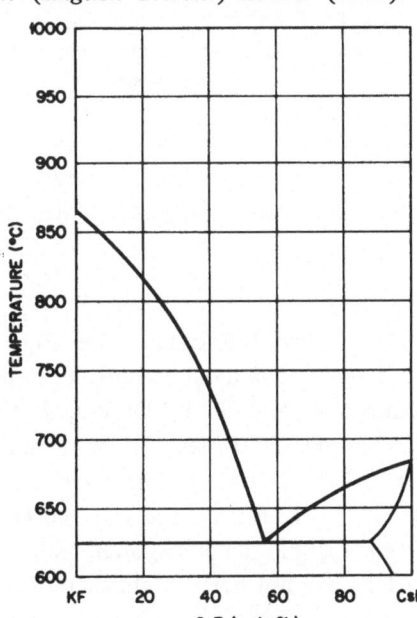

authors inferred that CsF can accommodate up to 15 mole % KF in solid solution at the eutectic temperature.

The System RbF–CsF. See Fig. 10. Source: R. G. Samuseva and V. E. Plyushchev, *Russ. J. Inorg. Chem. (English Transl.)* **10**:689 (1965). The authors deduced that RbF and CsF form a continuous series of solid

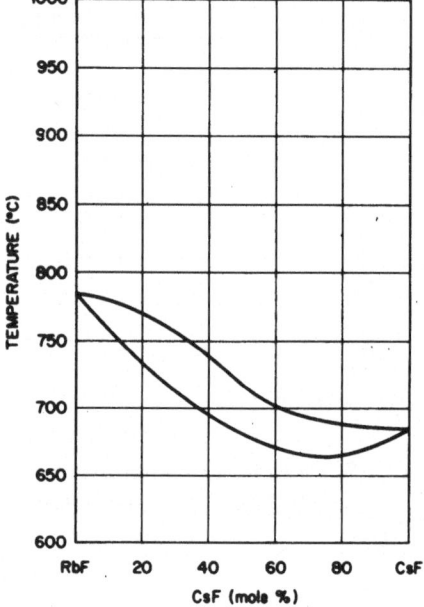

Fig. 10. The system RbF–CsF.

solutions with a temperature minimum in the liquidus. As constructed, the proposed phase diagram violates the phase rule.

The System LiF–BeF₂. See Fig. 11. Refinement of the LiF–BeF$_2$ phase diagram has most recently been accomplished by K. A. Romberger, J. Braunstein, and R. E. Thoma, *J. Phys. Chem.* **76**:1454 (1972) and by R. E. Thoma, H. Insley, H. A. Friedman, and G. M. Hebert, *J. Nucl. Mater.* **27**:166 (1968). In the most recent description of the structure of Li$_2$BeF$_4$, J. H. Burns and G. K. Gordon, *Acta Cryst.* **20**:135 (1966) reported Li$_2$BeF$_4$ as hexagonal, $R\bar{3}$, $a = 13.29$, $c = 8.91$. The structure of LiBeF$_3$ has not been established. Its powder pattern is listed in ORNL-3761.

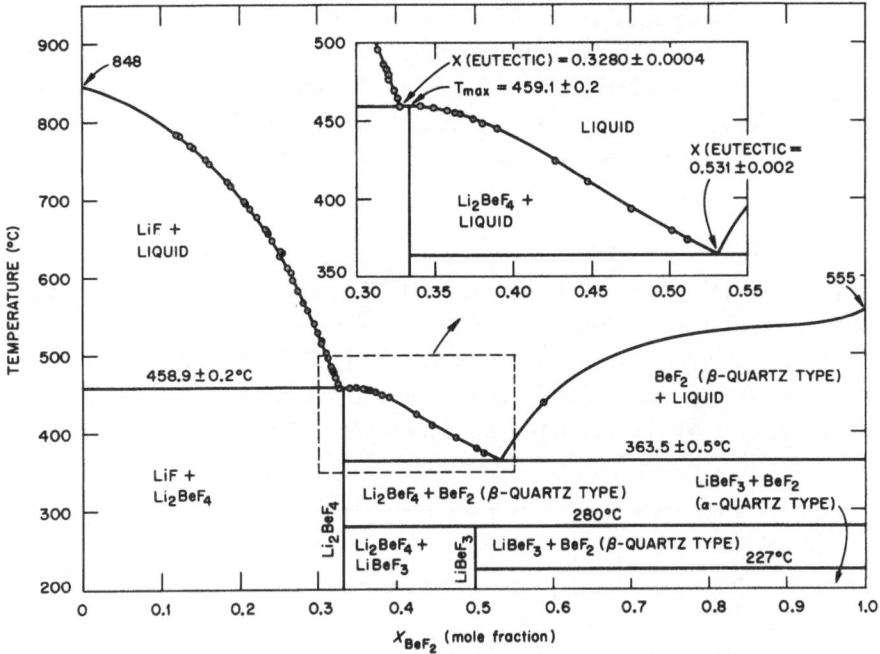

Fig. 11. The system LiF–BeF₂.

The System NaF–BeF₂. See Fig. 12. Source: D. M. Roy, R. Roy, and E. F. Osborn, *J. Am. Ceram. Soc.* **36**:185 (1953). These authors did

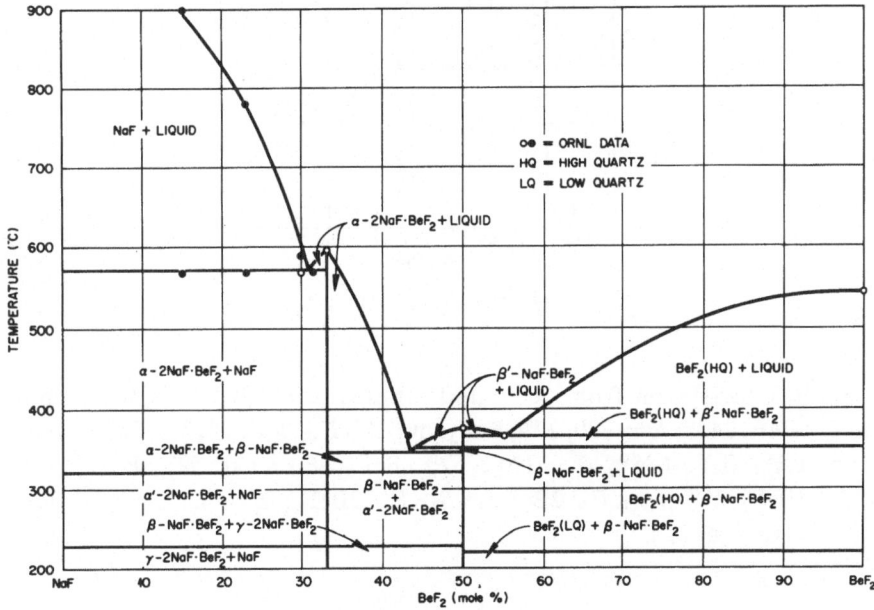

Fig. 12. The system NaF–BeF₂.

not list invariant equilibria; shown below are estimated values from the work reported and from experiments performed at ORNL. $E = 570°$, 31% BeF_2; m.p. Na_2BeF_4 600°; α-$Na_2BeF_4 \rightleftarrows \alpha'$-$Na_2BeF_4$ 320°; α'-$Na_2BeF_4 \rightleftarrows \gamma$-$Na_2BeF_4$ 225°; $E = 340°$, 43% BeF_2; m.p. $NaBeF_3$ 376 ± 5°; β-$NaBeF_3 \rightleftarrows \beta'$-$NaBeF_3 \approx$ 370°; $E = 365°$, 55% BeF_2.

The System KF–BeF₂. See Fig. 13. Source: R. E. Moore, C. J. Barton, T. N. McVay, and W. R. Grimes, ORNL (unpublished), 1955. Further characterization of the phase diagram has been reported by M. P. Borsenkova, A. V. Novoselova, Yu. P. Simanov, V. I. Chernikh, and E. I. Yarembash, *Zh. Neorg. Khim.* **1**:2071 (1956), and by A. V. Novoselova, Yu. M. Korenev, and M. P. Borzenkova, *Zh. Neorg. Khim.* **9**:2042 (1964), who

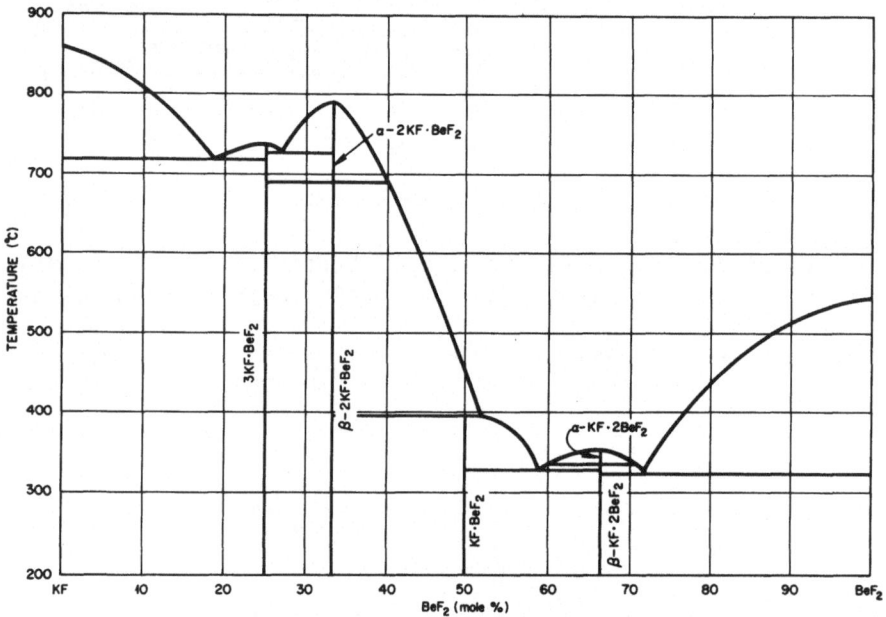

Fig. 13. The system KF–BeF₂.

described the system from 50 to 100% BeF_2. $E = 720°$, 19% BeF_2; m.p. 3KF · BeF_2 740°; $E = 730$, 27% BeF_2; m.p. 2KF · BeF_2 787°; α-2KF · $BeF_2 \rightleftarrows \beta$-2KF · BeF_2 685°; $P = 390°$, 52% BeF_2; $E = 330$, 59% BeF_2; m.p. KF · $2BeF_2$ 358°; α-KF · $2BeF_2 \rightleftarrows \beta$-KF · $2BeF_2$; $E = 323°$, 72.5% BeF_2 (Borzenkova *et al.*).

Structural data: β-2KF $\cdot$ BeF$_2$: orthorhombic, $a = 5.70$, $b = 9.90$, $c = 7.28$, H. O'Daniel and L. Tscheischwili, *Z. Krist.* **104**:350 (1942).

The System RbF–BeF$_2$. See Fig. 14. Sources: R. E. Moore, C. J. Barton, L. M. Bratcher, T. N. McVay, G. D. White, R. J. Sheil, W. R. Grimes, and R. E. Meadows, ORNL, unpublished work (1955–1956), and R. G. Grebenshchikov, *Dokl. Akad. Nauk SSSR* **114**:317 (1957).

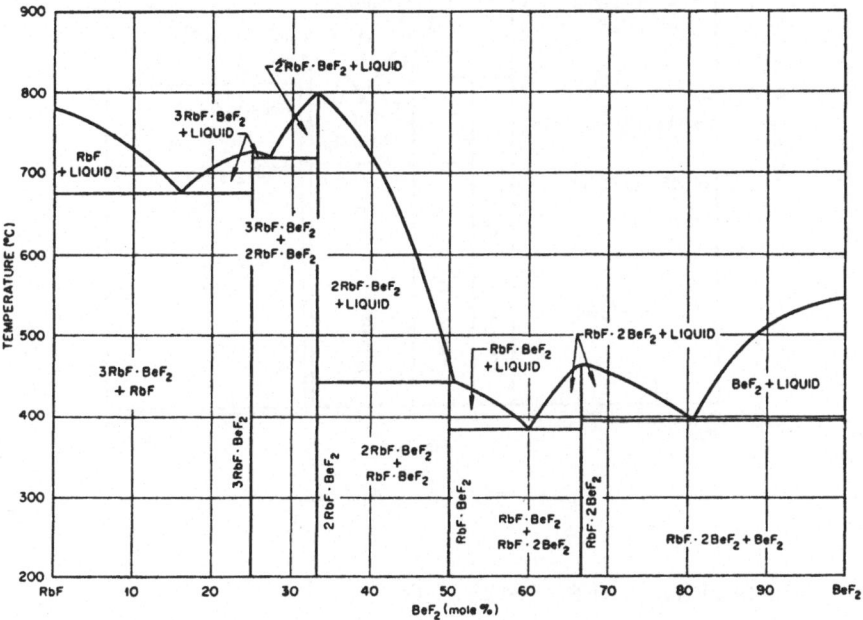

Fig. 14. The system RbF–BeF$_2$.

The two phase diagrams are in fair agreement, although the latter investigators have concluded that the solid-state relations are more complex than were deduced by the earlier group of workers. Invariant and singular equilibrium points found by Moore and co-workers are listed below.

The System RbF–BeF$_2$. See Fig. 15. $E = 675°$, 16% BeF$_2$; m.p. 3RbF $\cdot$ BeF$_2$ 725°; $E = 720°$, 27% BeF$_2$; m.p. 2RbF $\cdot$ BeF$_2$ 800°; $P = 442°$, 50.5% BeF$_2$; $E = 383°$, 61% BeF$_2$; m.p. RbF $\cdot$ 2BeF$_2$ 464°; $E = 397°$, 81% BeF$_2$. The phase diagram as reported by Grebenshchikova is shown in Fig. 15.

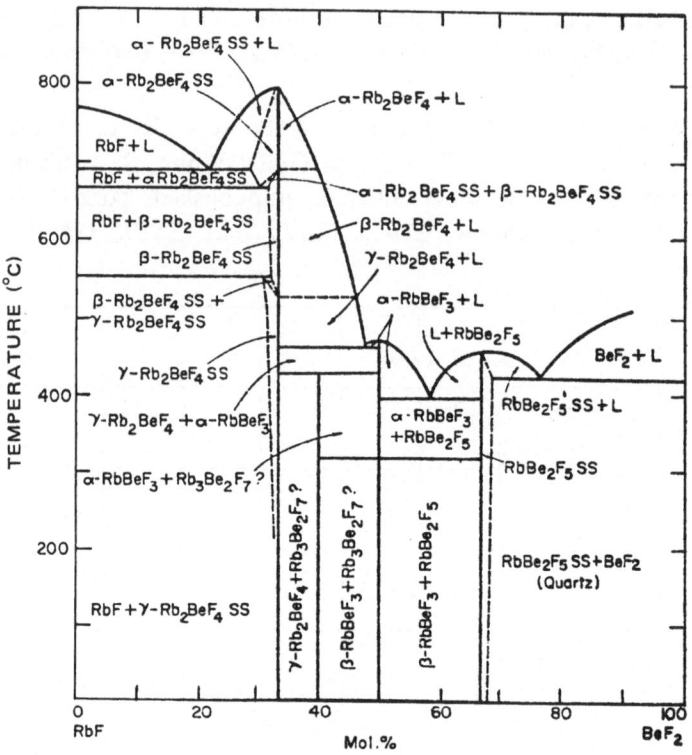

Fig. 15. The system RbF–BeF$_2$.

Structural data: RbF · BeF$_2$: triclinic, P_1 or P-1, $a = 4.69$, $b = 4.61$, $c = 6.12$, V. Ilyukhin and B. Belov, *Dokl. Akad. Nauk SSSR* **140**:1066 (1961).

The System CsF–BeF$_2$. See Fig. 16. Source: O. N. Breusov, A. V. Novoselova, and Yu. P. Simanov, *Dokl. Akad. Nauk SSSR* **118**:935 (1958). $E = 598°$, 14% BeF$_2$; $P = 659°$, 23.5% BeF$_2$; α-3CsF · BeF$_2 \rightleftarrows \beta$-3CsF · BeF$_2$ 617°; m.p. 2CsF · BeF$_2$ 793°; α-2CsF · BeF$_2 \rightleftarrows \beta$-2CsF · BeF$_2$ 404°; $E = 449°$, 48% BeF$_2$; m.p. CsF · BeF$_2$ 475°; α-CsF · BeF$_2 \rightleftarrows \beta$-CsF · BeF$_2$ 360°; β-CsF · BeF$_2 \rightleftarrows \alpha$-CsF · BeF$_2$ 140°; $E = 393°$, 58.4% BeF$_2$; m.p. CsF · 2BeF$_2$ 480°; α-CsF · 2BeF$_2 \rightleftarrows \beta$-CsF · 2BeF$_2$ 450°; $E = 367°$, 77.5% BeF$_2$.

Structural data: The compound CsBeF$_3$ was reported by H. Steinfink and G. D. Brunton, *Acta Cryst.* **B24**:807 (1968) to crystallize in the space group *Pnma* with $a = 4.828$, $b = 6.004$, and $c = 12.794$.

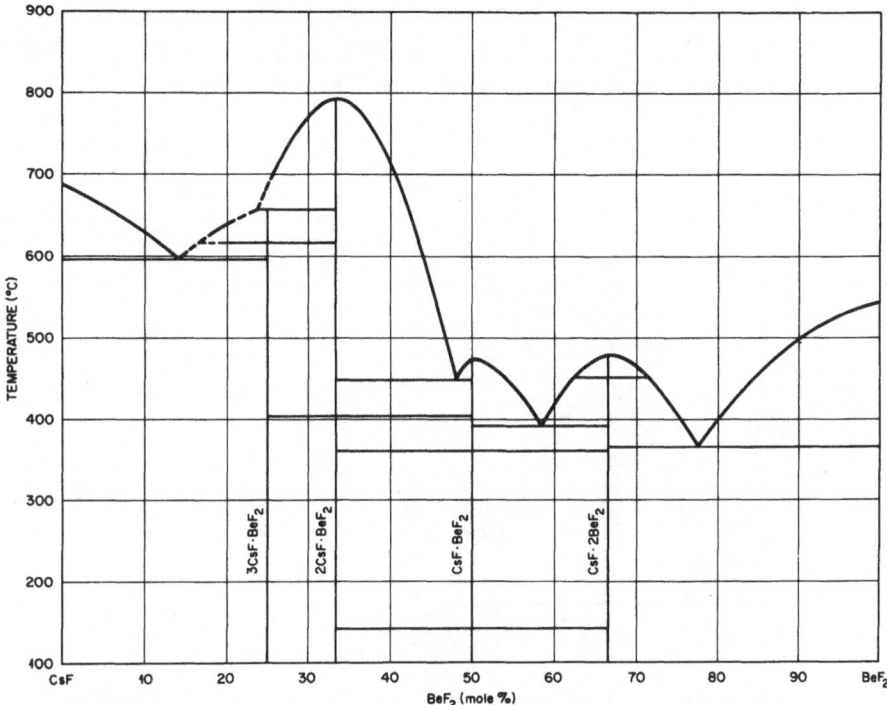

Fig. 16. The system CsF–BeF$_2$.

The System LiF–MgF$_2$. See Fig. 17. Source: W. E. Counts, Rustum Roy, and E. F. Osborn, *J. Am. Ceram. Soc.* **36**, 15 (1953). $E = 735°$, 36% MgF$_2$. The mutual solubility limits of the components were not given. The solidus between ~32 and ~38% MgF$_2$ should be regarded as horizontal so as to conform to the phase rule.

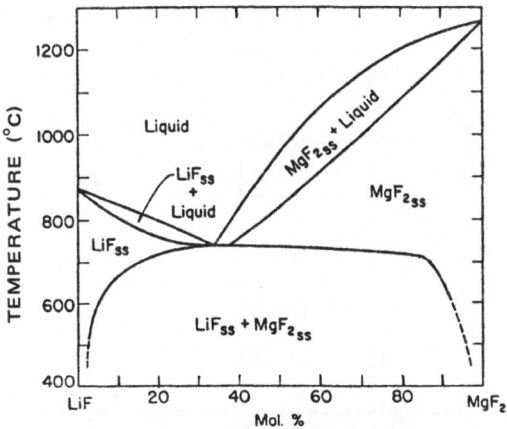

Fig. 17. The system LiF–MgF$_2$.

The System NaF–MgF$_2$. See Fig. 18. Source: A. G. Bergman and E. P. Dergunov, *Compt. Rend. Acad. Sci. URSS* **31**:755 (1941). $E = 830°$, 25% MgF$_2$; m.p. NaF · MgF$_2$ 1030°; $E = 1000°$, 64% MgF$_2$.

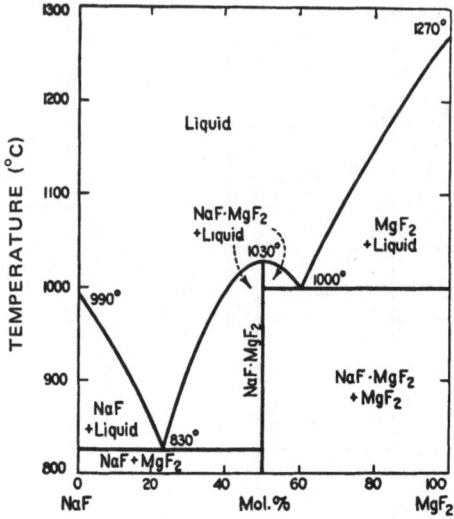

Fig. 18. The system NaF–MgF$_2$.

The System KF–MgF$_2$. See Fig. 19. Source: R. C. DeVries and Rustum Roy, *J. Am. Chem. Soc.* **75**:2841 (1953). $E = 778$, 15% MgF$_2$;

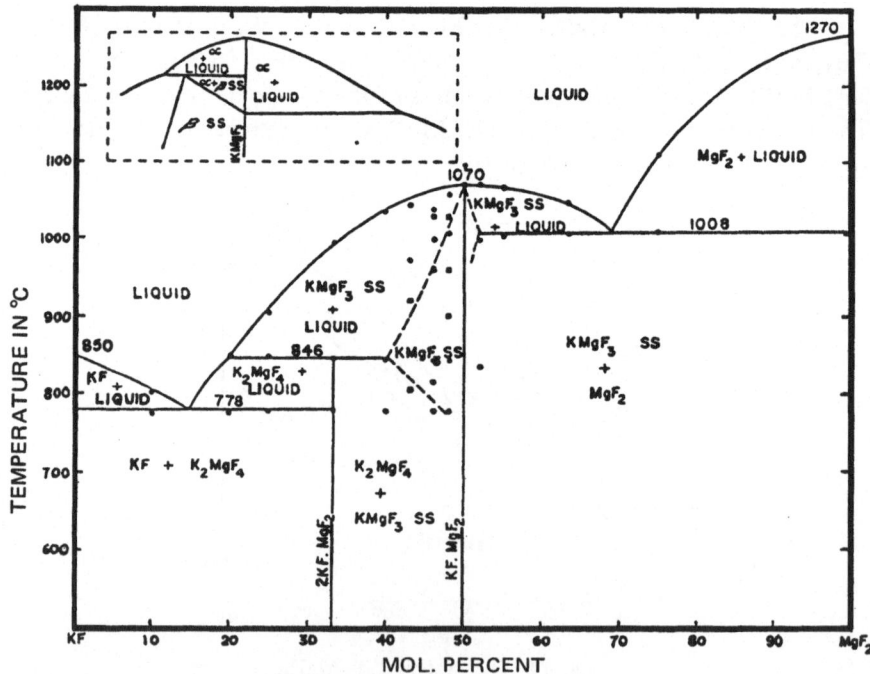

Fig. 19. The system KF–MgF$_2$.

$P = 846°$, 20.5% MgF_2; congruent m.p. $KF \cdot MgF_2$ $1070°$; $E = 1008°$, 69% MgF_2. A possible alternative to the phase relationships of $KF \cdot MgF_2$ is shown in the inset.

Structural data: The compound $KF \cdot MgF_2$ was reported by W. L. W. Ludekens and A. J. E. Welch, *Acta Cryst.* **5**:841 (1952), as a monoclinic structure, $a = b = c = 7.98$, $\beta = 91°18'$.

The System CsF–MgF₂.   The CsF–MgF_2 phase diagram has not been described. The structure of the intermediate compound, $4CsF \cdot 3MgF_2$, has been reported by H. Steinfink and G. D. Brunton, *Inorg. Chem.* **8**:1665 (1968), as orthorhombic *Cmca*, $a = 6.1333$, $b = 14.561$, $c = 13.653$. W. L. W. Ludekens and A. J. E. Welch have reported the structure of $CsF \cdot MgF_2$, *Acta Cryst.* **5**:841 (1952), as tetragonal, $a = 9.37$, $c = 8.70$.

The System LiF–CaF₂.   See Fig. 20. Source: D. L. Deadmore and J. S. Machin, *J. Phys. Chem.* **64**:824 (1960). The eutectic composition was not reported in the source reference; it is approximately 23 mole $\%$ CaF_2.

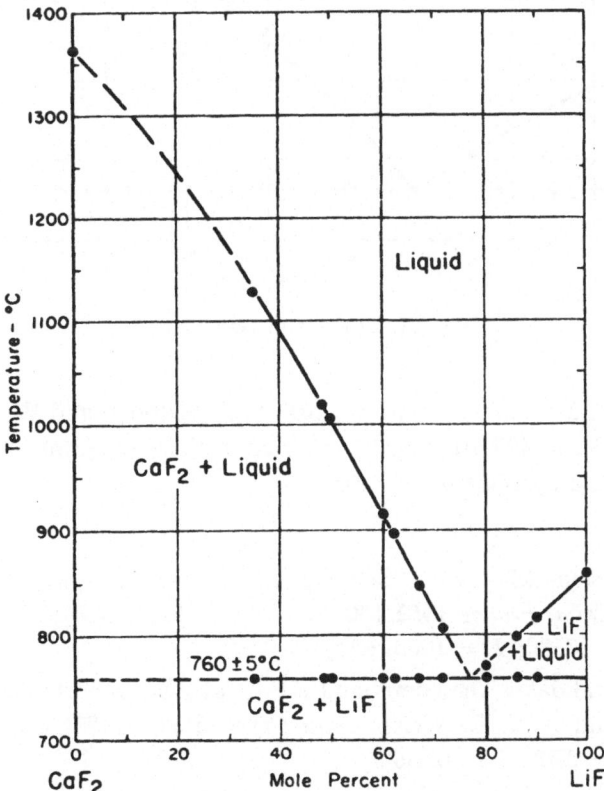

Fig. 20. The system LiF–CaF₂.

A phase diagram of the system was reported earlier by W. E. Roake, *J. Electrochem. Soc.* **104**:661 (1957). The reference diagram differs from the earlier version principally by adoption of $\sim$1360°C as the m.p. of CaF_2 rather than $\sim$1430°C as used by Roake.

The System NaF–CaF_2. See Fig. 21. Source: C. J. Barton, L. M. Bratcher and W. R. Grimes, ORNL, unpublished work (1955–1956).

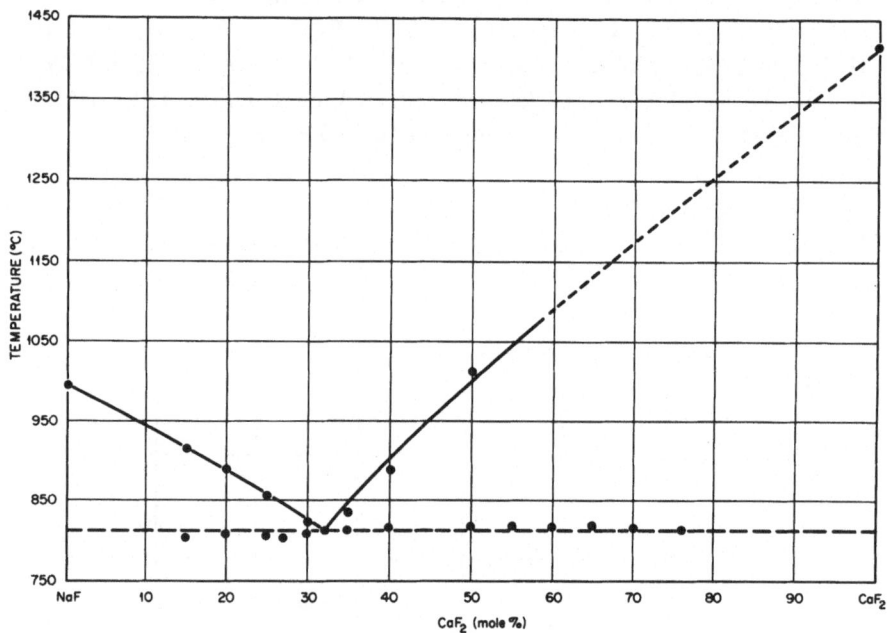

Fig. 21. The system NaF–CaF_2.

$E = 818°$, 32.5% CaF_2. In earlier work, P. P. Fedotieff and W. P. Iljinskii, *Z. Anorg. Chem.* **129**:101 (1923), reported a phase diagram of the system that is in close agreement with Fig. 21.

The System KF–CaF_2. See Fig. 22. Source: Mohammed Ishaque, *Bull. Soc. Chim. France* **1952**:130. $E = 782°$, 15.4% CaF_2; Cong. m.p. of KF · CaF_2, 1068°; $E = 1060$, 62% CaF_2.

Structural data: The compound KF · CaF_2 was reported by W. L. W. Ludekens and A. J. E. Welch, *Acta Cryst.* **5**:841 (1952), as monoclinic, $a = b = c = 8.78$, $\beta = 92°36'$.

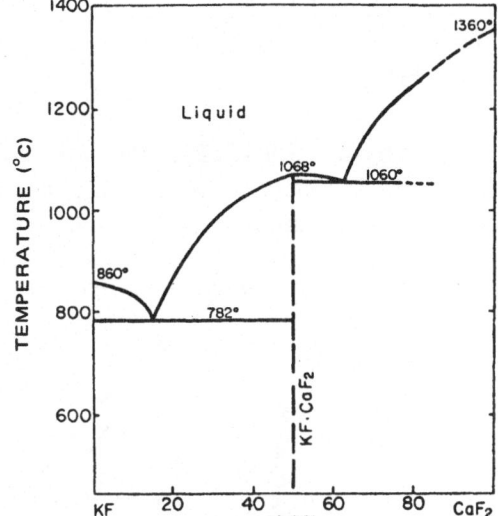

Fig. 22. The system KF–CaF$_2$.

The System RbF–CaF$_2$. See Fig. 23. Source: C. J. Barton, L. M. Bratcher, R. J. Sheil, and W. R. Grimes, ORNL, unpublished work (1956). $E = 7600$, 9% CaF$_2$; m.p. RbF · CaF$_2$ 1110°; $E = 1090°$, 57% CaF$_2$.

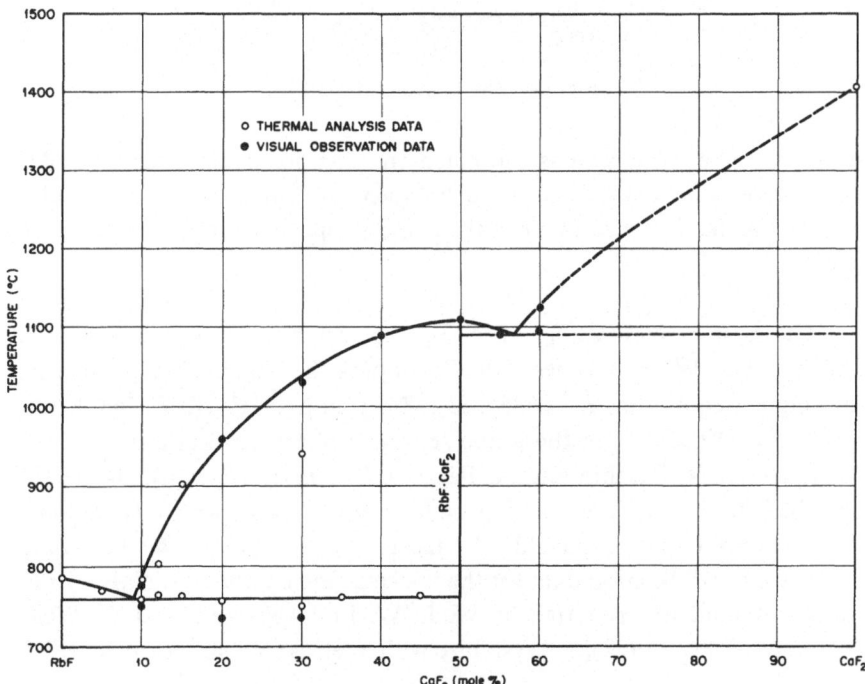

Fig. 23. The system RbF–CaF$_2$.

Structural data: The compound RbF · CaF$_2$ was reported by W. L. W. Ludekens and A. J. E. Welch, *Acta Cryst.* **5**:841 (1952), as cubic, with $a = 4.434$.

The System LiF–ZnF$_2$. See Fig. 24. Source: O. Schmitz-Dumont and Horst Bornefeld, *Z. Anorg. Allgem. Chem.* **287**:121 (1956). $E = 620°$, 64% ZnF$_2$. The minimum in the solid solution curve ("eutectic") is listed

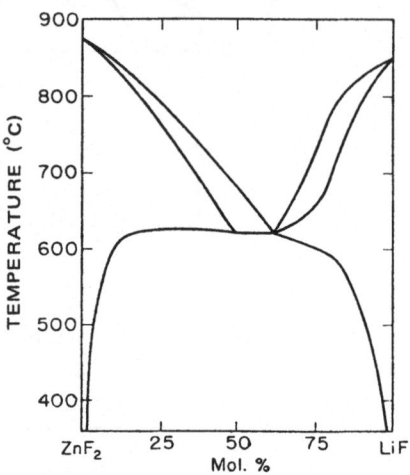

Fig. 24. The system LiF–ZnF$_2$.

as 620°C. The reviewer has estimated the compositions to be 64 mole % ZnF$_2$. The solid exsolution curve between ∼10 and 20% LiF should be regarded as having zero or negative slope so as to conform to the phase rule.

The System NaF–ZnF$_2$. See Fig. 25. Source: C. J. Barton, L. M. Bratcher, and W. R. Grimes, ORNL, unpublished work (1952). Invariant and singular equilibria: $E = 6350, 33\%$ ZnF$_2$; m.p. NaF · ZnF$_2$ 748 ± 10%; $E = 685°, 69\%$ ZnF$_2$. In the initial report of phase equilibria in the NaF–ZnF$_2$ system, N. Puschin and A. Baskow, *Z. Anorg. Chem.* **81**:347 (1913), described NaF · ZnF$_2$ as melting incongruently. A paper by O. Schmitz-Dumont and Horst Bornefeld, *Z. Anorg. Allgem. Chem.* **287**:120 (1956), includes x-ray diffraction data for the intermediate compound, NaF · ZnF$_2$. This compound was reported by W. L. W. Ludekens and A. J. E. Welch, *Acta Cryst.* **5**:841 (1952), to crystallize with a tetragonal structure, $a = 7.72$, $c = 8.11$.

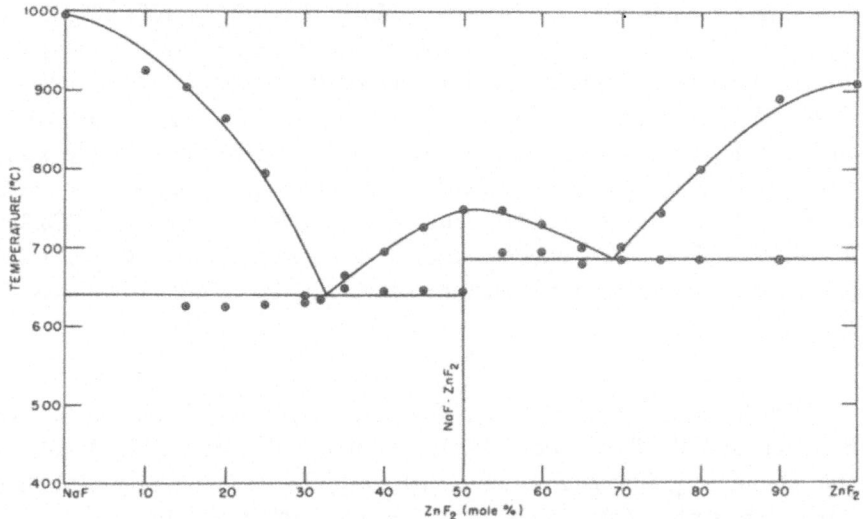

Fig. 25. The system NaF–ZnF$_2$.

The System KF–ZnF$_2$. See Fig. 26. Source: C. J. Barton, L. M. Bratcher, and W. R. Grimes, ORNL, unpublished work (1952). Invariant

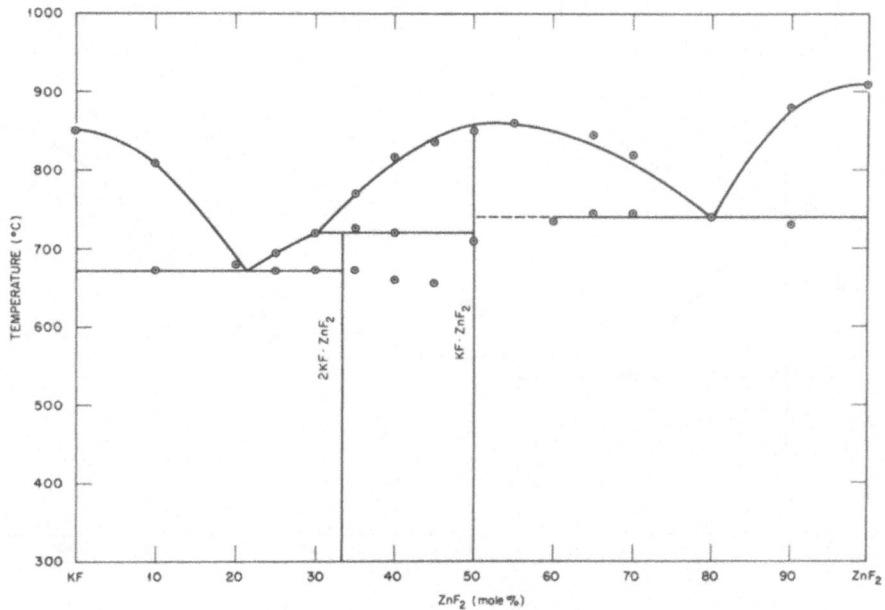

Fig. 26. The system KF–ZnF$_2$.

and singular equilibria: $E = 670$, 21% ZnF_2; $P = 720 \pm 10°$, 30% ZnF_2; m.p. $KF \cdot ZnF_2$ 850 $\pm$ 10°; $E = 740 \pm 5°$, 80% ZnF_2. In a subsequent investigation, O. Schmitz-Dumont and Horst Bornefeld, *Z. Anorg. Allgem. Chem.* **287**:122 (1956), concluded that $KZnF_3$ melts congruently at 870°C, and that K_2ZnF_4 melts incongruently at 737°C. They indexed the x-ray powder diffraction pattern of K_2ZnF_4 as tetragonal, $a = 4.009$, $c = 13.02$. G. D. Brunton *et al.* (ORNL-3761) reported $KZnF_3$ as cubic, *Pm3m*, $a = 4.056$. W. L. W. Ludekens and A. J. E. Welch, *Acta Cryst.* **5**:841 (1952), have reported the structure of $KZnF_3$ as tetragonal, $a = 8.47$, $c = 8.07$.

The System RbF–ZnF₂. See Fig. 27. Source: C. J. Barton, L. M. Bratcher, and W. R. Grimes, ORNL, unpublished work (1952). Invariant and singular equilibria: $E = 595 \pm 10°$, 21% ZnF_2; $P = 620 \pm 10°$, 32% ZnF_2; m.p. $RbF \cdot ZnF_2$ 730 $\pm$ 10%; $E = 650$, 70% ZnF_2. In a subsequent investigation, O. Schmitz-Dumont and Horst Bornefeld, *Z. Anorg. Allgem. Chem.* **287**:122 (1956), reported the melting temperatures of $2RbF \cdot ZnF_2$ and $RbF \cdot ZnF_2$ to be 640° and 740°, respectively. The liquidus profile reported by Dumont and Bornefeld is in close agreement with Fig. 27, with respect to compositions of the invariant equilibrium points.

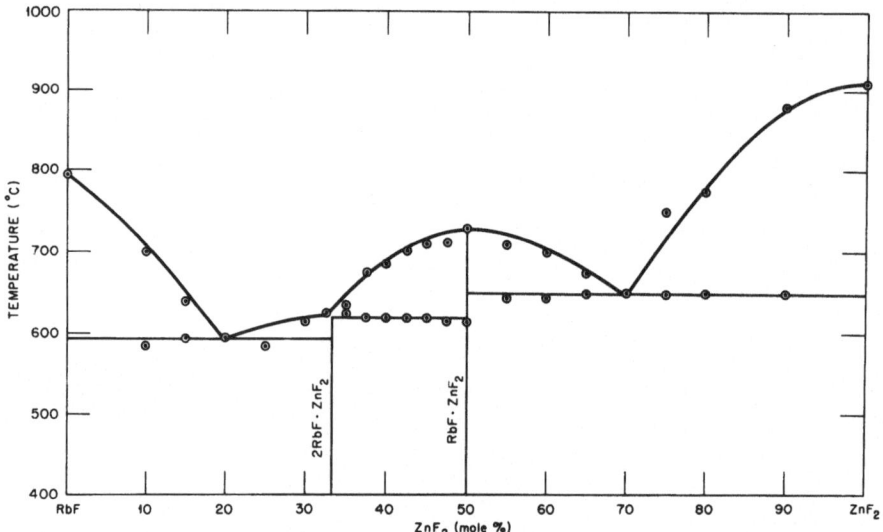

Fig. 27. The system RbF–ZnF₂.

Structural data: W. L. W. Ludekens and A. J. E. Welch, *Acta Cryst.* **5**:841 (1952), have reported the structure of $RbZnF_3$ as tetragonal, $a = 8.67$, $c = 9.01$.

The System CsF–ZnF$_2$. See Fig. 28. Source: O. Schmitz-Dumont and Horst Bornefeld, *Z. Anorg. Allgem. Chem.* **287**:122 (1956). Invariant equilibria described by the source reference include only the following: CsZnF$_3$–β-Cs$_2$ZnF$_4$ eutectic occurs at 43 mole % CsF and at 554°; incongruent melting point of CsZn$_2$F$_5$ 618°; Cs$_2$ZnF$_4$ melting point 633°; Cs$_2$ZnF$_4$ crystal transition temperature 633°.

Structure data: W. L. W. Ludekens and A. J. E. Welch, *Acta Cryst.* **5**:841 (1952), have reported the structure of CsZnF$_3$ as tetragonal, $a = 9.86$, $c = 9.01$.

Fig. 28. The system CsF–ZnF$_2$.

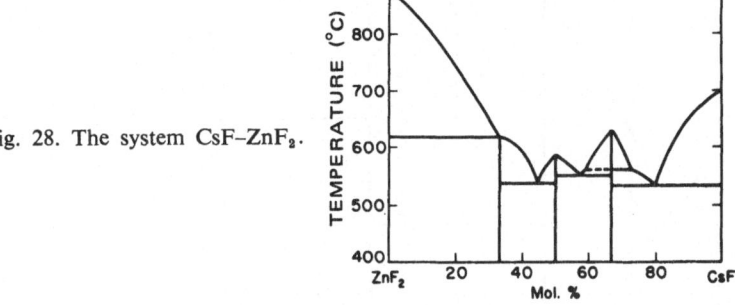

The System AgF–ZnF$_2$. See Fig. 29. Source: R. C. DeVries and Rustum Roy, *J. Am. Chem. Soc.* **75**:24 (1953). The minimum in the solid solution formed from AgF and AgF · ZnF$_2$ occurs at 380° and at 14 mole % ZnF$_2$. $E = 630$, 58% ZnF$_2$.

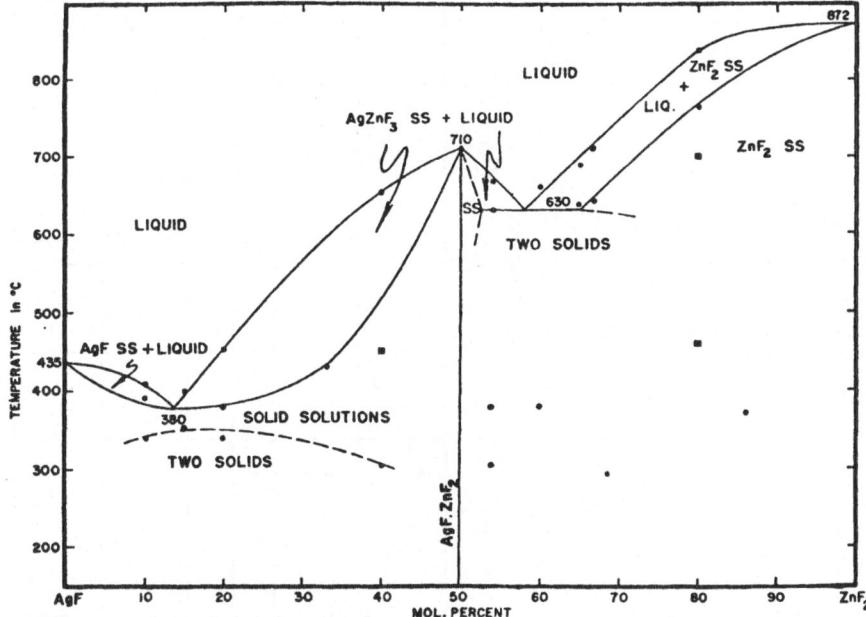

Fig. 29. The system AgF–ZnF$_2$.

The Systems RbF–SrF$_2$ and CsF–SrF$_2$. See Figs. 30 and 31. Source:
V. T. Berezhnaya and G. A. Bukhalova, *Russ. J. Inorg. Chem.* (*English*

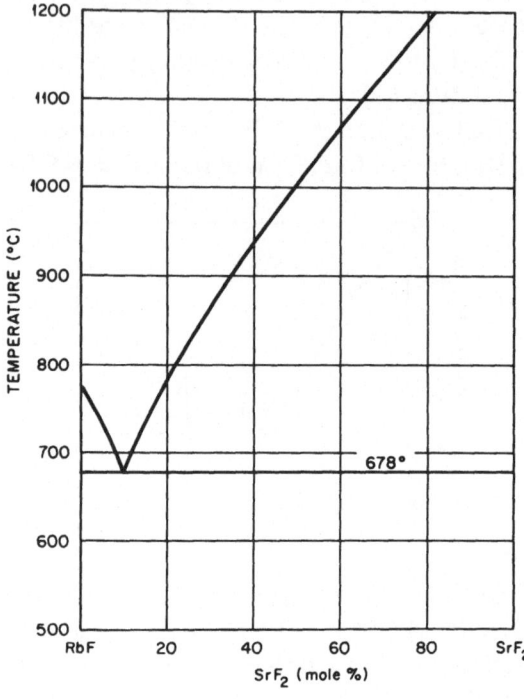

Fig. 30. The system RbF–SrF$_2$.

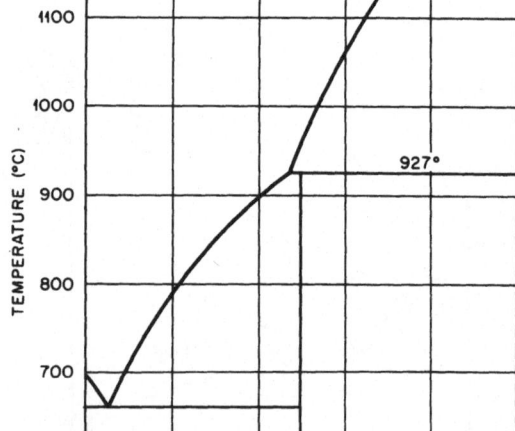

Fig. 31. The system CsF–SrF$_2$.

Transl.) **12**:1148 (1967). RbF–SrF$_2$: $E = 678°$, 10% SrF$_2$ (est.). CsF–SrF$_2$: $E = 660°$, 5% SrF$_2$, $P = 927°$, 48% SrF$_2$ (est.).

The System NaF–CdF$_2$. See Fig. 32. Source: N. A. Puschin and A. V. Baskow, *Z. Anorg. Chem.* **81**:359 (1913). $E = 660°$, 47.5% CdF$_2$.

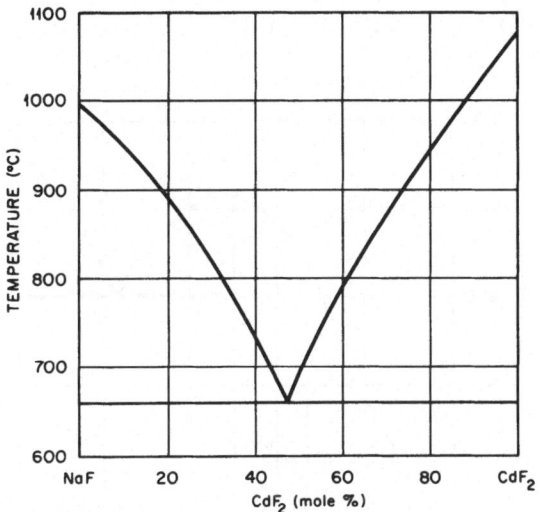

Fig. 32. The system NaF–CdF$_2$.

The System KF–CdF$_2$. See Fig. 33. Source: J. C. Cousseins, *Compt. Rend. Acad. Sci. Paris, Ser. C* **263**:585 (1966). $E = 694°$, 21% CdF$_2$;

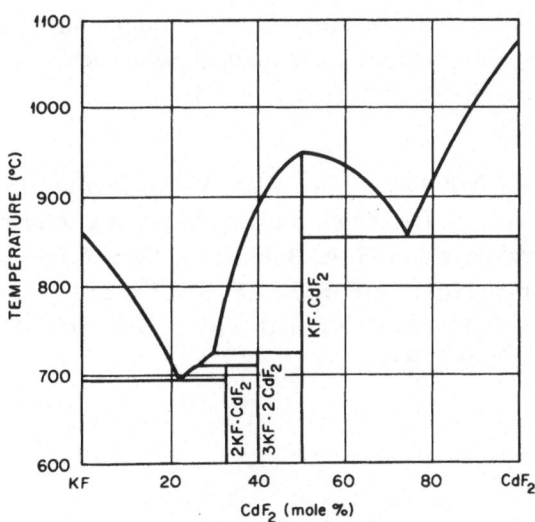

Fig. 33. The system KF–CdF$_2$.

$P = 710°$, 23% CdF_2; $P = 725°$, 25% CdF_2; m.p. KF · CdF_2 949°; $E =$ 855°, 77.5% CdF_2.

The System RbF–CdF$_2$. See Fig. 34. Source: J. C. Cousseins and C. P. Perez, *Compt. Rend. Acad. Sci. Paris, Ser. C* **264**:2060 (1967). $E =$ 713°, 12% CdF_2; $P = 792°$, 32% CdF_2; $P = 800°$, 32.5% CdF_2; m.p. RbF · CdF_2 1018°; $E = 903°$, 77% CdF_2. The phase diagram is based on

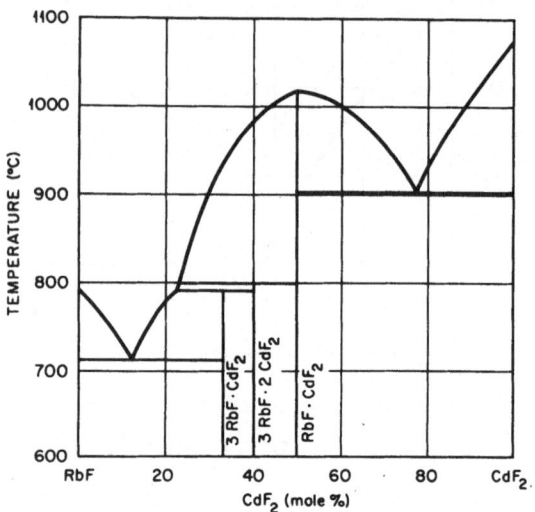

Fig. 34. The system RbF–CdF$_2$.

DTA and x-ray diffraction analysis. The compounds 2RbF · CdF_2 and 3RbF · 2CdF_2 are described as tetragonal, while RbF · CdF_2 is noted to be cubic (perovskite structure), $a = 4.398$.

The System MF–BaF$_2$. See Figs. 35–38. Source: G. A. Bukhalova and E. S. Yagub'yan, *Izv. Akad. Nauk SSSR, Neorg. Mater.* **3**:1096 (1967), and V. T. Berezhnaya and G. A. Bukhalova, *Russ. J. Inorg. Chem.* (*English Transl.*) **12**:1148 (1967). LiF–BaF$_2$: $E = 765°$, 16.5% BaF_2; NaF–BaF$_2$: $E = 812°$, 26% BaF_2 (est.); KF–BaF$_2$: $E = 729°$, 30% BaF_2 (est.); CsF–BaF$_2$: $E = 628°$, 16% BaF_2.

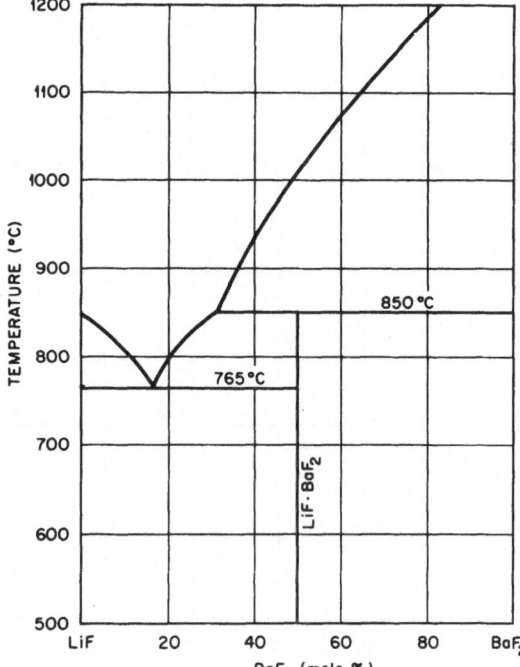

Fig. 35. The system LiF–BaF$_2$.

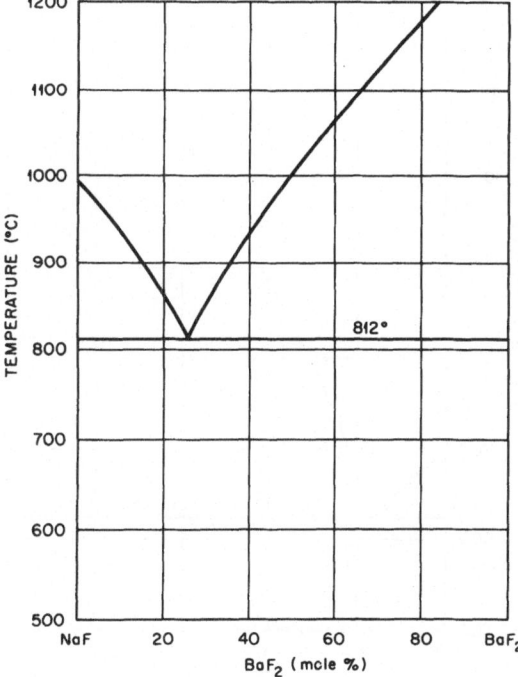

Fig. 36. The system NaF–BaF$_2$.

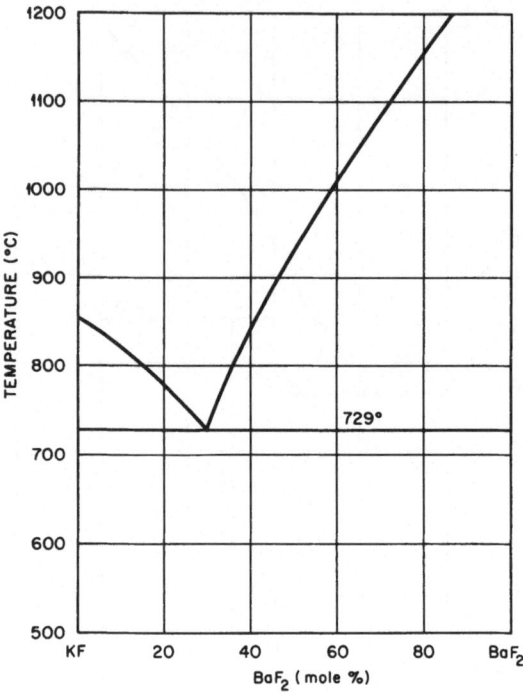

Fig. 37. The system KF–BaF$_2$.

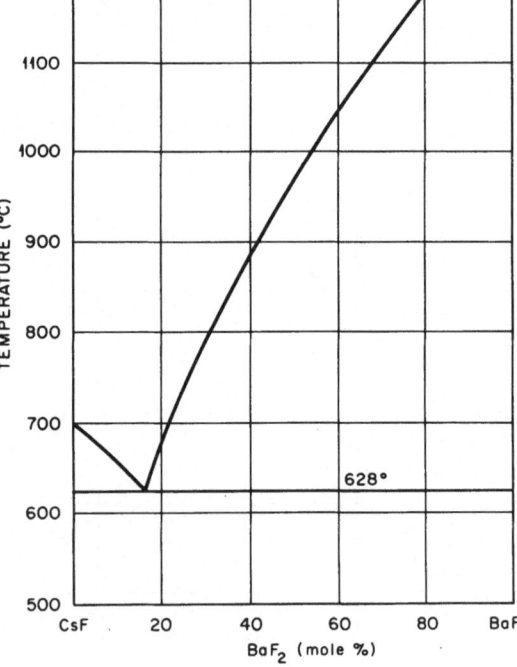

Fig. 38. The system CsF–BaF$_2$.

The System NaF–SnF₂. See Fig. 39. Source: J. D. Donaldson *et al.*, *J. Chem. Soc.* **1965**(714):3876. Invariant and singular equilibria were not

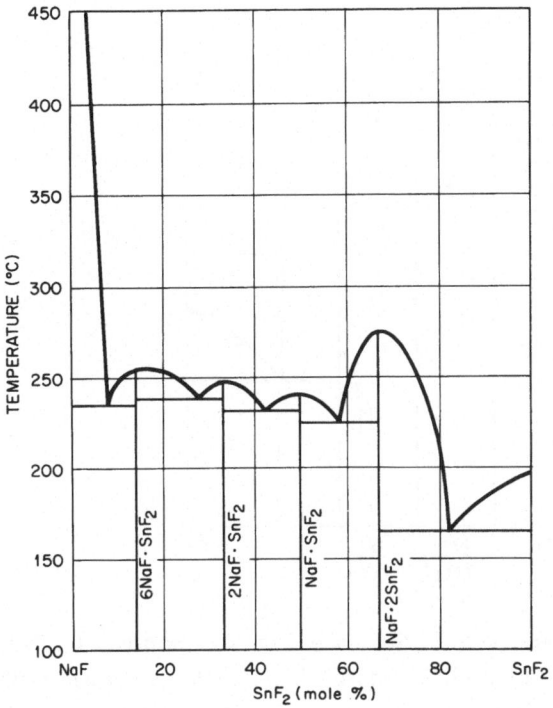

Fig. 39. The system NaF–SnF₂.

defined. Melting points of the intermediate compounds were given as follows: 6NaF · SnF₂ 279°; 2NaF · SnF₂ 273°; NaF · SnF₂ 265°; NaF · 2SnF₂ 299°.

The System KF–SnF₂. See Fig. 40. Source: J. D. Donaldson *et al.*, *J. Chem. Soc.*, **1965**(714):3876. Invariant and singular equilibria were not defined. Melting points of the intermediate compounds were given as follows: 6KF · SnF₂ 325°; 2KF · SnF₂ 323°; KF · SnF₂ 310°; KF · 2SnF₂ 354°.

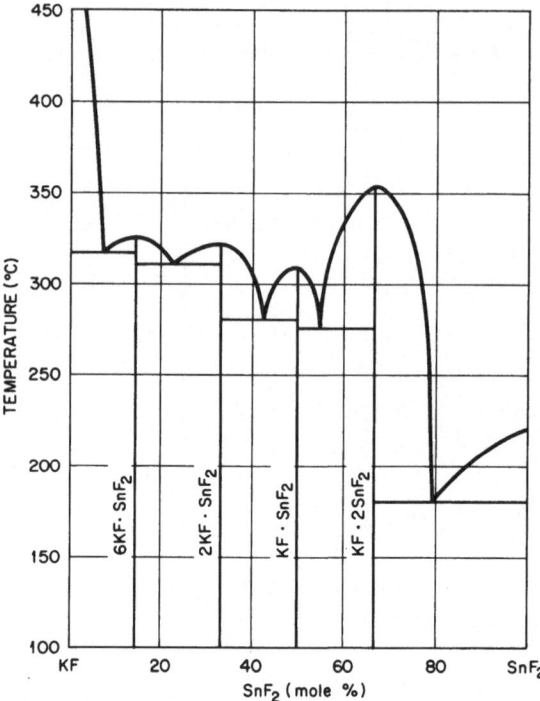

Fig. 40. The system KF–SnF$_2$.

The System NaF–PbF$_2$. See Fig. 41. Source: O. Schmitz-Dumont and Gunter Bergerhoff, *Z. Anorg. Allgem. Chem.* **283**:317 (1956). $E = 520°$, 66% PbF$_2$.

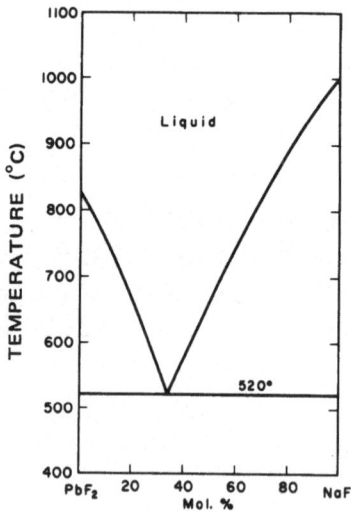

Fig. 41. The system NaF–PbF$_2$.

The System KF–PbF₂. See Fig. 42. Source: O. Schmitz-Dumont and Gunter Bergerhoff, *Z. Anorg. Allgem. Chem.* **283**:317 (1956). $P = 475°$, composition not given; $E = 465°$, 42 % PbF₂; $P = 645°$, composition not given. The maximum solubility of KF in crystalline PbF₂ is 15 mole %.

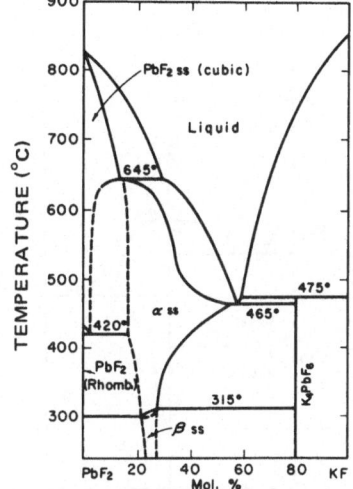

Fig. 42. The system KF–PbF₂.

The System RbF–PbF₂. See Fig. 43. Source: O. Schmitz-Dumont and Gunter Bergerhoff, *Z. Anorg. Allgem. Chem.* **283**:318 (1956). $P = 600°$, ∼70 mole % PbF₂; $E = 550°$, 55 % PbF₂; congruent m.p. of RbF · PbF₂ 560°; $E = 485°$, ∼30 % PbF₂.

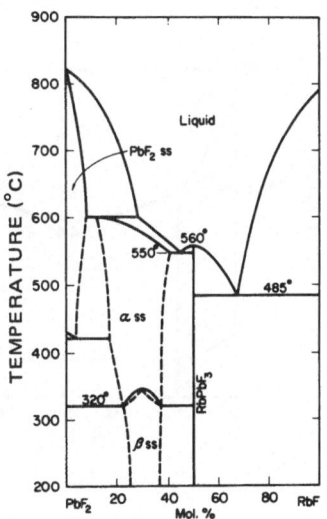

Fig. 43. The system RbF–PbF₂.

The System CsF–PbF₂. See Fig. 44. Source: O. Schmitz-Dumont and Gunter Bergerhoff, *Z. Anorg. Allgem. Chem.* **283**:319 (1956). $E = 500°$, 28% PbF₂; $P = 615°$, 40% PbF₂; congruent m.p. of CsF · PbF₂ 725°; $P = 615°$, 60% PbF₂; $E = 565°$, 71% PbF₂.

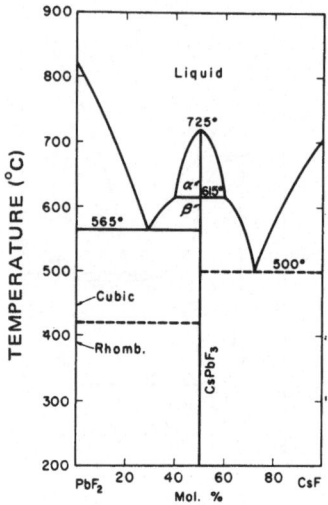

Fig. 44. The system CsF–PbF₂.

The System AgF–PbF₂. See Fig. 45. Source: C. Tubandt and S. Eggert, *Z. Anorg. Allgem. Chem.* **110**:223 (1920). The phase diagram as constructed in the source reference does not obey the phase rule.

Fig. 45. The system AgF–PbF₂.

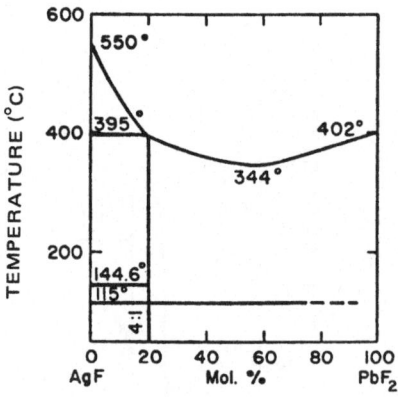

The Alkali Fluoride–MnF₂ Binary Systems. See Figs. 46–50. Source: I. N. Belyaev and O. Ya. Revina, *Russ. J. Inorg. Chem.* (*English Transl.*)

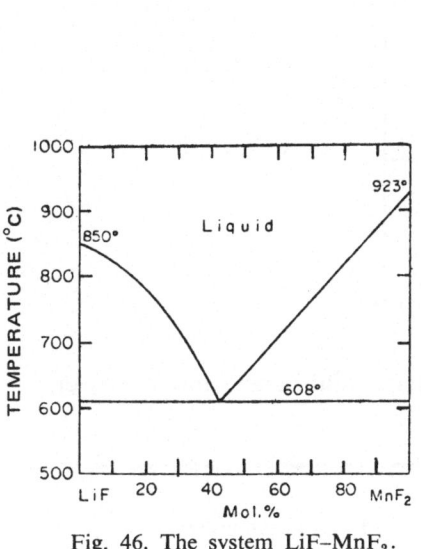

Fig. 46. The system LiF–MnF₂.

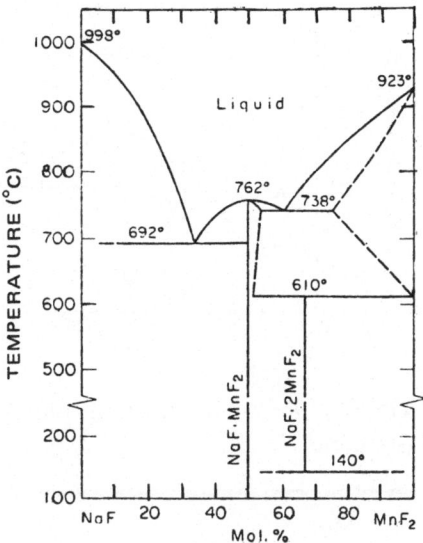

Fig. 47. The system NaF–MnF₂.

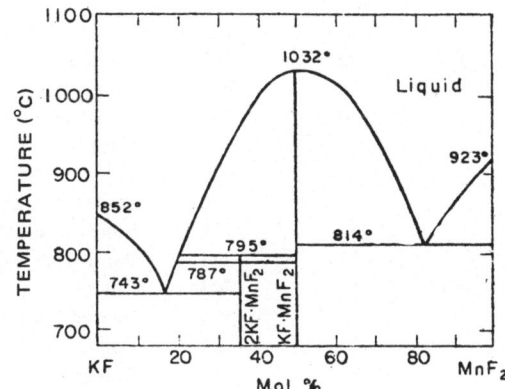

Fig. 48. The system KF–MnF₂.

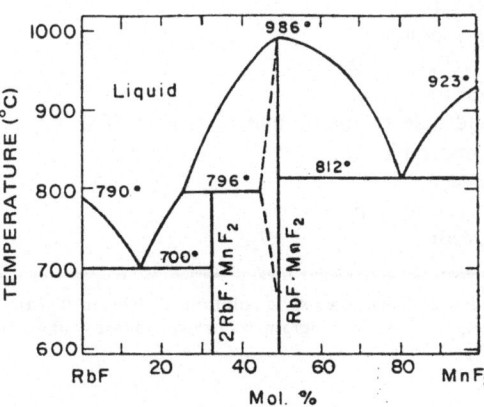

Fig. 49. The system RbF–MnF₂.

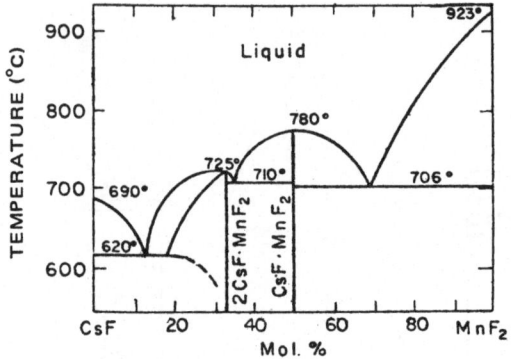

Fig. 50. The system CsF–MnF$_2$.

11:774 (1966). The invariant and singular equilibrium points are given in the table.

Mole %		Temperature	Type of equilibrium
MF	MnF$_2$		
LiF	43	608	Eutectic
NaF	34	692	Eutectic
	50	762	Congruent m.p.
	66	738	Eutectic
		610	Upper temp. of stability of NaF · 2MnF$_2$
		140	Solid-state inversion temp. of NaF · 2MnF$_2$
	55	738	Solid-state solubility limit of MnF$_2$ in NaF · MnF$_2$
	77	738	Solid-state solubility limit of NaF · MnF$_2$ in MnF$_2$
KF	17	743	Eutectic
		795	Incongruent melting temp. of 2KF · MnF$_2$
		787	Polymorphic transition temp. of 2KF · MnF$_2$
	84	814	Eutectic
RbF	16	700	Peritectic
	24	796	Peritectic
	(46)[a]	796	See footnote
	81	814	Eutectic
CsF	14	620	Eutectic
	20	620	Solid-state solubility limit of CsF in 2CsF · MnF$_2$
	33.3	725	Congruent m.p.
	35	710	Eutectic
	50	780	Congruent m.p.
	71	706	Eutectic

[a] The source report states that "a small region of restricted solid solutions of RbF in RbMnF$_3$ extends up to 54 mole % MnF$_2$," although the phase diagram, as drawn, indicates that the limit should be about 46 mole % MnF$_2$.

The System NaF–FeF$_2$. See Fig. 51. Source: R. E. Thoma, H. A. Friedman, B. S. Landau, and W. R. Grimes, ORNL, unpublished work (1957). $E = 680°$, 30% FeF$_2$; m.p. NaF · FeF$_2$, 783°; $E = 475°$, 63% FeF$_2$. The phase diagram was based on cooling-curve analysis, TGQ

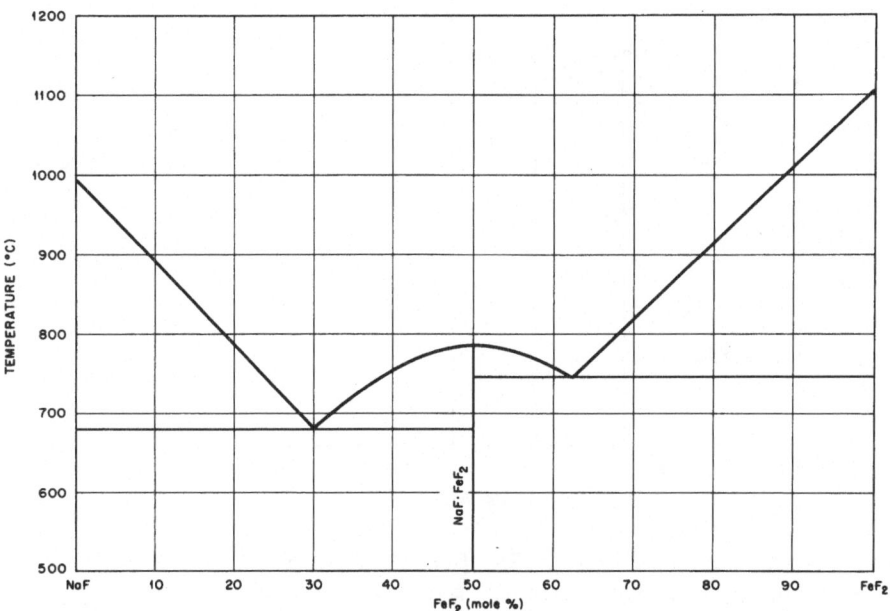

Fig. 51. The system NaF–FeF$_2$.

data, and x-ray diffraction analysis. A partial phase diagram of the system NaF–FeF$_3$ has been reported by N. Puschin and A. Baskow, *Z. Anorg. Chem.* **81**:361 (1913), showing a eutectic at ~890° and at ~35% FeF$_3$.

The System KF–CoF$_2$. See Fig. 52. Source: I. N. Belyaev and S. A. Shilov, *Russ. J. Inorg. Chem.* (*English Transl.*) **14**:1046 (1969). Invariant and singular equilibrium points: $E = 750°$, 19 mole % CoF$_2$; congruent m.p. of 2KF · CoF$_2$ 858°; $E = 848°$, 37.5 mole % CoF$_2$; congruent m.p. of KF · CoF$_2$ 1032°; $E = 945°$, 76 % CoF$_2$. Relative values of the heat evolved on cooling KF–CoF$_2$ mixtures are shown graphically on the phase diagram and should not be construed to signify phase transformations.

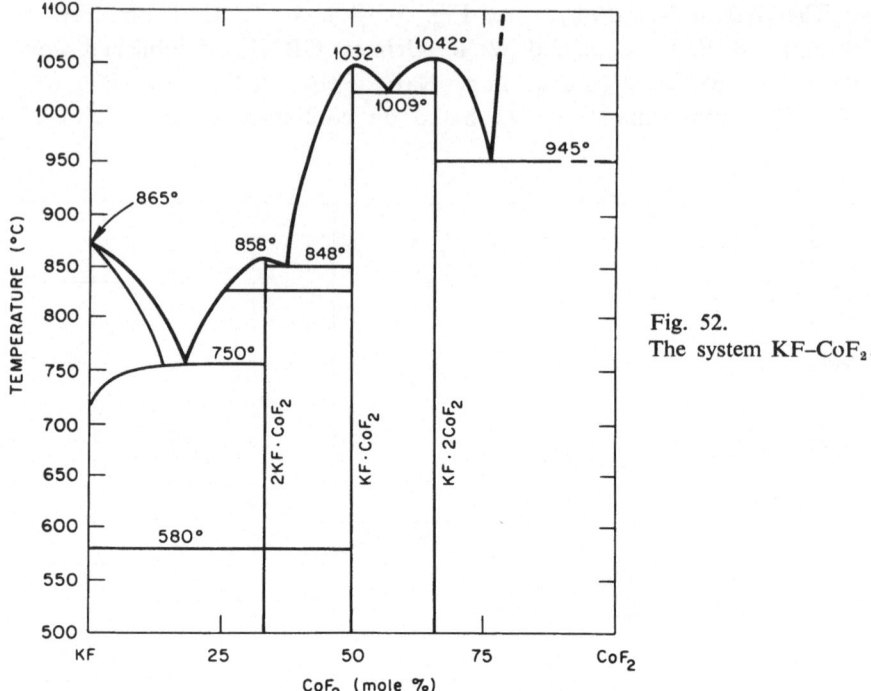

Fig. 52.
The system KF–CoF$_2$.

The System NaF–NiF$_2$. See Fig. 53. Source: R. E. Thoma, H. A. Friedman, B. S. Landau, and W. R. Grimes, ORNL, unpublished work

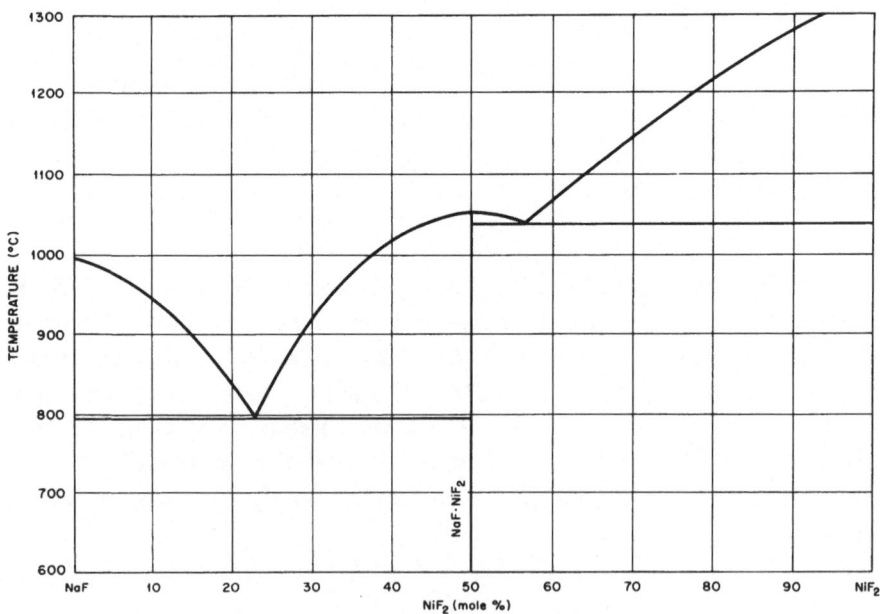

(1957). $E = 795°$; 23% NiF_2; m.p. $NaF \cdot NiF_2$ 1045°; $E = 1040°$, 57% NiF_2. The phase diagram was based on cooling-curve analysis, TGQ and x-ray diffraction analysis.

The System KF–NiF₂. See Fig. 54. Source: G. Wagner and D. Balz, *Z. Electrochem.* **56**:576 (1952). $E = 797°$, 9.1% NiF_2; $P = 930°$, 23.3% NiF_2; m.p. $KF \cdot NiF_2$ 1130 ± 2°; $E = 1084°$, 65.5% NiF_2.

Fig. 54. The system KF–NiF₂.

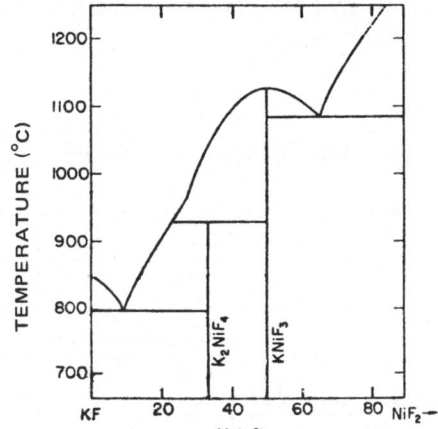

The System LiF–BF₃. See Fig. 55. Source: A. S. Dworkin and M. A. Bredig, in: Molten Salt Reactor Program Semiannual Progress Report for Period Ending August 31, 1972, ORNL-4832, p. 58. Dworkin and Bredig did not state the composition of the eutectic in their report, although they did list activity coefficients for LiF at mole fractions X_{BF_4} of 0.2, 0.4, and 0.6 as 1.16, 1.32, and 1.140, respectively.

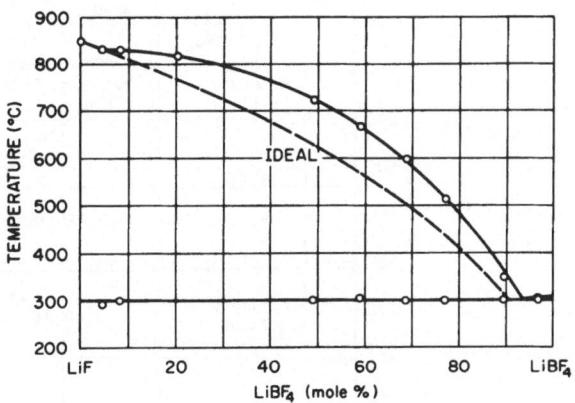

Fig. 55. The system LiF–BF₃.

The System NaF–BF$_3$. See Fig. 56. Source: C. J. Barton, L. O. Gilpatrick, J. A. Bornmann, H. H. Stone, T. N. McVay, and H. Insley, *J. Inorg. Nucl. Chem.* **33**:337 (1971). The eutectic is located at 92 ± 1 mole % NaBF$_4$ and melts at $384 \pm 2°$. An earlier report by V. G. Selivanov and V. V. Stender, *Zh. Neorg. Khim.* **3**:448 (1958), indicates the eutectic composition to be at 60 mole % NaBF$_4$ and the melting temperature to be 304°. The results are probably erroneous because of oxide impurities.

The structure of NaBF$_4$ has been reported by G. D. Brunton, *Acta Cryst.* **B24**:1703 (1968), as orthorhombic, *Cmcm*, $a = 6.8368$, $b = 6.2619$, $c = 6.7916$.

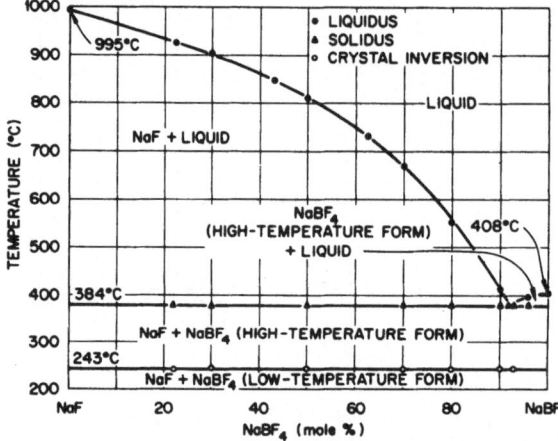

Fig. 56. The system NaF–BF$_3$.

The System KF–BF$_3$. See Fig. 57. Source: C. J. Barton *et al.*, *J. Inorg. Nucl. Chem.* **33**, 337 (1971). The eutectic formed by KF and KBF$_4$ occurs at 74.5 ± 1 mole % KBF$_4$, and at $460 \pm 2°$.

Fig. 57. The system KF–BF$_3$.

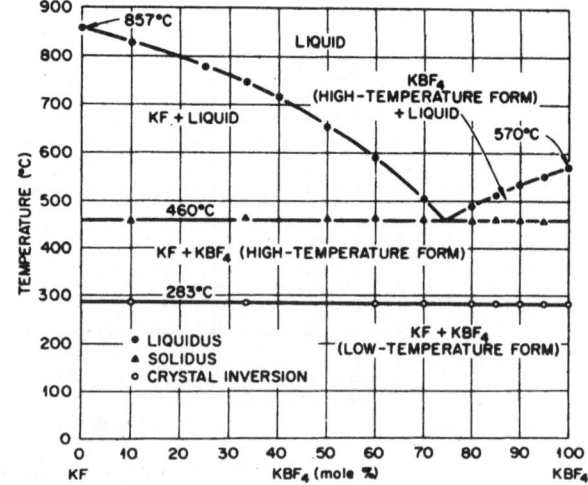

Structural data: KBF_4, orthorhombic, *Pnma*, $a = 8.6588$, $b = 5.4800$, $c = 7.0299$ [G. D. Brunton, *Acta Cryst.* **B25**:2161 (1969)].

The System RbF–BF$_3$. See Fig. 58. Source: L. O. Gilpatrick and C. J. Barton, *J. Inorg. Nucl. Chem.* **36**:725 (1974). The authors of the source reference deduced that the components, RbF and $RbBF_4$, form a simple eutectic system. $E = 442 \pm 2°$, 68.5% $RbBF_4$.

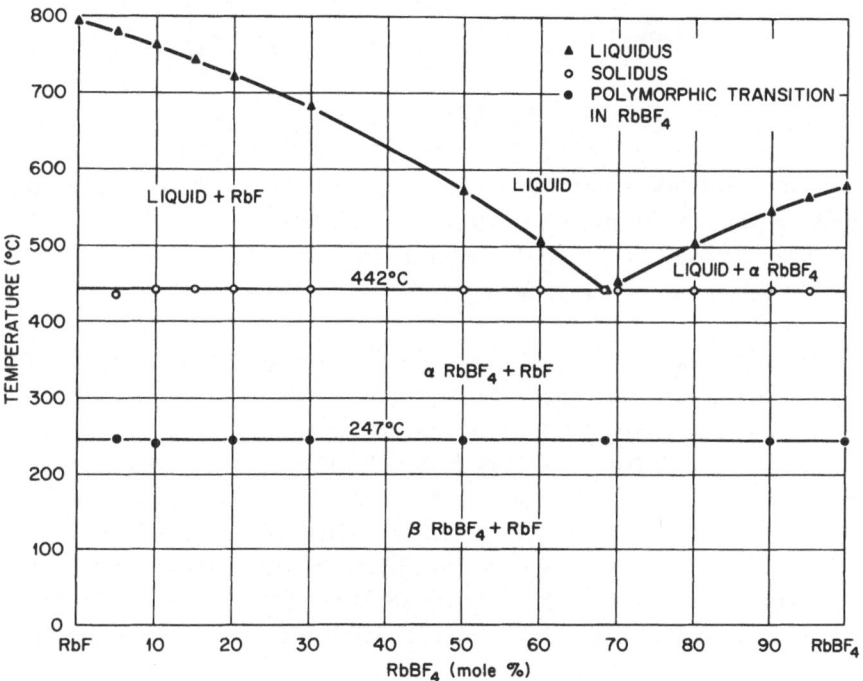

Fig. 58. The system RbF–BF$_3$.

Structural data: $RbBF_4$ crystallizes with an orthorhombic structure, *Pbnm*, $a = 7.29$, $b = 9.10$, $c = 5.13$ [M. J. R. Clark and H. Lynton, *Can. J. Chem.* **47**:2579 (1969)].

The System LiF–AlF$_3$. See Fig. 59. Source: E. P. Dergunov, *Dokl. Akad. Nauk, SSSR* **60**:1185 (1948). The system was described earlier, and more completely by P. P. Fedotieff and K. Timofieff, *Z. Anorg. Allgem. Chem.* **206**:266 (1937). Figure 59 is the phase diagram as proposed by Fedotieff and Timofieff.

Structural data: α-3LiF · AlF$_3$: orthorhombic, $a = 9.510$, $b = 8.2295$, $c = 4.8762$, *Pna*21 [J. H. Burns, A. C. Tennissen, and G. D. Brunton, *Acta Cryst.* **B24**, 225 (1968). The compound 3LiF · AlF$_3$ undergoes at

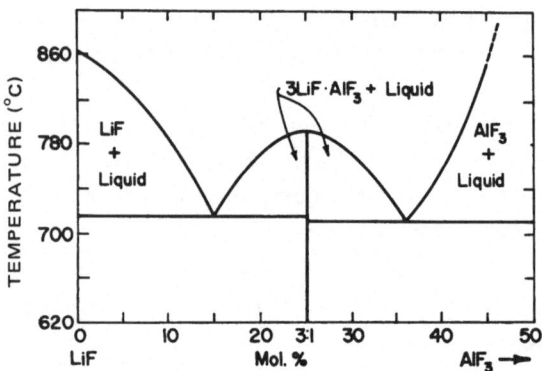

Fig. 59. The system LiF–AlF₃.

least one structural transition on heating and cooling. The structure given above is that of the room-temperature modification.

The viscosity of LiF–AlF₃ and LiF–NaF–AlF₃ melts was reported by M. M. Vetnikov and G. I. Sipniga, *Zh. Prikl. Khim.* **36**(9):905 (1963).

The System NaF–AlF₃. See Fig. 60. Source: K. Grojtheim, *Kgl. Norske Videnskap. Selskabs. Skrifter*, 5 (1956). Numerous examinations of the system NaF–AlF₃ have been reported, including phase diagrams by the following investigators: P. P. Fedotieff and W. P. Iljinskii, *Z. Anorg. Chem.* **80**:121 (1913); see also N. A. Puschin and A. V. Baskow, *ibid.*, **81**:350 (1913); E. P. Dergunov, *Dokl. Akad. Nauk SSSR* **60**(7):1185 (1948); Giiejiro Fuseya, Chuzo Sugihara, Nobuo Nagao, and Ichiro Teraoka, *J. Electrochem. Soc. Japan* **18**(2):66 (1950). A paper by P. A. Foster, Jr., *J. Am. Ceram. Soc.* **51**:107 (1968) discusses the melting point

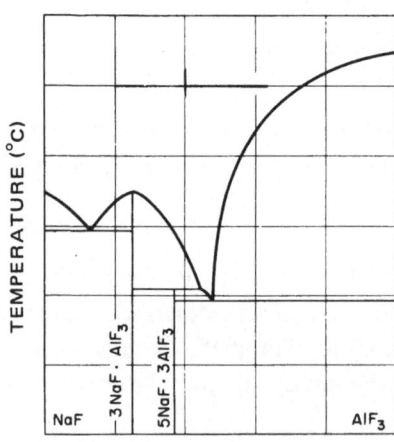

Fig. 60. The system NaF–AlF₃.

of Na_3AlF_6. The constitution of cryolite and $NaF-AlF_3$ melts has been discussed by W. B. Frank and L. M. Foster, *J. Phys. Chem.* **64**:310 (1960). A review of the structural modifications of Na_3AlF_6 may be found in A. F. Wells, *Structural Inorganic Chemistry*, 3rd ed., Oxford University Press, London (1962).

The System KF–AlF₃. See Fig. 61. Source: Bert Phillips, C. M. Warshaw, and J. Mockrin, *J. Am. Ceram. Soc.* **49**:633 (1966). The phase diagram as reported by these authors indicates that $KAlF_4$ is dimorphic and inverts on cooling from a cubic to an orthorhombic form at $\sim15°$. The authors do not list the compositions of the invariant equilibrium points in the systems, but do state that they are in good agreement with those listed by the previous investigators, P. P. Fedotieff and K. Timofieff, *Z. Anorg. Allgem. Chem.* **206**:263–66 (1932).

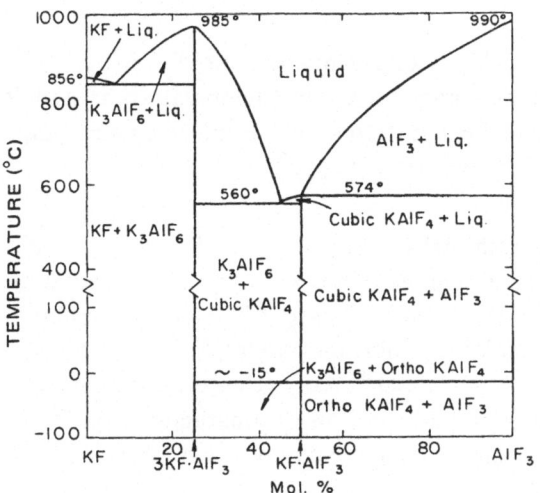

Fig. 61. The system KF–AlF₃.

Structural data: KF · AlF₃: tetragonal, $P4/mmm$, $a = 3.550$, $c = 6.139$ [G. D. Brunton *et al.*, ORNL-3761].

The System RbF–AlF₃. See Fig. 62. Sources: N. Puschin and A. Baskow, *Z. Anorg. Chem.* **81**:356 (1913), and E. P. Dergunov, *Dokl. Akad. Nauk SSSR* **50**:1185 (1948). Figure 62 is from Puschin and Baskow. $E = 790°$, 6.5 mole % AlF_3 (Puschin and Baskow); $E = 740°$, 7 % AlF_3 (Dergunov); congruent m.p. 3RbF · AlF₃, 985° (Puschin and Baskow);

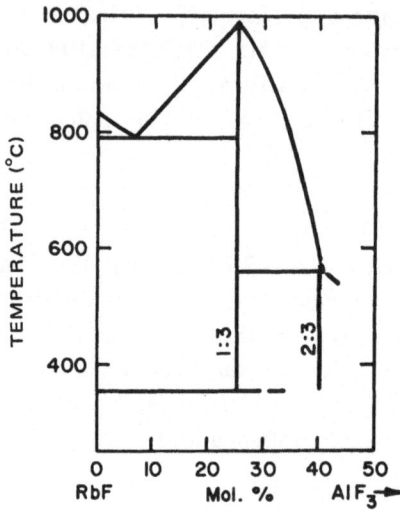

Fig. 62. The system RbF–AlF$_3$.

$P = 560°$, 41.5% AlF$_3$; congruent m.p. 3RbF · AlF$_3$, 914° (Dergunov). The source reference indicates a lower temperature of stability for 3RbF · AlF$_3$ of 350°, and the occurrence of the intermediate compound, 3RbF · 2AlF$_3$, melting incongruently at 560°.

The System CsF–AlF$_3$. See Fig. 63. Sources: N. Puschin and A. Baskow, *Z. Anorg. Chem.* **81**:356 (1913), and E. P. Dergunov, *Dokl. Akad. Nauk SSSR* **50**:1185 (1948). Figure 63 is from Puschin and Baskow. Invariant and singular equilibrium points: $E = 685°$, 5.5 % AlF$_3$ (Puschin and Baskow); $E = 663°$, ≃6 % AlF$_3$ (Dergunov); congruent m.p. 3CsF · AlF$_3$, 825° (Puschin and Baskow); congruent m.p. 3CsF · AlF$_3$, 808° (Dergunov). The eutectic at some composition richer in AlF$_3$ than 30 mole % melts at 490°.

Fig. 63. The system CsF–AlF$_3$.

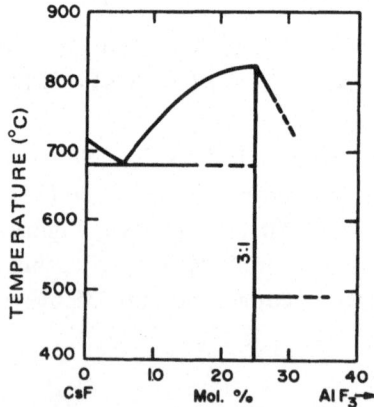

The System LiF–ScF₃. See Fig. 64. Source: E. P. Babaeva and G. A. Bukhalova, *Zh. Neorg. Khim.* **11**:648 (1966). $E = 606°$, 29% ScF_3.

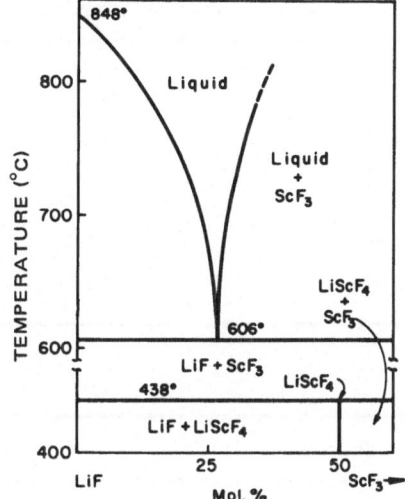

Fig. 64. The system LiF–ScF₃.

The System NaF–ScF₃. See Fig. 65. Source: R. E. Thoma and R. H. Karraker, *Inorg. Chem.* **5**:1933 (1966). Invariant and singular equi-

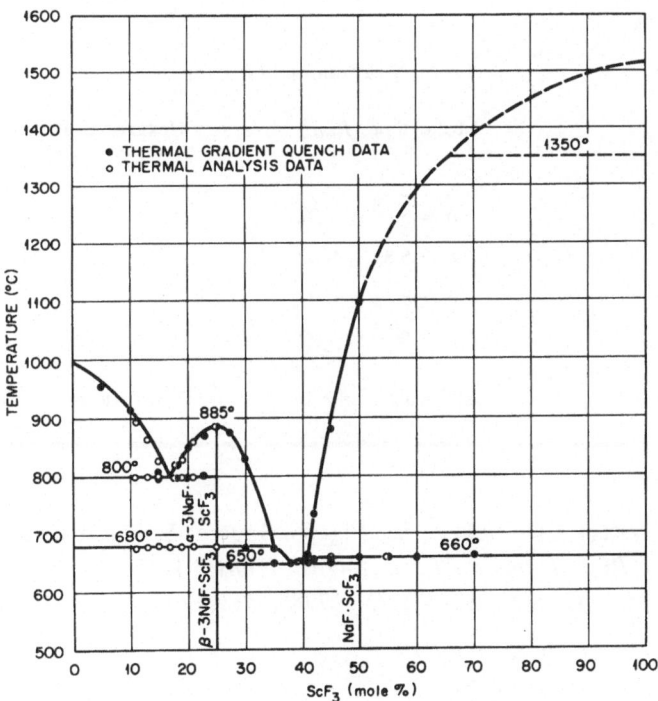

Fig. 65. The system NaF–ScF₃.

librium points and structural data are given in the tables.

Mole % ScF_3	Invariant temperature	Type of equilibrium
17	800	Eutectic
25	885	Congruent melting point
35	680	Peritectic inversion of $3NaF \cdot ScF_3$
38	650	Eutectic
42	660	Peritectic

Property	β-$3NaF \cdot ScF_3$	$NaF \cdot ScF_3$
Equilibrium stability range	<680°	<660°
Optical character	Uniaxial (+), $2V \simeq 0°$	Uniaxial (−)
Refractive indices	$N_\omega = 1.368$, $N_\varepsilon = 1.374$	$N_\omega = 1.431$
	From X-ray diffraction data	
Symmetry	Monoclinic, $P2_1/n$	Hexagonal, $P3_1 k2m$, $P3_2 k2m$, $P3_1 21$, or $P3_2 21$
Parameters	$a = 5.60 \pm 0.02$, $b = 5.81 \pm 0.02$, $c = 8.12 \pm 0.02$, $\beta = 90°45' \pm 5'$	$a = 12.97 \pm 0.03$, $c = 9.27 \pm 0.02$
Density	2.87 (calc.), $z = 2$	3.01 (calc.), $z = 18$

The System CsF–ScF$_3$. See Fig. 66. Source: E. P. Babaeva and G. A. Bukhalova, *Russ. J. Inorg. Chem.* (*English Transl.*) **11**:352 (1966). $E = 680°$, 4% ScF$_3$; m.p. Cs$_3$ScF$_6$ 1084°; $E = 798°$, 33% ScF$_3$.

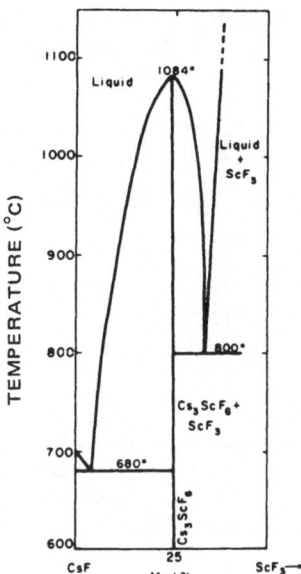

Fig. 66. The system CsF–ScF$_3$.

The System TlF–ScF$_3$. See Fig. 67. Source: J. Chassaing, *Rev. Chim. Min.* **9**:265 (1972). Invariant and singular equilibrium points:

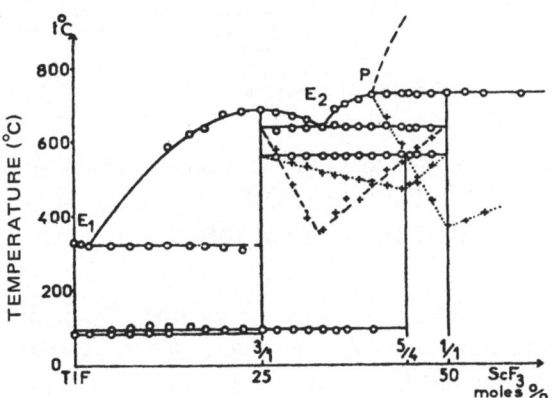

Fig. 67. The system TlF–ScF$_3$.

$E = 320°$, 2% ScF$_3$; m.p. Tl$_3$ScF$_6$ 686°; $E = 638°$, 33% ScF$_3$; $P = 732°$, 40% ScF$_3$; Tl$_5$Sc$_4$F$_{17}$ upper temperature of stability 563°; TlScF$_4$ upper temperature of stability 732°.

The System LiF–YF$_3$. See Fig. 68. Source: R. E. Thoma *et al.*, *J. Phys. Chem.* **65**:1096 (1961). Invariant equilibria: $E = 695°$, 19 % YF$_3$;

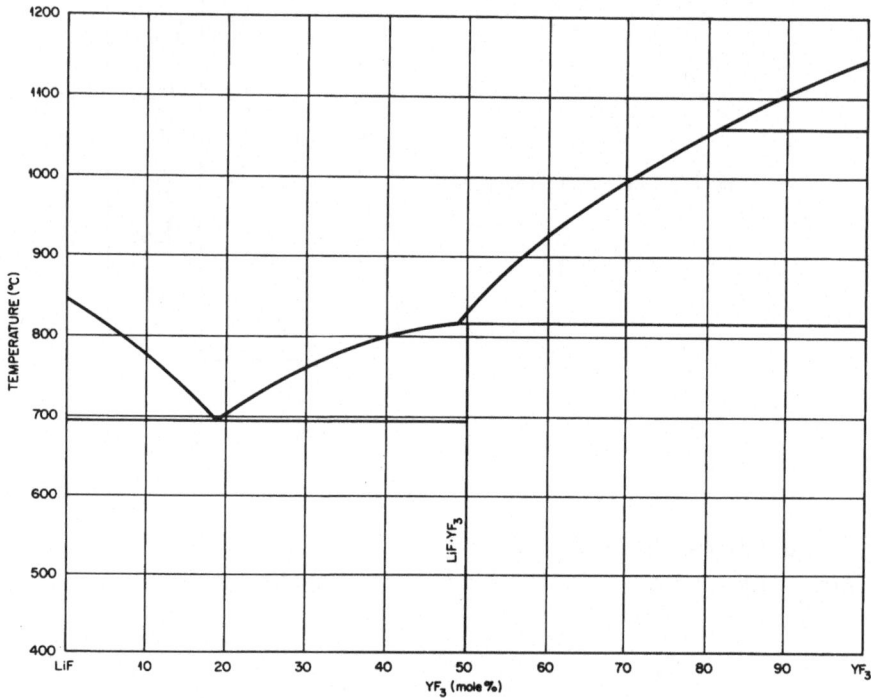

Fig. 68. The system LiF–YF$_3$.

$P = 819°$, 49%. The compound LiYF$_4$ is described in the source reference as tetragonal, $a = 5.26 \pm 0.03$, $c = 10.94 \pm 0.03$, S.G. $I4_1/a$. A solid state transition in YF$_3$ was shown to occur at 1052°.

The System NaF–YF$_3$. See Fig. 69. Source: R. E. Thoma, G. M. Hebert, H. Insley, and C. F. Weaver, *Inorg. Chem.* **2**:1005 (1963). A = liquid, B = NaF + liquid, C = NaF + hex. NaF · YF$_3$, D = hex. NaF · YF$_3$ + liquid, E = (fluorite) cubic NaF · YF$_3$–5NaF · 9YF$_3$ solid solution + liquid, F = (fluorite) cubic NaF · YF$_3$–5NaF · 9YF$_3$ ss, G = hex. NaF · YF$_3$ ss + cubic NaF · YF$_3$–5NaF · 9YF$_3$ ss, H = cubic NaF · YF$_3$–5NaF · 9YF$_3$ ss + ordered 5NaF · 9YF$_3$ ss, I = hex. NaF · YF$_3$ ss, J = hex. NaF · YF$_3$ ss + ordered 5NaF · 9YF$_3$ ss, K = ordered 5NaF · 9YF$_3$ ss, L = hex. NaF · YF$_3$ ss + orthorhombic YF$_3$, M = ordered 5NaF · 9YF$_3$ + ortho-YF$_3$, N = cubic 5 NaF · 9YF$_3$ + ortho-YF$_3$, O = ortho-YF$_3$ +

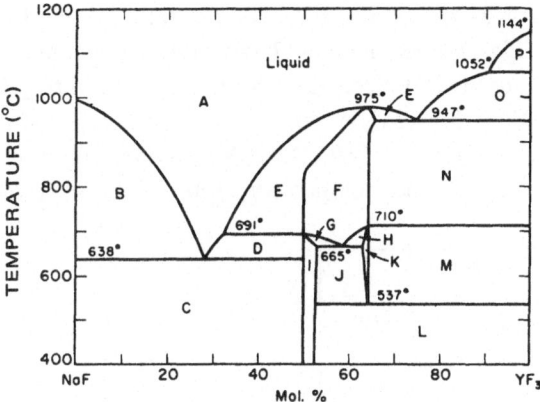

Fig. 69. The system NaF–YF$_3$.

liquid, and P = high-temperature YF$_3$ + liquid. Invariant and singular equilibrium points are given in the table.

Mole % YF$_3$	Temperature	Type of equilibrium	Equilibrium reaction
28	638	Eutectic	L $\rightleftharpoons$ NaF + hexagonal NaF · YF$_3$
32	691	Peritectic	Cubic NaF · YF$_3$ + L $\rightleftharpoons$ hexagonal NaF · YF$_3$ + L
58.5	665	Eutectoid	Fluoride ss $\rightleftharpoons$ hexagonal NaF · YF$_3$ ss + orthorhombic YF$_3$
64.28	975	Congruent m.p. for 5NaF · 9YF$_3$	L $\rightleftharpoons$ cubic 5NaF · 9YF$_3$
64.28	710	Inversion of 5NaF · 9YF$_3$	Disordered cubic 5NaF · 9YF$_3$ $\rightleftharpoons$ ordered 5NaF · 9YF$_3$
64.28	537	Dec. of 5NaF · 9YF$_3$	5NaF · 9YF$_3$ $\rightleftharpoons$ hexagonal NaF · YF$_3$ ss + YF$_3$ (orthorhombic)
75	947	Eutectic	L $\rightleftharpoons$ cubic 5NaF · 9YF$_3$ ss + orthorhombic YF$_3$
91	1052	Peritectic	L + high-temperature YF$_3$ $\rightleftharpoons$ L + orthorhombic YF$_3$

Structural data: NaF · YF$_3$ hexagonal, $a = 5.95$, $c = 3.52$, $P63/m$ [J. H. Burns, ORNL-3262 (1962)]. Lattice constants for the cubic solid solutions range from 5.447 to 5.530 at the 5NaF · 9YF$_3$ composition limit.

The System KF–YF₃. See Fig. 70. Source: G. A. Bukhalova and E. P. Babaeva, *Zh. Neorg. Khim.* **11**:644 (1966). The binary system, as reported in the source reference, was adopted as the limiting binary system diagram for the NaF–KF–YF₃ system shown in Fig. 183. Invariant and singular equilibrium points: $E = 760°$, 13 % YF₃; congruent m.p. K₃YF₆ 980°; $E = 750°$, 43 % YF₃. Upper temperature limit of stability of KYF₄ 702°.

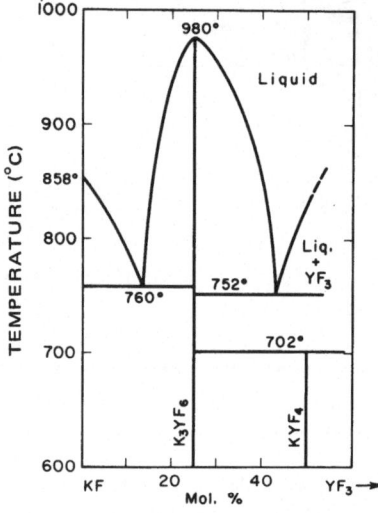

Fig. 70. The system KF–YF₃.

The System RbF–YF₃. See Fig. 71. Source: E. P. Dergunov, *Dokl. Akad. Nauk SSSR* **60**:1186 (1948). Compositions of the invariant equilibrium points were not given in the source reference. The eutectics are estimated to occur at 9 and 40 mole % YF₃.

Fig. 71. The system RbF–YF₃.

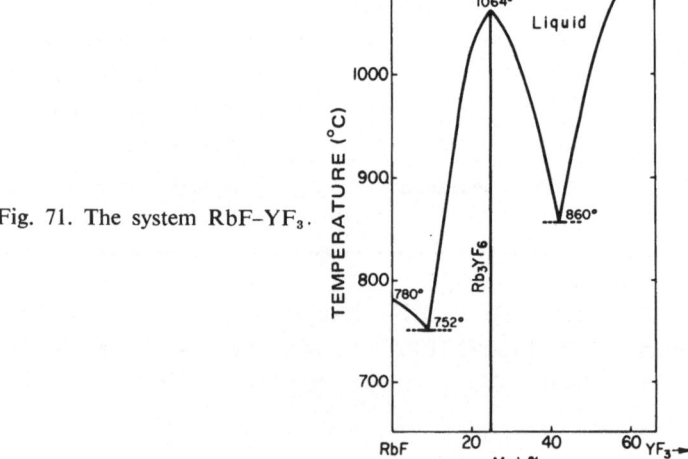

The System CsF–YF$_3$. See Fig. 72. Source: G. A. Bukhalova and E. P. Babaeva, *Russ. J. Inorg. Chem.* (*English Transl.*) **11**:219 (1966). The phase diagram reported in the source reference agrees with that of E. P. Dergunov, *Dokl. Akad. Nauk SSSR* **60**:1185 (1948). The eutectics occur at 4 mole % YF$_3$, 670°; and at 37.5 mole % YF$_3$, 890°.

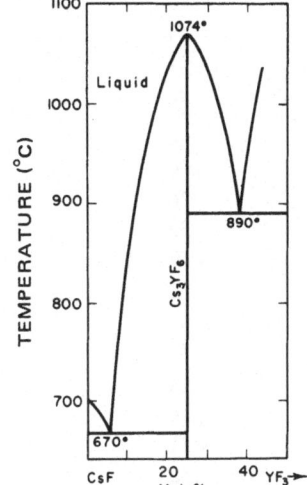

Fig. 72. The system CsF–YF$_3$.

The LiF–LnF$_3$ Systems. See Fig. 73. Source: R. E. Thoma, G. D. Brunton, R. A. Penneman and T. K. Keenan, *Inorg. Chem.* **9**:1096 (1970).

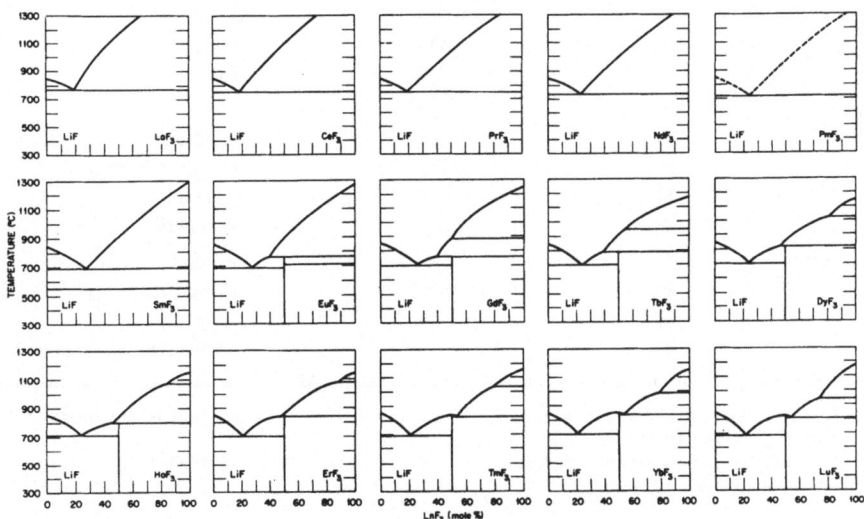

Fig. 73. The LiF–LnF$_3$ systems.

Invariant and singular equilibrium points are given in the table.

Ln	Mole % LnF$_3$	Temperature	Type of equilibrium
La	20[a]	770	Eutectic
Ce	19[b]	755	Eutectic
Pr	19	750	Eutectic
Nd	23	738	Eutectic
Sm	27	698	Eutectic
Eu	27	688	Eutectic
	30	710	Peritectic
	40	760	Peritectic
Gd	26	700	Eutectic
	39	755	Peritectic
	50	875	Peritectic
Tb	24	700	Eutectic
	39	790	Peritectic
	(54)[c]	950	Peritectic
Dy	(24)	700	Eutectic
	46	820	Peritectic
	(81)	1030	Peritectic
Ho	(24)	(710)	Eutectic
	46	798	Peritectic
	(84)	1070	Peritectic
Er	21	700	Eutectic
	48	840	Peritectic
	(88)	1075	Peritectic
Tm	21	700	Eutectic
	50	(835)	Congruent m.p.
	53	824	Eutectic
	(78)	1030	Peritectic
Yb	21	700	Eutectic
	50	850	Congruent m.p.
	53.5	840	Eutectic
	(79)	985	Peritectic
Lu	22	695	Eutectic
	50	(825)	Congruent m.p.
	54	810	Eutectic
	74	945	Peritectic

[a] Reported as LiF–LaF$_3$ (86–14 mole %), 758°, by G. A. Bukhalova and E. P. Babaev, *Zh. Neorgan. Khim.* **10**:1883 (1965).
[b] From C. J. Barton and R. A. Strehlow, *J. Inorg. Nucl. Chem.* **20**:45 (1961).
[c] Estimated value.

The compounds of the $LiLnF_4$ class are all isostructural and isomorphous with $LiYF_4$ [R. E. Thoma *et al.*, *Inorg. Chem.* **65**:1096 (1961)]. Details of the structure were first established through the work of Brunton, in a determination of the structure of $LiYbF_4$ from single crystals (see source reference).

The NaF–LnF$_3$ Systems. See Fig. 74. Source: R. E. Thoma, H. Insley, and G. M. Hebert, *Inorg. Chem.* **5**:1222 (1966). Invariant and sin-

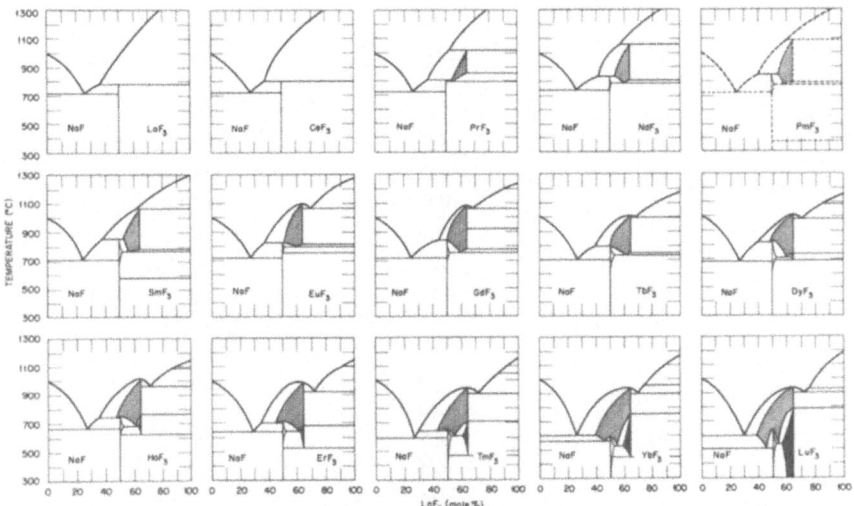

Fig. 74. The NaF–LnF$_3$ systems.

gular equilibrium points and structural data are given in the tables.

Ln	Mole % LnF$_3$	Temperature	Type of equilibrium
La	29	730[a]	Eutectic
	43	810[a]	Peritectic
Ce[b]	28	726	Eutectic
	38	810	Peritectic
Pr	27	733	Eutectic
	37	810	Peritectic
	52	1025	Peritectic
Nd	26	732	Eutectic

(*Continued*)

Ln	Mole % LnF$_3$	Temperature	Type of equilibrium
Nd	42	850	Peritectic
	58	1056	Peritectic
Sm	25	725	Eutectic
	38	834	Peritectic
	62	1060	Peritectic
Eu	25	718	Eutectic
	39	825	Peritectic
	64.3	1091	Congruent m.p.
	70	1060	Eutectic
Gd	25	712	Eutectic
	43	835	Peritectic
	64.3	1080	Congruent m.p.
	68	1058	Eutectic
Tb	27	696	Eutectic
	39	789	Peritectic
	64.3	1031	Congruent m.p.
	68	994	Eutectic
Dy	27	680	Eutectic
	38	780	Peritectic
	64.3	1020	Congruent m.p.
	70	982	Eutectic
Ho	27	663	Eutectic
	35	745	Peritectic
	64.3	1010	Congruent m.p.
	71	962	Eutectic
Er	27	643	Eutectic
	35	702	Peritectic
	64.3	975	Congruent m.p.
	72	920	Eutectic
Tm	27	592	Eutectic
	29	632	Peritectic
	64.3	940	Congruent m.p.
	~70	~905	Eutectic
Yb	26	594	Eutectic
	64.3	928	Congruent m.p.
	~70	893	Eutectic
Lu	29	595	Eutectic
	64.3	930	Congruent m.p.
	72	873	Eutectic

[a] Eutectic and peritectic temperatures were reported by E. Mathes and S. Holz, *Z. Chem.* **2**:22 (1962), to be about 630 and 770°C, respectively.
[b] C. J. Barton, J. D. Redman and R. A. Strehlow, *J. Inorg. Nucl. Chem.* **20**:45 (1961).

Compound formula	Symmetry	Space group or type	Lattice constants (Å)		Refractive indices	
			a_0	c_0	N_ε	N_ω
$NaF \cdot LaF_3$	Hexagonal	$P\bar{6}$	6.157	3.822	1.500	1.486
$NaF \cdot CeF_3$	Hexagonal	$P\bar{6}$	6.131	3.776	1.514	1.493
$NaF \cdot PrF_3{}^a$	Cubic	$Fm3m$	5.695–5.702			
$NaF \cdot PrF_3$	Hexagonal	$P\bar{6}$	6.123	3.822	1.516	1.494
$NaF \cdot NdF_3$	Cubic	$Fm3m$	5.670–5.678			
$NaF \cdot NdF_3$	Hexagonal	$P\bar{6}$	6.100	3.711	1.515	1.493
$NaF \cdot SmF_3$	Cubic	$Fm3m$	5.605–5.628			
$NaF \cdot SmF_3$	Hexagonal	$P\bar{6}$	6.051	3.640	1.516	1.492
$NaF \cdot EuF_3$	Cubic	$Fm3m$	5.575–5.605			
$NaF \cdot EuF_3$	Hexagonal	$P\bar{6}$	6.044	3.613	1.516	1.492
$NaF \cdot GdF_3$	Cubic	$Fm3m$	5.552–5.583			
$NaF \cdot GdF_3$	Hexagonal	$P\bar{6}$	6.020	3.601	1.507	1.483
$NaF \cdot TbF_3$	Cubic	$Fm3m$	5.525–5.563			
$NaF \cdot TbF_3$	Hexagonal	$P\bar{6}$	6.008	3.580	1.506	1.486
$NaF \cdot DyF_3$	Cubic	$Fm3m$	5.508–5.545			
$NaF \cdot DyF_3$	Hexagonal	$P\bar{6}$	5.985	3.554	1.510	1.486
$NaF \cdot HoF_3$	Cubic	$Fm3m$	5.480–5.526			
$NaF \cdot HoF_3$	Hexagonal	$P\bar{6}$	5.991	3.528	1.510	1.486
$NaF \cdot ErF_3$	Cubic	$Fm3m$	5.455–5.510			
$NaF \cdot ErF_3$	Hexagonal	$P\bar{6}$	5.959	3.514	1.504	1.482
$NaF \cdot TmF_3$	Cubic	$Fm3m$	5.435–5.495			
$NaF \cdot TmF_3$	Hexagonal	$P\bar{6}$	5.953	3.494	1.496	1.476
$NaF \cdot YbF_3$	Cubic	$Fm3m$	5.418–5.480			
$NaF \cdot YbF_3$	Hexagonal	$P\bar{6}$	5.929	3.471	~1.490	
$NaF \cdot LuF_3$	Cubic	$Fm3m$	5.400–5.465			
$NaF \cdot LuF_3$	Hexagonal	$P\bar{6}$	5.912	3.458	1.428	1.484

[a] J. H. Burns, *Inorg. Chem.* **4**:881 (1965). X-ray data for the hydrated double fluorides of NaF and the lanthanide trifluorides were reported by L. R. Batsanova, L. M. Yankovskaya, and L. V. Lukina, *Russ. J. Inorg. Chem.* (*English Transl.*) **17**:654 (1972).

The KF–LnF₃ Systems. See Fig. 75. Sources: E. P. Dergunov, *Dokl. Akad. Nauk SSSR* **60**:1185 (1948); E. P. Dergunov, *ibid.*, **85**:1025 (1952); C. J. Barton, L. O. Gilpatrick, G. D. Brunton, D. Hsu, and H. Insley, *J. Inorg. Nucl. Chem.* **33**:325 (1971); B. S. Zakharova, L. P. Reshetnikova, and A. V. Novoselova, *Vestn. Mosk Univ., Ser. II* **22**:102 (1967); and

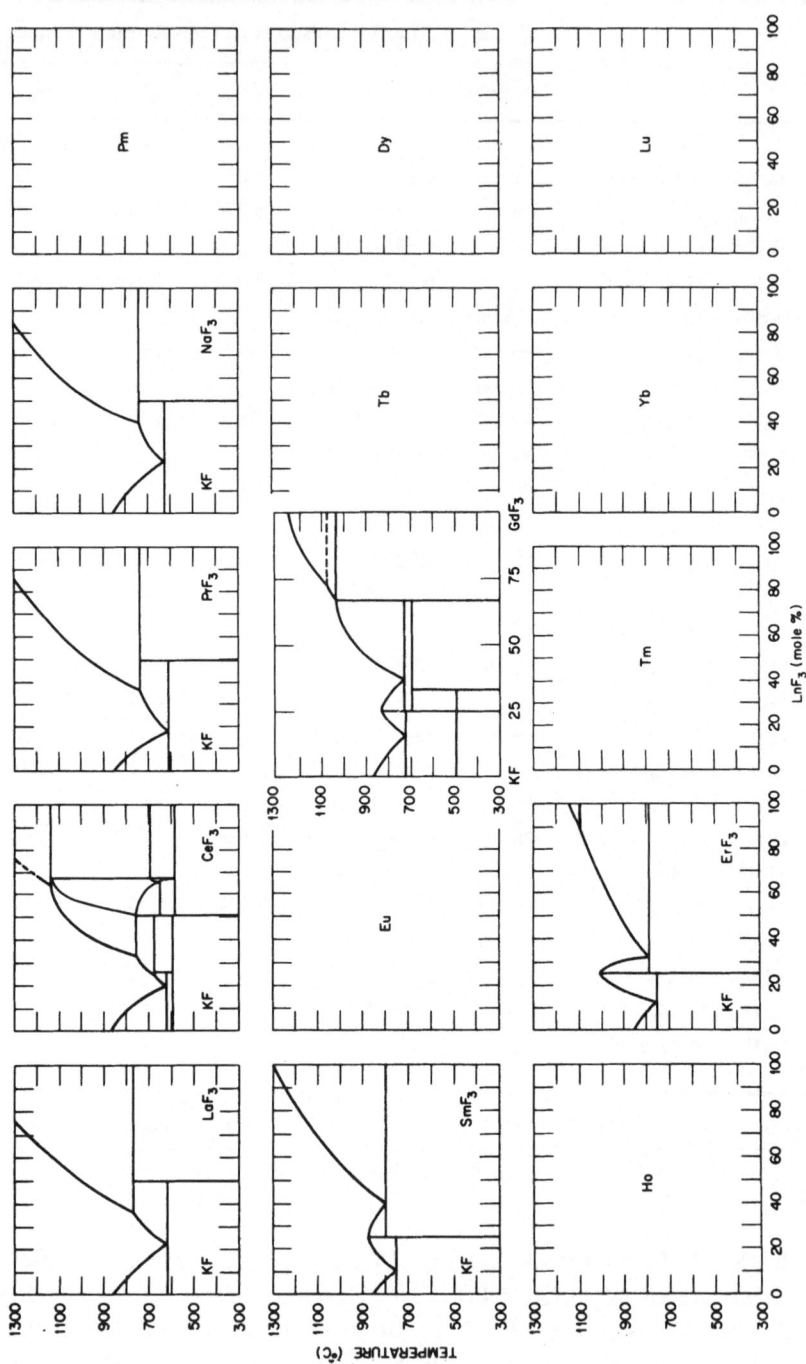

Fig. 75. The KF–LnF₃ systems.

A. Kozak, M. Samouel and A. Chretien, *Rev. de Chim. Min.* **10**:259 (1973). Invariant and singular equilibrium points are given in the table.

Ln	Mole % LnF$_3$	Temperature	Type of equilibrium
La[a]	22	620	Eutectic [Dergunov (1948)]
	36	770	Peritectic
Ce	19	620 ± 3	Eutectic [Barton *et al.* (1971)]
	24	765 ± 10	Peritectic
	32	755 ± 5	Peritectic
	63	1135 ± 15	Peritectic
	65	650 ± 12	Eutectoid
Pr[a]	23	610	Eutectic [Dergunov (1952)]
	63	740	Peritectic
Nd	23	625	Eutectic [Zakharov *et al.* (1967)]
		550	Upper limit of stability of 3KF · NdF$_3$
	40	740	Peritectic
Sm	10	758	Eutectic [Dergunov (1952)]
	25	882	Congruent m.p.
	40	800	Eutectic
Gd	5	672	Eutectic [Kozak *et al.* (1973)]
	25	997	Congruent m.p.
	37	880	Eutectic
	66.7	1160	Congruent m.p.
	74	1128	Eutectic
Er[a]	12	756	Eutectic [Dergunov (1952)]
	25	1012	Congruent m.p.
	32	790	Eutectic

[a] Compositions were not listed in the source reference. Values given below are estimated.

Structural data: 3KF · CeF$_3$: orthorhombic, *Pnma*, $a = 6.2895$, $b = 3.8040$, $c = 15.596$ [G. D. Brunton, *Acta Cryst.* **B25**:600 (1969). KF · LaF$_3$: hexagonal, *P-62m*, $a = 6.526$, $c = 3.791$ [W. H. Zachariasen, *J. Am. Chem. Soc.* **70**:2147 (1948).

The RbF–LnF$_3$ Systems. See Fig. 76. Sources: See KF–LnF$_3$ systems. Invariant and singular equilibrium points are given in the table.

Structural data: RbF · LaF$_3$: tetragonal, $a = 8.11$, $c = 6.5$ [E. P. Dergunov, *Dokl. Akad. Nauk SSSR* **60**:1185 (1948).

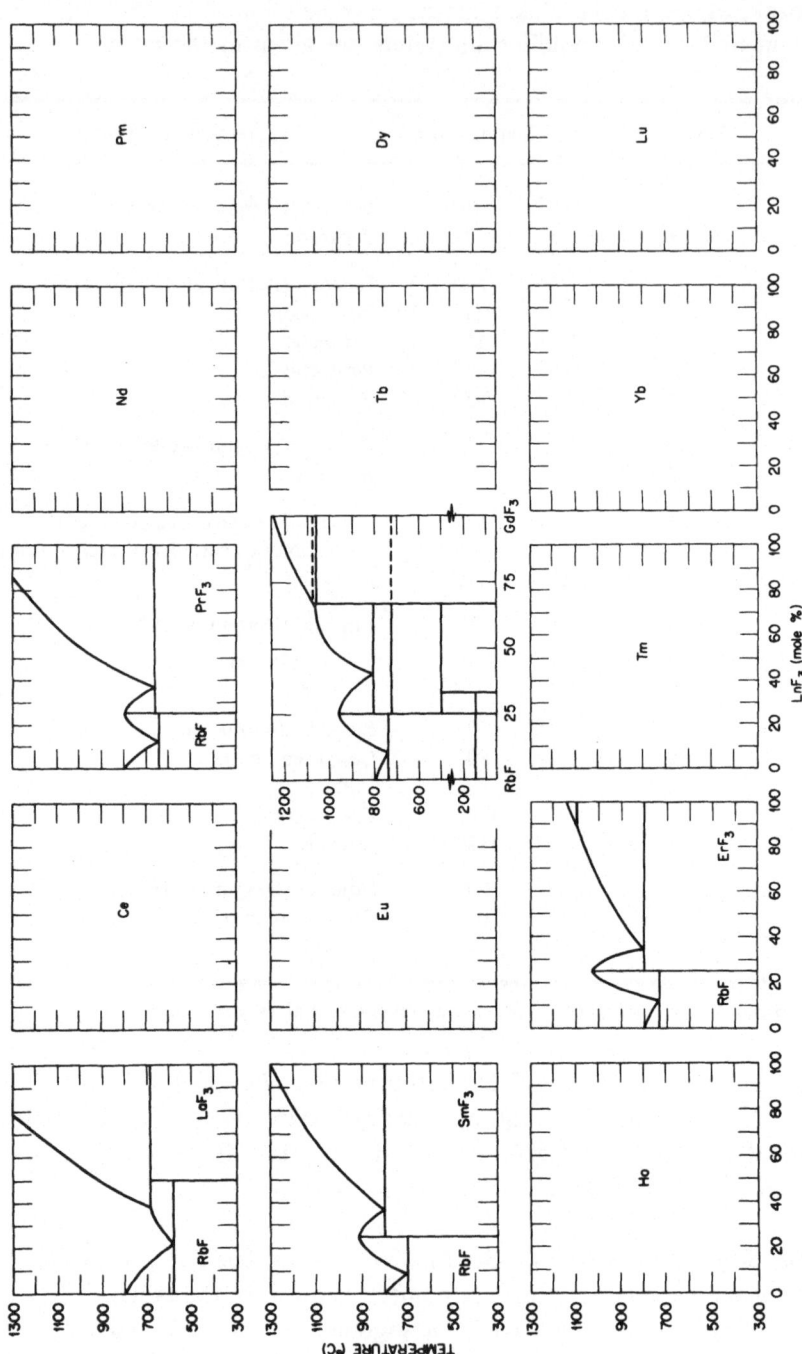

Fig. 76. The RbF–LnF₃ systems.

Ln	Mole %	Temperature	Type of equilibrium
La[a]	22	582	Eutectic [Dergunov (1948)]
	38	684	Peritectic
Pr[a]	22	640	Eutectic [Dergunov (1952)]
	25	797	Congruent m.p.
	37	660	Eutectic
Sm[a]	9	700	Eutectic [Dergunov (1952)]
	25	916	Congruent m.p.
	37	800	Eutectic
Gd	10	796	Eutectic [Kozak et al. (1973)]
	25	962	Congruent m.p.
		500	Upper temperature limit of stability of $2RbF \cdot GdF_3$
	40	803	Eutectic
	65	1060	Peritectic
Er[a]	12	732	Eutectic [Dergunov (1952)]
	25	1034	Congruent m.p.
	33.3	800	Eutectic

[a] Compositions were not listed in the source reference. Values listed here are estimated.

The $CsF–LnF_3$ Systems. See Fig. 77. Sources: See $KF–LnF_3$ systems. Invariant and singular equilibrium points are given in the table.

Ln	Comp. mole %	Temperature	Type of equilibrium
La[a]	12	600	Eutectic [Dergunov (1952)]
	25	795	Congruent m.p.
	33	726	Eutectic
Pr[a]	8	654	Eutectic [Dergunov (1952)]
	25	920	Congruent m.p.
	41	783	Eutectic
Gd	5	672	Eutectic [Kozak et al. (1973)]
	25	997	Congruent m.p.
	37	880	Eutectic
	66.7	1160	Congruent m.p.
	76	1128	Eutectic
Er[a]	5	634	Eutectic [Dergunov (1952)]
	25	1048	Congruent m.p.
	42	800	Eutectic

[a] Compositions were not listed in the source reference. Values listed here are estimated.

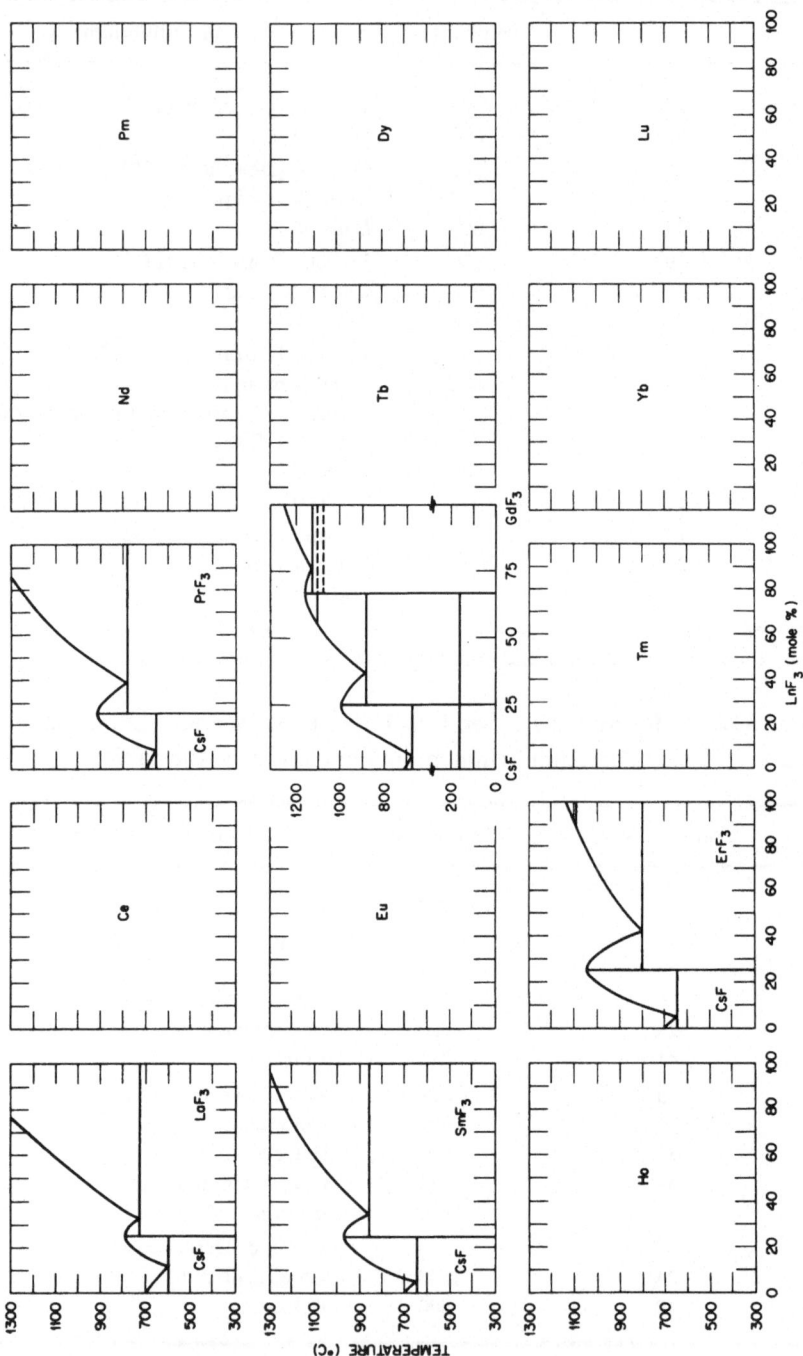

Fig. 77. The CsF–LnF₃ systems.

The System RbF–VF₃. See Fig. 78. Source: J. C. Cousseins and J. C. Cretenet, *Compt. Rend.* **265**:1464 (1967). Invariant and singular equilibrium

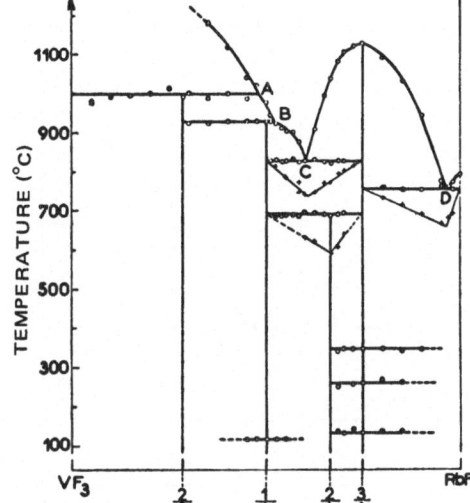

Fig. 78. The system RbF–VF₃.

points are given in the table.

Mole % ThF₄	Temperature	Type of equilibrium
4.5	755	Eutectic
25	1128	m.p. 3RbF · VF₃
	345	3RbF · VF₃ inversion temp.
	258	" " "
	130	" " "
	690	Upper temp. of stability of 2RbF · VF₃
39.5	828	Eutectic
48	928	Peritectic
	115	RbF · VF₂ inversion temp.
52	996	Peritectic

The source paper lists the x-ray powder diffraction data for four of the intermediate compounds formed in the system.

The System LiF–UF₃. See Fig. 79. Source: C. J. Barton, V. S. Coleman, L. M. Bratcher, and R. R. Grimes, ORNL, unpublished work (1953–1954). $E = 770°$, 27% UF₃.

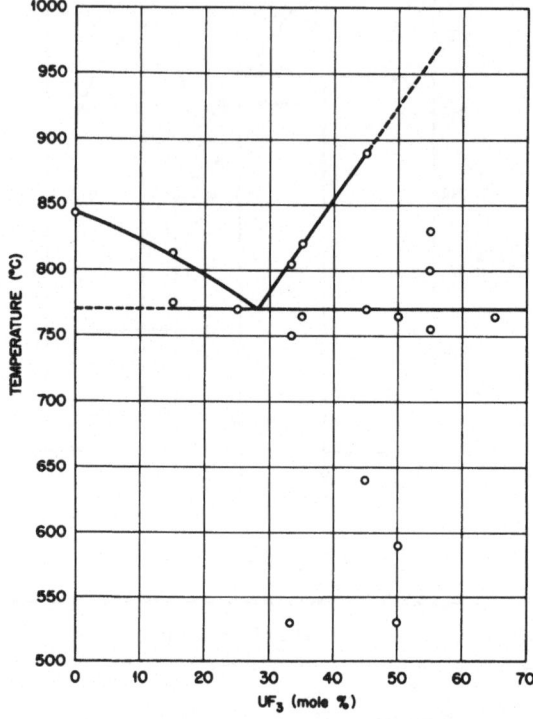

Fig. 79. The system LiF–UF₃.

The System LiF–PuF₃. See Fig. 80. Source: C. J. Barton and R. A. Strehlow, *J. Inorg. Nucl. Chem.* **18**:143 (1961). $E = 743°$, 19.5% PuF₃.

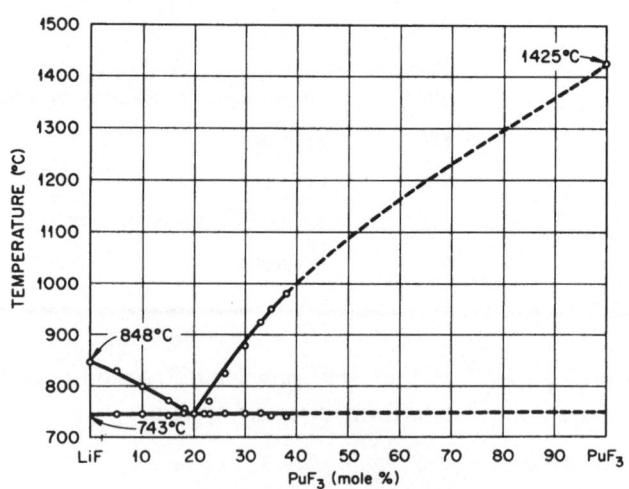

Fig. 80. The system LiF–PuF₃.

The System NaF–UF$_3$. See Fig. 81. Source: C. J. Barton, V. S. Coleman, T. N. McVay, and W. R. Grimes, ORNL, unpublished work (1953–1954). $E = 715°$, 27 % UF$_4$; $P = 775°$, 35 % UF$_4$; α-NaF · UF$_3 \rightleftarrows$ β-NaF · UF$_3$, 595°.

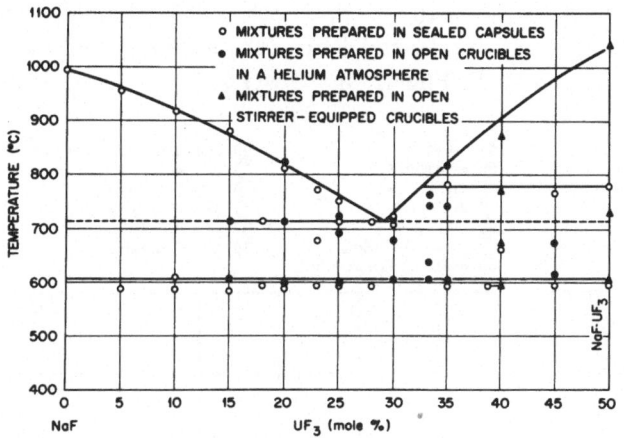

Fig. 81. The system NaF–UF$_3$.

Structural data: NaF · UF$_3$: hexagonal, $P\bar{6}2m$, $a = 6.167$, $c = 3.77$ [W. H. Zachariasen, *J. Am. Chem. Soc.* **70**:2147 (1948)].

The System NaF–PuF$_3$. See Fig. 82. Source: C. J. Barton, J. D. Redman, and R. A. Strehlow, *J. Inorg. Nucl. Chem.* **20**:45 (1961). Invariant equilibrium points: $E = 727°$, 23% PuF$_3$; $P = 842°$, 39% PuF$_3$.

Structural data: NaF · PuF$_3$: hexagonal, $P\bar{6}2m$, $a = 6.117$, $c = 3.746$ [W. H. Zachariasen, *J. Am. Chem. Soc.* **70**:2147 (1948)].

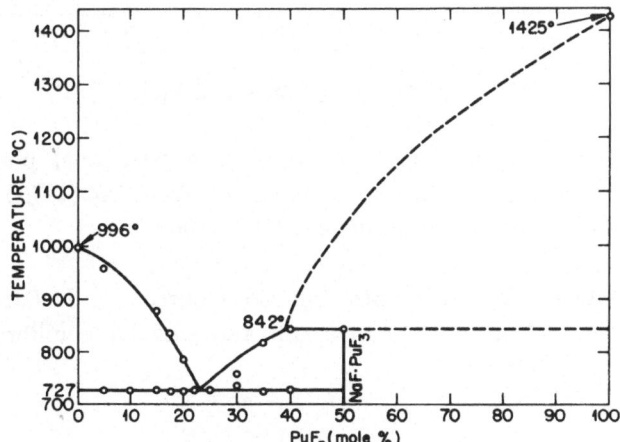

Fig. 82. The system NaF–PuF$_3$.

The System LiF–ZrF$_4$.   See Fig. 83. Source: R. E. Thoma, H. Insley, H. A. Friedman, and G. M. Hebert, *J. Chem. Eng. Data* **10**:219 (1965). $E = 598°$, 21 % ZrF$_4$; m.p. 3LiF · ZrF$_4$ 662°; α-3LiF · ZrF$_4 \rightleftarrows \beta$-3LiF · ZrF$_4$, 475°; β-3LiF · ZrF$_4 \rightleftarrows$ LiF + 2LiF · ZrF$_4$, 470°; $E = 570°$, 29.5 % ZrF$_4$; m.p. 2LiF · ZrF$_4$ 596°; $E = 507°$, 49 % ZrF$_4$; $P = 520°$, 51.5 % ZrF$_4$; 3LiF · 4ZrF$_4 \rightleftarrows$ 2LiF · ZrF$_4$ + ZrF$_4$, 466°. Liquidus curves of the binary systems LiF–, NaF–, and RbF–ZrF$_4$ were reported by K. A. Sense, R. W. Stone, and R. B. Filbert, Jr., USAEC Report BMI-1199, 31 (1957).

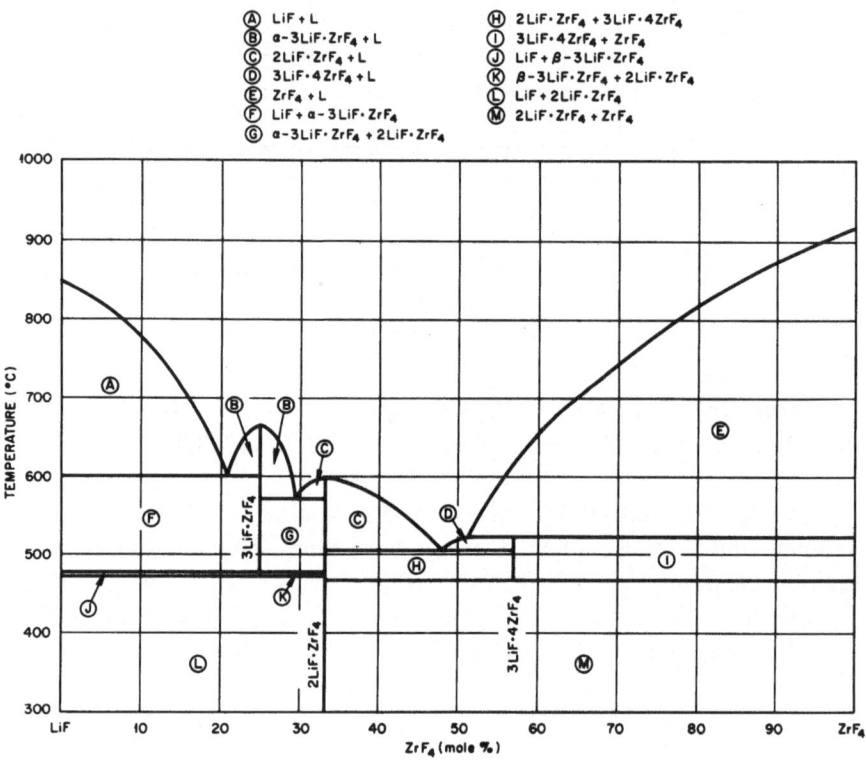

Fig. 83. The system LiF–ZrF$_4$.

Structural data: Li$_2$ZrF$_6$ crystallizes as a hexagonal phase, $P\bar{3}1m$, $a = 4.9733$, $c = 4.658$ [G. Brunton, *Acta Cryst.* **B29**:2294 (1973); R. Hoppe and W. Dahne, *Naturwissenschaften* **47**:397 (1960)].

The System NaF–ZrF$_4$.   See Fig. 84. Source: C. J. Barton *et al.*, *J. Phys. Chem.* **62**:665 (1958). Invariant and singular equilibrium points are given in the table.

Mole % ZrF$_4$ in liquid	Temperature	Type of equilibrium
20	747	Eutectic
25	850	Congruent m.p.
	523	Inversion
	500	Eutectoid
34	640	Peritectic
39.5	544	Peritectic
40	533	Inversion
	487	Decomposition
40.5	505	Inversion
40.5	500	Eutectic
46.2	525	Congruent m.p.
49.5	512	Eutectic
56.5	537	Peritectic

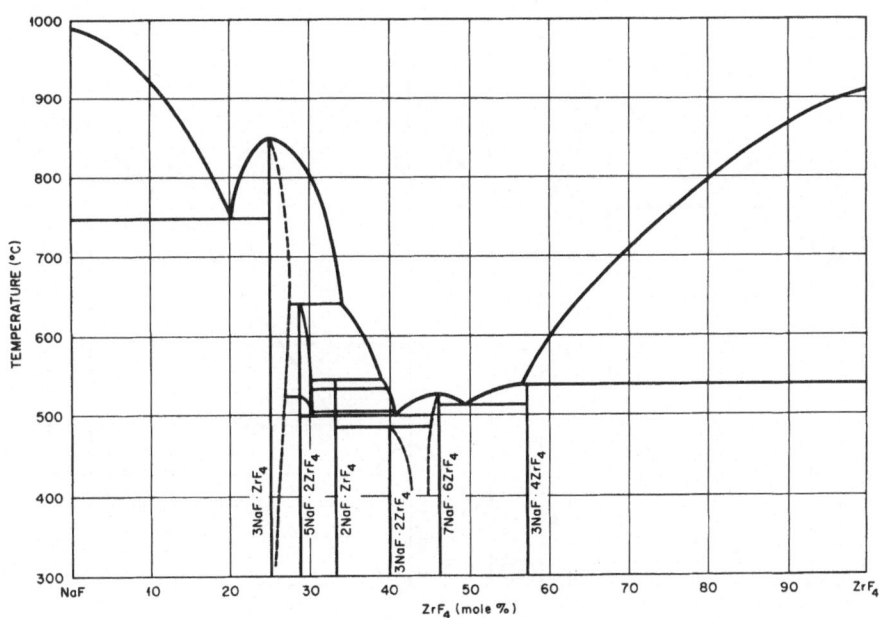

Fig. 84. The system NaF–ZrF$_4$.

A phase diagram of the system was reported by R. Winand, *Compt. Rend. Congr. Intern. Chimie Ind., 31ᵉ, Liege*, 744 (1958), approximately concurrently with the publication of the source reference. It bears only gross resemblance to Fig. 84 and has not been evaluated. A liquidus curve for the ZrF_4 primary phase field was computed by K. A. Sense, R. W. Stone, and O. B. Filbert, Jr., USAEC Report BMI-1199, 31 (1957); see Fig. 1509, *Phase Diagrams for Ceramists* (1964).

Structural data: $3NaF \cdot ZrF_4$: tetragonal, *I4/mmm*, $a = 5.31, c = 10.50$ [L. A. Harris, *Acta Cryst.* **12**:172 (1959)]; $\beta\text{-}2NaF \cdot ZrF_4$: tetragonal, $a = 13.85$, $c = 1023$ [G. D. Brunton *et al.*, ORNL-3761]; $\gamma\text{-}NaF \cdot ZrF_4$: monoclinic, *P21/c*, $a = 5.5562$, $b = 5.4069$, $c = 16.073$, $\beta = 95.886°$ [G. D. Brunton *et al.*, ORNL-3761]; $7NaF \cdot 6ZrF_4$: hexagonal, *R-3*, $a = 13.807$, $c = 9.429$ [J. H. Burns, R. D. Ellison, and H. A. Levy, *Acta Cryst.* **B24**:230 (1968); $5NaF \cdot 2ZrF_4$: monoclinic, *C2/m*, $a = 11.62$, $b = 5.49$, $c = 8.44$, $\beta = 97.7°$ (corresponds to low-temperature form shown in phase diagram) [R. M. Herak *et al.*, *Bilten Inst. Boris Kidric* **14**:21 (1963)].

The System $KF–ZrF_4$. See Figs. 85 and 86. Sources: C. J. Barton, H. Insley, R. P. Metcalf, R. E. Thoma, and W. R. Grimes, ORNL, unpublished work (1951–1955), and A. V. Novoselova, Yu. M. Korenev, and

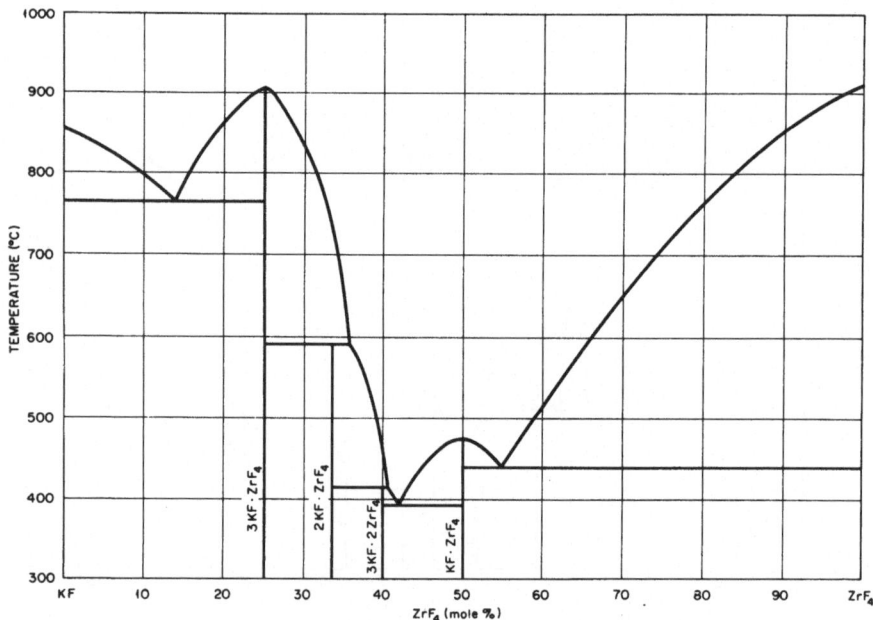

Fig. 85. The system $KF–ZrF_4$.

Yu. P. Simanov, *Dokl. Akad. Nauk SSSR* **139**:893 (1961). Invariant equilibria are listed below from the ORNL work. $E = 765$, 14% ZrF_4; congruent m.p. $3KF \cdot ZrF_4$ 910°; $P = 590°$, 36% ZrF_4; $P = 412°$, 40.5% ZrF_4: $E = 390°$, 42% ZrF_4; congruent m.p. $KF \cdot ZrF_4$ 475°; $E = 440°$, 55% ZrF_4. The complex diagram reported by Novoselova *et al.* has not been evaluated by subsequent investigators.

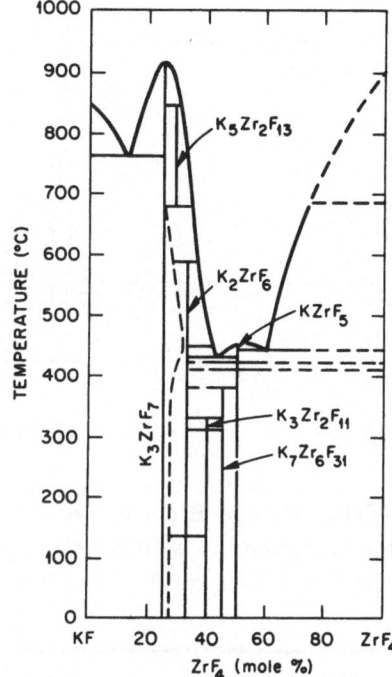

Fig. 86. The system KF–ZrF_4.

Structural data: $3KF \cdot ZrF_4$: cubic, *Fm3m*, $a = 8.988$ [G. D. Brunton *et al.*, ORNL-3761]. $2KF \cdot ZrF_4$: orthorhombic, *Immm*, $a = 6.58$, $b = 11.42$, $c = 6.94$ [V. Bode and G. Teufer, *Z. Anorg. Allgem. Chem.* **283**:18 (1956).

The System RbF–ZrF$_4$. See Fig. 87. Source: R. E. Moore, R. E. Thoma, C. J. Barton, W. R. Grimes, B. S. Landau, and H. A. Friedman, ORNL, unpublished work (1955–1956). Invariant and singular equilibrium points are given in the table.

Vapor–liquid equilibria were described by K. A. Sense, R. W. Stone, and R. B. Filbert, USAEC Report BMI-1199, 17 (1957).

Structural data: $3RbF \cdot ZrF_4$: cubic, *Fm3m*, $a = 9.31$ [G. D. Brunton *et al.*, ORNL-3761]. $2RbF \cdot ZrF_4$: hexagonal, *P3m*, $a = 6.16$, $c = 4.82$ [V. Bode and G. Teufer, *Z. Anorg. Allgem. Chem.* **283**:18 (1956)]. $5RbF \cdot$

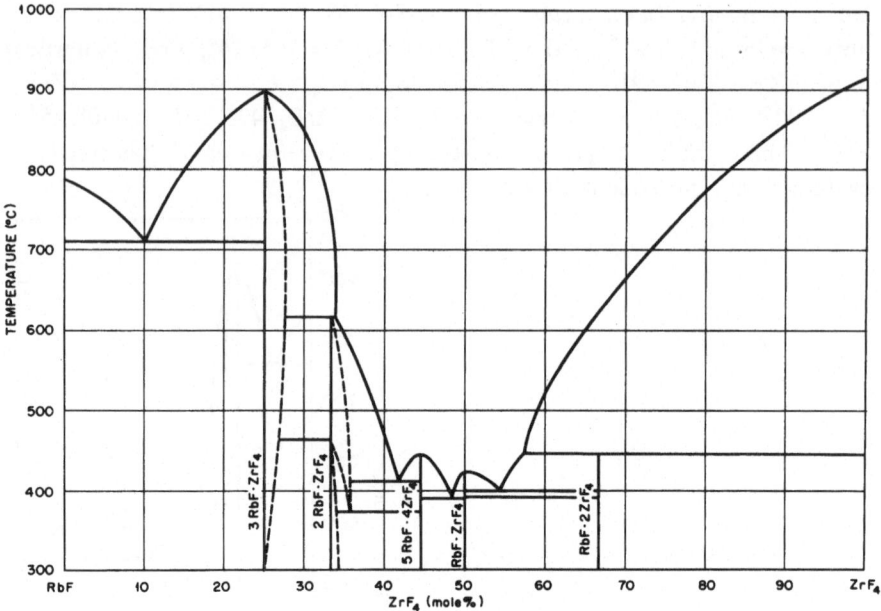

Fig. 87. The system RbF–ZrF$_4$.

4ZrF$_4$: George Brunton, *Acta Cryst.* **B27**:1944 (1971), has reported that the compound crystallizes with space group $P2_1$; $a = 11.520$, $b = 11.222$, $c = 7.868$, and $\cos \beta = 0.1445$.

Mole % ZrF$_4$ in liquid	Temperature	Type of equilibrium
10	710	Eutectic
25	897	Congruent melting point
34	620	Peritectic
	460	Inversion
	370	Lowered inversion and decomposition
42	410	Eutectic
44.4	445	Congruent melting point
48	390	Eutectic
50	423	Congruent melting point
	391	Inversion
54	400	Eutectic
57	447	Peritectic

The System CsF–ZrF₄.   See Fig. 88. Source: G. D. Robbins, R. E. Thoma, and H. Insley, *J. Inorg. Nucl. Chem.* **27**:559 (1965). Invariant and

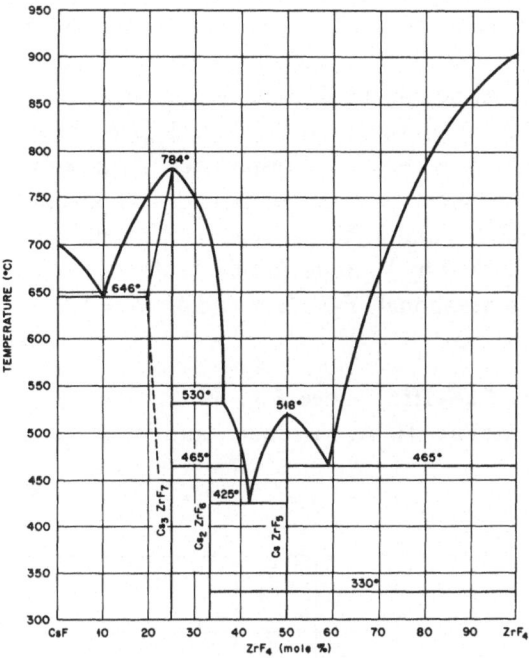

Fig. 88. The system CsF–ZrF₄.

singular equilibrium points are given in the table.

Mole % ZrF_4	Temperature	Type of equilibrium
10	646	Eutectic
25	784	Congruent melting point
37	530	Peritectic
	465	Inversion of Cs_2ZrF_6
42	425	Eutectic
50	518	Congruent melting
	330	Inversion of $CsZrF_5$
55	465	Eutectic

The compound Cs_3ZrF_7 was observed to vary in composition as shown in the phase diagram, accommodating slightly more than 20% CsF into its lattice at the solidus. Refractive indices of the solid solution were observed to vary in this composition range from 1.470 at the stoichiometric composition to 1.480 near the CsF-rich terminus of the solid solution. Cs_3ZrF_7 is face-centered cubic with $a = 9.70 \pm 0.02$; the space group is $F\bar{4}3m$, $F432$, or $Fm3m$. Although the fluoride-ion arrangement is disordered, the Zr^{4+} and Cs^+ are at or very near $(0, 0, 0)$ + face centering and $(\frac{1}{2}, \frac{1}{2}, \frac{1}{2}; \frac{1}{4}, \frac{1}{4}, \frac{1}{4}; \frac{3}{4}, \frac{3}{4}, \frac{3}{4})$ + face centering, respectively. The x-ray density of the compound is 4.53 g/cm^{-3}. This compound is apparently isotypic with K_3ZrF_7, $(NH_4)_3ZrF_7$, and K_3UF_7 as regards the metal ions. The compound Cs_2ZrF_6 was described by V. Bode and G. Teufer, *Z. Anorg. Allgem. Chem.* **283**:18 (1956) as hexagonal, *P-3m1*, $a = 6.41$, $c = 5.01$.

The System NaF–HfF$_4$. See Fig. 89. Source: R. E. Thoma, C. F. Weaver, T. N. McVay, H. A. Friedman, and W. R. Grimes, ORNL, un-

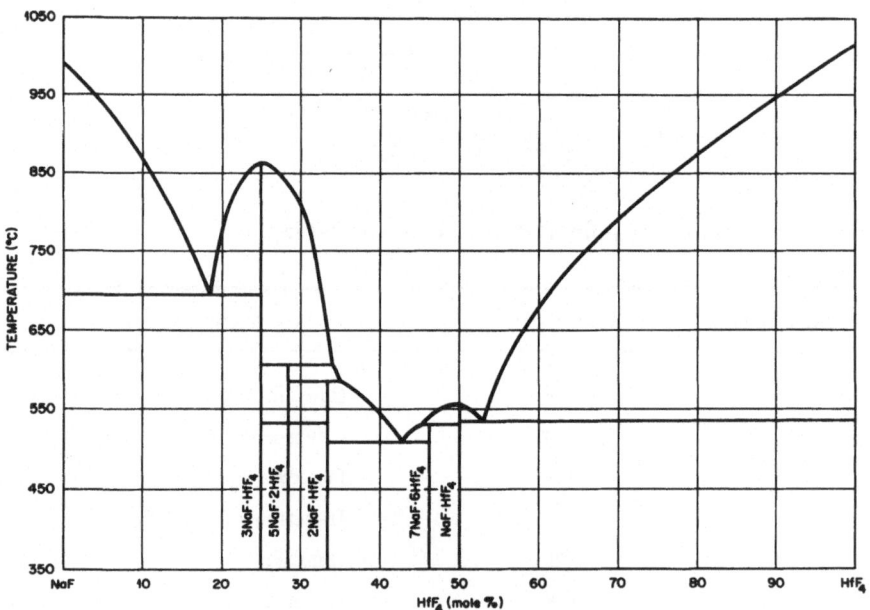

Fig. 89. The system NaF–HfF$_4$.

published work (1965–1968). Invariant and singular equilibrium points are given in the table.

Mole % HfF$_4$ in liquid	Temperature	Type of equilibrium
18.5	695	Eutectic
25	863	Congruent melting point
34	606	Peritectic
	535	Inversion
35	586	Peritectic
	Below 350	Inversion
	Below 350	Inversion
43	510	Eutectic
45	531	Peritectic
50	557	Congruent melting point
53	535	Eutectic

Structural data: $3NaF \cdot HfF_4$, hexagonal, $a = 6.1$, $c = 10.84$ [G. D. Brunton *et al.*, ORNL-3761 (1965)].

A determination of the crystal structure of $3NaF \cdot HfF_4$ has been reported by L. A. Harris *et al.*, *Acta Cryst.* **12**:172 (1959).

The System NaF–HfF$_4$. See Fig. 90. A phase diagram of the NaF–HfF$_4$ system was reported by I. N. Sheiko, V. T. Mal'tsev, and G. A. Bukhalova, *Ukr. Khim. Zh.* **32**:1292 (1966), as shown in Fig. 90. The disparity between this phase diagram and that shown in Fig. 89 indicates that further investigations of the system would be useful.

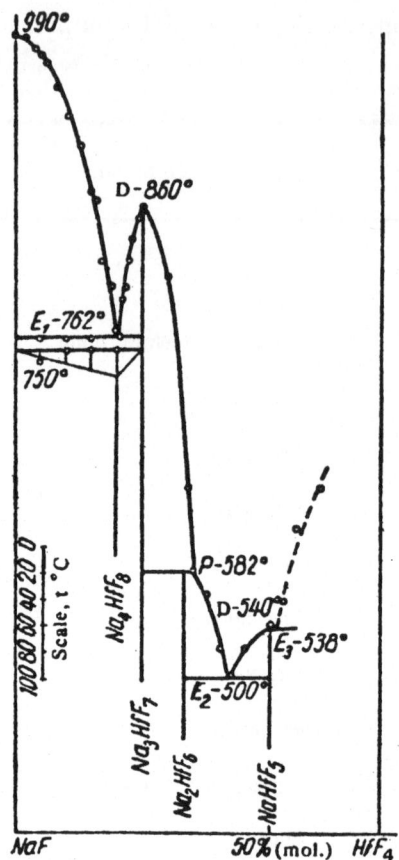

Fig. 90. The system NaF–HfF₄.

The System KF–HfF₄. See Fig. 91. Source: I. N. Sheiko, V. T. Mal'tsev, and G. A. Bukhalova, *Ukr. Khim. Zh.* **32**:1292 (1966). $E = 766°$, 13% HfF₄; m.p. $3KF \cdot HfF_4$ 923°; $P = 608°$, 35% HfF₄; $E = 408°$, 40% HfF₄; m.p. $KF \cdot HfF_4$ 432°; $E = 420°$, 57% HfF₄.

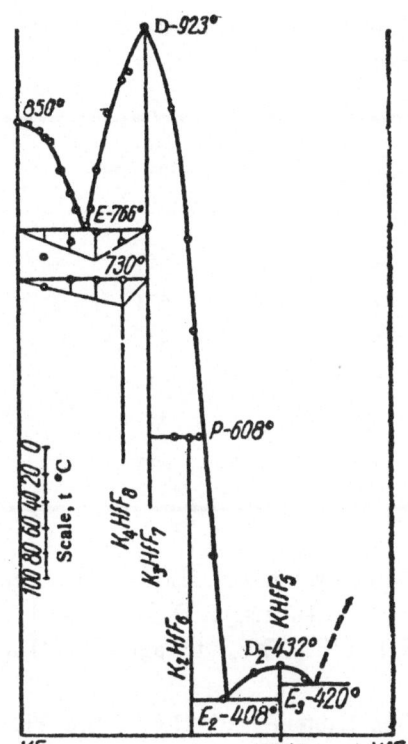

Fig. 91. The system KF–HfF$_4$.

The System LiF–ThF$_4$. See Fig. 92. Source: R. E. Thoma, H. Insley, B. S. Landau, H. A. Friedman, and W. R. Grimes, *J. Phys. Chem.*

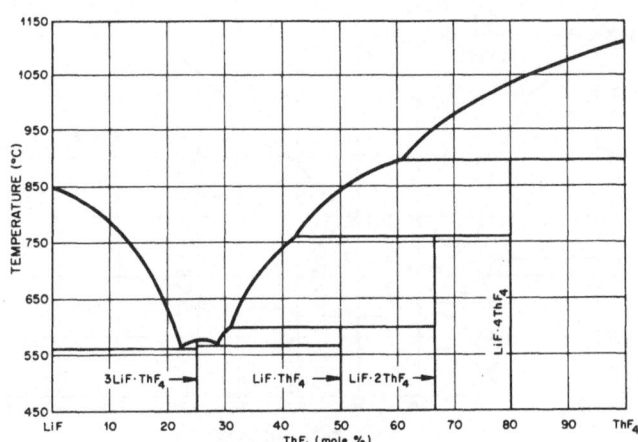

Fig. 92. The system LiF–ThF$_4$.

63, 1266 (1959). Invariant and singular equilibrium points are given in the table.

Mole % ThF$_4$ in liquid	Temperature	Type of equilibrium
23	568	Eutectic
25	573	Congruent melting point
29	565	Eutectic
30.5	597	Peritectic
42	762	Peritectic
62	897	Peritectic

Structural data: $3LiF \cdot ThF_4$: tetragonal, $P4/nmm$, $a = 6.206$, $c = 6.470$ [L. A. Harris, G. D. White, and R. E. Thoma, *J. Phys. Chem.* **63**:1974 (1959)]. $LiF \cdot ThF_4$: tetragonal, $141/a$, $a = 15.20$, $c = 6.60$.

The System NaF–ThF$_4$. See Fig. 93. Source: R. E. Thoma *et al.*, *J. Phys. Chem.* **63**:1266 (1959), and Roy E. Thoma, *J. Inorg. Nucl. Chem.*

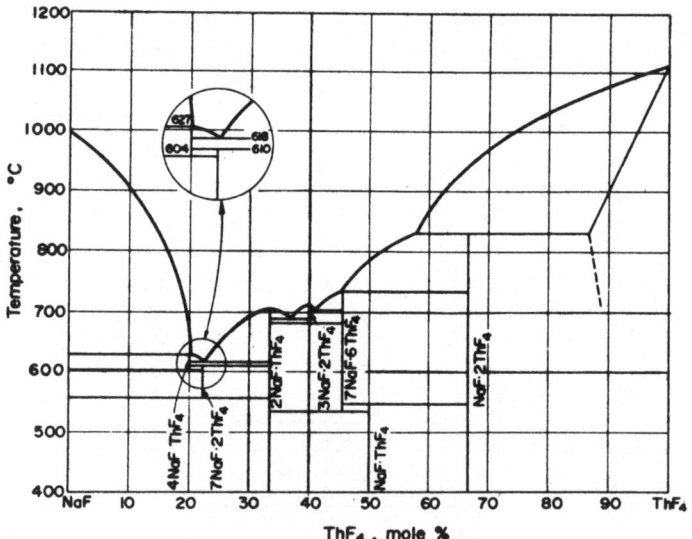

Fig. 93. The system NaF–ThF$_4$.

34:2747 (1972). Invariant and singular equilibrium points are given in the table.

Mole % ThF$_4$ in liquid	Temperature	Type of equilibrium
21.5	645	Peritectic
22.5	618	Eutectic
	604	Inversion
	558	Decomposition
33.3	705	Congruent melting point
37	690	Eutectic
40	712	Congruent melting point
41	705	Eutectic
	683	Decomposition
45.5	730	Peritectic
58	831	Peritectic

Structural data: 4NaF · ThF$_4$: cubic, $a = 11.04$ [W. H. Zachariasen, *J. Am. Chem. Soc.* **70**:2147 (1948). 7NaF · 2ThF$_4$: hexagonal, $a = 12.713$, $c = 10.377$ [G. D. Brunton *et al.*, ORNL-3761 (1965)]. $\beta_2 = 2$NaF · ThF$_4$: hexagonal, $P32$, $a = 5.99$, $c = 3.81$ [W. H. Zachariasen, *ibid.*]. $\delta = 2$NaF · ThF$_4$: hexagonal, $a = 6.14$, $c = 7.36$ [W. H. Zachariasen, *ibid.*]. 3NaF · 2ThF$_4$: cubic, $a = 5.62$ [G. D. Brunton, *ibid.*]. 7NaF · 6ThF$_4$: hexagonal, $R\bar{3}$, $a = 14.96$, $c = 9.912$ [G. D. Brunton, *ibid.*]. NaF · 2ThF$_4$: cubic, $I\bar{4}3M$, $a = 8.722$ [W. H. Zachariasen, *Acta Cryst.* **2**:390 (1949)].

The System KF–ThF$_4$. See Fig. 94. Source: W. J. Asker, E. R. Segnit, and A. W. Wylie, *J. Chem. Soc.* **1952**:4470. The system has also been described by A. G. Bergman and E. P. Dergunov, *Dokl. Akad. Nauk SSSR* **53**:753 (1941), and by V. S. Emelyanov and A. J. Evstyukhin, *J. Nucl. Energy* **5**:108 (1957). From independent examination at ORNL, the invariant equilibria listed by Asker *et al.* are considered to be the most reliable. This examination indicated that K$_2$ThF$_6$ melts congruently. Invariant and singular equilibrium points are given in the table.

Mole % ThF$_4$ in liquid	Temperature	Type of equilibrium
14	694	Eutectic
	635	Inversion
	712	Peritectic
25	865	Congruent melting point
	570	Decomposition
31	691	Eutectic
	747	Peritectic
	645	Inversion
50[a]	905	Congruent melting point
56	875	Eutectic
66	930	Peritectic
75	990	Congruent melting point
78	980	Eutectic

[a] It has been shown that the compound labeled by Asker *et al.* has the actual formula 7KF · 6ThF$_4$; cf. R. E. Thoma, Crystal Structures of Some Compounds of UF$_4$ and ThF$_4$ with Alkali Fluorides, ORNL-CF-58-12-40 (December 11, 1958).

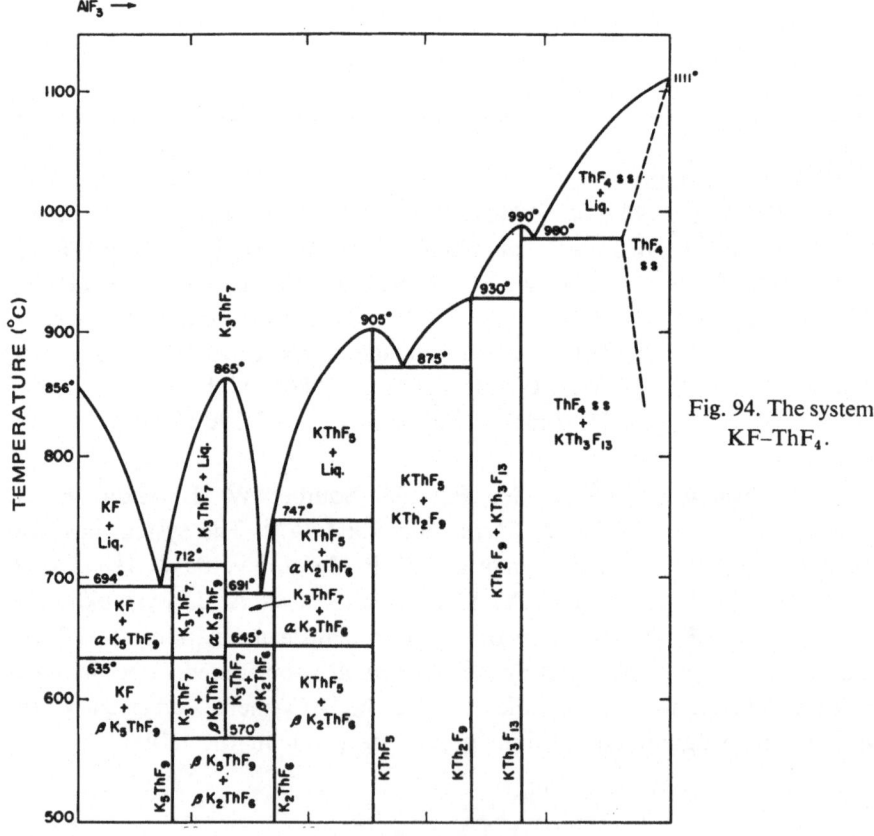

Fig. 94. The system KF–ThF$_4$.

Structural data: $5KF \cdot ThF_4$: orthorhombic, $Cmc2_1$, $a = 7.848$, $b = 12.840$, $c = 10.785$ [R. R. Ryan and R. A. Penneman, *Acta Cryst.* **B27**:829 (1971). $2KF \cdot ThF_4$: hexagonal, $P\bar{6}2m$, $a = 6.565$, $c = 3.815$ [W. H. Zachariasen, *J. Am. Chem. Soc.* **70**:2157 (1948)].

The System RbF–ThF$_4$. See Fig. 95. Source: R. E. Thoma, H. Insley, G. M. Hebert, and G. D. Brunton, ORNL, unpublished work (1961–1965).

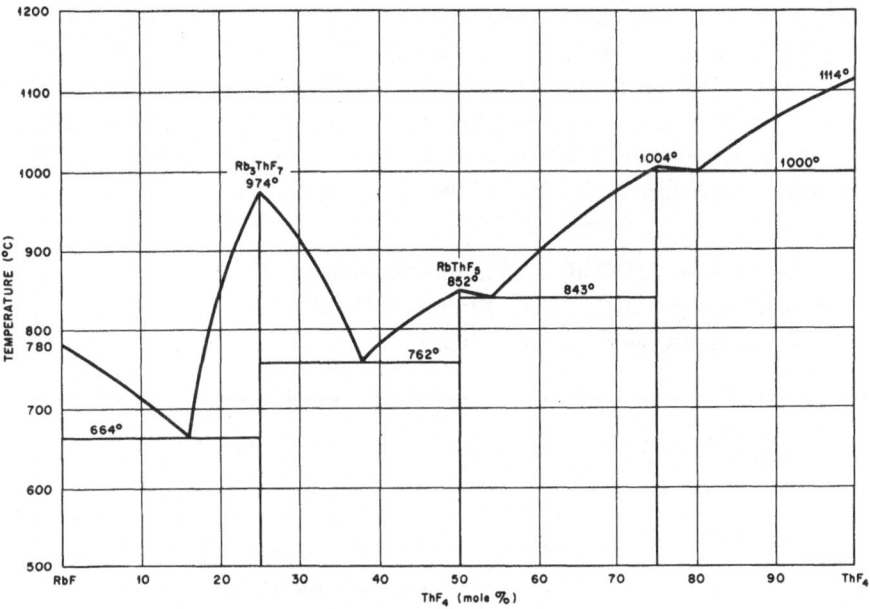

Fig. 95. The system RbF–ThF$_4$.

Invariant and singular equilibrium points are given in the table.

Mole % ThF$_4$	Temperature	Type of equilibrium
15.5	665	Eutectic
	545	Upper temp. of stability of $5RbF \cdot ThF_4$
20.5	895	Peritectic
25	973	m.p. $3RbF \cdot ThF_4$
	488	Lower temp. of stability of $3RbF \cdot ThF_4$
31	845	Eutectic
33.3	862	m.p. $2RbF \cdot ThF_4$
39.5	803	Eutectic
46.2	835	m.p. $7RbF \cdot 6ThF_4$
51.5	800	Eutectic
65.5	925	Peritectic
75	995	Peritectic

An earlier description of the system was given by E. P. Dergunov and A. G. Bergman, *Dokl. Akad. Nauk SSSR* **60**:391 (1948). See Fig. 1505, *Phase Diagrams for Ceramists* (1964). The phase diagram shown here was constructed from experimental data obtained as a means of evaluating the Dergunov and Bergman report.

Structural data: $3RbF \cdot ThF_4$: cubic, $a = 9.62$ [E. P. Dergunov and A. G. Bergman, *ibid.*]. $2RbF \cdot ThF_4$: hexagonal, $P\bar{6}2m$, $a = 6.85$, $c = 3.83$ [L. A. Harris, *Acta Cryst.* **13**:502 (1960)]. $7RbF \cdot 6ThF_4$: rhombohedral, $R\bar{3}$, $a = 9.58$; hexagonal parameters: $a = 15.39$, $c = 10.73$ [G. D. Brunton *et al.*, ORNL-3761 (1965). $RbF \cdot 3ThF_4$: $P2_1ma$; $a = 8.6490$, $b = 8.176$, $c = 7.4453$ [George Brunton, *Acta Cryst.* **B27**:1823 (1971). $RbF \cdot 6ThF_4$: hexagonal, $R3m$, $a = 8.33$, $c = 25.40$ [G. D. Brunton, *Acta Cryst.* **B28**:144 (1972)]. X-ray powder diffraction data are listed in ORNL-3761 for all of the intermediate phases shown in the phase diagram.

The System CsF–ThF₄. See Fig. 96. Source: R. E. Thoma and T. S. Carlton, *J. Inorg. Nucl. Chem.* **17**:88 (1961). Invariant and singular equilibrium points are given in the table.

Mole % ThF₄ in liquid	Invariant temperature	Type of equilibrium
9	615	Eutectic
25	980	Congruent melting point
30.5	838	Eutectic
33.3	869	Congruent melting point
40	775	Eutectic
50	839	Congruent melting point
53	797	Eutectic
55	842	Peritectic
56.5	860	Peritectic
	813	Lower limit of stability of CsF · 2ThF₄
75	830	Upper limit of stability of CsF · 3ThF₄
84	1010	Peritectic

Structural data: $3CsF \cdot ThF_4$: cubic, $a = 10.04$.

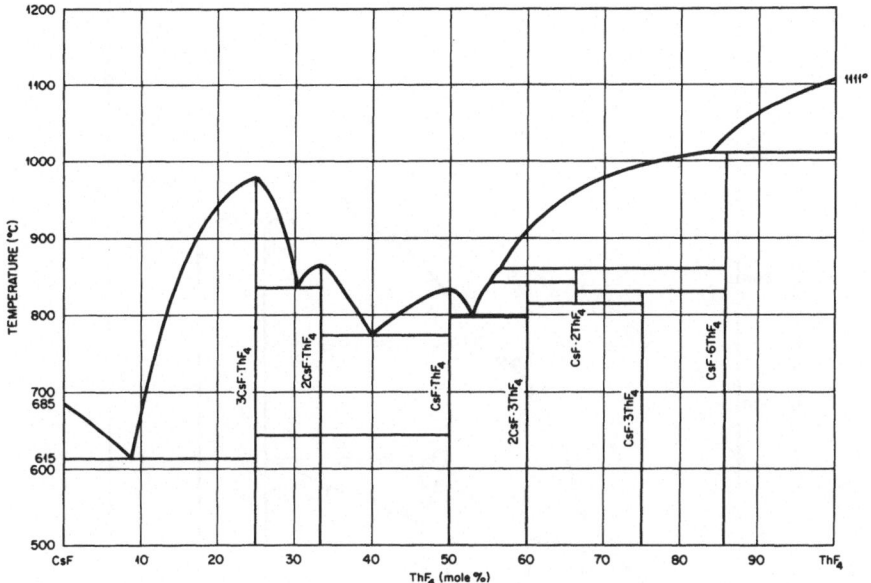

Fig. 96. The system CsF–ThF$_4$.

The System TlF–ThF$_4$. See Fig. 97. Source: D. Avignant and J. C. Cousseins, *Compt. Rend. Acad. Sci., Ser. C* **272**:2151 (1971). Invariant and singular equilibrium points are given in the table.

Mole % ThF$_4$	Temperature	Type of equilibrium
10	285	Eutectic
23	378	Peritectic
24	385	Peritectic
42.5	701	Peritectic
	519	Upper temp. of stability of TlF · ThF$_4$
60.5	902	Peritectic
65	956	Peritectic

On the basis of x-ray powder diffraction data, it was inferred that 3TlF · ThF$_4$ is cubic; at low temperatures TlF · UF$_6$ is orthorhombic;

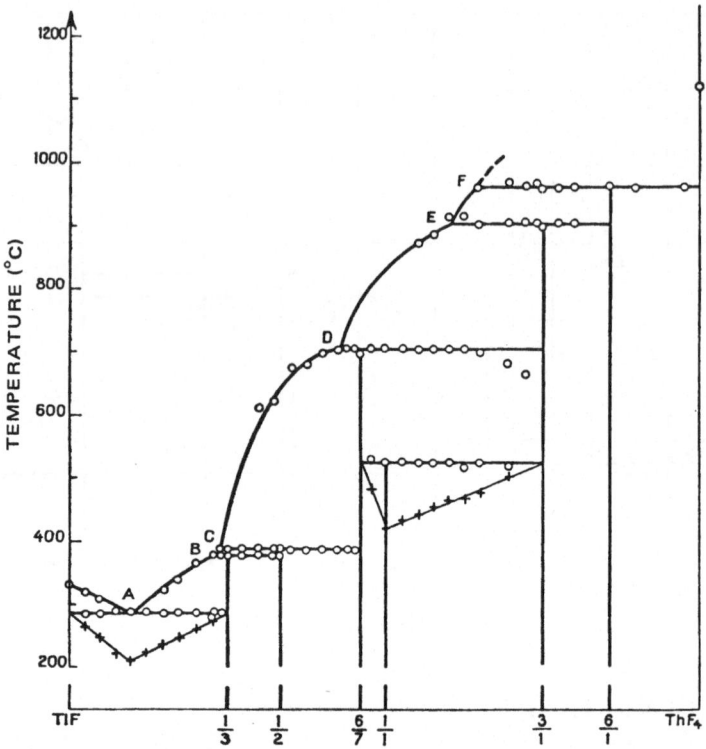

Fig. 97. The system TlF–ThF$_4$.

7TlF · 6ThF$_4$ is rhombohedral and isostructural with 7NaF · 6ZrF$_4$, with hexagonal parameters $a = 15.60$, $c = 10.84$; and TlF · 6ThF$_4$ is hexagonal, $P6_3/mmc$, $a = 8.31$, $c = 16.86$.

The System LiF–UF$_4$. See Fig. 98. Source: C. J. Barton *et al.*, *J. Am. Ceram. Soc.* **41**:63 (1958). Lower stability limit for 4LiF · UF$_4$ 470°; $P = 500°$, 26% UF$_4$; $E = 490°$, 27% UF$_4$; $P = 610°$, 40% UF$_4$; $P = 775°$, 57% UF$_4$.

Structural data: 4LiF · UF$_4$, orthorhombic, *Pnma*, $a = 9.960$, $b = 9.883$, $c = 5.986$ [G. D. Brunton, *J. Inorg. Nucl. Chem.* **29**:1631 (1967)]. 3LiF · UF$_4$ (metastable phase): tetragonal, $a = 6.13$, $c = 6.39$ [G. D. Brunton *et al.*, ORNL-3761]. LiF · UF$_4$: tetragonal, $I4_1/c$, $a = 14.884$, $c = 6.547$ [G. D. Brunton, *Acta Cryst.* **21**:814 (1966)].

The compound LiF · UF$_4$ was reported in the source reference to have the formula 7LiF · 6UF$_4$. The correct stoichiometry was established by determination of the crystal structure.

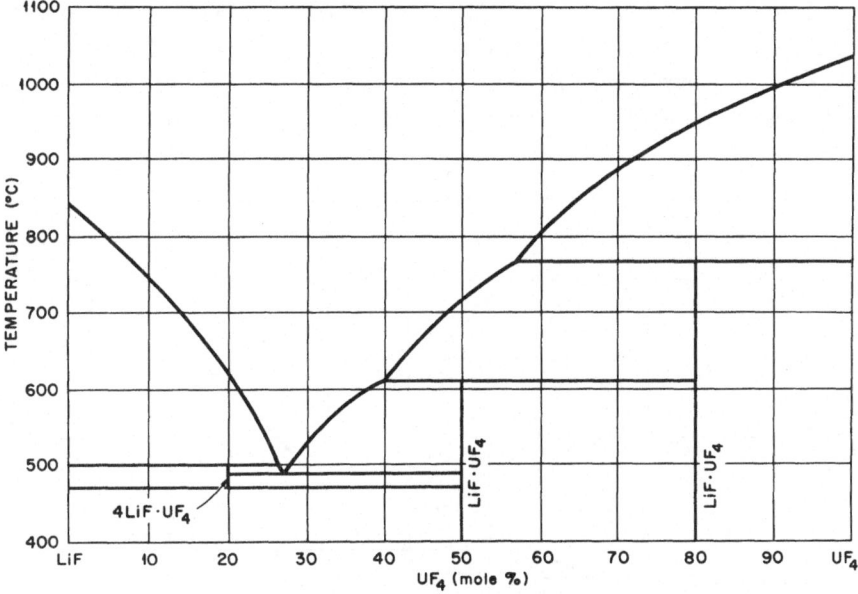

Fig. 98. The system LiF–UF$_4$.

The System NaF–UF$_4$. See Fig. 99. Source: C. J. Barton *et al.*, *J. Am. Ceram. Soc.* **41**:63 (1958). Invariant and singular equilibrium points are given in the table.

Mole % UF$_4$ in liquid	Temperature	Type of equilibrium	Coexisting phases
21.5	618	Eutectic	$L \rightleftarrows \alpha \cdot 3NaF \cdot UF_4 + NaF$
25	629	Congruent melting point	$L \rightleftarrows \alpha \cdot 3NaF \cdot UF_4$
	528	Inversion	$\alpha \cdot 3NaF \cdot UF_4 \rightleftarrows \beta \cdot 3NaF \cdot UF_4$
	497	Decomposition	$\beta \cdot 3NaF \cdot UF_4 \rightleftarrows NaF + 2NaF \cdot UF_4$
28	623	Eutectic	$L \rightleftarrows \alpha \cdot 3NaF \cdot UF_4 + 2NaF \cdot UF_4$
32.5	648	Peritectic	$L + 5NaF \cdot 3UF_4 \rightleftarrows 2NaF \cdot UF_4$
37	673	Peritectic	$L + 7NaF \cdot 6UF_4 \rightleftarrows 5NaF \cdot 3UF_4$
	630	Decomposition	$5NaF \cdot 3UF_4 \rightleftarrows 2NaF \cdot UF_4$ $+ 7NaF \cdot 6UF_4$
46.2	718	Congruent melting point	$L \rightleftarrows 7NaF \cdot 6UF_4$
56	680	Eutectic	$L \rightleftarrows 7NaF \cdot 6UF_4 + UF_4$
	660	Upper stability limit of NaF · 2UF$_4$	$7NaF \cdot 6UF_4 + UF_4 \rightleftarrows NaF \cdot 2UF_4$

Structural data: 7NaF · 2UF$_4$: orthorhombic, *Fmmm*, *Fmmm*2, or *F*222, $a = 17.7$, $b = 29.8$, $c = 12.7$ [G. D. Brunton *et al.*, ORNL-3761]. 3NaF · UF$_4$: tetragonal, *I4mmm*, $a = 5.448$, $c = 10.896$ [G. D. Brunton *et al.*, *ibid.*; see also W. H. Zachariasen, *J. Am. Chem. Soc.* **70**:2147 (1948)].

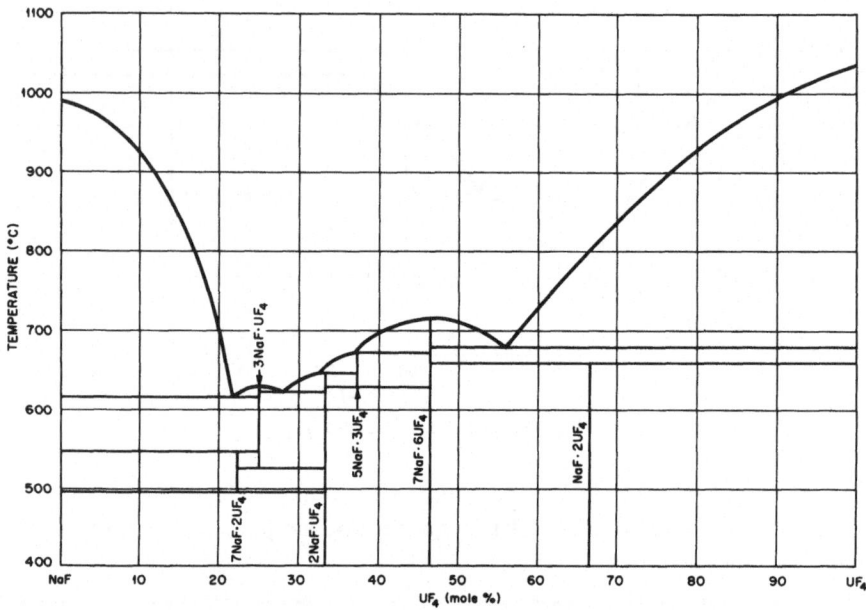

Fig. 99. The system NaF–UF$_4$.

δ-2NaF · UF$_4$: hexagonal, $a = 6.11$, $c = 7.25$ [W. H. Zachariasen, *ibid.*]. γ-2NaF · UF$_4$: orthorhombic, *Immm*, $a = 5.56$, $b = 4.01$, $c = 11.64$ [W. H. Zachariasen, *ibid.*]. β$_2$-2NaF · UF$_4$: hexagonal, *P*32, $a = 5.95$, $c = 3.73$ [W. H. Zachariasen, *ibid.*]. 7NaF · 6UF$_4$: hexagonal, *R*3, $a = 14.72$, $c = 9.84$ [G. D. Brunton *et al.*, *ibid.*].

The System KF–UF$_4$. See Fig. 100. Source: R. E. Thoma, H. Insley, B. S. Landau, H. A. Friedman, and W. R. Grimes, *J. Am. Ceram. Soc.* **41**:539 (1958). Invariant and singular equilibrium points are given in the table.

Mole % UF$_4$ in liquid	Temperature	Type of equilibrium
15	735	Eutectic
25	957	Congruent melting point
	608	Decomposition
35	755	Peritectic
38.5	740	Eutectic
46.2	789	Congruent melting point
54.5	735	Eutectic
65	765	Peritectic

Structural data; 3KF · UF$_4$: orthorhombic, *Pnmm* or *Pnm2$_1$*, $a = 6.59$, $b = 8.30$, $c = 7.20$ [J. H. Burns, ORNL-3262 (1962)]. β_1-2KF · UF$_4$: hexagonal, *P$\bar{6}$2m*, $a = 6.5528$, $c = 3.749$ [W. H. Zachariasen, *J. Am. Chem. Soc.* **70**:2147 (1948)]. 7KF · 6UF$_4$: rhombohedral, *R$\bar{3}$*, $a = 15.09$, $c = 10.38$ [G. D. Brunton *et al.*, ORNL-3761]. KF · 2UF$_4$: orthorhombic, *Pnam*, $a = 8.7021$, $b = 11.477$, $c = 7.0350$ [G. D. Brunton, *Acta Cryst.* **B25**:1919 (1969)]. KF · 6UF$_4$: hexagonal, *P6/mmc*, $a = 8.18$, $c = 16.42$ [W. H. Zachariasen, *J. Am. Chem. Soc.* **70**:2147 (1948).

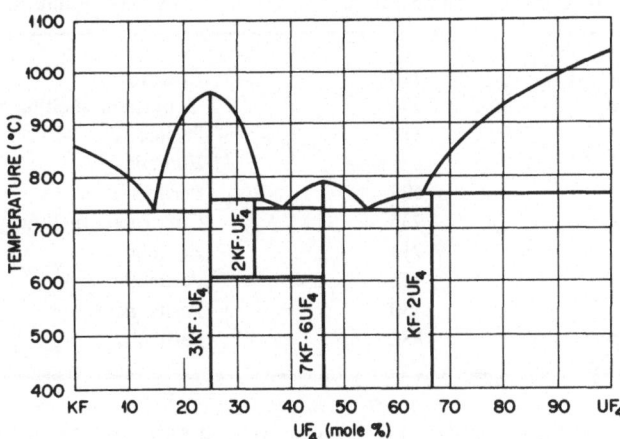

Fig. 100. The system KF–UF$_4$.

The System RbF–UF₄. See Fig. 101. Source: R. E. Thoma, H. Insley, B. S. Landau, H. A. Friedman, and W. R. Grimes, *J. Am. Ceram. Soc.*

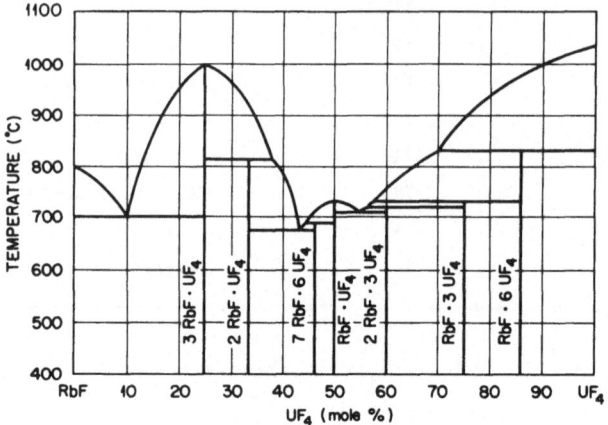

Fig. 101. The system RbF–UF₄.

41:538 (1958). Invariant and singular equilibrium points are given in the table.

Mole % UF₄ in liquid	Temperature	Type of equilibrium
10	710	Eutectic
25	995	Congruent melting point
38	818	Peritectic
43.5	675	Eutectic
44	693	Peritectic
50	735	Congruent melting point
55	714	Eutectic
56.5	722	Peritectic
57	730	Peritectic
70.5	832	Peritectic

Structural data: $3RbF \cdot UF_4$: cubic, $a = 9.5667$ [G. D. Brunton *et al.*, ORNL-3761]. $2RbF \cdot UF_4$: orthorhombic, *cmcm*, $a = 6.998$, $b = 12.098$, $c = 7.669$ [T. K. Keenan, *Inorg. Nucl. Chem. Letters* 3:463 (1967). $7RbF \cdot 6UF_4$: rhombohedral, $a = 9.595$; hexagonal parameters: $a = 15.49$, $c = 10.42$]. $RbF \cdot 6UF_4$: hexagonal, $P6_3/mmc$, $a = 8.195$, $c = 16.37$ [G. D. Brunton *et al.*, ORNL-3761].

The System CsF–UF₄. See Fig. 102. Source: C. J. Barton, L. M. Bratcher, T. N. McVay, and R. E. Thoma, ORNL, unpublished work

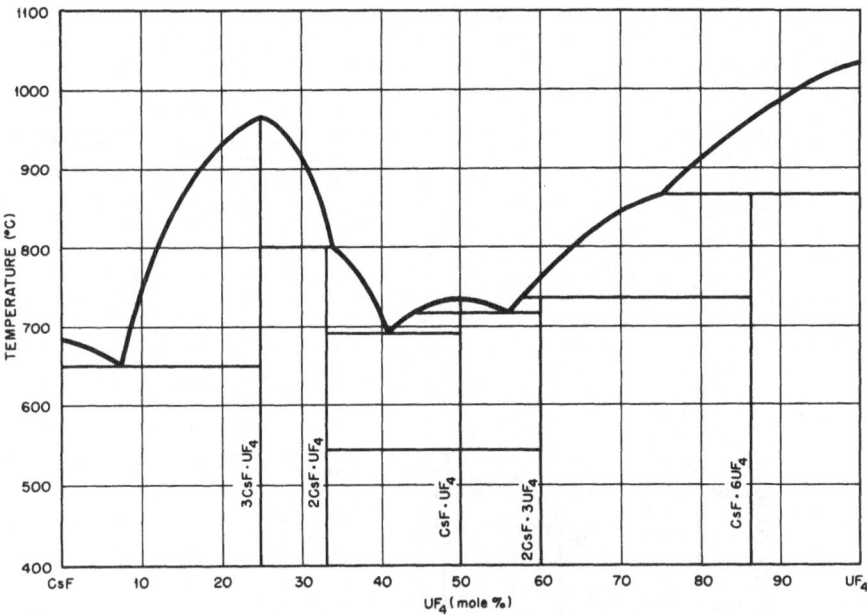

Fig. 102. The system CsF·UF₄.

(1951–1953). Invariant and singular equilibrium points are given in the table.

Mole % UF₄ in liquid	Temperature	Type of equilibrium
7.5	650	Eutectic
25	970	Congruent melting point
34	800	Peritectic
41	695	Eutectic
50	735	Congruent melting point
	550	Inversion
53	725	Eutectic

Structural data: $2CsF \cdot 3UF_4$, monoclinic, $P21/M$, $a_0 = 18.316$, $b_0 = 8.455$, $c_0 = 8.234$, $\beta = 96.12°$ [ORNL-3761]; $CsF \cdot 6UF_4$ hexagonal,

$P63/mmc$, $a_0 = 8.2424$, $c_0 = 16.412$ [G. D. Brunton, *Acta Cryst.* **B27**:225 (1971)].

The System TlF–UF$_4$. See Fig. 103. Source: D. Avignant and J. C. Cousseins, *Compt. Rend.* **272**:2151 (1971). $E = 274°$, 87.5 TlF; Tl$_4$UF$_8$ inversion at 227°; $P = 542°$, 77 TlF; Tl$_3$UF$_7$ lower temperature of stability, 285°; m.p. Tl$_2$UF$_6$ 620°; $E = 598°$, 61 TlF; Tl$_7$U$_6$F$_3$, upper temperature of stability 548°; $P = 640$, 53.5 TlF; $P = 674°$, 40 TlF; $P = 717°$, composition undefined.

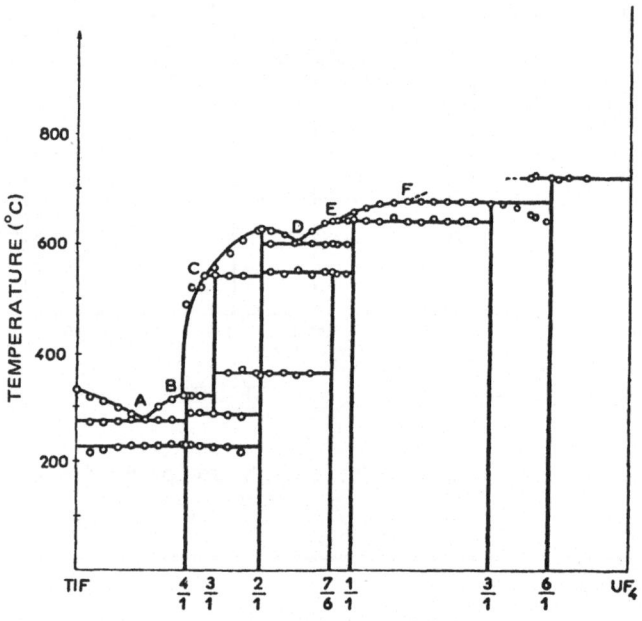

Fig. 103. The system TlF–UF$_4$.

Crystallographic data: Tl$_3$UF$_7$: cubic, $a = 9.434$; Tl$_2$UF$_6$: orthorhombic, $a = 4.07$, $b = 6.97$, $c = 10.80$; TlU$_6$F$_{25}$: hexagonal, $a = 8.18$, $c = 16.46$.

The System KF–NbF$_5$. See Fig. 104. Source: Ahmed Mukhtar and Rene Winand, *Compt. Rend.* **260**:3675 (1965). $E = 723°$, 23% K$_2$NbF$_7$; congruent m.p. of KF · K$_2$NbF$_7$ 780°; $E = 706°$, 80% K$_2$NbF$_7$.

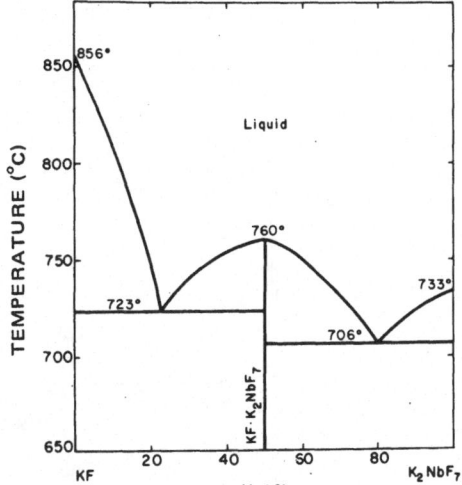

Fig. 104. The system KF–NbF$_5$.

The System KF–TaF$_5$. See Fig. 105. Source: Ts'ui Ping-hsin, N. P. Luzhnaya, and V. I. Konstantinov, *Zh. Neorg. Khim.* **8**:389 (1963). $E = 717°$, 21.5 % KF; $E = 727°$, 73% KF (est.); liquidus curves for the two

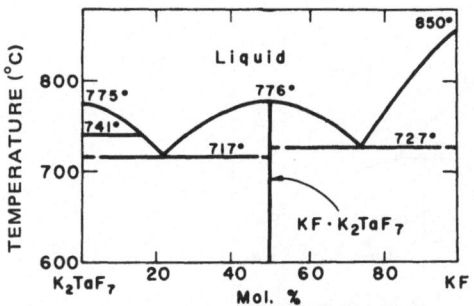

Fig. 105. The system KF–TaF$_5$.

crystalline modifications of K$_2$TaF$_7$ intersect at the transformation point located at 741° and at 18% KF.

The System BeF$_2$–MgF$_2$. See Fig. 106. Source: W. E. Counts, Rustum Roy, and E. F. Osborn, *J. Am. Ceram. Soc.* **36**:14 (1953). $E = 528 \pm 5°$, 5% MgF$_2$.

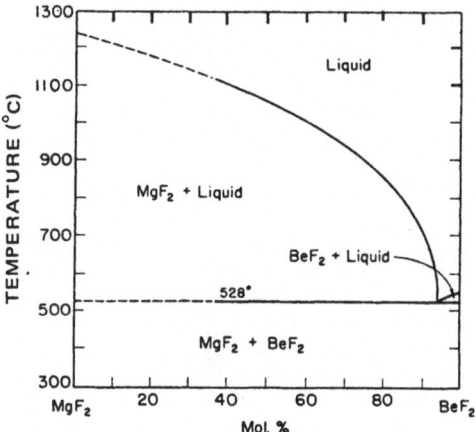

Fig. 106. The system BeF_2–MgF_2.

The System BeF_2–CaF_2. See Fig. 107. Source: W. E. Counts, Rustum Roy, and E. F. Osborn, *J. Am. Ceram. Soc.* **36**:12 (1953). $E = 495°$, 11% CaF_2; $P = 890°$, 43% CaF_2.

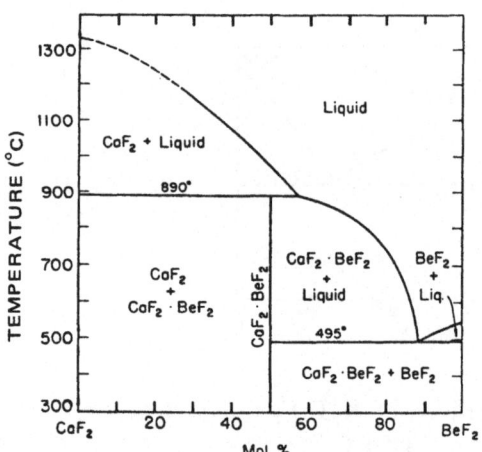

Fig. 107. The system BeF_2–CaF_2.

The System BeF_2–SrF_2. See Fig. 108. Source: O. N. Breusov, G. Trapp, A. V. Novoselova, and Yu. P. Simanov, *Zh. Neorg. Khim.* **4**:672 (1959). Invariant and singular equilibrium points are given in the table.

Mole % BeF$_2$	Temperature	Type of equilibrium
~100	582 ± 29	Eutectic
100	220	Polymorphic transition in crystalline BeF$_2$
92.5	774	Monotectic
75	774	Monotectic
50	954 ± 10	Congruent m.p.
	923	Polymorphic transition in crystalline BeF$_2$ · SrF$_2$
60	883	Eutectic

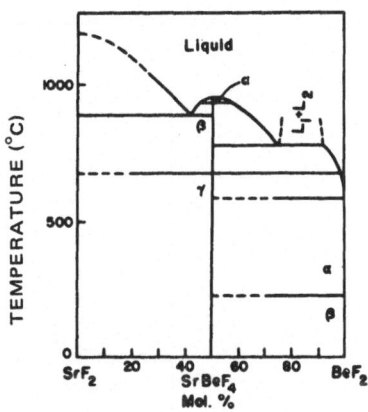

Fig. 108. The system BeF$_2$–SrF$_2$.

The System BeF$_2$–BaF$_2$. See Fig. 109. Source: D. F. Kirkina, A. V. Novoselova, and Yu. P. Simanov, *Zh. Neorg. Khim.* **1**:128 (1956). $E =$

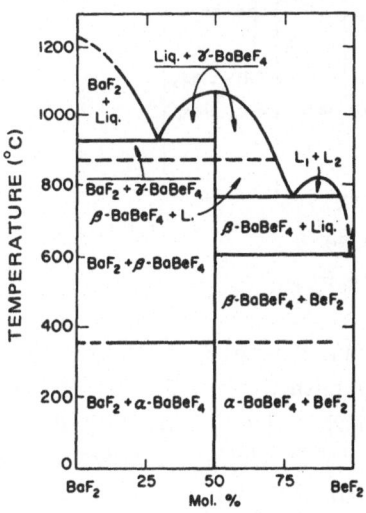

Fig. 109. The system BeF$_2$–BaF$_2$.

$600°$, $1\text{–}2\%$ BaF_2; intersections of the liquidus and binodal curves at 7 and 20 mole $\%$ BaF_2; congruent m.p. $BeF_2 \cdot BaF_2$ $1080°$; $E = 920$, 72% BaF_2. $BeF_2 \cdot BaF_2$ polymorphic transformation temperatures occur at 870 and $360°$.

The System $BeF_2\text{–}PbF_2$. See Fig. 110. Source: D. M. Roy, R. Roy, and E. F. Osborn, *J. Am. Ceram. Soc.* **37**:300 (1954). $E = 525°$, 90% BeF_2; m.p. $BeF_2 \cdot PbF_2$ $585 \pm 5°$; $E = 465°$, 21% BeF_2; m.p. $BeF_2 \cdot 3PbF_2$ $482 \pm 5°$; $E = 477°$, 79% PbF_2 (est.).

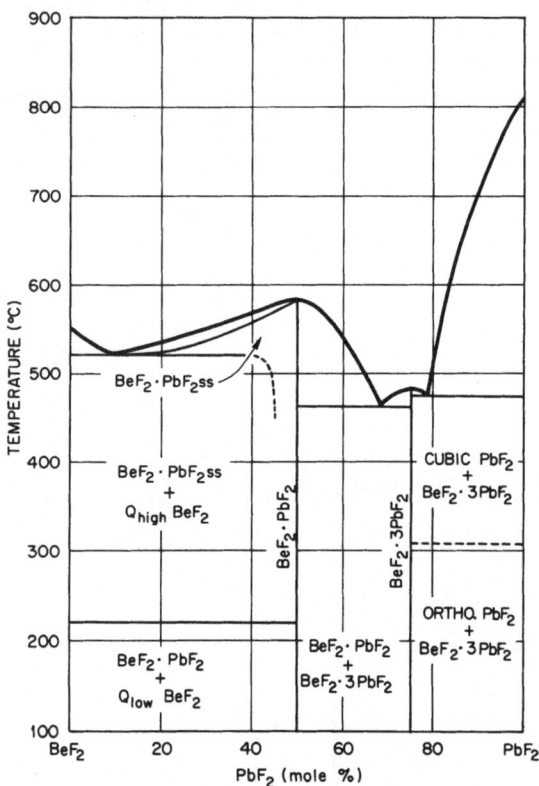

Fig. 110. The system $BeF_2\text{–}PbF_2$.

The System $MgF_2\text{–}CaF_2$. See Fig. 111. Source: C. J. Barton, L. M. Bratcher, and J. P. Blakely, ORNL, unpublished work (1955–1956). The eutectic occurs at 48 mole $\%$ CaF_2, and at $985°$. The initial phase diagram of the $MgF_2\text{–}CaF_2$ system was reported by E. Beck, *Metallurgie* **5**:504 (1908). Phase diagrams of the ternary systems $LiF\text{–}KF\text{–}MgF_2$ A. G. Berg-

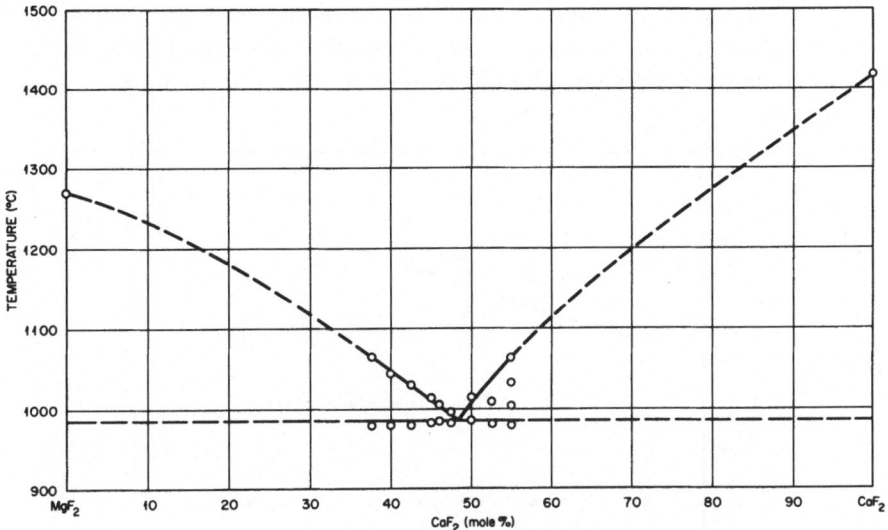

Fig. 111. The system MgF_2–CaF_2.

man and E. P. Dergunov, *Compt. Rend. Acad. Sci. URSS* **31**:75 (1941)] and NaF–KF–MgF_2 [A. G. Bergman and E. P. Dergunov, *op. cit.*, **48**:330 (1945)] include information pertaining to the binary system.

The System MgF_2–BaF_2. See Fig. 112. Source: M. Okamoto and U. Nisioka, *Sci. Rept. Tohoku Univ., First Ser.* **24**:142 (1935–1936). $E = 910°$, 15% MgF_2; $P = 930°$, 19% MgF_2.

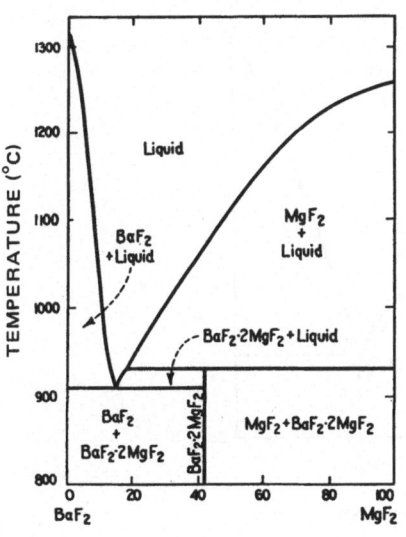

Fig. 112. The system MgF_2–BaF_2.

The System CaF_2–CdF_2. See Fig. 113. Source: M. P. O'Horo and W. B. White, *J. Am. Ceram. Soc.* **54**:588 (1971). CaF_2 and CdF_2 form a continuous series of solid solutions without a temperature minimum.

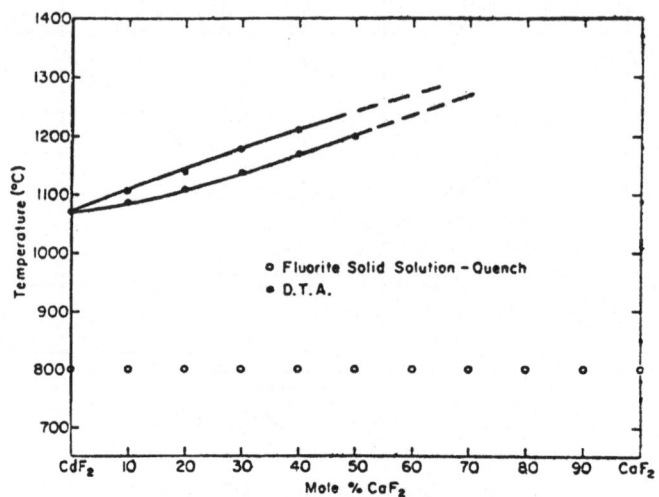

Fig. 113. The system CaF_2–CdF_2.

The System NiF_2–BaF_2. See Fig. 114. Source: J. C. Cousseins and M. Samouel, *Compt. Rend.* **265**:1121 (1967). $P = 1000°$, 39.5% BaF_2;

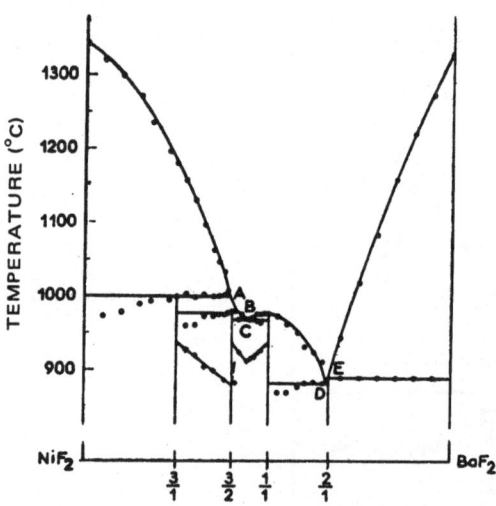

Fig. 114. The system NiF_2–BaF_2.

$P = 977°$, 42% BaF_2; m.p. $NiF_2 \cdot BaF_2$ 977°; $E = 882°$, 60% BaF_2; $P = 890°$, 65% BaF_2.

Structural data: $NiF \cdot BaF_2$: orthorhombic, $a = 14.46$, $b = 4.15$, $c = 5.80$.

The System CuF_2–BaF_2. See Fig. 115. Source: M. Samouel, *Compt. Rend.* **270**:1805 (1970). $E = 616°$, 22.5% BaF_2; $P = 624°$, 26% BaF_2; $E =$

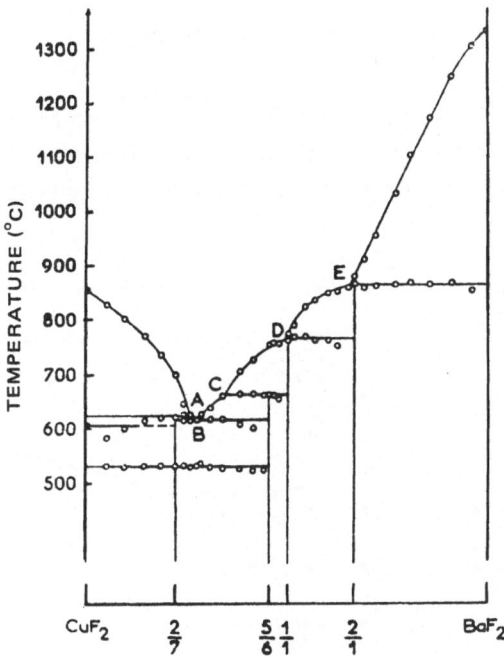

Fig. 115. The system CuF_2–BaF_2.

616°, 27.5% BaF_2; $P = 663°$, 34.5% BaF_2; $P = 764$, 49.4% BaF_2; $P = 863$, 65% BaF_2.

The System SrF_2–BaF_2. See Fig. 116. Source: Ralph H. Nafziger, *J. Am. Ceram. Soc.* **54**:467 (1971). SrF_2 and BaF_2 form a continuous series of solid solutions with a shallow minimum in the liquidus curve at 27 ± 2 wt. % SrF_2 and 1305°. Length of each vertical line indicates one standard deviation for replicate liquidus and solidus temperature measurements for each composition studied.

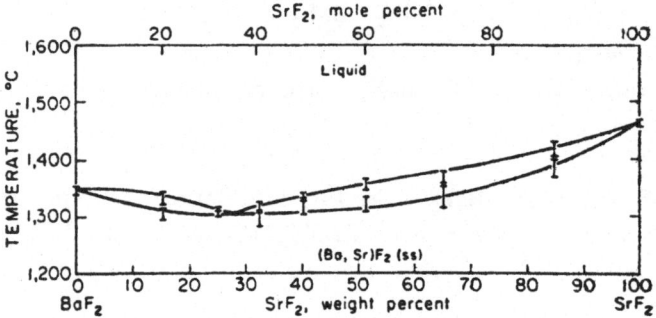

Fig. 116. The system SrF_2–BaF_2.

The System CdF_2–PbF_2. See Fig. 117. Source: M. P. O'Horo and W. B. White, *J. Am. Ceram. Soc.* **54**:588 (1971). The minimum temperature in the solid solution occurs at 60 mole % PbF_2. Its exact composition was not stated.

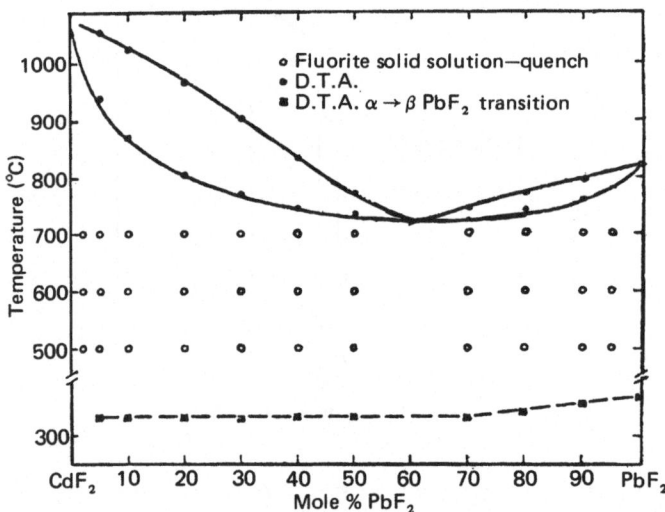

Fig. 117. The system CdF_2–PbF_2.

The System ZnF_2–CdF_2. See Fig. 118. Source: M. P. O'Horo and W. B. White, *J. Am. Ceram. Soc.* **54**:588 (1971). The invariant equilibrium points in the CdF_2–ZnF_2 system were not given in the source reference.

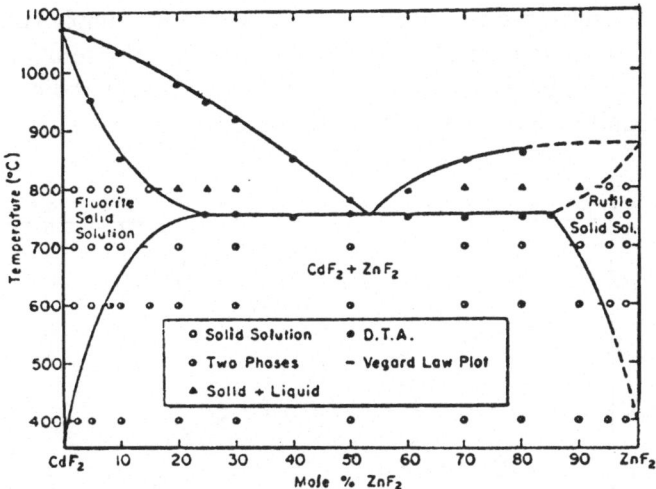

Fig. 118. The system ZnF_2–CdF_2.

The System MgF_2–ScF_3. See Fig. 119. Source: L. N. Komissarova and B. I. Pokrovskii, *Dokl. Akad. Nauk SSSR* **149**:300 (1963). $E = 1095°$, 34% ScF_3. The eutectoid invariant point at $840°$ occurs at 5 mole $\%$ ScF_3.

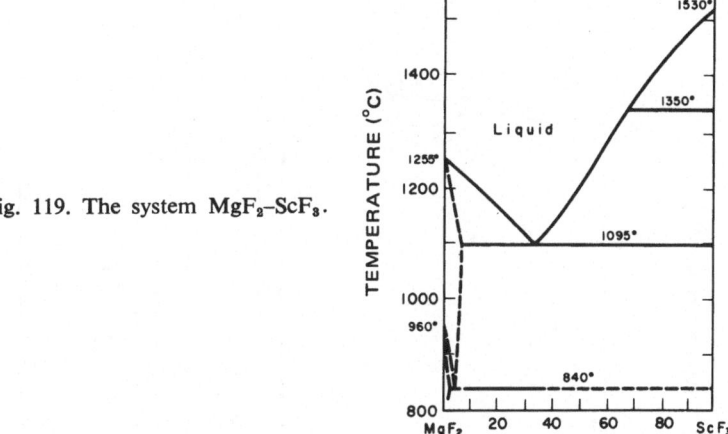

Fig. 119. The system MgF_2–ScF_3.

The System MgF_2–YF_3. See Fig. 120. Source: R. H. Nafziger, R. L. Lincoln, and R. Riazance, *J. Inorg. Nucl. Chem.* **35**:421 (1973). The authors of the source reference describe the system as "of the simple eutectic type with a very small amount of orthorhombic (α-YF$_3$) solid solution and perhaps a limited amount of YF$_3$ solid solubility in MgF$_2$, but did not list the exact temperature nor composition of the eutectic.

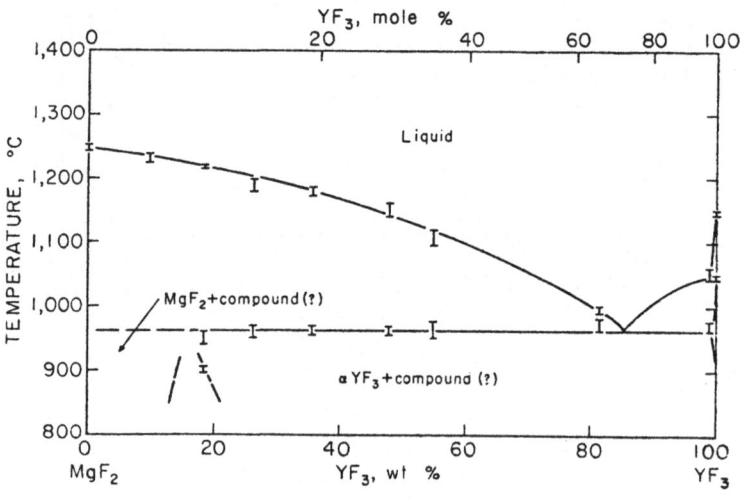

Fig. 120. The system MgF$_2$–YF$_3$.

The System CaF$_2$–YF$_3$. See Fig. 121. Source: J. Short and Rustum Roy, *J. Phys. Chem.* **67**:1861 (1963). Invariant equilibria were not fixed

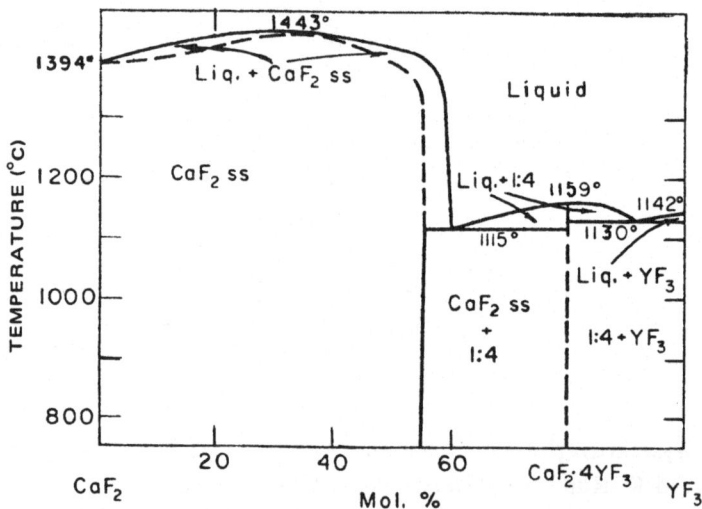

Fig. 121. The system CaF$_2$–YF$_3$.

precisely by Short and Roy. They state that the 1443° maximum in the liquidus curve occurs at about 33 mole % YF_3 and that the limit of solid solution of YF_3 in CaF_2 is about 55 mole % YF_3. The 1115° eutectic is located at approximately 58 mole % YF_3. The phase melting congruently at 1159° was tentatively assigned the formula $CaF_2 \cdot 4YF_3$, although the authors recognized that it could represent the limit of CaF_2 solubility in a "hitherto unreported hexagonal form of YF_3."

The System SrF_2–YF_3. See Fig. 122. Source: R. H. Nafziger, R. L. Lincoln, and N. Riazance, *J. Inorg. Nucl. Chem.* **35**:421 (1973). The maximum in the liquidus curve at 12 wt. % (8.05 mole %) YF_3 occurs at 1462°.

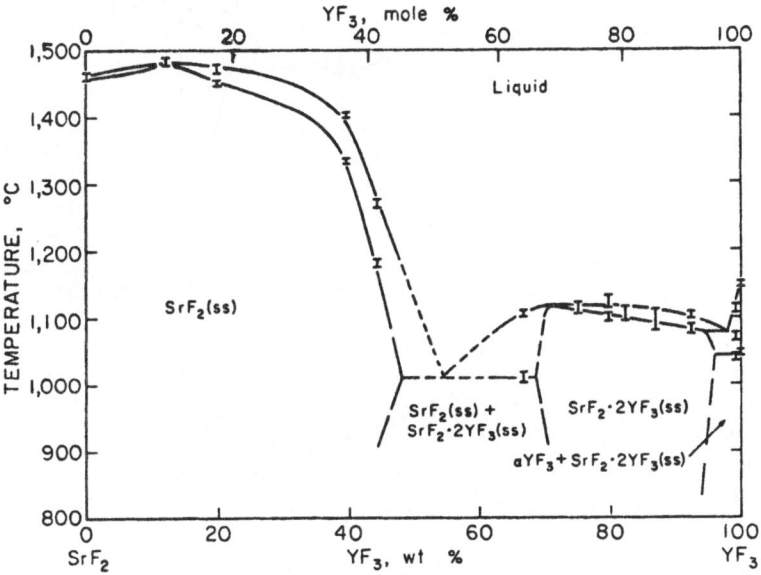

Fig. 122. The system SrF_2–YF_3.

Compositions and temperatures associated with the saturated solid solutions near the mid-composition range were not delineated in the source reference. The temperature of the eutectic at 98 wt. % (96.9 mole %) YF_3 is 1075°.

The Systems MF_2–LaF_3. See Fig. 123a–d. Source: R. H. Nafziger and Riazance, *J. Am. Ceram. Soc.* **55**:130 (1972). Composition and temperature of the invariant equilibrium points in the alkaline earth–LaF_3 systems were listed only for MgF_2–LaF_3, where the eutectic temperature

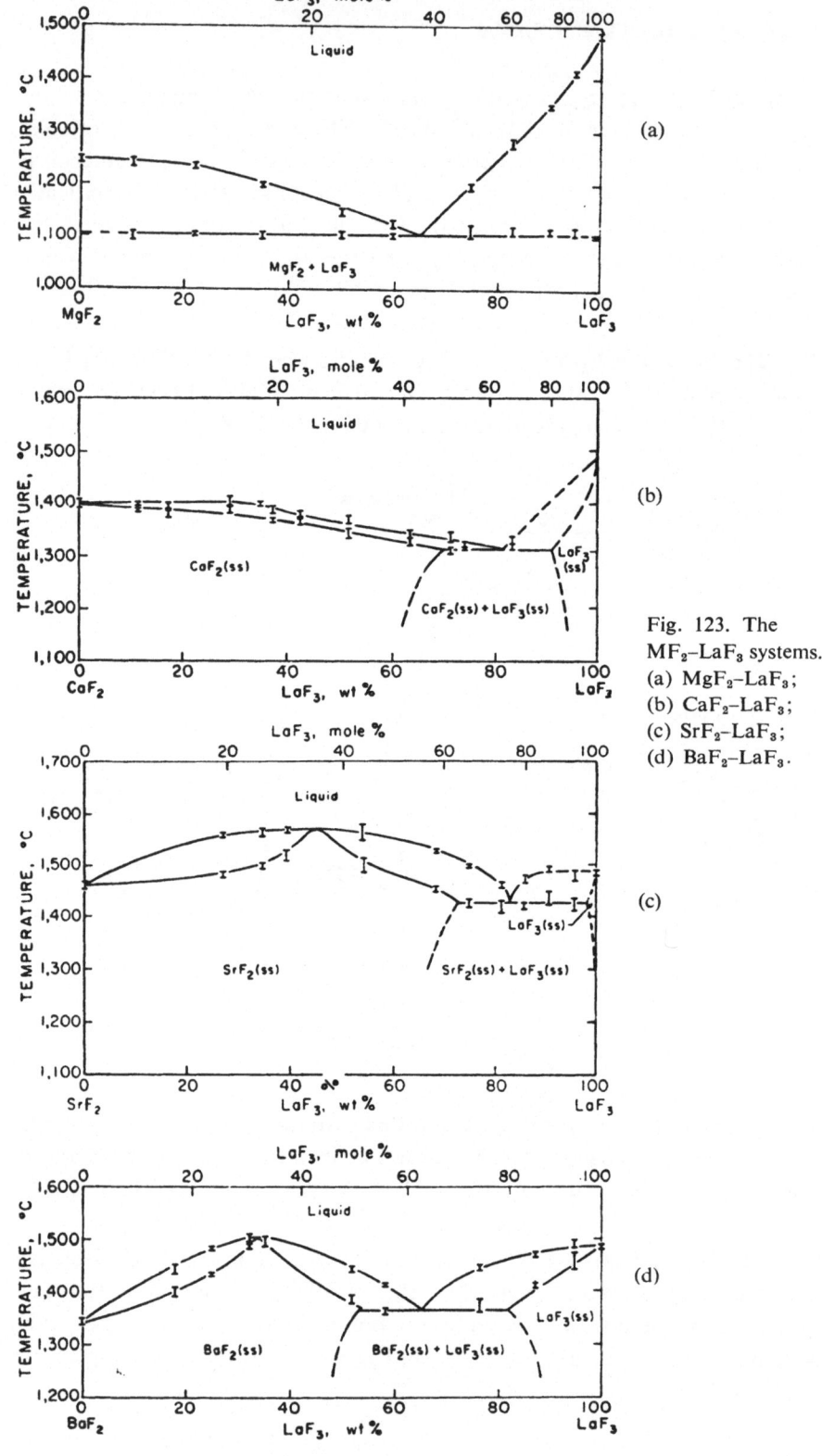

Fig. 123. The MF$_2$–LaF$_3$ systems. (a) MgF$_2$–LaF$_3$; (b) CaF$_2$–LaF$_3$; (c) SrF$_2$–LaF$_3$; (d) BaF$_2$–LaF$_3$.

was reported as 1105°, at 35 ± 1 wt. % MgF_2. Evidence of solid solution formation was observed in each of the binary systems except for MgF_2–LaF_3.

The System CaF_2–LaF_3. See Fig. 124. Source: E. G. Ippolitov, N. G. Gogadze, and B. M. Zhigarnovskii, *Russ. J. Inorg. Chem.* (*English Transl.*) **15**:1729 (1970). The eutectic in the limited solid solution series occurs at 60 mole % LaF_3, 1300°.

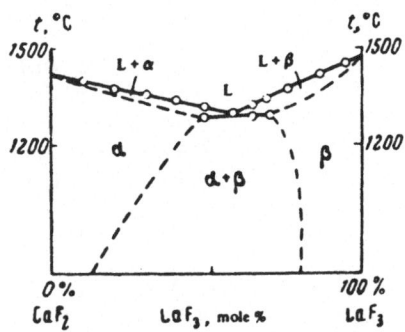

Fig. 124. The system CaF_2–LaF_3. (L) Single-phase liquid field; (α) fluorite solid solution; (β) tysonite solid solution.

The System CaF_2–NdF_3. See Fig. 125. Source: E. G. Ippolitov, N. G. Gogadze, and B. M. Zhigarnovskii, *Russ. J. Inorg. Chem.* (*English Transl.*) **15**:1729 (1970). The eutectic in the limited solid solution series occurs at 58 mole % NdF_3, 1275°.

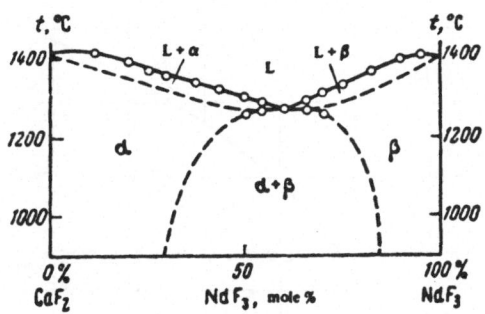

Fig. 125. The system CaF_2–NdF_3.

The System CaF_2–GdF_3. See Fig. 126. Source: N. G. Gogadze, E. G. Ippolitov, and B. M. Zhigarnovskii, *Russ. J. Inorg. Chem.* (*English Transl.*) **12**:600 (1972). $E = 1210°$, 58% GdF_3; maximum on liquidus, 1300°, 75–80% GdF_3.

Structural data: The cubic solid solution for the composition 75 mole % GdF_3 has a lattice constant $a = 5.56$: At 1020°C the cubic phase undergoes a polymorphic transformation, and below this temperature there is a hexagonal structure with lattice parameters $a = 6.84$, $c = 7.03$.

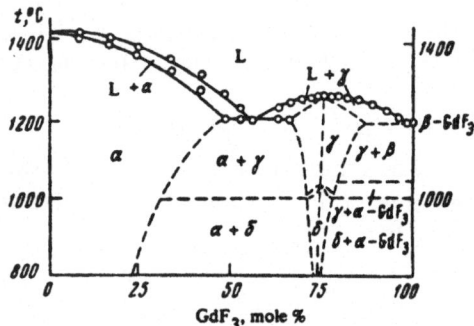

Fig. 126. The system CaF_2–GdF_3. (α) Solid solution with fluorite structure; (γ) solid solution with cubic structure; (δ) hexagonal solid solution; (β-GdF_3) high-molecular (hexagonal) modification of GdF_3; (α-GdF_3) orthorhombic modification of GdF_3.

The System CrF_2–CrF_3. See Fig. 127. Source: B. J. Sturm, *Inorg. Chem.* **1**:665 (1962). Invariant equilibrium points: $E = 831 \pm 5°$, 14% CrF_2; $P = 997 \pm 5°$, 29% CrF_2.

Structural data: $CrF_2 \cdot CrF_3$: monoclinic, $C2/c$, $a = 7.773$, $b = 7.540$, $c = 7.440$, $\beta = 124.25°$ (H. Steinfink and J. H. Burns, *Acta Cryst.* **17**:823 (1964).

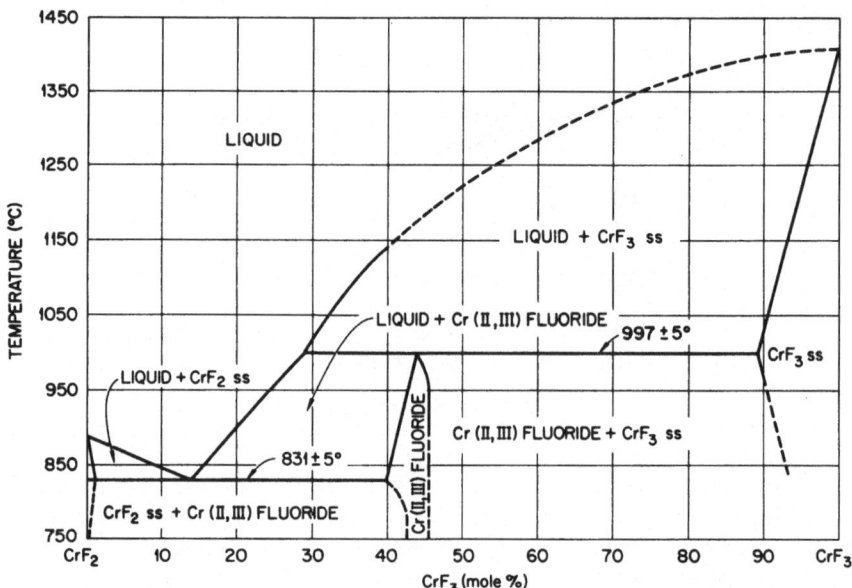

Fig. 127. The system CrF_2–CrF_3.

The System BaF_2–UF_3. See Fig. 128. Source: R. W. M. D'Eye and F. S. Martin, *J. Chem. Soc.* **1957**:1851. Authors of the source reference found that the limit of solubility of UF_3 in BaF_2 is roughly 50 mole % and that the solubility limit of BaF_2 in UF_3 is about 20 mole %. The peritectic temperature was inferred to be about 1400°, although its composition was not deduced.

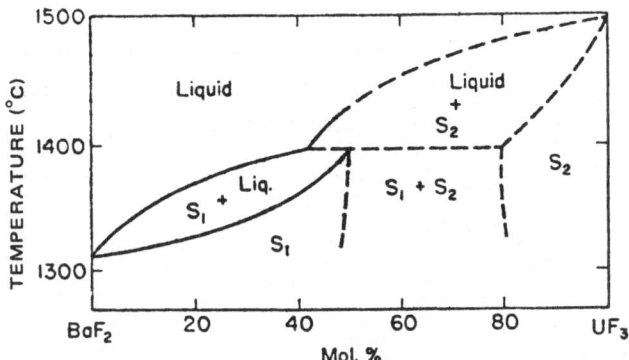

Fig. 128. The system BaF_2–UF_3.

The System PbF_2–AlF_3. See Fig. 129. Source: R. G. Shore and B. M. Wanklyn, *J. Am. Ceram. Soc.* **52**:79 (1969). $E = 570°$, 76.5% PbF_2. The composition of the 649° peritectic was not given. It occurs at approximately 69% PbF_2. The limit of solid-state solubility of AlF_3 in PbF_2 was reported to be 15%. Structure data: The compound $3PbF_2 \cdot 2AlF_3$ was reported by Shore and Wanklyn to have tetragonal structure, $I4/m$, $a = 14.23$, $c = 7.20$.

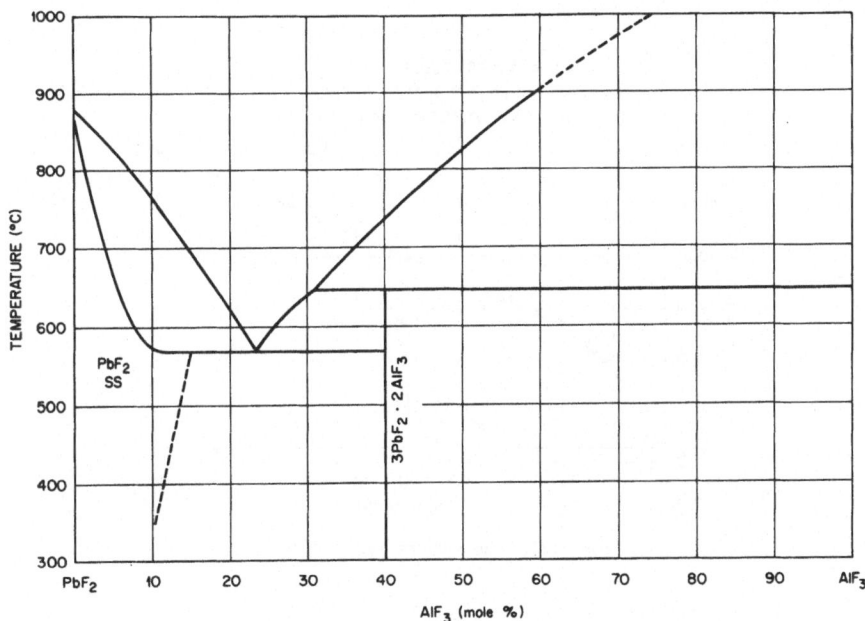

Fig. 129. The system PbF_2–AlF_3.

The System BeF_2–ZrF_4. See Fig. 130. Source: R. E. Thoma, H. Insley, H. A. Friedman, and G. M. Hebert, *J. Nucl. Mater.* **27**:166 (1968). $E = 535°$, $7.3\% ZrF_4$, immiscibility limits at $645°$, 14, and $26\% ZrF_4$; critical solution point $\sim745°$, $20\% ZrF_4$. A phase diagram of the system was reported earlier by U. M. Korenev and A. V. Novoselova, *Dokl. Akad. Nauk SSSR* **149**:5 (1963), which indicates the existence of an intermediate compound, $BeZrF_6$. No evidence of such a compound was discovered in the experiments reported in the source reference.

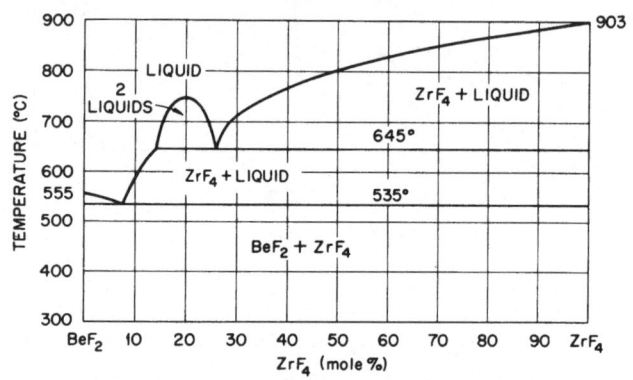

Fig. 130. The system BeF_2–ZrF_4.

The System BeF_2–ThF_4. See Fig. 131. Source: R. E. Thoma, H. Insley, H. A. Friedman, and C. F. Weaver, *J. Phys. Chem.* **64**:865 (1960). $E = 527 \pm 3°$, $2.0\% ThF_4$.

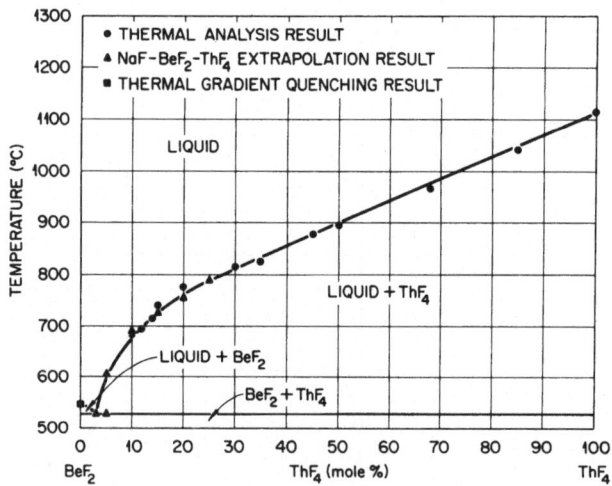

Fig. 131. The system BeF_2–ThF_4.

The System BeF$_2$–UF$_4$. See Fig. 132. Source: L. V. Jones *et al.*, *J. Am. Ceram. Soc.* **45**:79 (1962). $E = 535 \pm 2°$, 0.5% UF$_4$.

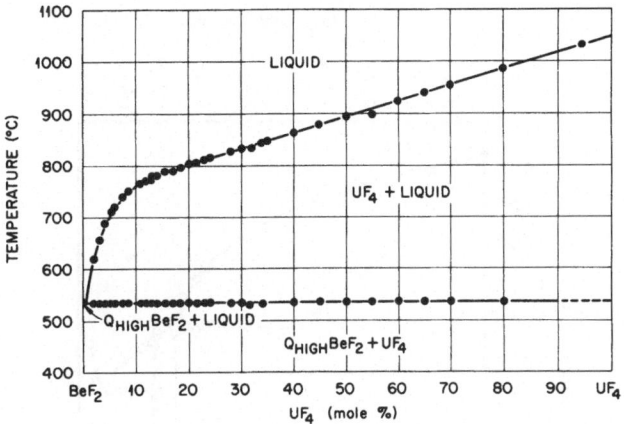

Fig. 132. The system BeF$_2$–UF$_4$.

The System MgF$_2$–ThF$_4$. See Fig. 133. Source: J. O. Blomeke, ORNL-1030 (declassified) (1951). $E = 915°$, 25% ThF$_4$; congruent m.p. MgTh$_2$F$_{10}$ 937°, $E = 925°$, 40% ThF$_4$.

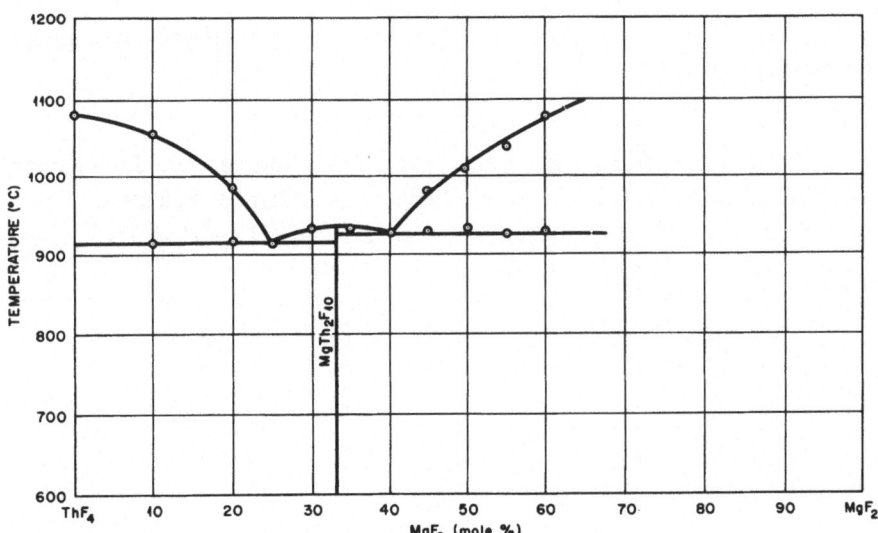

Fig. 133. The system MgF$_2$–ThF$_4$.

The System CaF$_2$–UF$_4$. See Fig. 134. Source: N. P. Nekrasova *et al.*, *At. Energ.* (*USSR*) **22**:293 (1967). Invariant and singular equilibrium

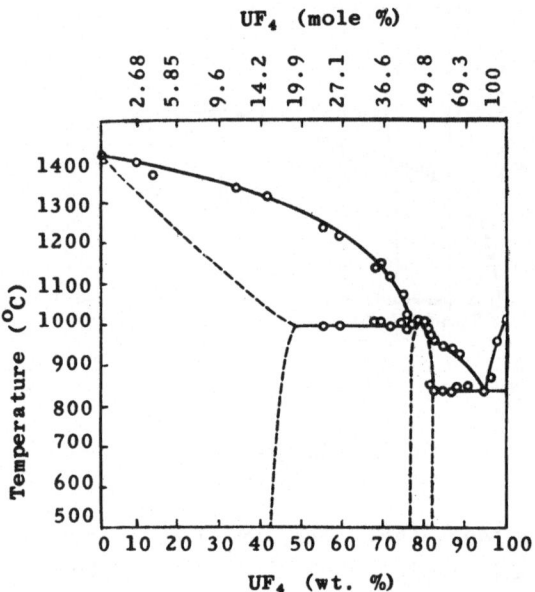

Fig. 134. The system CaF$_2$–UF$_4$.

points: $E = 997°$, 76.7% UF$_4$; congruent m.p. CaUF$_6$ 1005°; $E = 1008°$, 94.5% UF$_4$.

The System SnF$_2$–UF$_4$. See Fig. 135. Source: B. J. Thamer, USAEC Report LA-2286, July, 1959, Los Alamos Scientific Laboratory. $E = 212°$, 0.5% MF$_4$; $P = 340°$, 4.5% UF$_4$; $P = 371°$, 7.8% UF$_4$.

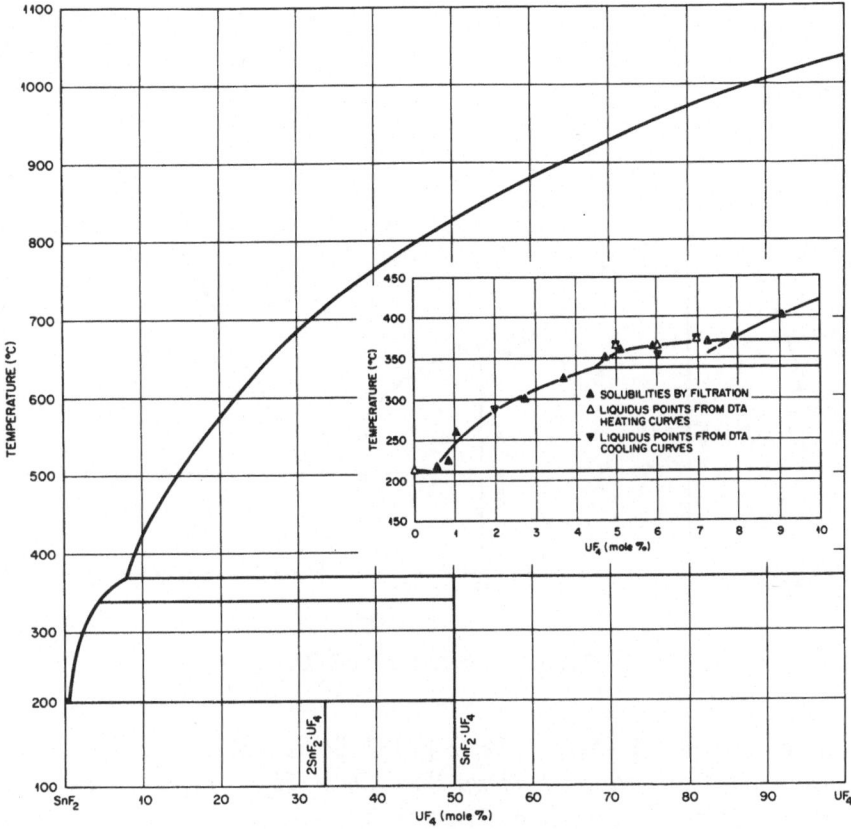

Fig. 135. The system SnF_2-UF_4.

The System PbF_2-UF_4. See Fig. 136. Source: C. J. Barton, L. M. Bratcher, J. P. Blakely, G. J. Nessle, and W. R. Grimes, ORNL, unpublished work (1950–1951). Congruent m.p. $6PbF_2 \cdot UF_4$ 920°; $E = 835°$, 35% UF_4; congruent m.p. $3PbF_2 \cdot 2UF_4$ 840°; $E = 762 \pm 10°$, 62% UF_4. The eutectic at quite high PbF_2 concentrations is not shown on this diagram. The authors examined thermal data but found no evidence of a eutectic thermal effect.

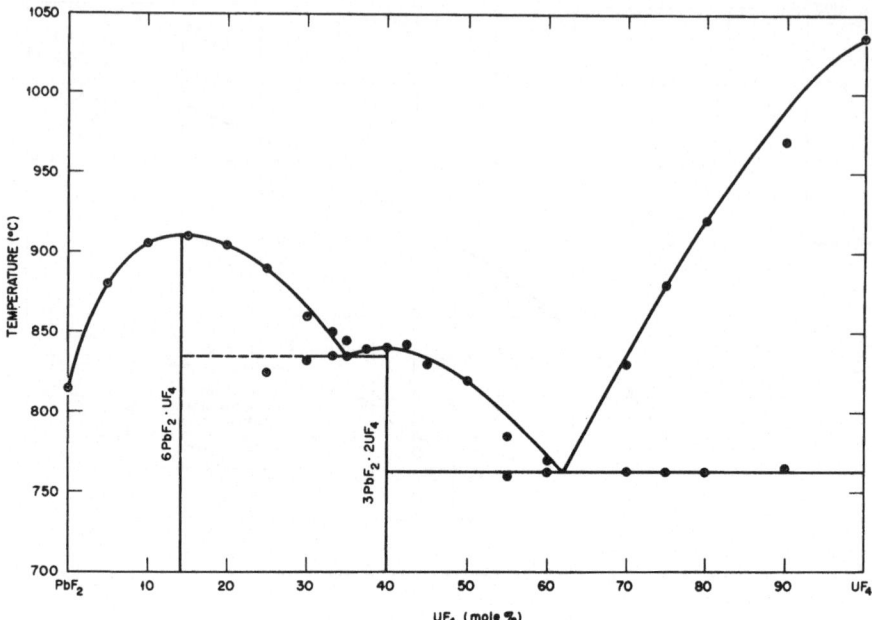

Fig. 136. The system PbF_2–UF_4.

The System YF_3–LaF_3. See Fig. 137. Source: R. H. Nafziger, R. L. Lincoln and R. Riazance, *J. Inorg. Nucl. Chem.* **35**:421 (1973). $E = 1037°$, 81 wt. % (15 mole %) YF_3.

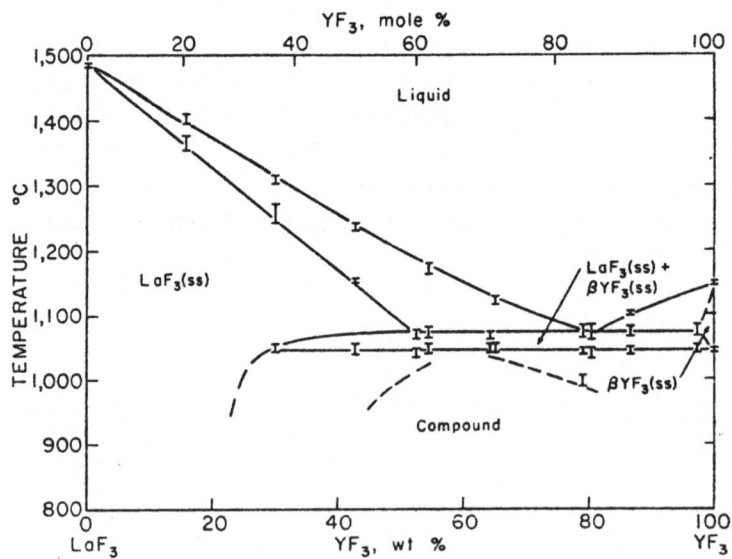

The System CeF_3–ThF_4. See Fig. 138. Source: L. O. Gilpatrick, H. Insley, and C. J. Barton, ORNL-4622, MSR Program Semiannual Progress Report for Period Ending Aug. 31, 1970, p. 91. $E = 950°$, 75% ThF_4; $P = 960°$, 73% ThF_4; $P = 1155°$, approx. 52% ThF_4.

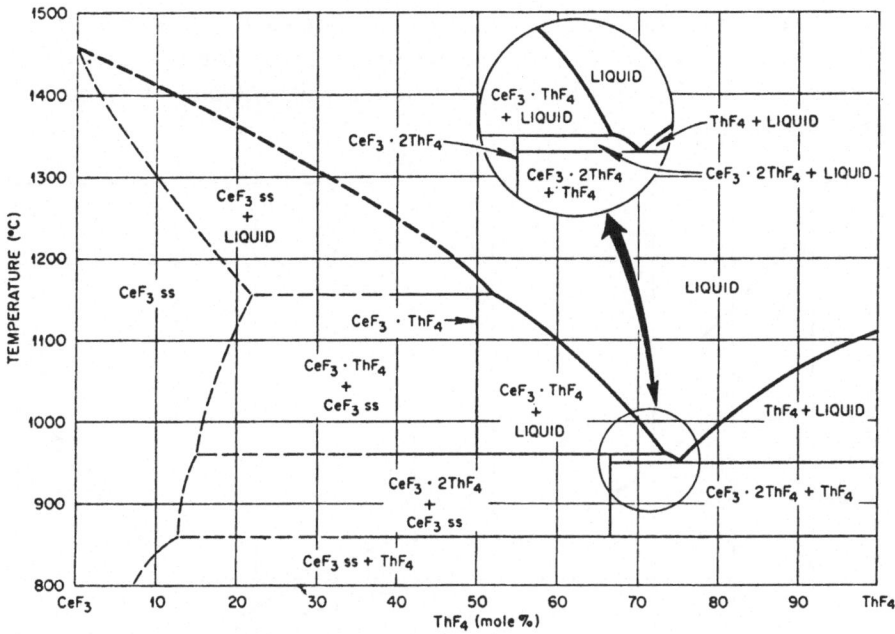

Fig. 138. The system CeF_3–ThF_4.

The System UF_3–UF_4. See Fig. 139. Source: R. E. Thoma, G. D. Brunton, and H. Insley, *J. Inorg. Nucl. Chem.* **36**:1095 (1974). $E = 865 \pm 4°$,

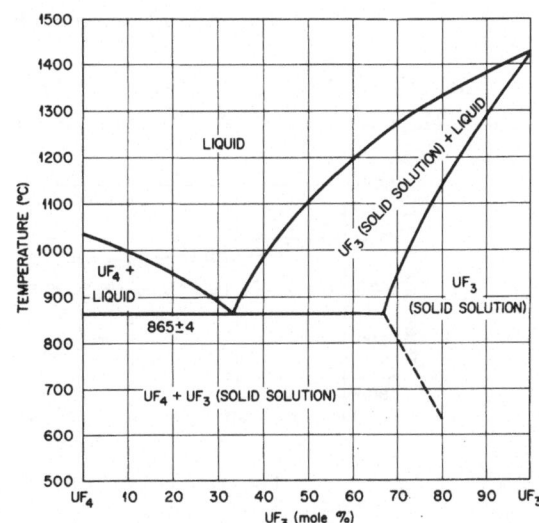

Fig. 139. The system UF_3–UF_4.

32% UF_3. The results of the investigation described in the source reference
are in disagreement with those reported earlier by L. A. Khripin, S. A.
Poduzova, and G. M. Zadneprovskii, *Russ. J. Inorg. Chem.* (*English
Transl.*) **13**:1439 (1968).

The System ZrF_4–UF_4. See Fig. 140. Source: C. J. Barton *et al.*,
J. Phys. Chem. **62**:665 (1958). The system ZrF_4–UF_4 forms a continuous
series of solid solutions having a minimum melting temperature of 765°
at 23 mole % UF_4.

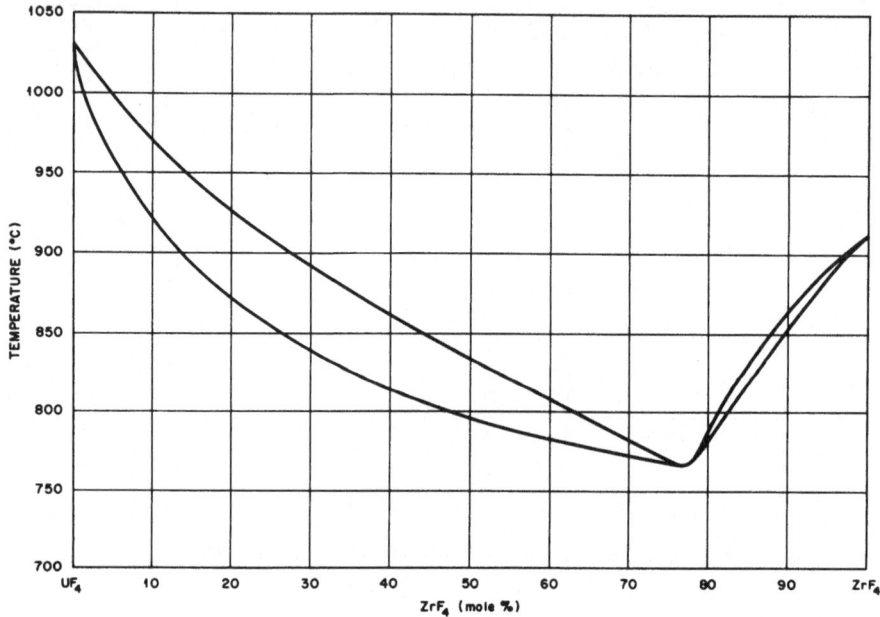

Fig. 140. The system ZrF_4–UF_4.

The System ThF_4–UF_4. See Fig. 141. Source: C. F. Weaver, R. E.
Thoma, H. Insley, and H. A. Friedman, *J. Am. Ceram. Soc.* **43**:213 (1960).

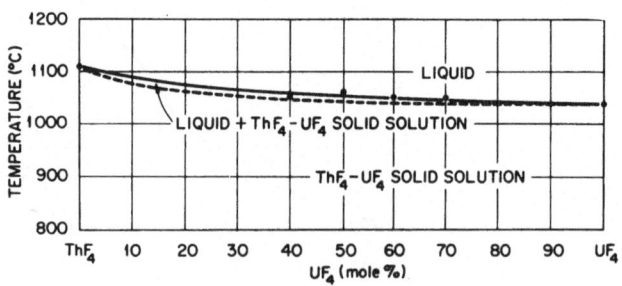

Fig. 141. The system ThF_4–UF_4.

The system ThF_4–UF_4 forms a continuous series of solid solutions without a temperature minimum.

The System LiF–SbF₅. The phase diagram for the system has not yet been described. However, the structure of an intermediate compound, $LiSbF_5$, has been determined by J. H. Burns, *Acta. Cryst.* **15**:1098 (1962), as hexagonal, $R\bar{3}$, $a = 5.18$, $c = 13.60$.

The System XeF₂–VF₅. See Fig. 142. Source: V. A. Legasov and A. S. Marinin, *Russ. J. Phys. Chem.* (*English Transl.*) **46**:1388 (1972). The intermediate compound $XeF_2 \cdot VF_5$ melts congruently at $38 \pm 1°$;

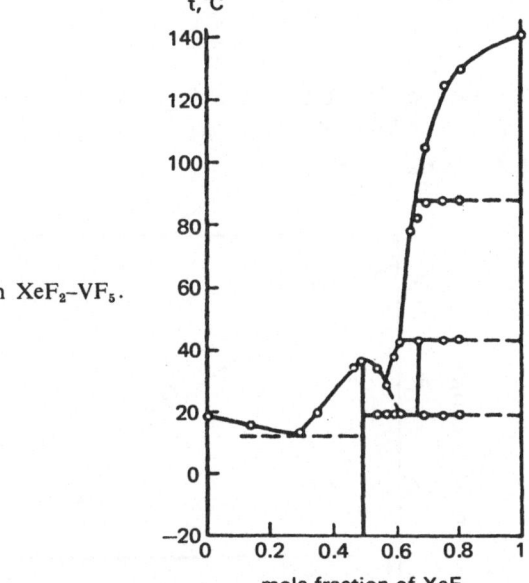

Fig. 142. The system XeF₂–VF₅.

mole fraction of XeF₂

the peritectic associated with the upper limit of stability of the intermediate compound, $2XeF_2 \cdot VF_5$, occurs at $44°$. Compositions of the invariant equilibrium points were not given in the source reference.

The System XeF₂–IF₅. See Fig. 143. Source: V. A. Legasov, V. B. Sokolov, and B. B. Chaivanov, *Russ. J. Phys. Chem.* (*English Transl.*) **43**:1650 (1969). The composition of the invariant equilibrium points was not given by the authors of the source reference. The m.p. of $XeF_2 \cdot IF_5$ is $102 \pm 0.7°$.

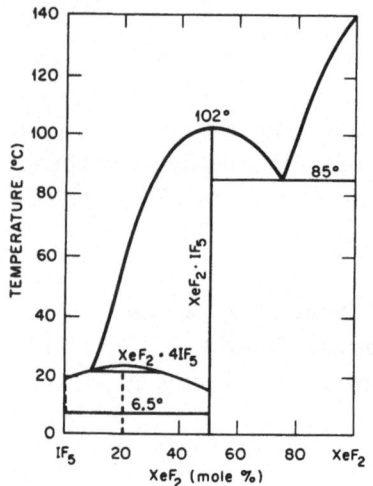

Fig. 143. The system XeF$_2$–IF$_5$.

The System BF$_3$–UF$_6$. See Fig. 144. Source: J. Fischer and R. C. Vogel, *J. Am. Chem. Soc.* **76**:4381 (1954). The eutectic occurs at $-63.5°$ and at $3.2 \pm 0.5\%$ UF$_6$.

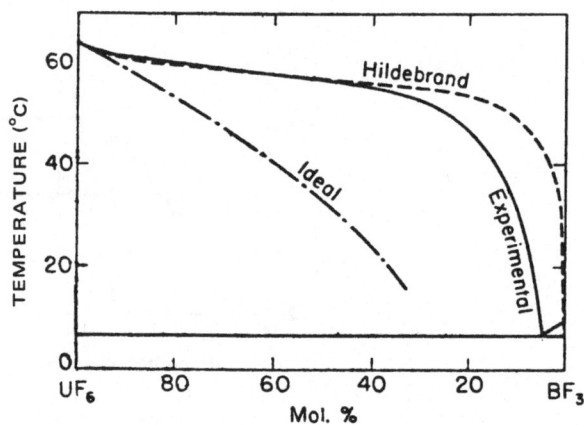

Fig. 144. The system BF$_3$–UF$_6$.

The System BF$_5$–UF$_6$. See Fig. 145. Source: J. Fischer and R. C. Vogel, *J. Am. Chem. Soc.* **76**:4829 (1954). $E = 6.4°$, 4.1% UF$_6$.

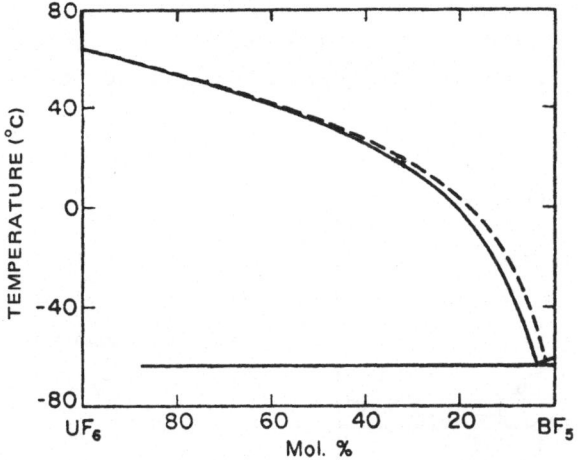

Fig. 145. The system BF_5–UF_6.

The System MoF_5–SbF_5. See Fig. 146. Source: V. K. Ezhov, *Russ. J. Inorg. Chem. (English Transl.)* **17** (1972). Invariant equilibrium data were not described quantitatively in the reference source, which cites the original reference as: V. N. Prusakov and V. K. Ezhov, Materialy Sim-

Fig. 146. The system MoF_5–SbF_5.

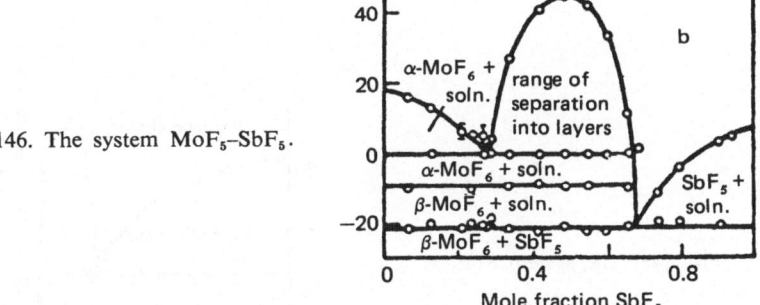

poziuma SEV "Issledovaniya v Oblasti Pererabotki Obluchennogo Topliva" (Proceedings of the SEV Symposium "Studies in the Processing of Irradiated Fuel"), KAE SSSR (1968), p. 331.

The System VF$_5$–UF$_6$. See Fig. 147. Source: V. K. Ezhov, *Russ. J. Inorg. Chem. (English Transl.)* **17**:1059 (1972). The temperature and composition of the eutectic were not given in the source reference.

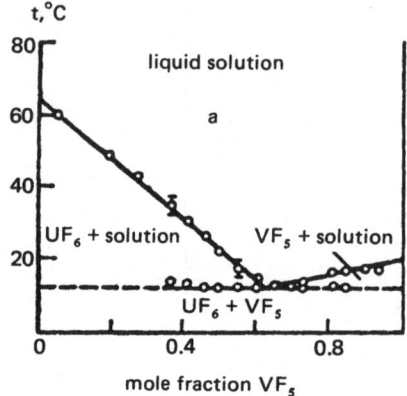

Fig. 147. The system VF$_5$–UF$_6$.

The System SbF$_5$–UF$_6$. See Fig. 148. Source: V. K. Ezhov, *Russ. J. Inorg. Chem.* **17**:1059 (1972). The region of liquid–liquid immiscibility extends from 23.3 to 39.0 mole % UF$_6$. The temperature of the monotectic was not specified in the source reference. The upper critical temperature of separation is 73.5°. The eutectic occurs at −5.5° and at 19.5 mole % UF$_6$.

Fig. 148. The system SbF$_5$–UF$_6$.

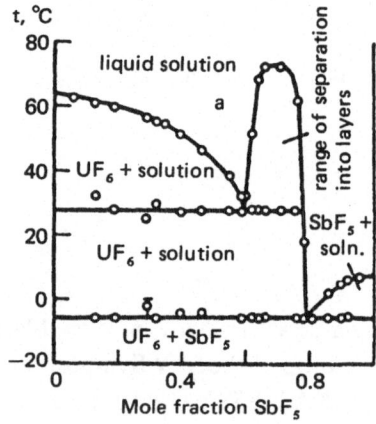

The System TaF₅–UF₆. See Fig. 149. Source: V. K. Ezhov, *Russ. J. Inorg. Chem. (English Transl.)* **17**:1059 (1972). The composition and temperature of the invariant equilibrium points were not given in the source reference.

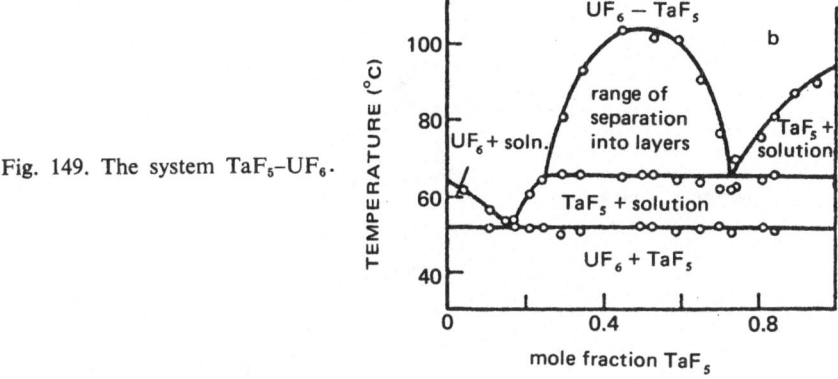

Fig. 149. The system TaF₅–UF₆.

The System WF₆–UF₆. See Fig. 150. Source: V. N. Prusakov and V. K. Eshov, *At. Energ. (USSR)* **25**:64 (1968). The system forms a single eutectic, at $-1.0 \pm 0.7°$ and 96 ± 2 mole %. No evidence of solid solution was observed.

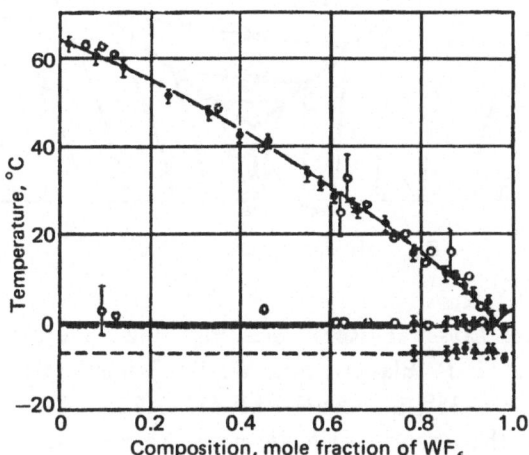

Fig. 150. The system WF₆–UF₆.

TERNARY SYSTEMS

The System LiF–NaF–KF.   See Fig. 151. Source: A. G. Bergman and
E. P. Dergunov, *Compt. Rend. Acad. Sci. URSS* **31**:753 (1941). $E = 454°$,
LiF–NaF–KF (46.5–11.5–42.0%). The isotherms shown in Fig. 151 were
verified to $\pm 5°$ for 0 to 60% KF by C. J. Barton *et al.*, ORNL (1951–1954).

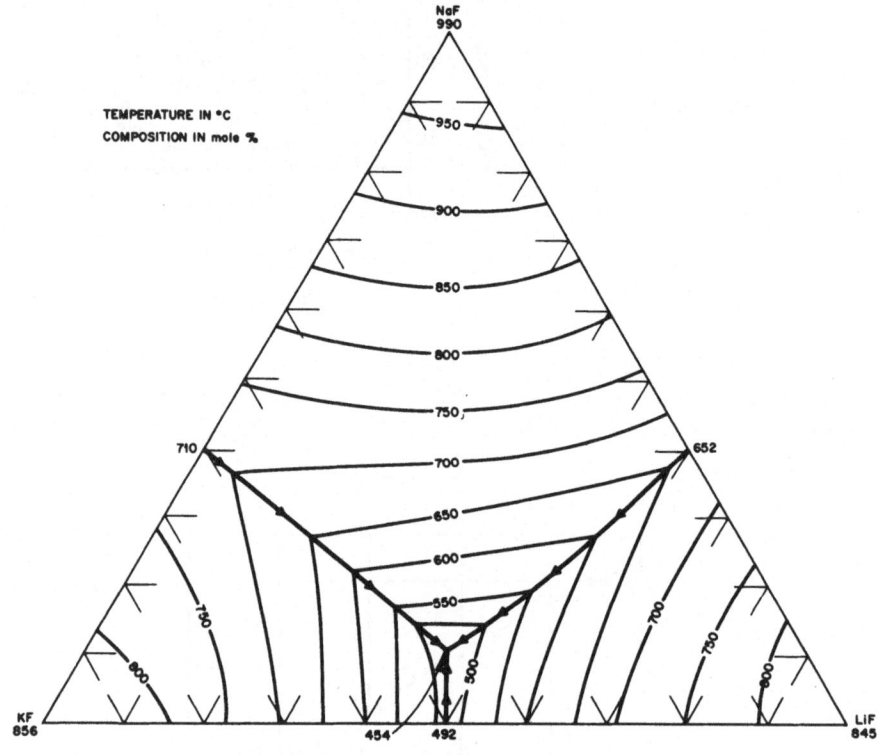

Fig. 151. The system LiF–NaF–KF.

The System LiF–NaF–RbF.   See Fig. 152. Source: C. J. Barton,
L. M. Bratcher, J. P. Blakely, and W. R. Grimes, ORNL, unpublished
work (1951). $E = 435°$, LiF–NaF–RbF (42–6–52%). The composition and
temperature of the ternary peritectic point were not determined. General

characteristics of the system were reported by E. P. Dergunov, *Dokl. Akad. Nauk SSSR* **58**: 1369 (1947). According to this work, the composition and temperature of the eutectic is LiF–NaF–RbF (46.5–6.5–47%) and 426°.

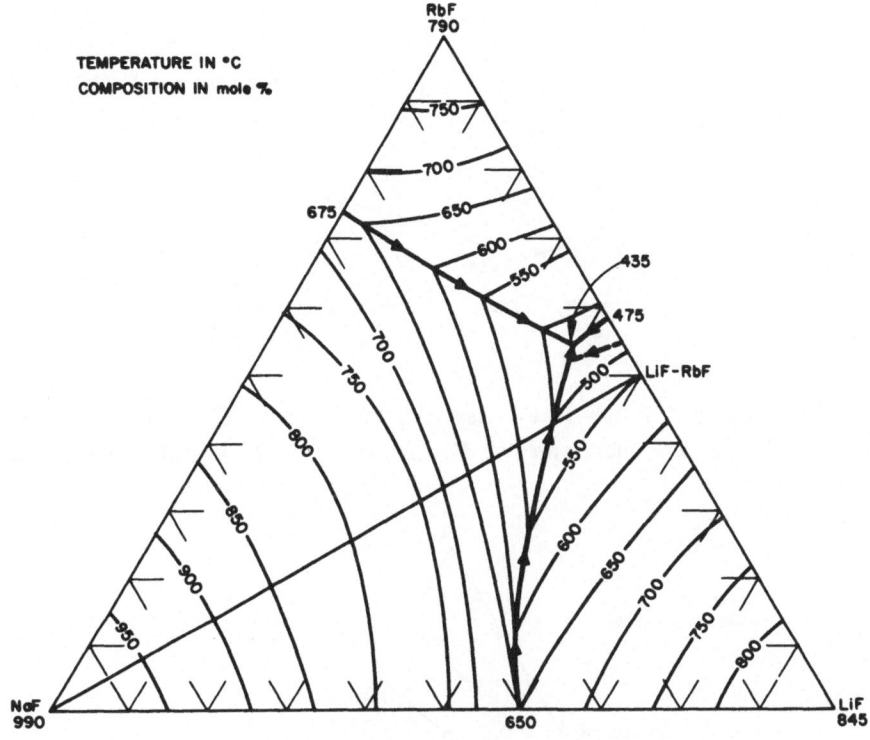

Fig. 152. The system LiF–NaF–RbF.

The System LiF–NaF–CsF. See Fig. 153. Source: G. A. Bukhalova and D. V. Sementsova, *Russ. J. Inorg. Chem.* (*English Transl.*) **10**:1880 (1965). $E = 448°$, LiF–NaF–CsF (48–11–41%) $E = 446°$, LiF–NaF–CsF (9–37–54%). The authors of the source reference emphasize that the results of their investigation of the LiF–NaF–CsF system clearly indicate that the compound LiF–CsF melts congruently.

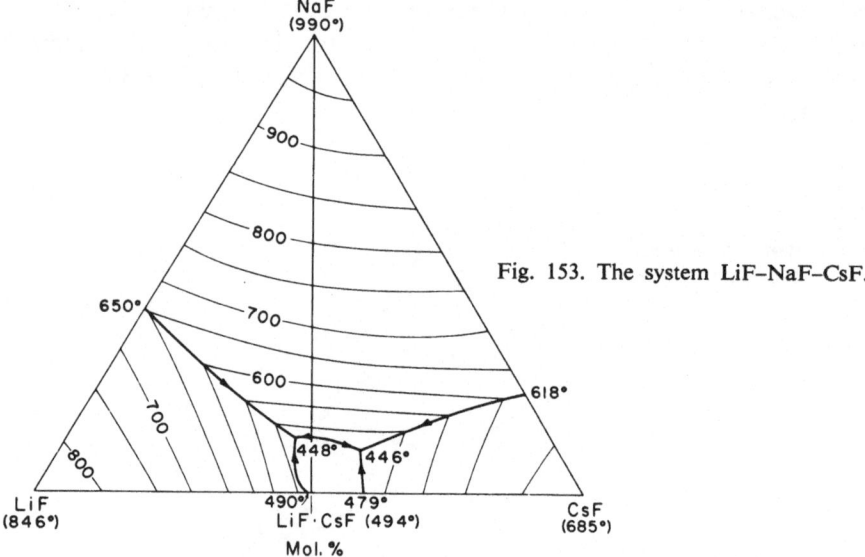

Fig. 153. The system LiF–NaF–CsF.

The System LiF–KF–RbF. See Fig. 154. Source: C. J. Barton, J. P. Blakely, L. M. Bratcher, and W. R. Grimes, ORNL, unpublished work

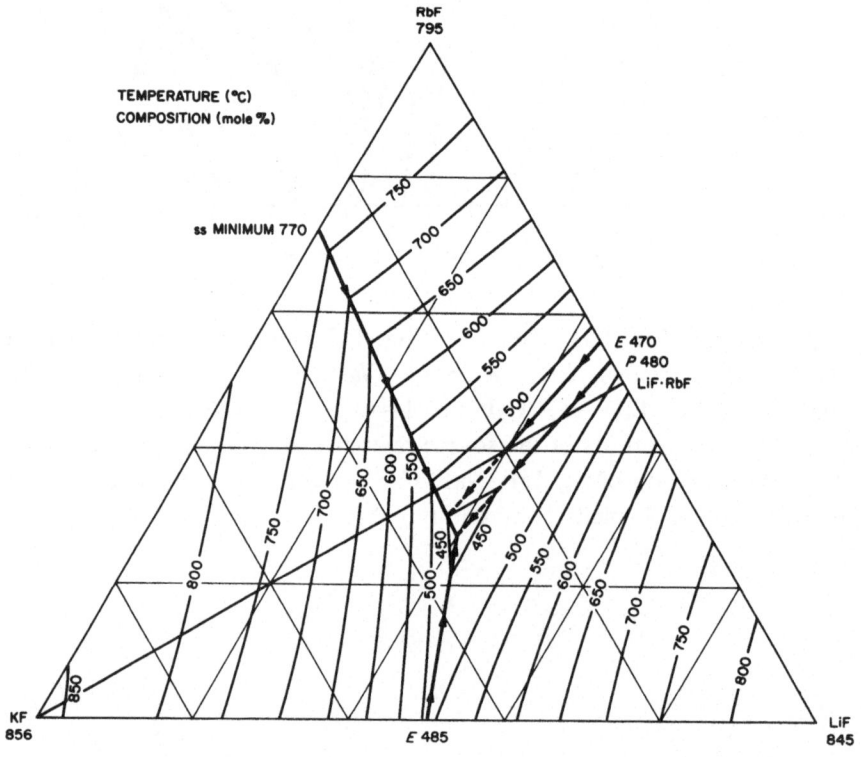

(1951). A single invariant point occurs within the system, $E = 440 \pm 10°$, LiF–KF–RbF (40–32.5–27.5%), m.p. $440 \pm 10°$. General characteristics of the system were reported in earlier work by E. P. Dergunov, *Dokl. Akad. Nauk SSSR* **58**:1369 (1947).

The System NaF–KF–RbF. See Fig. 155. Source: C. J. Barton, L. M. Bratcher, J. P. Blakely, and W. R. Grimes, ORNL, unpublished work (1951). $E = 621 \pm 10°$, NaF–KF–RbF (21–5–74%).

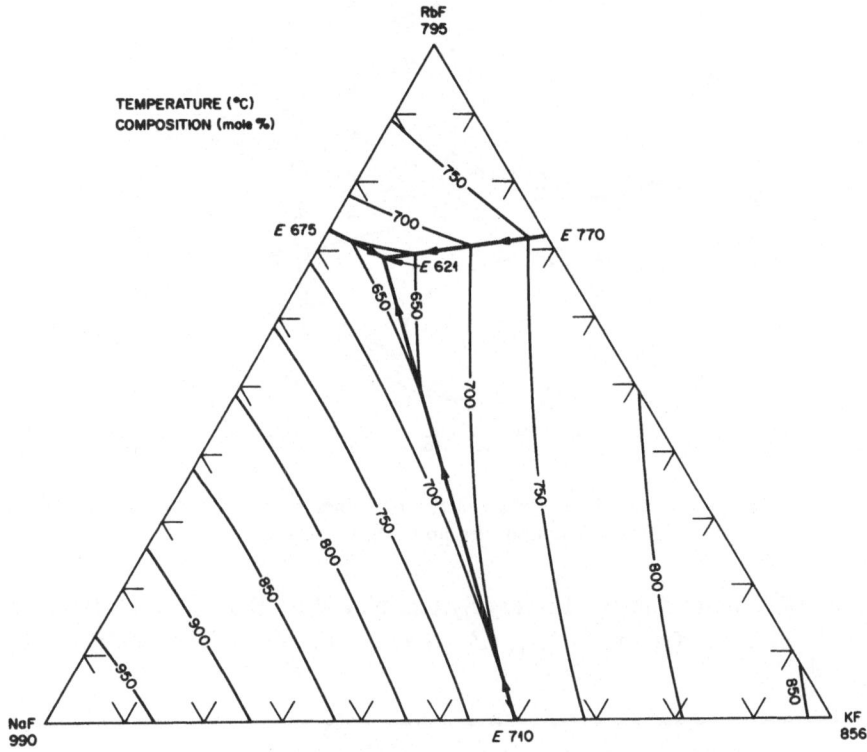

Fig. 155. The system NaF–KF–RbF.

The System LiF–NaF–BeF$_2$. See Fig. 156. Source: R. E. Moore, C. J. Barton, W. R. Grimes, R. E. Meadows, L. M. Bratcher, G. D. White, and T. N. McVay, ORNL, unpublished work (1951–1958). Invariant equilibria shown in the phase diagram by the intersections of dotted-line boundary curves have not been determined with sufficient precision to be listed in the table. Phase relationships of two of the ternary compounds in this system have been reported by W. Jahn [Silicate Models. V. NaLi-(BeF$_4$), a Model Substance for Monticellite, CaMg(SiO$_4$), *Z. Anorg. Allgem. Chem.* **276**:113–127 (1954); Silicate Models. VI. Na$_3$Li(BeF$_4$)$_2$,

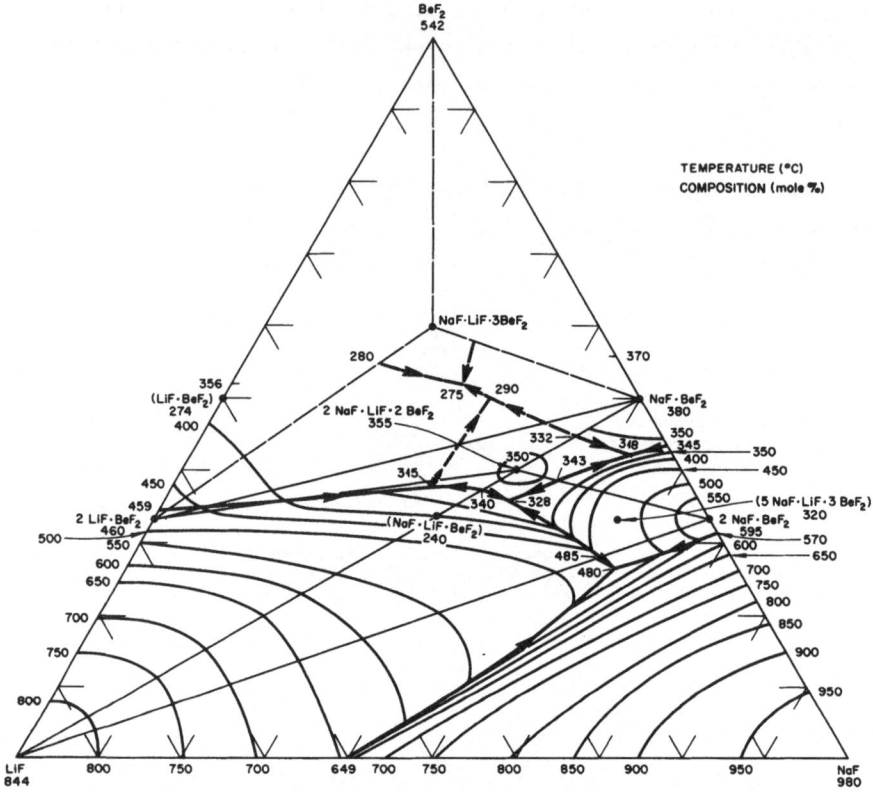

Fig. 156. The system LiF–NaF–BeF$_2$. The dotted lines represent incompletely defined phase boundaries and Alkemade lines.

a New Compound in the Ternary System NaF–LiF–BeF$_2$, and Its Relation to Merwinite, Ca$_3$Mg(SiO$_4$)$_2$, *Z. Anorg. Allgem. Chem.* **277**:274–286 (1954)].

	Mole %		Temperature	Type of equilibrium
LiF	NaF	BeF$_2$		
15	58	27	480	Eutectic
23	41	36	328	Eutectic
20	40	40	355	Congruent melting point
5	53	42	318	Eutectic
31.5	31	37.5	315	Eutectic

(Continued)

	Mole %			
			Minimum temperature on Alkemede lines	
			Temperature	Alkemede line
LiF	NaF	BeF₂		
16	56	28	485	2NaF · BeF₂–LiF
26	37	37	340	2NaF · LiF · 2BeF₂–LiF
11	44	45	332	NaF · BeF₂–2NaF · LiF · 2BeF
16	45	39	343	2NaF · BeF₂–2NaF · LiF · 2BeF
30.5	31	38.5	316	2LiF · BeF₂–2NaF · LiF · 2BeF₂

The System LiF–RbF–BeF₂. See Fig. 157. Source: T. B. Rhineham-mer, D. E. Etter, C. R. Hudgens, N. E. Rogers, and P. A. Tucker, unpub-

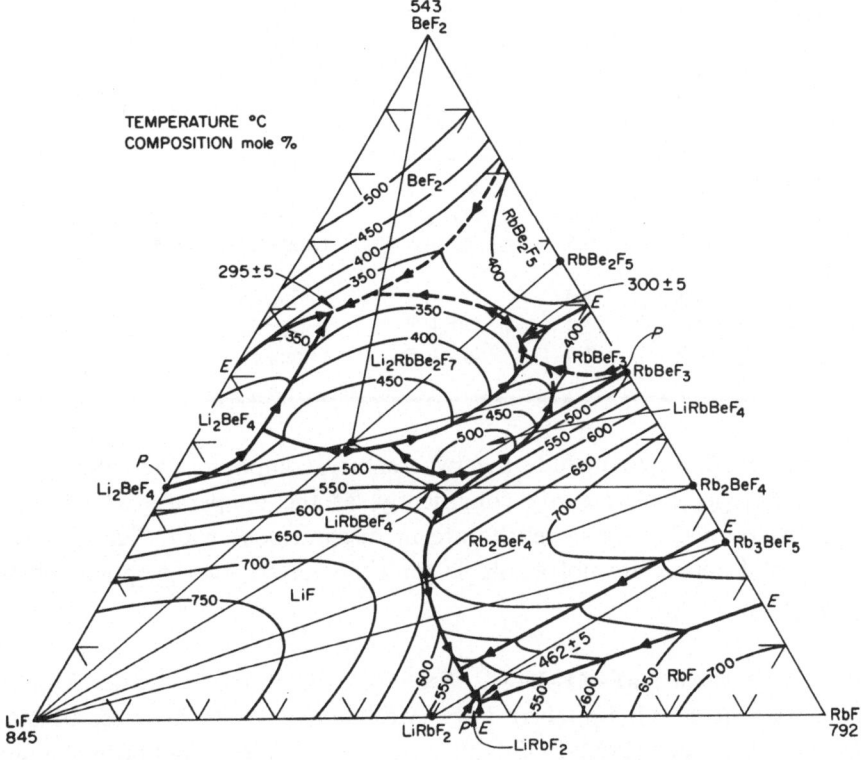

Fig. 157. The system LiF–RbF–BeF₂.

lished work, Mound Laboratory, Miamisburg, Ohio (1956–1958). Invariant and singular equilibrium points are given in the table. In a study of isotypic fluoroberyllates of the formula $Li_3Me(BeF_4)_2$, J. LeRoy, L. Pontonnier, and S. Aleonard, *Mater. Res. Bull.* **6**:267 (1971), described the section

Mole %			Temperature	Type of equilibrium
LiF	RbF	BeF_2		
40.0	20.0	40.0	485 ± 5	Congruent melting point
33.3	33.3	33.3	544 ± 5	Incongruent melting point of $LiF \cdot RbF \cdot BeF_2$
33.0	8.0	59.0	295 ± 5	Eutectic
26.0	12.0	62.0	305–320	Peritectic
12.0	29.0	59.0	334 ± 5	Quasi-binary eutectic
11.0	34.0	55.0	300 ± 5	Eutectic
11.5	35.5	53.0	335 ± 5	Peritectic
9.0	40.0	51.0	380 ± 5	Peritectic
50.0	8.0	42.0	426 ± 5	Peritectic
41.5	19.5	39.0	473 ± 5	Quasi-binary eutectic
35.0	25.5	39.5	470 ± 5	Peritectic
23.0	40.0	37.0	530 ± 5	Peritectic
30.0	35.0	35.0	544 ± 5	Maximum on the boundary
41.0	39.5	19.5	640 ± 5	Quasi-binary eutectic
43.0	50.0	7.0	527 ± 5	Peritectic
43.0	54.0	3.0	470 ± 5	Peritectic
43.0	55.0	2.0	462 ± 5	Eutectic

Li_2BeF_4–Rb_2BeF_4 to include the two ternary compounds $LiRbBeF_4$ and $Li_3Rb(BeF_4)_2$, each melting congruently, at the temperatures 560° and 500°, respectively. These results are in disagreement with Fig. 157. Experimental details are not available for either of the two proposed phase diagrams.

The System NaF–KF–BeF_2. See Fig. 158. Source: R. E. Moore, C. J. Barton, L. M. Bratcher, T. N. McVay, and W. R. Grimes, ORNL, unpublished work (1952–1955). Invariant and singular equilibrium points were not established with sufficient precision to list here.

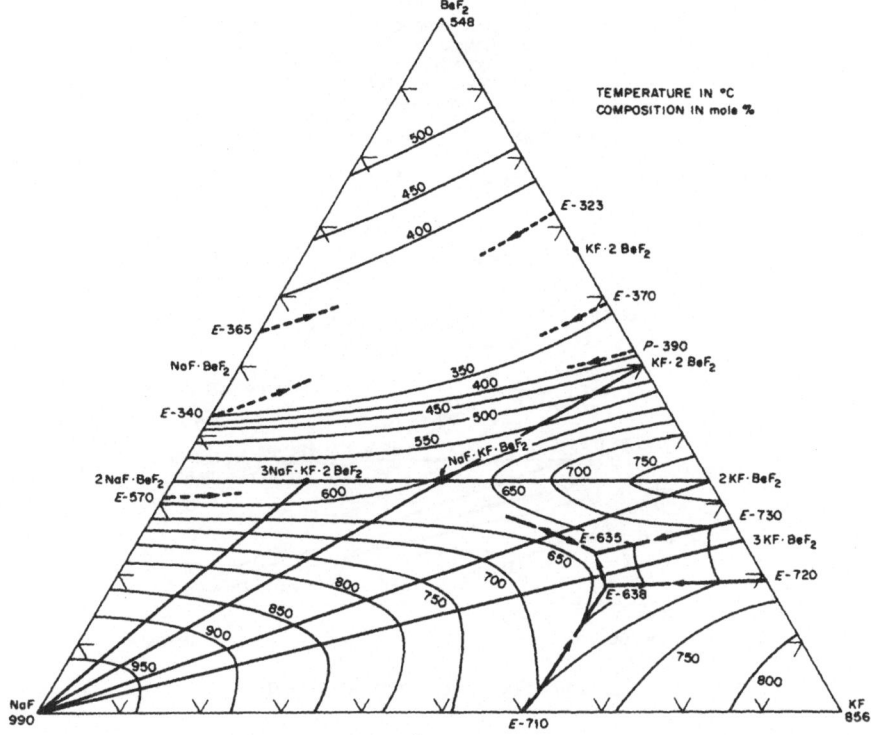

Fig. 158. The system NaF–KF–BeF$_2$.

The System NaF–RbF–BeF$_2$. See Fig. 159. Source: R. E. Moore, C. J. Barton, L. M. Bratcher, and W. R. Grimes, ORNL, unpublished work (1954–1957). Invariant and singular equilibrium points are given in the table.

Mole %			Temperature	Type of equilibrium
NaF	RbF	BeF$_2$		
20.5	68.5	11.0	610	Eutectic
50	37.5	12.5	640	Quasi-binary eutectic
46	38	16	635	Eutectic
45	37	18	650	Quasi-binary eutectic
37	33	30	570	Peritectic
52	15	33	477	Eutectic
33	33.7	33.3	580	Peritectic
51.7	15	33.3	480	Quasi-binary eutectic

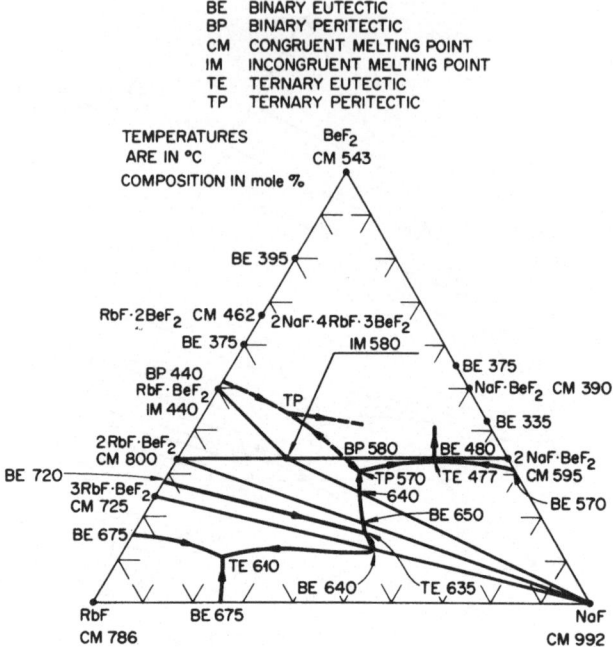

BE   BINARY EUTECTIC
BP   BINARY PERITECTIC
CM   CONGRUENT MELTING POINT
IM   INCONGRUENT MELTING POINT
TE   TERNARY EUTECTIC
TP   TERNARY PERITECTIC

Fig. 159. The system NaF–RbF–BeF₂.

The System LiF–NaF–MgF₂. See Fig. 160. Source: A. G. Bergman and E. P. Dergunov, *Compt. Rend. Acad. Sci. URSS* **31**:755 (1941). $E =$ 630°, LiF–NaF–MgF₂ (47–43–10%); $E = 684°$, LiF–NaF–MgF₂ (59–12–29%).

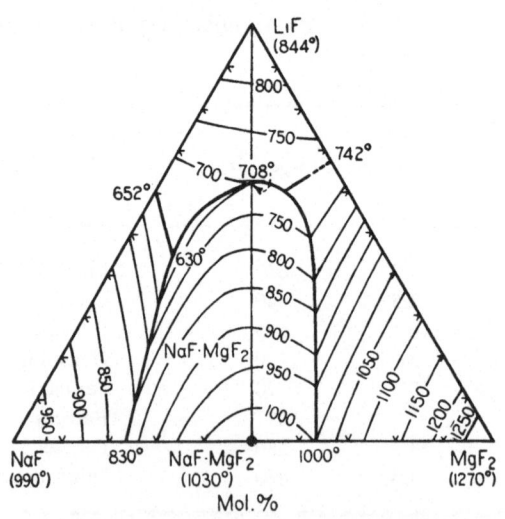

Fig. 160. The system LiF–NaF–MgF₂.

The System LiF–KF–MgF$_2$. See Fig. 161. Source: A. G. Bergman and S. P. Pavlenko, *Compt. Rend. Acad. Sci. URSS* **30**:818 (1941). The authors reported evidence of both a peritectic and a eutectic in the LiF–KF–2KF · MgF$_2$ subsystem. The eutectic, solidifying at 488° (2° lower than that for LiF–KF), occurs close to the LiF–KF binary system. Its composition and the temperature and composition of the peritectic in the subsystem were not established. The eutectic in the LiF–MgF$_2$–2KF · MgF$_2$ subsystem occurs at 713°, LiF–KF–MgF$_2$ (63.5–6–30.5%).

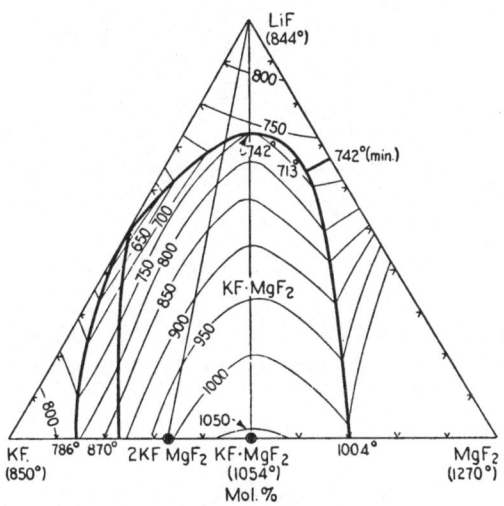

Fig. 161. The system LiF–KF–MgF$_2$.

The System NaF–KF–MgF$_2$. See Fig. 162. Source: A. G. Bergman and E. P. Dergunov, *Compt. Rend. Acad. Sci. URSS* **48**:330 (1945). Invariant and singular equilibrium points are given in the table.

Mole %			Temperature	Type of equilibrium
NaF	KF	MgF$_2$		
33.0	13.5	53.5	975	Eutectic
62.5	15.0	22.5	798	Eutectic
34.5	59.0	6.5	685	Eutectic
39.0	50.0	11.0	710	Peritectic

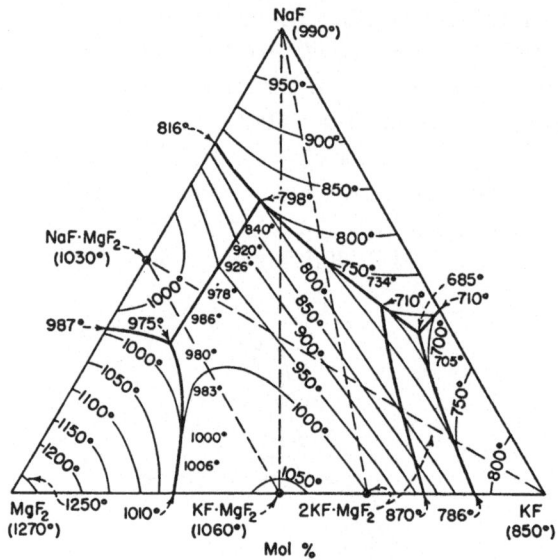

Fig. 162. The system NaF–KF–MgF$_2$.

The System LiF–NaF–CaF$_2$. See Fig. 163. Source: C. J. Barton, L. M. Bratcher, and W. R. Grimes, ORNL, unpublished work (1955–

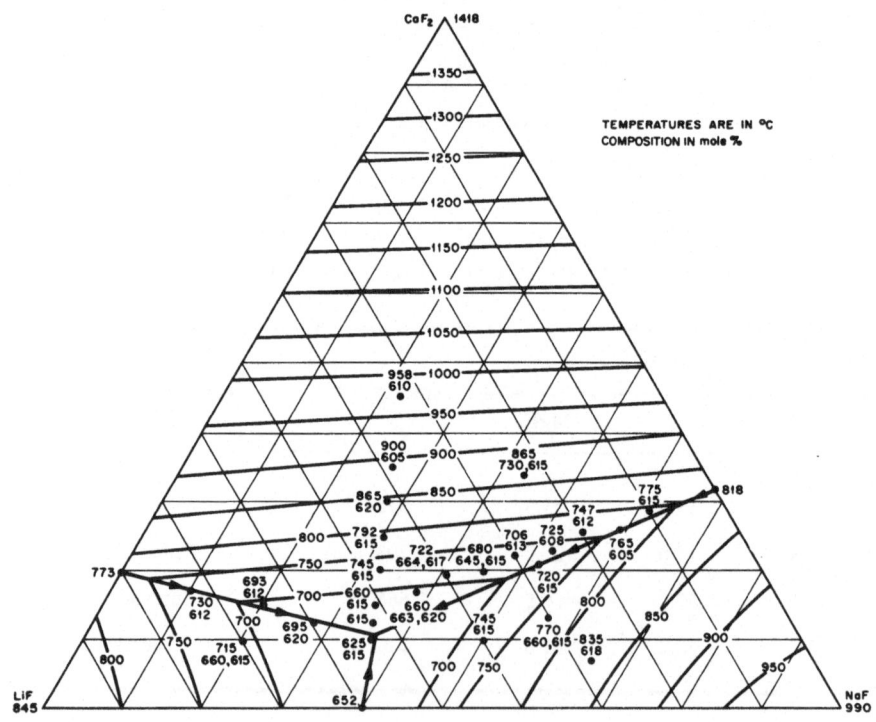

1956). $E = 616°$, LiF–NaF–CaF$_2$ (53–36–11%) phase diagram of the system was reported by G. A. Bukhalova, K. Sulaimankulov, and A. K. Bastandzhiyan, *Zh. Neorg. Khim.* **4**:1138 (1959).

The System LiF–CsF–CaF$_2$. See Fig. 164. Source: D. V. Sementsova, G. A. Bukhalova, and Z. A. Mateiko, *Russ. J. Inorg. Chem.* (*English Transl.*)

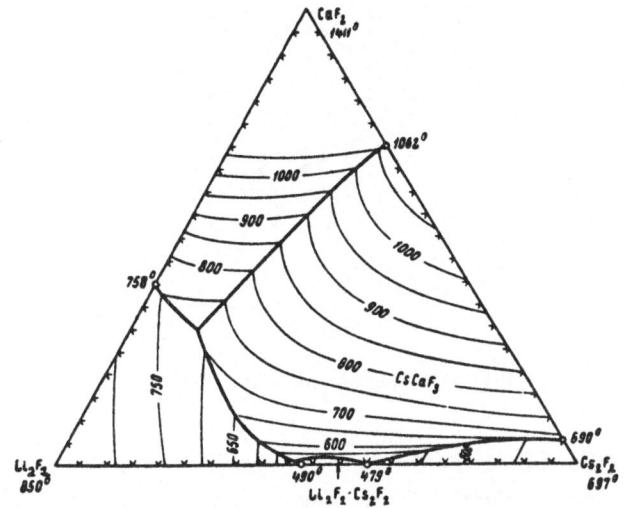

Fig. 164. The system LiF–CsF–CaF$_2$.

12:865 (1967). Invariant and singular equilibrium points are given in the table.

Mole %			Temperature	Type of equilibrium
Li$_2$F$_2$	Cs$_2$F$_2$	CaF$_2$		
57	13	30	708	Eutectic
54.5	44.0	1.5	476	Eutectic
39.0	60.5	0.5	462	Eutectic

The System NaF–KF–CaF$_2$. See Fig. 165. Source: G. A. Bukhalova and V. T. Berezhnaya, *Russ. J. Inorg. Chem.* (*English Transl.*) **4**:1197 (1959). $E = 759°$, Na$_2$F$_2$–K$_2$F$_2$–BaF$_2$ (31–34–35%); $E = 621°$, Na$_2$F$_2$–K$_2$F$_2$–BaF$_2$ (51–36–13%).

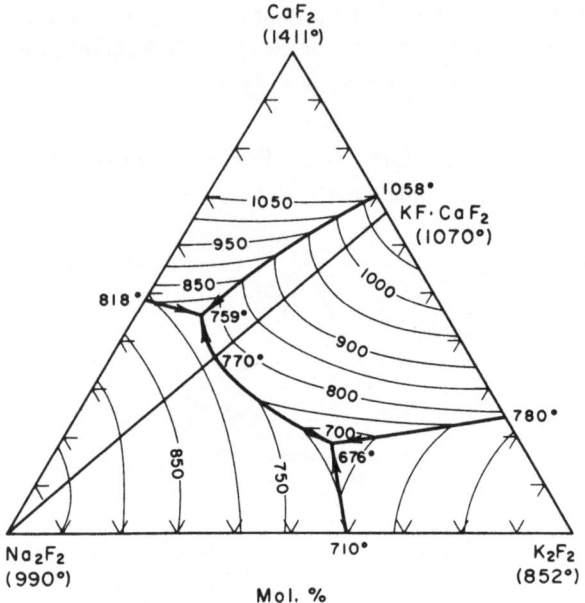

Fig. 165. The system NaF–KF–CaF₂.

The System LiF–CsF–MnF₂. See Fig. 166. Source: I. N. Belyaev and O. Ya. Revina, *Russ. J. Inorg. Chem.* (*English Transl.*) **11**:1041 (1966).

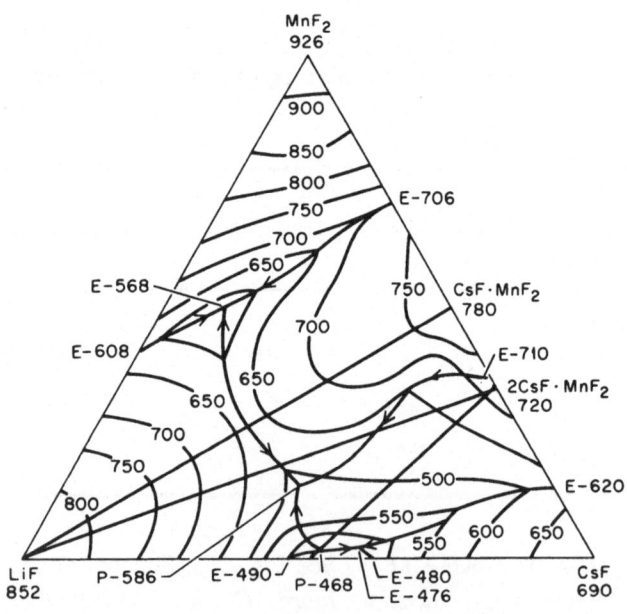

Fig. 166. The system LiF–CsF–MnF₂.

Invariant and singular equilibrium points are given in the table.

	Mole %		Temperature	Type of equilibrium
LiF	CsF	MnF$_2$		
40	10	50	568	Eutectic
40	58	2	476	Eutectic
47	52	1	488	Peritectic
45	40	15	586	Peritectic

The System NaF–CsF–MnF$_2$. See Fig. 167. Source: I. N. Belyaev and O. Ya. Revina, *Russ. J. Inorg. Chem.* (*English Transl.*) **11**:1041 (1966). Invariant and singular equilibrium points are given in the table.

	Mole %		Temperature	Type of equilibrium
NaF	CsF	MnF$_2$		
18	74	8	570	Eutectic
25	44	31	634	Eutectic
41	20	39	640	Eutectic
17	16	67	650	Eutectic

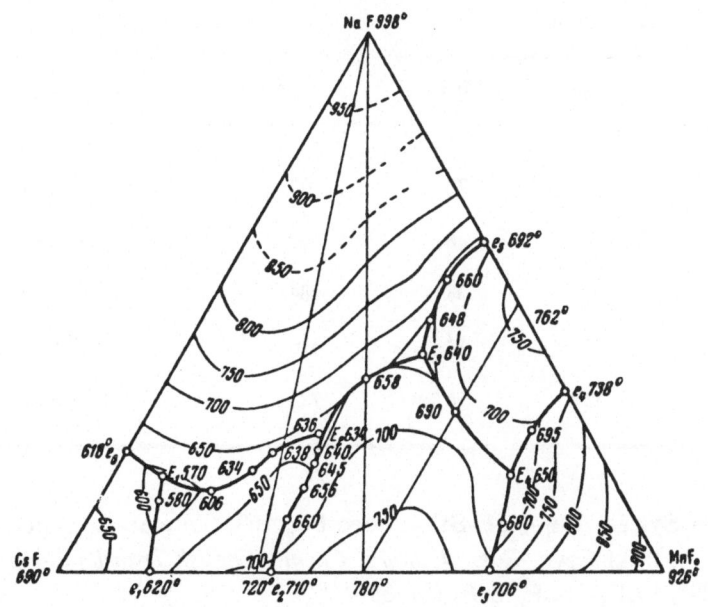

Fig. 167. The system NaF–CsF–MnF$_2$.

The System KF–CsF–MnF$_2$. See Fig. 168. Source: I. N. Belyaev and O. Ya. Revina, *Russ. J. Inorg. Chem.* (*English Transl.*) **11**:1041 (1966).

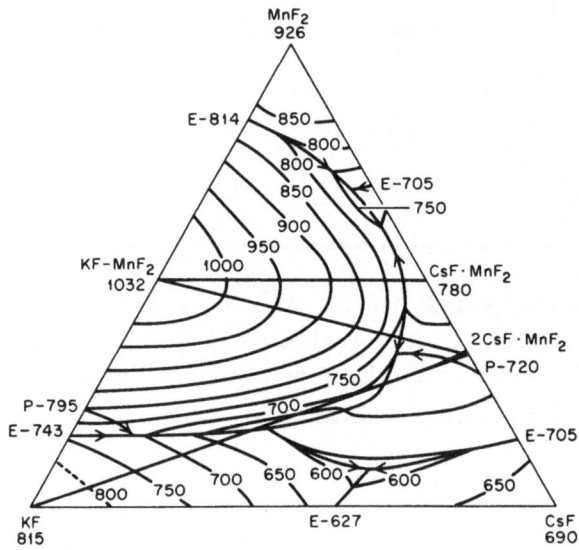

Fig. 168. The system KF–CsF–MnF$_2$.

Invariant and singular equilibrium points are given in the table.

Liquid mole %			Temperature	Type of equilibrium
KF	CsF	MnF$_2$		
3	27	70	702	Eutectic
32	60	8	560	Eutectic
13	54	38	684	Peritectic
46	37	17	610	Peritectic
73	11	16	725	Peritectic

The System LiF–NaF–SrF$_2$. See Fig. 169. Source: V. T. Berezhnaya and G. A. Bukhalova, *Russ. J. Inorg. Chem.* (*English Transl.*) **5**:445 (1960). $E = 624°$, Li$_2$F$_2$–Na$_2$F$_2$–SrF$_2$ (51–33–16%).

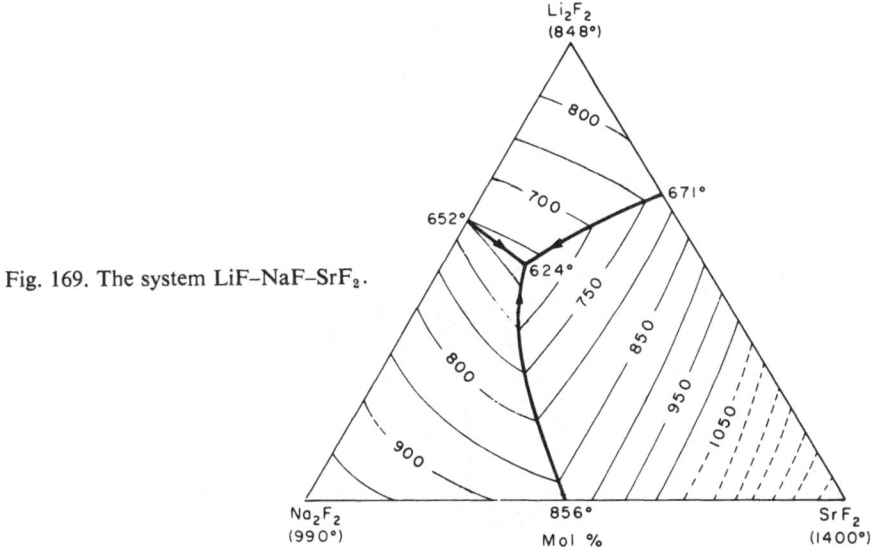

Fig. 169. The system LiF–NaF–SrF$_2$.

The System NaF–KF–SrF$_2$. See Fig. 170. Source: V. T. Berezhnaya and G. A. Bukhalova, *Russ. J. Inorg. Chem.* (*English Transl.*) **5**:925 (1960). $E = 664°$, Na$_2$F$_2$–K$_2$F$_2$–SrF$_2$ (31–40–29%).

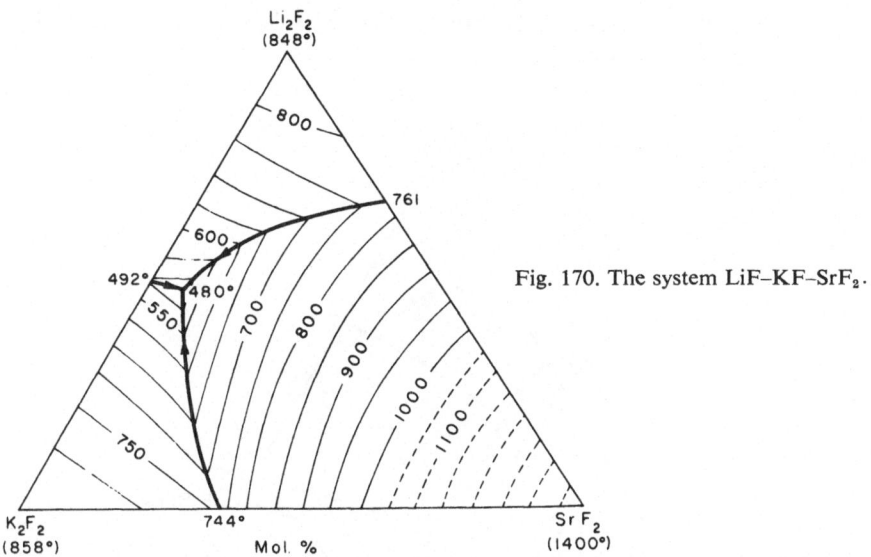

Fig. 170. The system LiF–KF–SrF$_2$.

The System NaF–KF–SrF$_2$. See Fig. 171. Source: V. T. Berezhnaya and G. A. Bukhalova, *Russ. J. Inorg. Chem.* (*English Transl.*) **5**:925 (1960). $E = 664°$, Na$_2$F$_2$–K$_2$F$_2$–SrF$_2$ (31–40–29%).

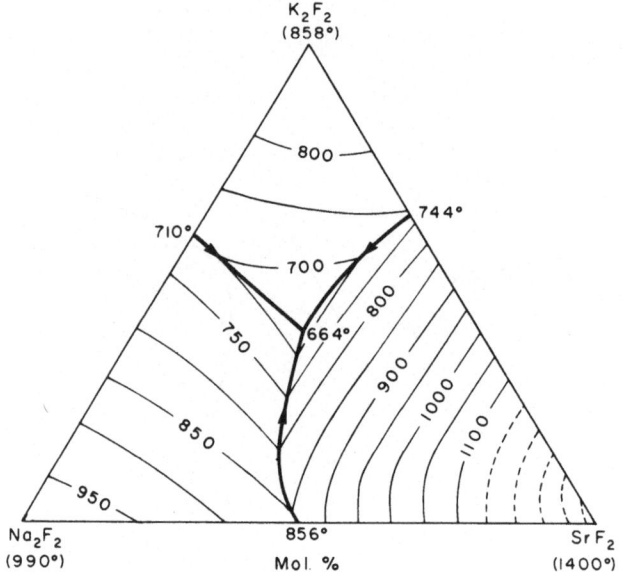

Fig. 171. The system NaF–KF–SrF₂.

The System Li₂F₂–Na₂F₂–BaF₂. See Fig. 172. Source: V. T. Berezhnaya and G. A. Bukhalova, *Russ. J. Inorg. Chem. (English Transl.)* **4**:1200 (1959). $E = 621°$, Li₂F₂–Na₂F₂–BaF₂ (51–36–13%); $P = 694°$, Li₂F₂–Na₂F₂–BaF₂ (31–34–35%).

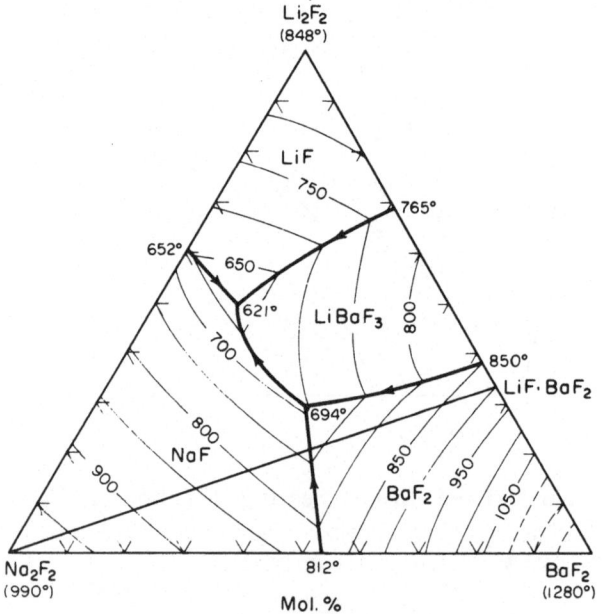

Fig. 172. The system LiF₂–Na₂F₂–BaF₂.

The System LiF–KF–BaF₂. See Fig. 173. Source: V. T. Berezhnaya and G. A. Bukhalova, *Russ. J. Inorg. Chem.* (*English Transl.*) **4**:1199 (1959). $E = 472°$, Li_2F_2–K_2F_2–BaF_2 (49–46–5%); $P = 588°$, Li_2F_2–K_2F_2–BaF_2 (29–49–22%).

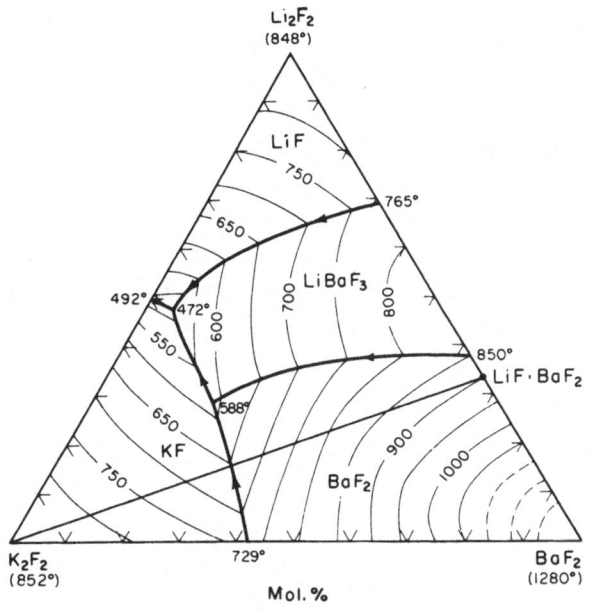

Fig. 173. The system LiF–KF–BaF₂.

The System NaF–KF–BF₃. See Fig. 174. Source: C. J. Barton, L. O. Gilpatrick, J. A. Bornmann, T. N. McVay, and H. Insley, *J. Inorg. Nucl.*

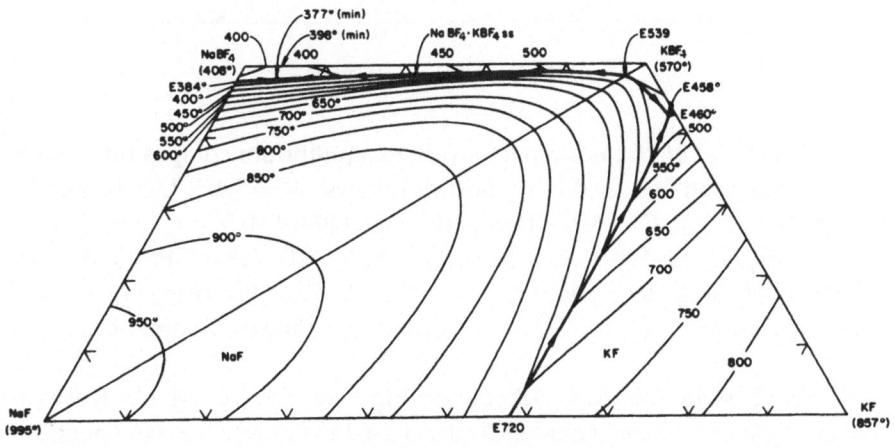

Fig. 174. The system NaF–KF–BF₃.

Chem. **33**:345 (1971). The composition of the ternary eutectic in the NaF–KBF$_4$–KF subsystem was not determined. It is described as being very close to the KF–KBF$_4$ binary eutectic in composition and melts at 458°. The shallow minimum at the liquidus, corresponding to the minimum in the NaBF$_4$–KBF$_4$ binary system, has the composition NaF–KF–BF$_3$ (48.4–3.5–48.1 %); the melting temperature is 377 $\pm$ 2°. Evidence of the existence of a subsolidus compound, 3NaBF$_4$ · KBF$_4$, stable between approximately 193 and 280°, was noted in the reference source.

The System LiF–NaF–AlF$_3$. See Fig. 175. Source: R. E. Thoma, B. J. Sturm, and E. H. Guinn, ORNL-3594 (1964). See also Fig. 3439 in *Phase Diagrams for Ceramists* (1969). A phase diagram of the system has also been reported by M. A. Kuvakin, *Russ. J. Inorg. Chem. (English Transl.)* **14**:146 (1969), that is in close agreement with the system as shown

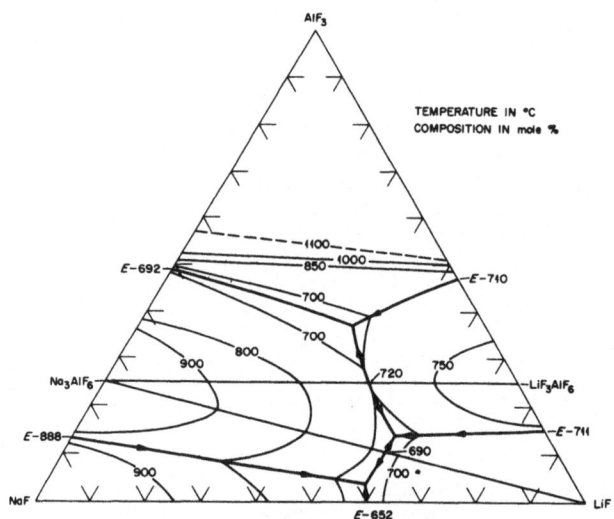

Fig. 175. The system LiF–NaF–AlF$_3$.

here. Thoma *et al.* did not report invariant equilibrium composition points. They were given in Kuvakin's report as follows: $E = 580°$, LiF–NaF–AlF$_3$ (51.4–38–10.6%); $E = 600°$, LiF–NaF–AlF$_3$ (40–17–43%); $P = 635°$, LiF–NaF–AlF$_3$ (22–19–59%); $E = 565°$, LiF–NaF–AlF$_3$ (18–17–65%). A phase diagram of the system was reported by M. A. Kuvakin (Fig. 176). In this source, composition–temperatures sections are shown in some detail.

The System LiF–NaF–AlF$_3$. See Fig. 176. Source: M. A. Kuvakin, *Russ. J. Inorg. Chem. (English Transl.)* **14**:146 (1969). See comments on Fig. 175.

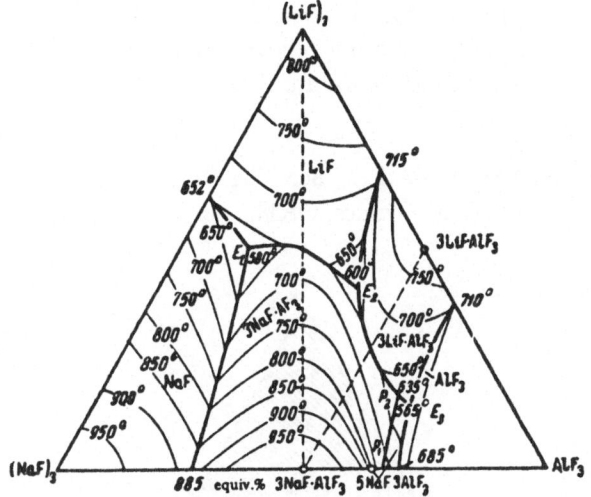

Fig. 176. The system LiF–NaF–AlF₃.

The System LiF–KF–AlF₃.

See Fig. 177. Source: R. E. Thoma, B. J. Sturm, and E. H. Guinn, Molten-Salt Solvents for Fluoride Volatility

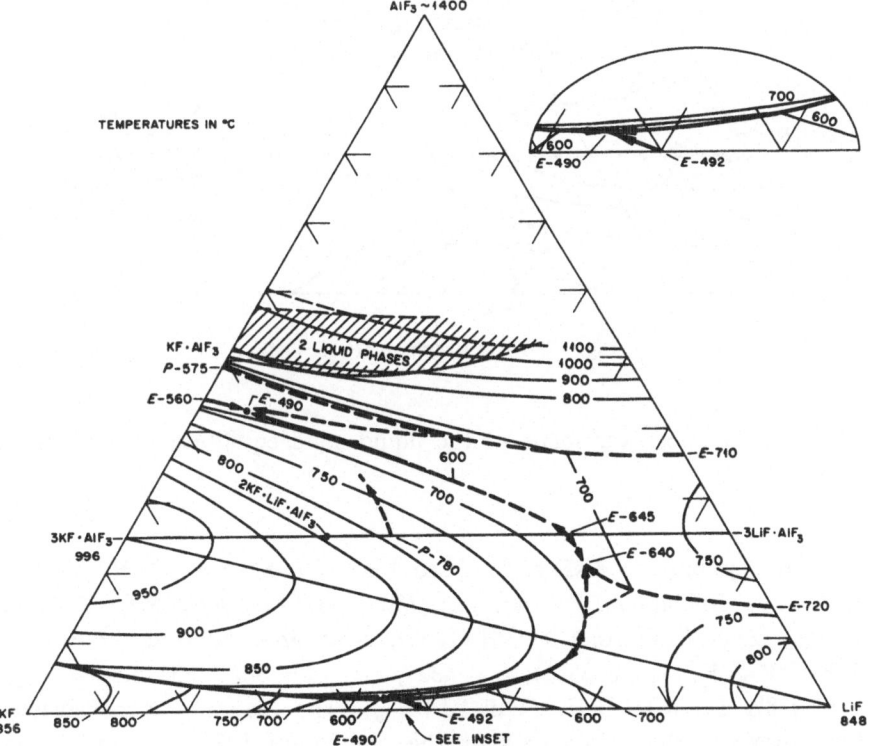

Fig. 177. The system LiF–KF–AlF₃.

Processing of Aluminum-Matrix Nuclear Fuel Elements, ORNL-3594, August, 1964. The authors did not report composition and temperature of invariant points. All primary data are listed in source report.

The System NaF–KF–AlF$_3$. See Fig. 178. Source: C. J. Barton, L. M. Bratcher, and W. R. Grimes, ORNL, unpublished work (1951–1952).

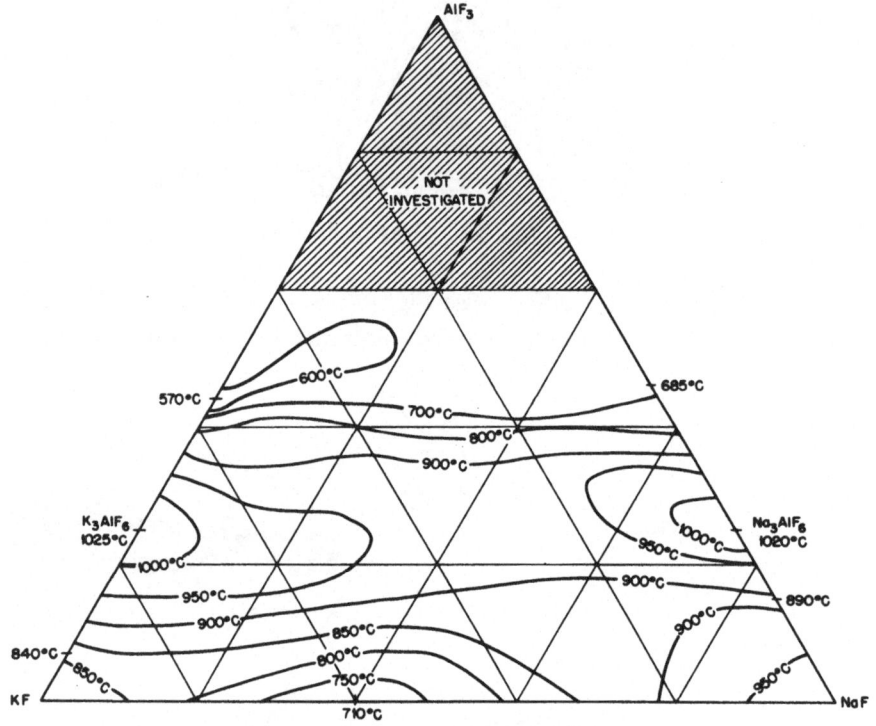

Fig. 178. The system NaF–KF–AlF$_3$.

No determination was made of the liquid–solid equilibria in the ternary system.

The System LiF–CsF–ScF$_3$. See Fig. 179. Source: E. P. Babaeva and G. A. Bukhalova, *Zh. Neorgan. Khim.* **11**:648 (1966). *Russ. J. Inorg. Chem.* (*English Transl.*) **11**:351 (1966). Invariant equilibria: $E = 584°$, LiF–CsF–ScF$_3$ (51–26–23%); $E = 432°$, LiF–CsF–ScF$_3$ (53–46.88–0.12%); $E = 438°$, LiF–CsF–ScF$_3$ (39.9–60–0.1%). The diagram as constructed here supports the inference that the compound LiF · CsF melts congruently.

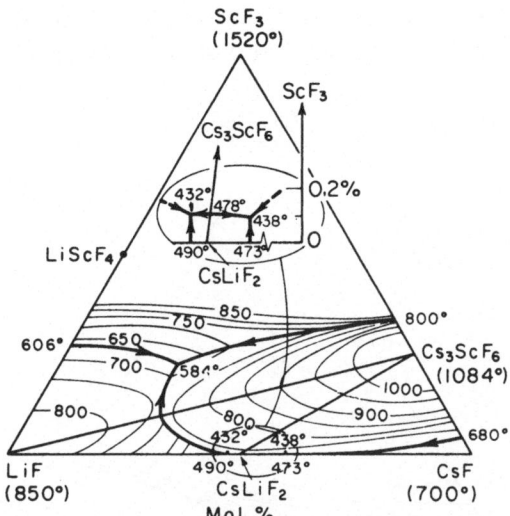

Fig. 179. The system LiF–CsF–ScF₃.

The System LiF–CsF–YF₃. See Fig. 180. Source: G. A. Bukhalova and E. P. Babaeva, *Russ. J. Inorg. Chem.* (*English Transl.*) **11**:220 (1966). Invariant and singular equilibria: $E = 650°$, LiF–CsF–YF₃ (47.5–31–21.5%); $E = 460°$, LiF–CsF–YF₃ (51.5–48–0.5%); $E = 430°$, LiF–CsF–YF₃ (39.8–60–0.2); saddle point, 710° LiF–CsF–YF₃ (57.5–32–10.5).

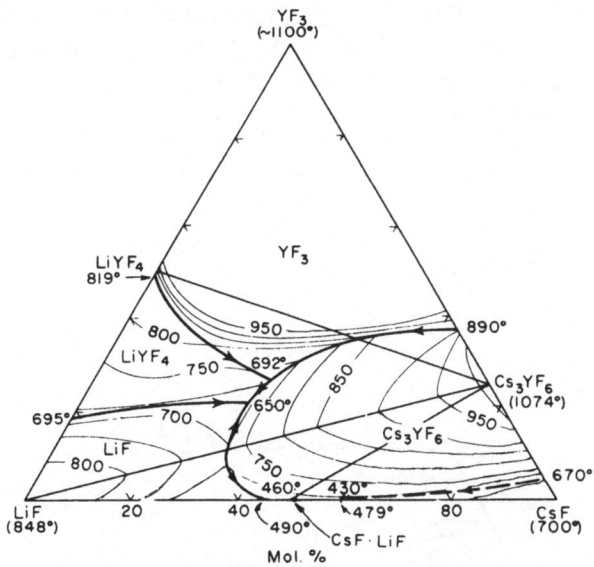

Fig. 180. The system LiF–CsF–YF₃.

The System LiF–CsF–LaF₃. See Fig. 181. Source: G. A. Bukhalova and E. P. Babaeva, *Russ. J. Inorg. Chem.* (*English Transl.*) **10**:1027 (1965).

Invariant equilibria: $E = 417°$, LiF–CsF–LaF$_3$ (33–61–6%); E–443°, LiF–CsF–LaF$_3$ (39–52–9%); $E = 480°$, LiF–CsF–LaF$_3$ (35–45–20%).

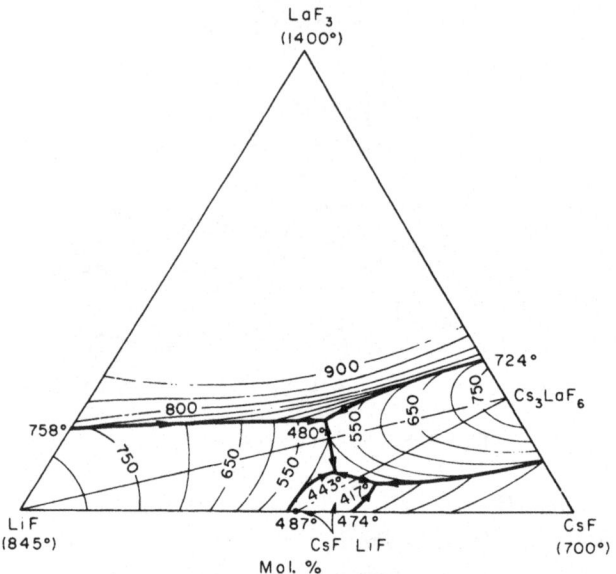

Fig. 181. The system LiF–CsF–LaF$_3$.

The System NaF–KF–ScF$_3$. See Fig. 182. Source: E. P. Babaeva and G. A. Bukhalova, *Russ. J. Inorg. Chem.* (*English Transl.*) **10**:793 (1965). The solid-solution field of 3NaF · ScF$_3$–3KF · ScF$_3$ meets the NaF and KF crystallization fields at $E = 674°$, NaF–KF–ScF$_3$ (35–60–5%) or to K$_3$F$_3$–Na$_3$F$_3$–3NaF · ScF$_3$ (62.5–8.5–29%).

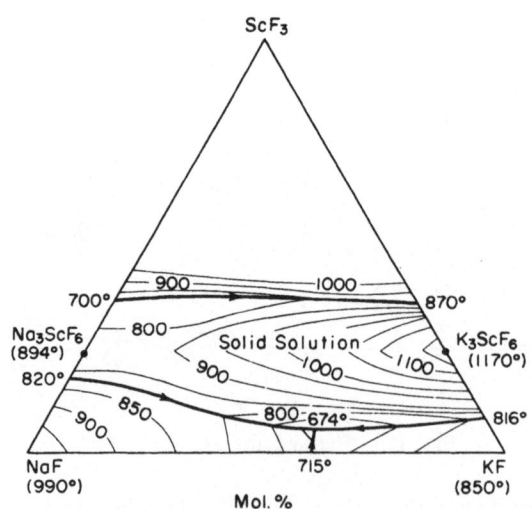

Fig. 182. The system NaF–KF–ScF$_3$.

The System NaF–KF–YF$_3$. See Fig. 183. Source: G. A. Bukhalova and E. P. Babaeva, *Russ. J. Inorg. Chem.* (*English Transl.*) **11**:350 (1966). $E = 554°$, NaF–KF–YF$_3$ (44–21–35%) $E = 648°$, NaF–KF–YF$_3$ (32–58–10%).

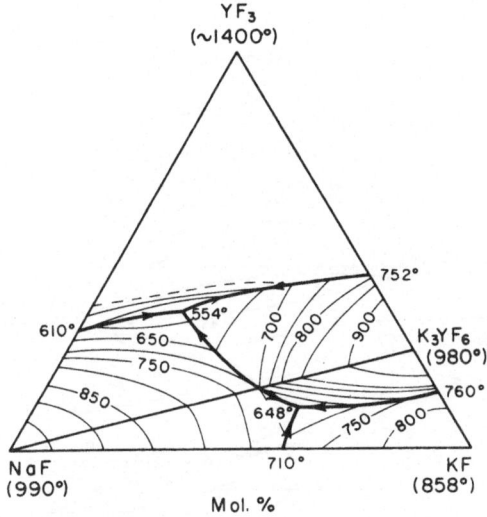

Fig. 183. The system NaF–KF–YF$_3$.

The System NaF–KF–LaF$_3$. See Fig. 184. Source: G. A. Bukhalova, E. P. Babaev, and T. M. Khliyan, *Russ. J. Inorg. Chem.* (*English Transl.*) **10**:1160 (1965). $E = 563°$, NaF–KF–LaF$_3$ (25–58–17%).

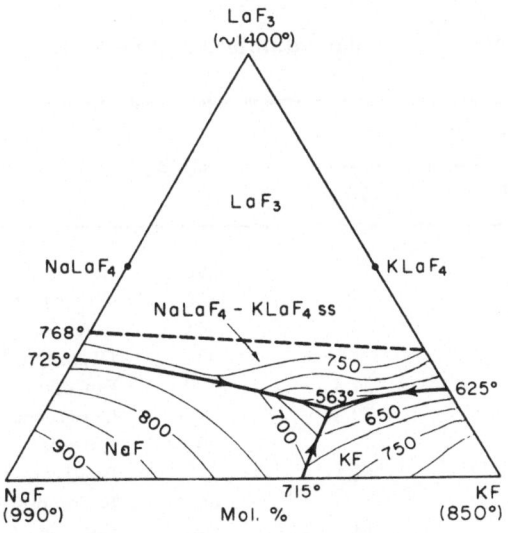

Fig. 184. The system NaF–KF–LaF$_3$.

The System LiF–NaF–ZrF₄. See Fig. 185. Source: R. E. Thoma, H. Insley, H. A. Friedman, and G. M. Hebert, *J. Chem. Eng. Data* **10**:219

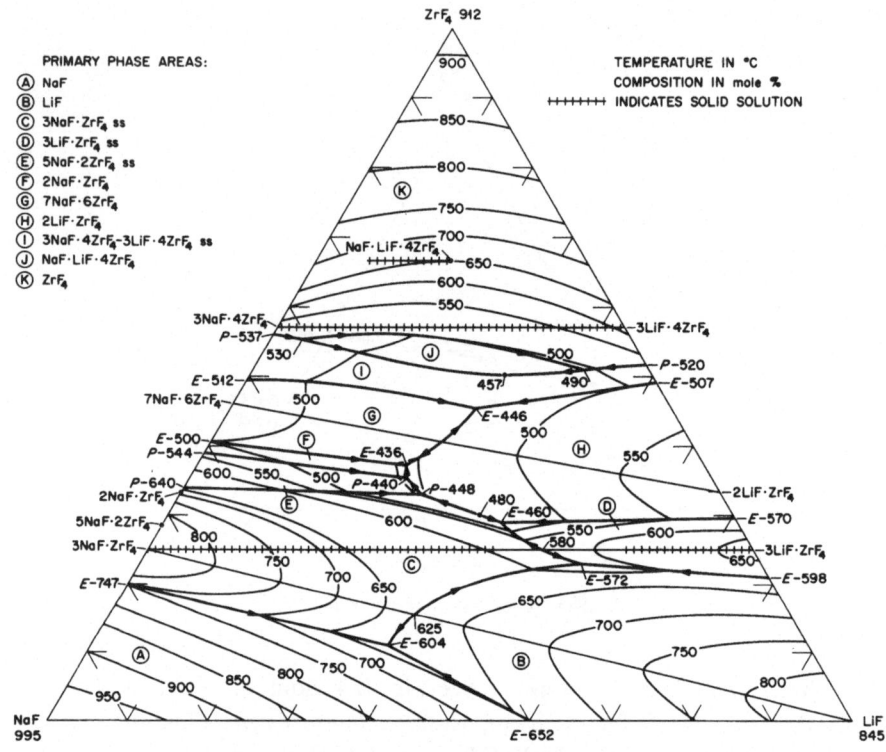

Fig. 185. The system LiF–NaF–ZrF₄.

(1965). Invariant and singular equilibrium points are given in the table.

Mole %			Temperature	Type of equilibrium
LiF	NaF	ZrF₄		
79		21	598	Eutectic
75		25	662	Congruent melting point
			475	Inversion
			470	Decomposition
70.5		29.5	570	Eutectic
66.7		33.3	596	Congruent melting point
51		49	507	Eutectic
48.5		51.5	520	Peritectic
			466	Decomposition
37	52	11	694	Eutectic

(Continued)

Mole %			Temperature	Type of equilibrium
LiF	NaF	ZrF$_4$		
38	46.5	15.5	625	Boundary curve maximum
55	22	23	572	Eutectic
49	26	25	580	Boundary curve maximum
42	29	29	460	Eutectic
38.5	31.5	30	480	Boundary curve maximum
30	37	33	448	Peritectic
27	37.5	35.5	440	Peritectic
26	37	37	436	Eutectic
29	30.5	40.5	470	Boundary curve maximum
30.5	24	45.5	446	Eutectic
32	18	50	457	Boundary curve minimum
41	8	51	490	Peritectic
4	40.5	55.5	530	Peritectic

The System NaF–KF–ZrF$_4$. See Fig. 186. Source: R. E. Thoma, C. J. Barton, H. Insley, H. A. Friedman, and W. R. Grimes, ORNL,

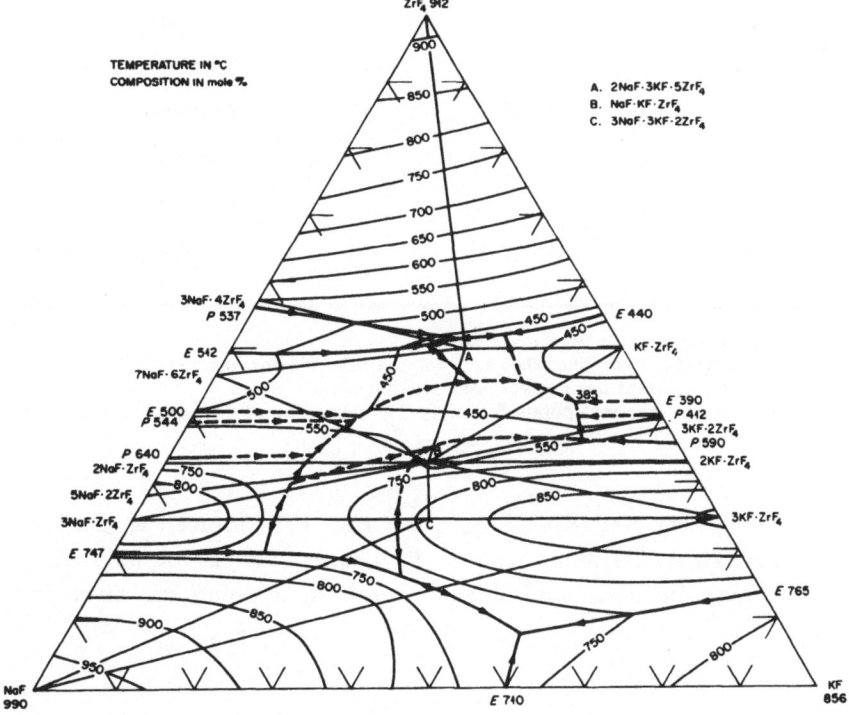

Fig. 186. The system NaF–KF–ZrF$_4$.

unpublished work (1951–1955). Invariant and singular equilibrium points are given in the table.

Mole %[b]			Temperature	Type of equilibrium
NaF	KF	ZrF_4		
34	58	8	695	Eutectic
45	38	17	720	Eutectic
61	19	20	710	Eutectic
52	17	31		Peritectic
48	18	34		Peritectic
33	32	35		Peritectic
12	51	37		Peritectic
40	21	39		Peritectic
39	21	40		Peritectic
11	49	40		Peritectic
10	48	42	385	Eutectic
15	40	45		Eutectic
22	33	45		Peritectic
25	25	50		Eutectic
21	29	51		Eutectic
20	30	50	432	Congruent melting point

[a] Three solid phases have been observed routinely in the composition region 30–50 mole % ZrF_4 which have not been identified as to composition. Whether these exhibit primary phases at the liquidus surface has not yet been determined.

[b] Compositions of invariant points shown by the intersection of dotted boundary curves are approximate.

The System NaF–RbF–ZrF$_4$. See Fig. 187. Source: R. E. Thoma, H. Insley, H. A. Friedman, and W. R. Grimes, ORNL, unpublished work (1951–1956). Invariant and singular equilibrium points are given in the table.

Mole %			Temperature	Type of equilibrium
NaF	RbF	ZrF_4		
23	72	5	643	Eutectic
50	27	23	720	Eutectic
37	32	31	605	Eutectic
39	26	35	545	Peritectic
36.3	24.2	39.5	427	Peritectic
34.5	24	41.5	424	Peritectic
34	23.5	42.5	422	Peritectic
33	23.5	43.5	420	Eutectic
28.5	21.5	50	443	Eutectic
28	21.5	50.5	446	Peritectic
23.5	39.5	37	470	Peritectic
21	40	39	438	Peritectic
8	50	42	400	Eutectic
8.5	47	44.5	395	Eutectic
6.2	45.8	48	380	Eutectic
5	42	53	398	Eutectic
6.5	39	54.5	423	Peritectic
33.3	33.3	33.3	642	Congruent melting point
25	25	50	462	Congruent melting point

Minimum temperatures on Alkemede lines

Mole %			Temperature	Alkemede line
NaF	RbF	ZrF_4		
24.75	24.75	51.5	455	$NaF \cdot RbF \cdot 2ZrF_4$–$ZrF_4$
6	44	50	402	$NaF \cdot RbF \cdot 2ZrF_4$–$RbF \cdot ZrF_4$
7.5	46.5	46	405	$NaF \cdot RbF \cdot 2ZrF_4$–$5RbF \cdot 4ZrF_4$
8.5	48.5	43	405	$3NaF \cdot RbF \cdot 4ZrF_4$–$5RbF \cdot 4ZrF_4$
28	28	44	435	$3NaF \cdot RbF \cdot 4ZrF_4$–$NaF \cdot RbF \cdot 2ZrF_4$
22	40	38	442	$3NaF \cdot 3RbF \cdot 4ZrF_4$–$2RbF \cdot ZrF_4$
29	41.5	49.5	450	$NaF \cdot RbF \cdot 2ZrF_4$–$7NaF \cdot 6ZrF_4$
37	24.5	38.5	610	$NaF \cdot RbF \cdot ZrF_4$–$3NaF \cdot ZrF_4$
47	28	25	732	$3NaF \cdot ZrF_4$–$3RbF \cdot ZrF_4$
36	48	16	777	NaF–$3RbF \cdot ZrF_4$
30	36.7	33.3	598	$NaF \cdot RbF \cdot ZrF_4$–$3RbF \cdot ZrF_4$

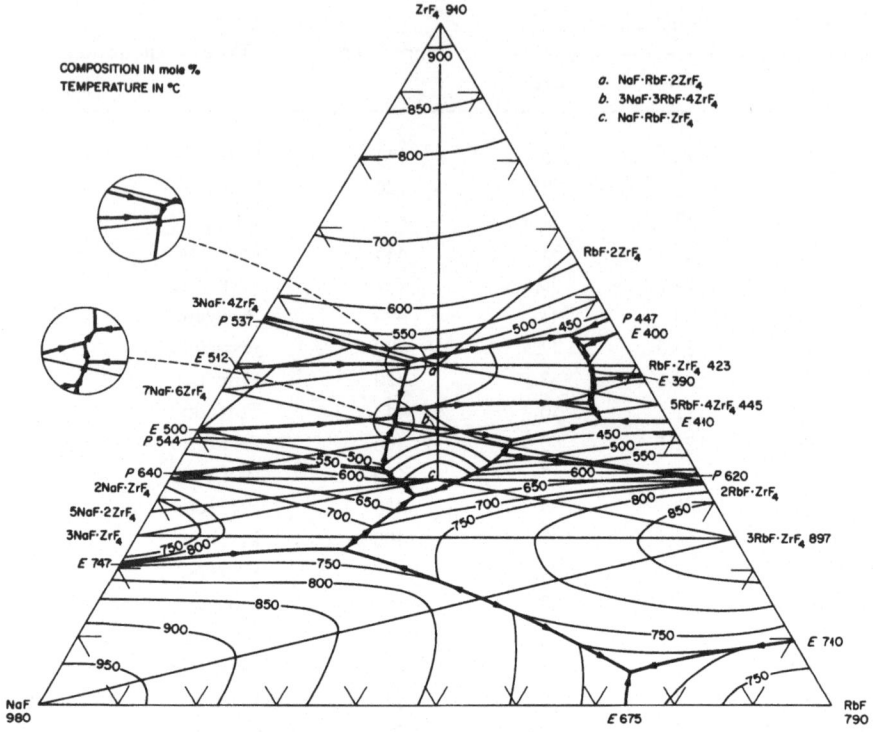

Fig. 187. The system NaF–RbF–ZrF$_4$.

The System LiF–NaF–ThF$_4$. See Fig. 188. Source: R. E. Thoma, *J. Inorg. Nucl. Chem.* **34**:2747 (1972). Invariant and singular equilibrium points are given in the table.

Mole %			Temperature	Type of equilibrium
LiF	NaF	ThF$_4$		
30	56.5	13.5	505	Eutectic
18	65	17	558	Peritectic
46.5	35.5	18	536	Max. on boundary curve
7	72.5	20.5	604	Peritectic
2	77.5	20.5	610	Peritectic
43.5	32.5	24	509	Eutectic
53	17	30	531	Max. on boundary curve
54.5	13.5	32	525	Eutectic
54.5	12.5	33	540	Peritectic
6	56.5	37.5	683	Peritectic
50	12	38	585	Peritectic

(*Continued*)

	Mole %		Temperature	Type of equilibrium
LiF	NaF	ThF$_4$		
48	12	40	643	Peritectic
44	12	44	725	Peritectic
16	27	57	825	Peritectic

Structure data: 7(Na, Li)F · 6ThF$_4$: hexagonal, *P*3Cl, $a = 9.9056$, $c = 13.282$ [G. D. Brunton and D. R. Sears, *Acta Cryst.* **B25**:2519 (1969)]. This phase was regarded as a ternary compound of variable composition when the structure was determined. In the final examination of phase relations in the LiF–NaF–ThF$_4$ ternary system, it was concluded that the structure was a new crystal modification of 7NaF · 6ThF$_4$ that could accommodate extensive solid solution of Li$^+$.

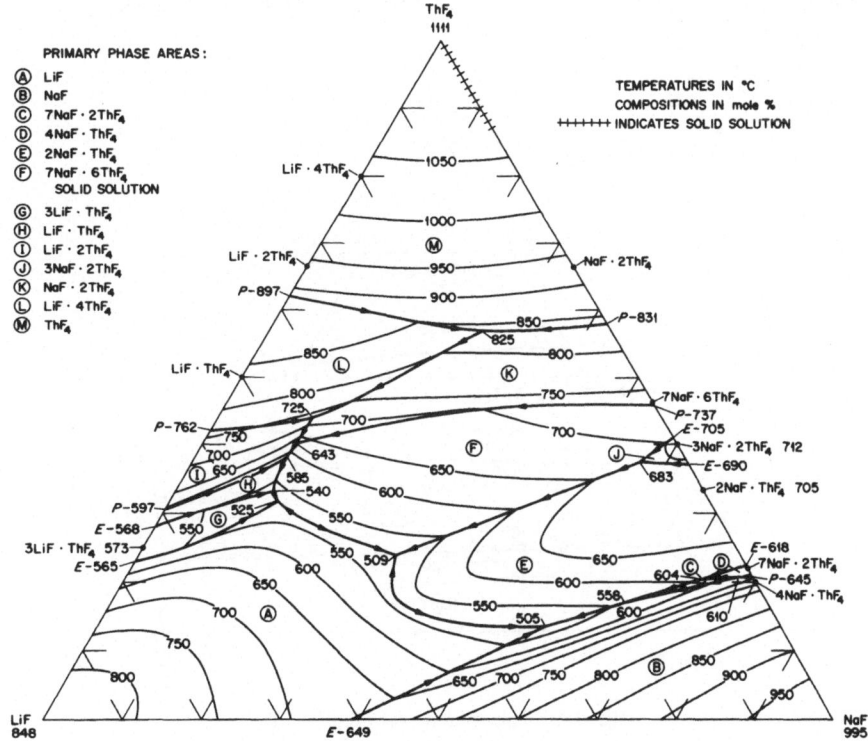

Fig. 188. The system LiF–NaF–ThF$_4$.

The System NaF–KF–ThF$_4$. Source: R. E. Thoma, H. A. Friedman, and H. Inslev. ORNL, unpublished work (1955–1958). Investigations with the system NaF–KF–ThF$_4$ were not carried out to the extent that a phase diagram could be constructed. Within the system, a single ternary compound NaF · KF · ThF$_4$ (isostructural with NaF · KF · UF$_4$) was found to exist. It forms extensive solid solutions with 2NaF · ThF$_4$. The quasi-binary system, NaF–3KF · ThF$_4$, is a limiting system of the subsystem NaF–KF–3KF · ThF$_4$, in which a single eutectic occurs near the composition NaF–KF–ThF$_4$ (36–57–7 mole %). Within the subsystem NaF–NaF · KF · ThF$_4$–2NaF · ThF$_4$, a temperature minimum occurs near the composition NaF–KF–ThF$_4$ (11–67–22 mole %), and at 535°.

Structural data: NaF · KF · ThF$_4$: hexagonal, P-3, a = 3.63073, c = 7.8907 [G. D. Brunton et al., ORNL-3761]. Crystals of the ternary compound NaF · KF · ThF$_4$ are hexagonal, P-3, a = 6.32, c = 8.18 [Brunton, ibid.].

The System KF–RbF–ThF$_4$. See Fig. 189. Source: E. P. Dergunov and A. G. Bergman, *Dokl. Akad. Nauk SSSR* **60**:394 (1948). No ternary invariant equilibria are found in the KF–RbF–ThF$_4$ system. The compounds indicated as KThF$_5$ and RbThF$_5$ are probably 7KF · 6ThF$_4$ and 7RbF · 6ThF$_4$.

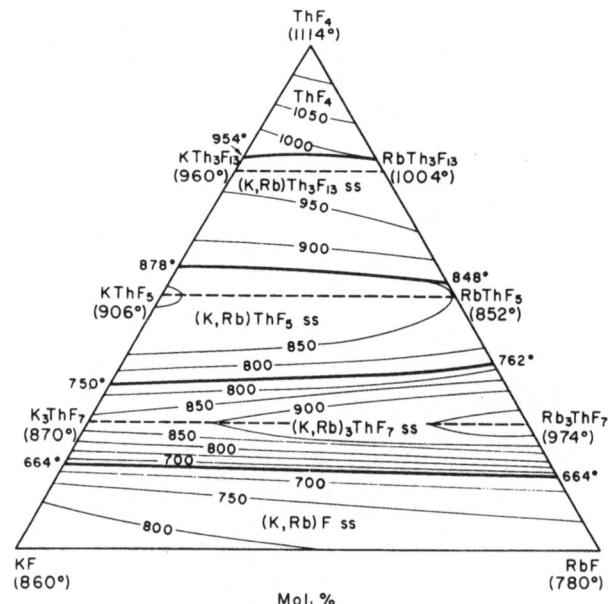

Fig. 189. The system KF–RbF–ThF$_4$.

The System LiF–NaF–UF$_4$. See Fig. 190. Source: R. E. Thoma *et al., J. Am. Ceram. Soc.* **42**:21 (1959). Invariant and singular equilibrium points are given in the table.

Mole %			Temperature	Type of equilibrium	Coexisting solid phases
NaF	LiF	UF$_4$			
60.0	21.0	19.0	480	Eutectic	2NaF · UF$_4$, NaF, LiF
65.0	13.0	22.0	497	Peritectic	β-3NaF · UF$_4$, 2NaF · UF$_4$, NaF
7.0	65.5	27.5	470	Peritectic	7LiF · 6UF$_4$ ss,[a] 4LiF · UF$_4$, LiF
35.0	37.0	28.0	480	Eutectic	2NaF · UF$_4$, 7NaF · 6UF$_4$ ss, LiF
57.0	13.0	30.0	630	Peritectic	2NaF · UF$_4$, 5NaF · 3UF$_4$, 7NaF · 6UF$_4$ ss
24.3	43.5	32.2	445	Eutectic	7NaF · 6UF$_4$ ss, LiF, 7LiF · 6UF$_4$ ss
24.5	29.0	46.5	602	Eutectic	NaF · 2UF$_4$, 7LiF · 6UF$_4$ ss, 7NaF · 6UF$_4$ ss
24.3	28.7	47.0	605	Peritectic	NaF · 2UF$_4$, LiF · 4UF$_4$, 7LiF · 6UF$_4$ ss
23.5	28.0	48.5	640	Peritectic	UF$_4$, NaF · 2UF$_4$, LiF · 4UF$_4$
37.5	10.5	52.0	660	Peritectic	UF$_4$, NaF · 2UF$_4$, 7NaF · 6UF$_4$ ss

[a] The compounds 7LiF · 6UF$_4$ and 7NaF · 6UF$_4$ as they occur in the ternary system are members of the solid solution 7LiF · 6UF$_4$–7NaF · 6UF$_4$.

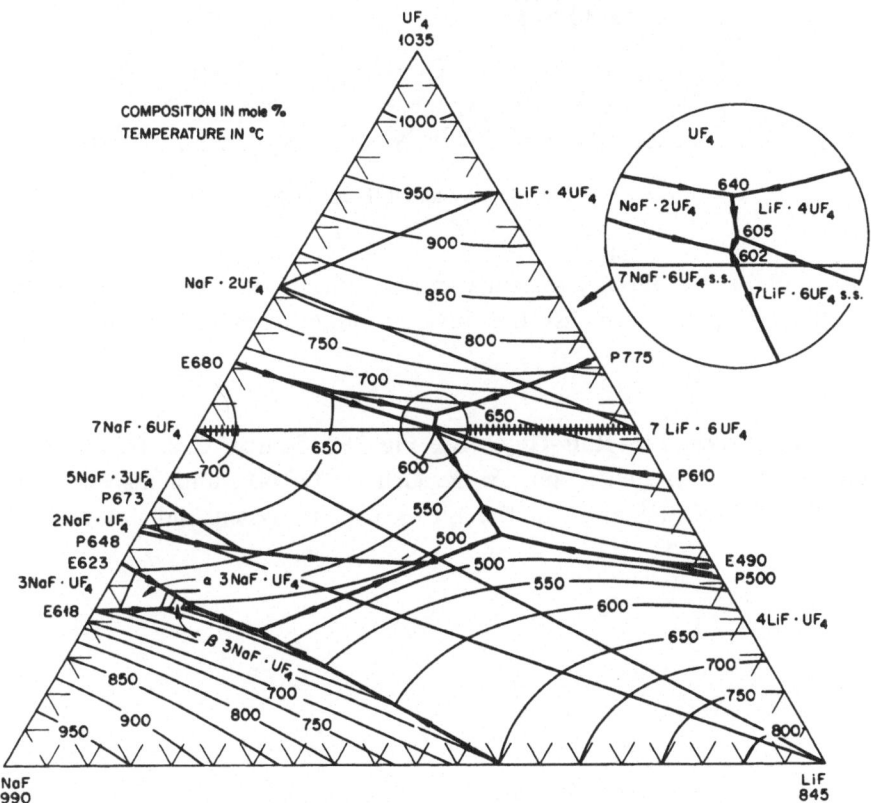

Fig. 190. The system LiF–NaF–UF$_4$. –|–|–|–|–|–|– indicates solid solution.

The System LiF–KF–UF₄. See Fig. 191. Source: C. J. Barton, J. P. Blakely, L. M. Bratcher, and W. R. Grimes, ORNL, unpublished work

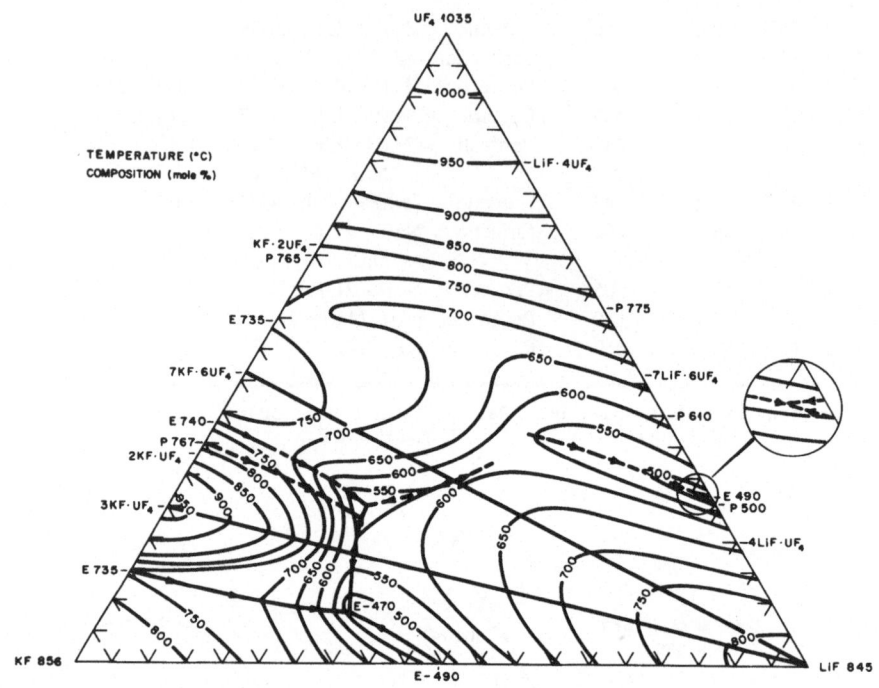

Fig. 191. The system LiF–KF–UF₄.

(1950–1951). The ternary system was not defined with sufficient precision in the work cited to list the invariant-point compositions and temperatures.

The System LiF–RbF–UF₄. See Fig. 192. Source: C. J. Barton, J. P. Blakely, L. M. Bratcher, and W. R. Grimes, ORNL, unpublished work, (1950–1951). Examination of the LiF–RbF–UF₄ system did not proceed to the point that invariant equilibria were determined, except for the eutectic at 450°, LiF–RbF–UF$_4$ (39–58–3%).

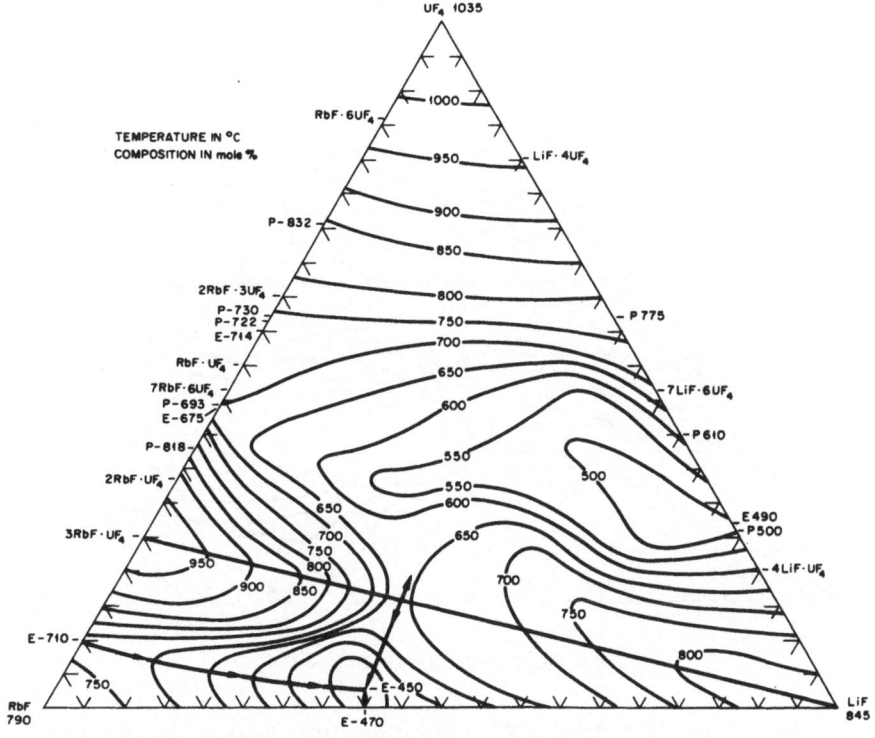

Fig. 192. The system LiF–RbF–UF$_4$.

The System NaF–KF–UF$_4$. See Fig. 193. Source: R. E. Thoma, C. J. Barton, J. P. Blakely, R. E. Moore, H. Insley, and H. A. Friedman, ORNL, unpublished work (1950–1958). Investigation of the phase diagram did not proceed to the extent that invariant equilibrium points were established accurately.

Structural data: NaF · KF · ThF$_4$: hexagonal, P-3, $a = 6.24$, $c = 7.80$ [G. D. Brunton *et al.*, ORNL-3761].

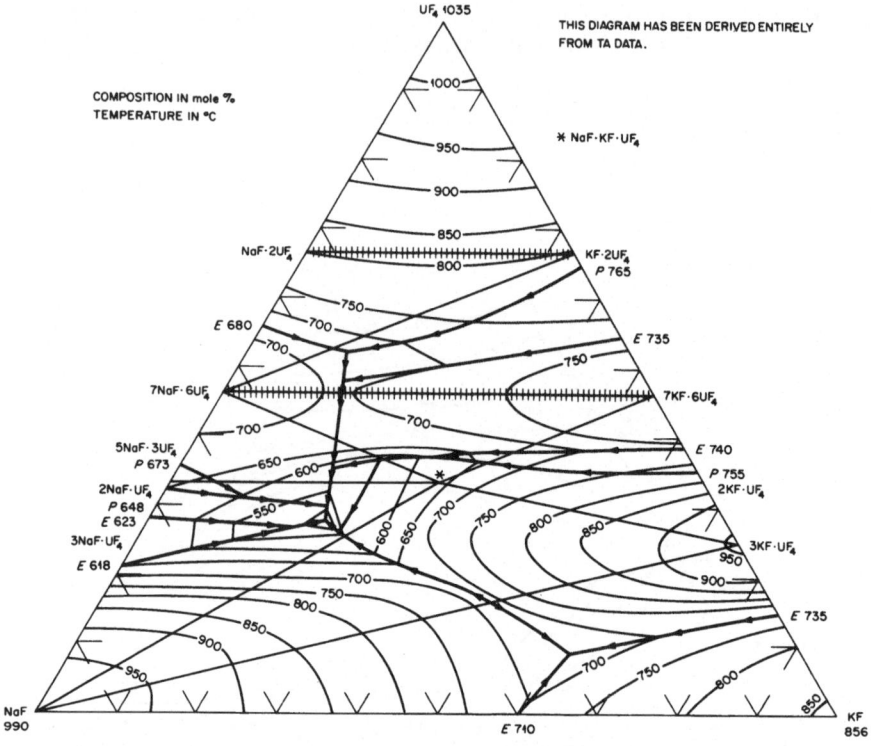

Fig. 193. The system NaF–KF–UF$_4$.

The System NaF–RbF–UF$_4$. See Fig. 194. Source: R. E. Thoma, H. Insley, H. A. Friedman, and W. R. Grimes, ORNL, unpublished work (1955–1956). Equilibrium relationships of the ternary compound along several composition–temperature sections were given in ORNL-2548, p. 105–107. Invariant and singular equilibrium points are given in the table.

Mole %			Temperature	Type of equilibrium
NaF	RbF	UF$_4$		
18	73	9	670	Eutectic
46	33	21	470	Eutectic
44	33	23	500	Peritectic
45	30	25	485	Peritectic
52	23	25	500	Decomposition
41	26	33	555	Peritectic
57	8	35	630	Decomposition

(Continued)

Mole %			Temperature	Type of equilibrium
NaF	RbF	UF$_4$		
33	30	37	535	Eutectic
32	29	39	540	Peritectic
26	27	47	620	Eutectic
25	25	50	630	Eutectic
27	22	51	633	Peritectic
33	14	53	655	Peritectic
42	3	55	678	Peritectic

Structural data: NaF · RbF · UF$_4$: hexagonal, P-3, $a = 6.29$, $c = 8.13$ [G. D. Brunton *et al.*, ORNL-3761 (1965)].

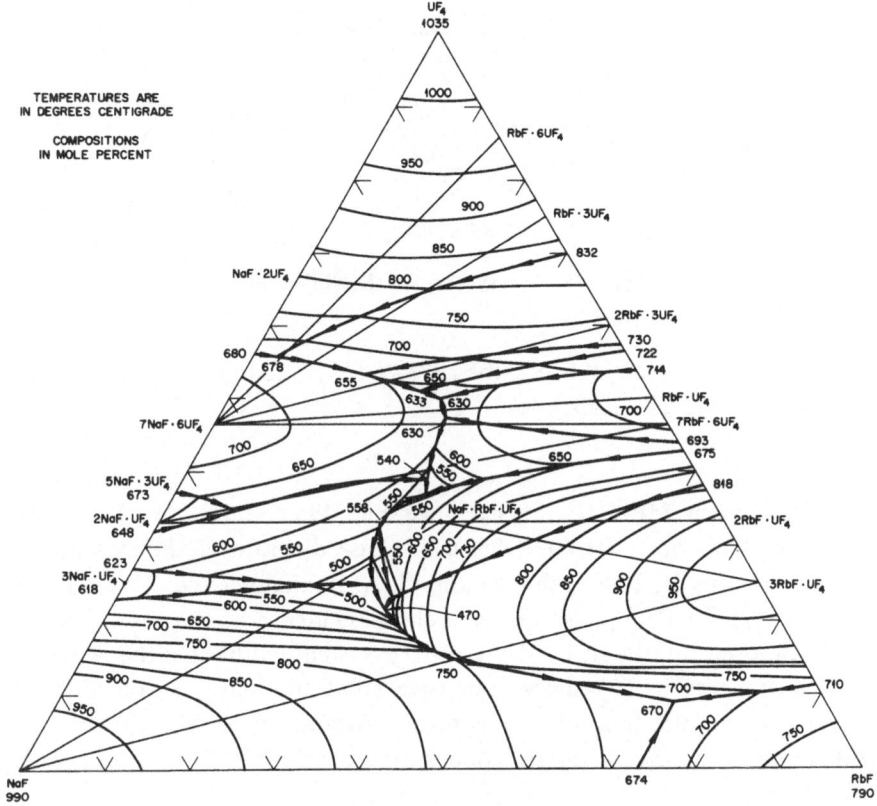

Fig. 194. The system NaF–RbF–UF$_4$.

The System $KF-K_2NbF_7-NaF$. See Fig. 195. Source: L. A. Kamenskaya, V. I. Konstantinov, and A. M. Matveev, *Russ. J. Inorg. Chem. (English Transl.)* **17**:1343 (1972). $E = 655°$, $NaF-KF-K_2NbF_7$ (25.5–54.5–

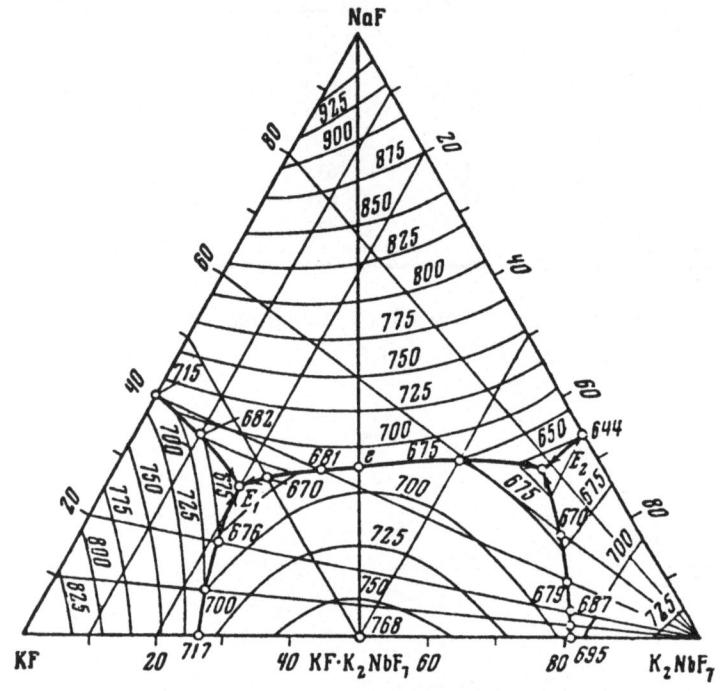

Fig. 195. The system $KF-K_2NbF_7-NaF$.

20%); $E = 636°$, $NaF-KF-K_2NbF_7$ (28–9–63%); saddle point, $E = 688°$, $NaF-K_2NbF_7$ (27.5–72.5%).

The System $LiF-CaF_2-MgF_2$. See Fig. 196. Source: W. E. Roake, *J. Electrochem. Soc.* **104**:409 (1957). A phase diagram of the system was also reported by V. T. Berezhnaya and G. A. Bukhalova, *Zh. Neorg. Khim.* **4**:903 (1959). Neither phase diagram shows evidence of the occurrence of solid solutions in the $LiF-MgF_2$ binary system, although the work of Counts *et al.* (Fig. 17) shows extensive solid solubility. Roake did not report the composition of the eutectic. According to Berezhnaya and Bukhalova, the eutectic in the system occurs at 676°, $Li_2F_2-MgF_2-CaF_2$ (47–35–18%).

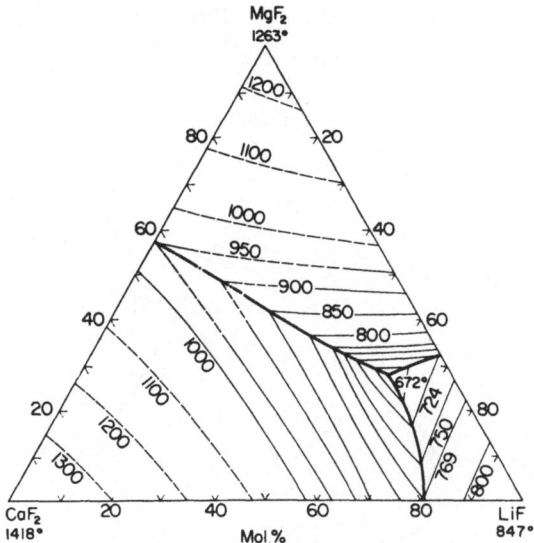

Fig. 196. The system LiF–CaF₂–MgF₂.

The System LiF–MgF₂–SrF₂. See Fig. 197. Source: Z. A. Mateiko and G. A. Bukhalova, *Russ. J. Inorg. Chem.* (*English Transl.*) 7:84 (1962). $E = 646°$, $Li_2F_2–MgF_2–SrF_2$ (36–39–25%).

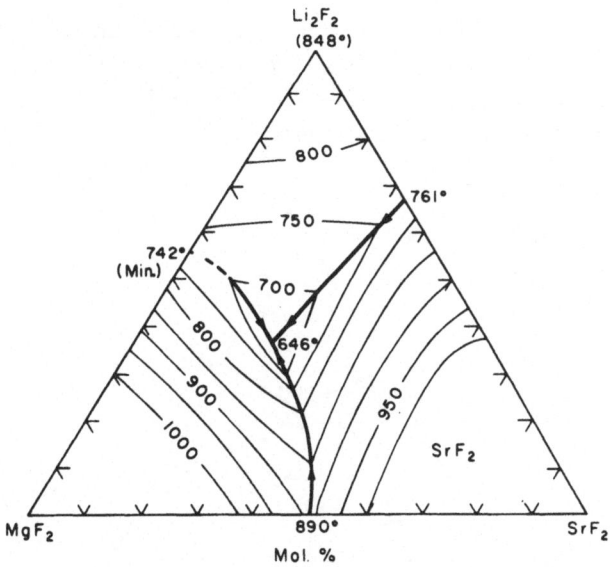

Fig. 197. The system LiF–MgF₂–SrF₂.

The System LiF–MgF₂–BaF₂. See Fig. 198. Source: G. A. Bukhalova and V. T. Berezhnaya, *Russ. J. Inorg. Chem.* (*English Transl.*) 4:51 (1959).

$E = 748°$, 17.0 Li_2F_2–29.0 MgF_2–54.0 BaF_2; $P = 741°$, 25.0 Li_2F_2–46.5 MgF_2–28.5 BaF_2; $E = 654°$, Li_2F_2–MgF_2–BaF_2 (35–35–30%).

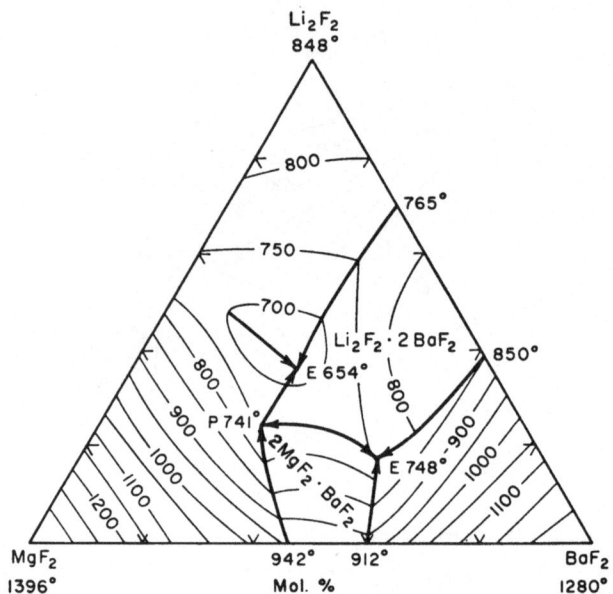

Fig. 198. The system $LiF–MgF_2–BaF_2$.

The System $LiF–CaF_2–SrF_2$. See Fig. 199. Source: Z. A. Mateiko and G. A. Bukhalova, *Russ. J. Inorg. Chem. (English Transl.)* **6**:882 (1961). $E = 740°$, Li_2F_2–CaF_2–SrF_2 (57–21–22%). The authors note that CaF_2 and SrF_2 form a continuous solid solution.

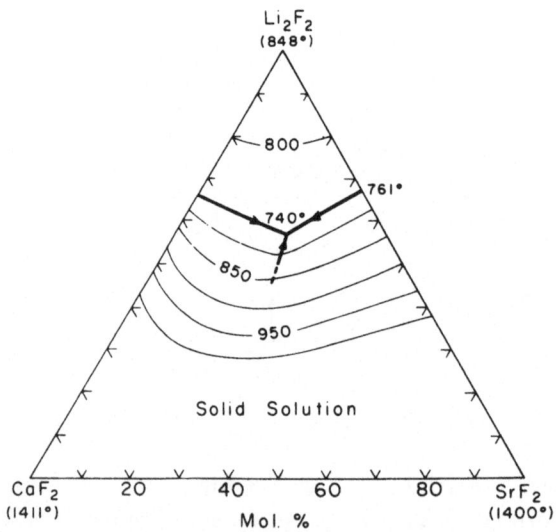

Fig. 199. The system $LiF–CaF_2–SrF_2$.

The System LiF–CaF₂–BaF₂. See Fig. 200. Source: G. A. Bukhalova and V. T. Berezhnaya, *Russ. J. Inorg. Chem.* (*English Transl.*) **2**:1409 (1957). $E = 710°$, Li_2F_2–CaF_2–BaF_2 (50.5–25.5–24%); $E = 758°$, Li_2F_2–CaF_2–BaF_2 (30.0–32.8–37.2%). The authors of the source reference note that CaF_2 and BaF_2 form continuous solid solutions with a temperature minimum at 1277° and at 50 mole % BaF_2.

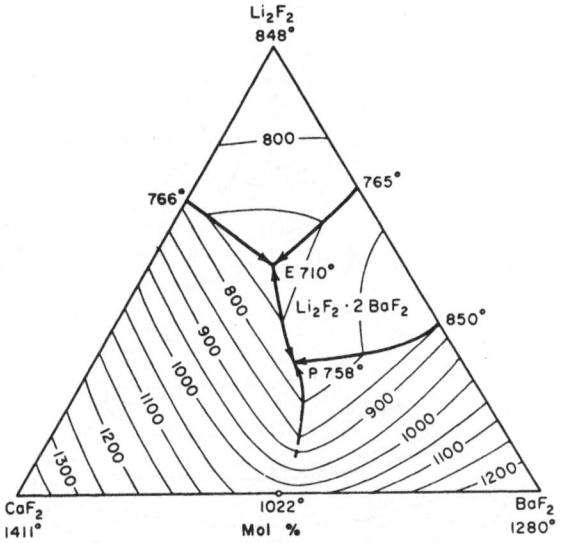

Fig. 200. The system LiF–CaF₂–BaF₂.

The System LiF–SrF₂–BaF₂. See Fig. 201. Source: Z. A. Mateiko and G. A. Bukhalova, *Russ. J. Inorg. Chem.* (*English Transl.*) **7**:85 (1962).

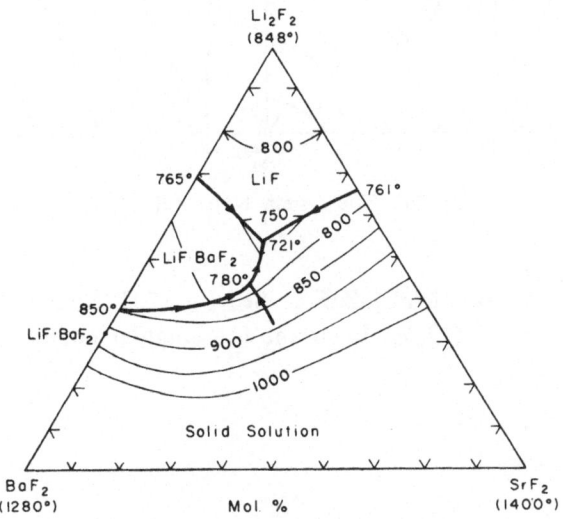

Fig. 201. The system LiF–SrF₂–BaF₂.

$E = 721°$, $Li_2F_2–SrF_2–BaF_2$ (53–21–26%); $P = 780°$, $Li_2F_2–SrF_2–BaF_2$ (43–24–33%).

The System $NaF–MgF_2–CaF_2$. See Fig. 202. Source: C. J. Barton, L. M. Bratcher, J. P. Blakely, and W. R. Grimes, ORNL, unpublished work (1956). $E = 745°$, $NaF–MgF_2–CaF_2$ (65–12–23%); $E = 905°$, $NaF–MgF_2–CaF_2$ (35–37–28%).

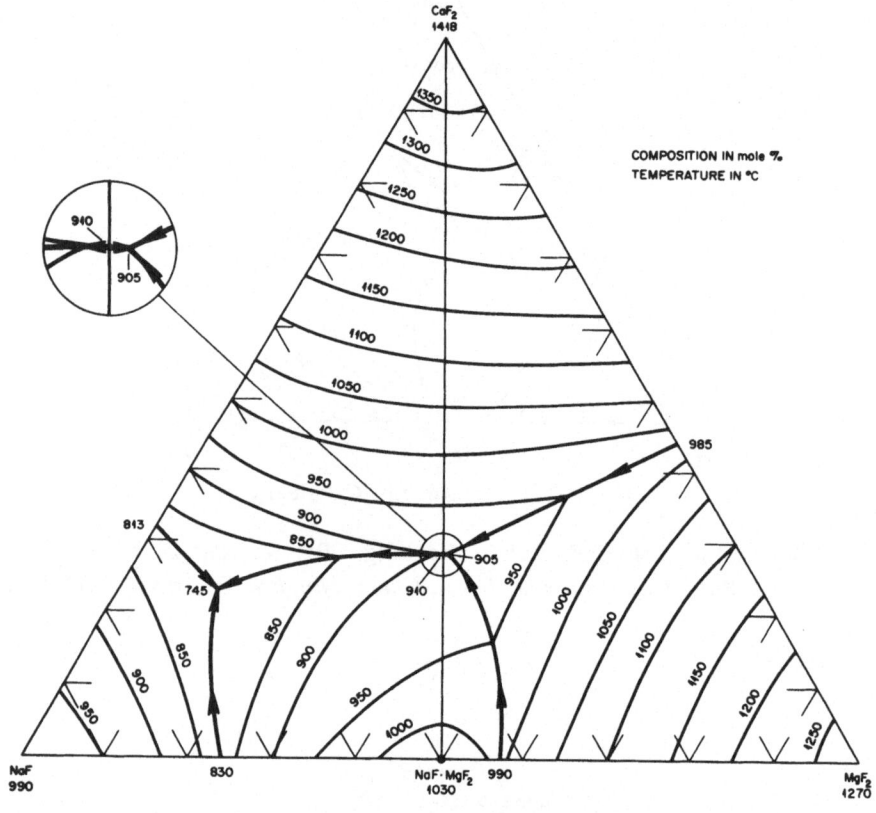

Fig. 202. The system $NaF–MgF_2–CaF_2$.

The System $NaF–MgF_2–BaF_2$. See Fig. 203. Source: G. Grube, *Z. Elektrochem.* **33**:483 (1927). Invariant and singular equilibrium points are given in the table.

	Mole %		Temperature	Type of invariant equilibrium
NaF	MgF_2	BaF_2		
32	18	50	750	Eutectic
8	21	71	800	Eutectic
8	36	56	850	Eutectic

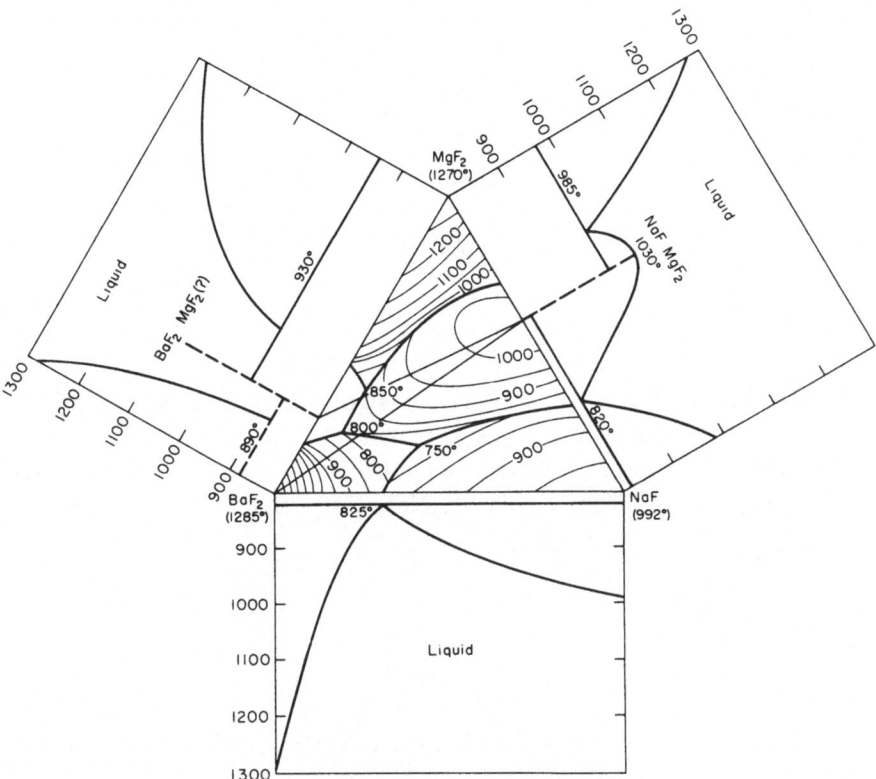

Fig. 203. The system $NaF–MgF_2–BaF_2$.

The System $NaF–CaF_2–SrF_2$. See Fig. 204. Source: G. A. Bukhalova, V. T. Berezhnaya, and Z. A. Mateiko, *Russ. J. Inorg. Chem.* (*English Transl.*) 7:1156 (1962). The eutectic invariant point separating the primary phase field of NaF from that of the continuous solid solutions formed by CaF_2 and SrF_2 occurs at the composition Na_2F_2–CaF_2–SrF_2 (62–12.5–25.5%) and at 699°.

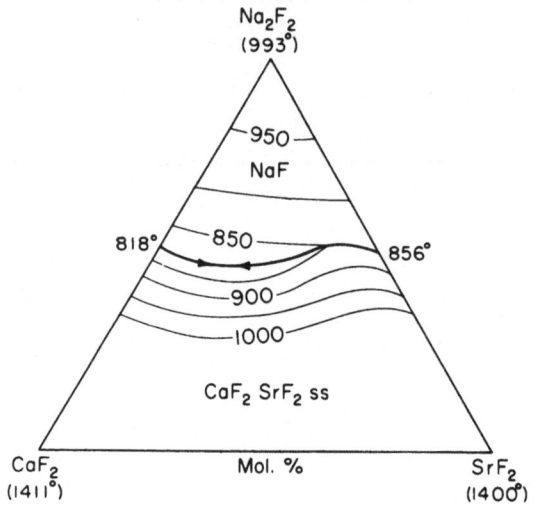

Fig. 204. The system NaF–CaF_2–SrF_2.

The System NaF–CaF_2–BaF_2. See Fig. 205. Source: G. A. Bukhalova, V. T. Berezhnaya, and A. G. Bergman, *Russ. J. Inorg. Chem.* (*English Transl.*) **6**:1197 (1961). Invariant equilibrium points: $E = 680°$, NaF–CaF_2–BaF_2 (53–14–33%); $P = 867°$, NaF–CaF_2–BaF_2 (18–35–47%).

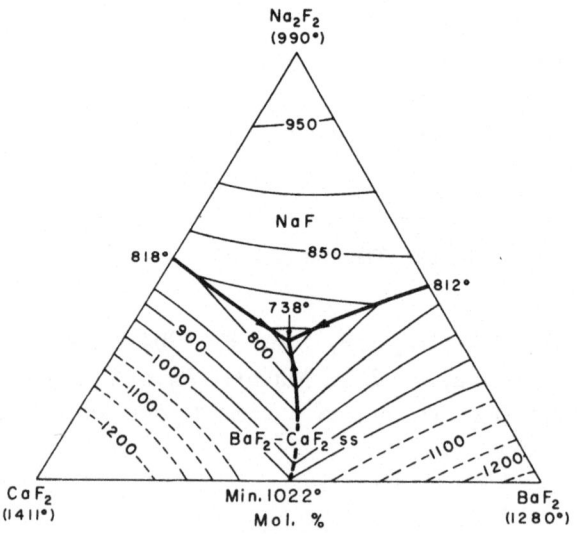

Fig. 205. The system NaF–CaF_2–BaF_2.

The System NaF–SrF$_2$–BaF$_2$. See Fig. 206. Source: G. A. Bukhalova, Z. A. Mateiko, and V. T. Berezhnaya, *Russ. J. Inorg. Chem.* (*English Transl.*) 7:856 (1962). Continuous solid solutions are formed by BaF$_2$ and SrF$_2$. The boundary separating the crystallization fields has a minimum at 804°, Na$_2$F$_2$–SrF$_2$–BaF$_2$ (44–26–30%).

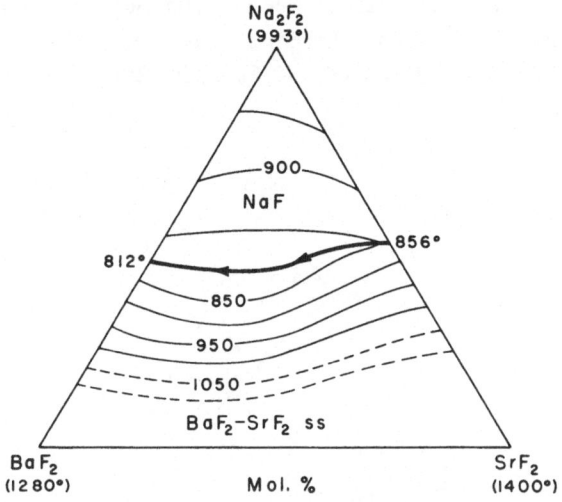

Fig. 206. The system NaF–SrF$_2$–BaF$_2$.

The System KF–CaF$_2$–SrF$_2$. See Fig. 207. Source: G. A. Bukhalova, V. T. Berezhnaya, and Z. A. Mateiko, *Russ. J. Inorg. Chem.* (*English*

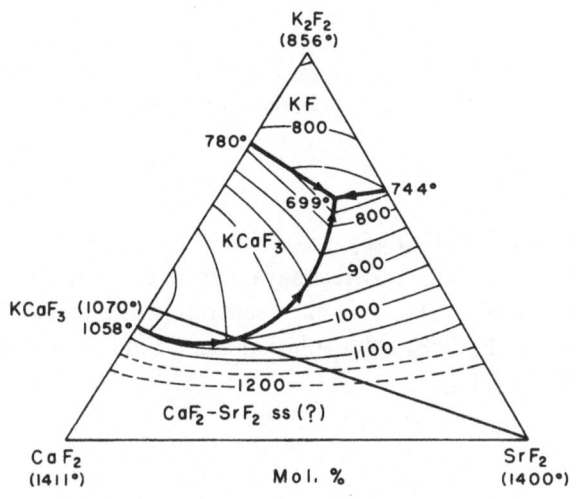

Fig. 207. The system KF–CaF$_2$–SrF$_2$.

Transl.) **7**:1156 (1962). The boundary separating the primary phase field of NaF from that of the continuous solid solutions formed by CaF_2 and SrF_2 has a minimum melting point of 804° at Na_2F_2–CaF_2–SrF_2 (48–36–16%). The eutectic at 699° occurs at K_2F_2–CaF_2–SrF_2 (62–12.5–25.5%).

The System KF–BaF$_2$–CaF$_2$. See Fig. 208. Source: G. A. Bukhalova, V. T. Berezhnaya, and A. G. Bergman, *Russ. J. Inorg. Chem.* (*English Transl.*) **6**:1197 (1961). $E = 680°$, K_2F_2–CaF_2–BaF_2 (53–14–33%); $P =$

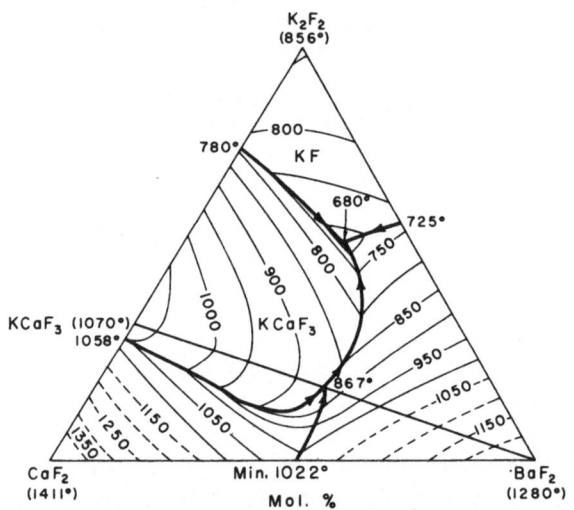

Fig. 208. The system KF–BaF$_2$–CaF$_2$.

867°, K_2F_2–CaF_2–BaF_2 (18–35–47%). It is noteworthy that the congruently melting compound $KCaF_3$ also becomes incongruently melting in the K, Ca, F, Cl system.

The System KF–SrF$_2$–BaF$_2$. See Fig. 209. Source: G. A. Bukhalova, Z. A. Mateiko, and V. T. Berezhnaya, *Russ. J. Inorg. Chem.* (*English Transl.*) **7**:1655 (1962). The boundary separating the crystallization fields of NaF and the BaF_2–SrF_2 solid solution has a minimum at 804°, Na_2F_2–SrF_2–BaF_2 (44–26–30%).

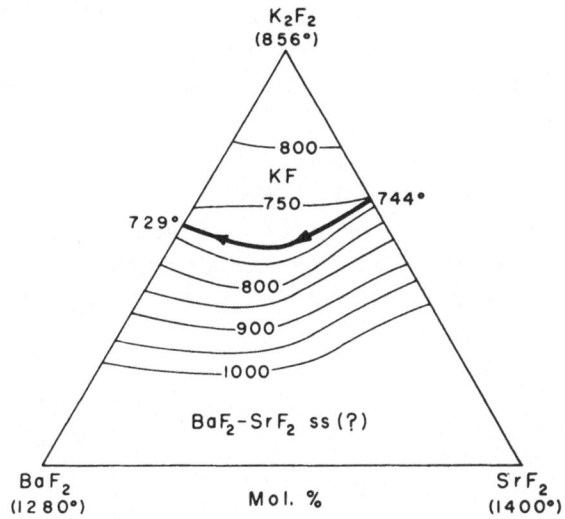

Fig. 209. The system KF–SrF₂–BaF₂.

The System LiF–BeF₂–AlF₃. See Fig. 210. Source: R. L. Boles and R. E. Thoma, Volatility Process Phase Studies—A Survey of Molten

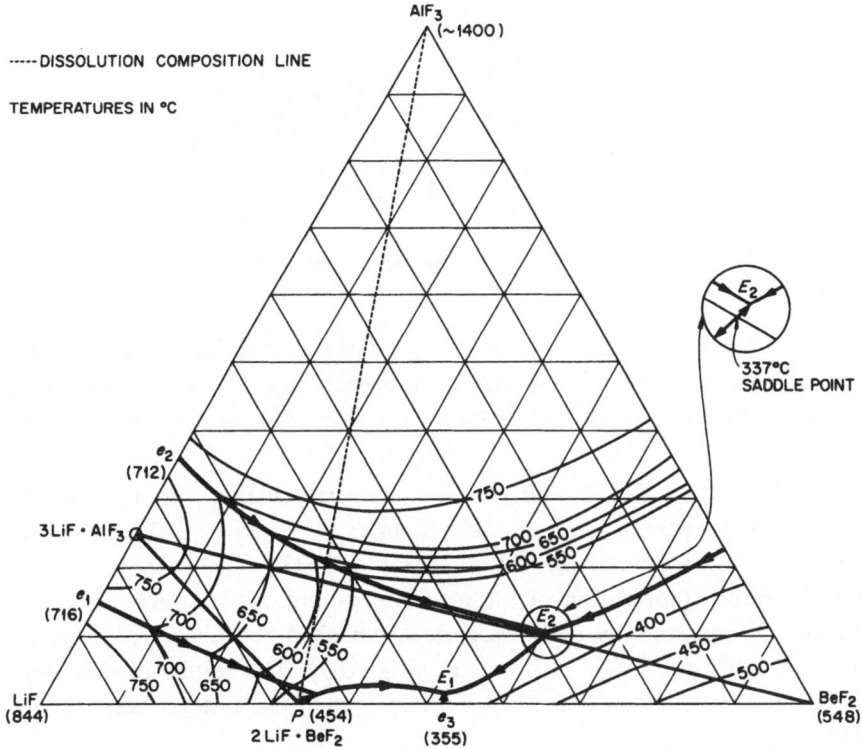

Fig. 210. The system LiF–BeF₂–AlF₃.

Fluoride Solvents Suitable for Dissolution of AlF₃, ORNL-TM-400, October 22, 1962. Invariant and singular equilibrium composition points were not listed by the authors. All the experimental data are included in the source reference.

The System NaF–MgF₂–AlF₃. See Fig. 211. Source: E. Vatslavik and A. I. Belyaev, *Russ. J. Inorg. Chem.* (*English Transl.*) **3**:324 (1958). No invariant equilibrium points were found to exist in the section of the phase diagram reported in the source reference. The section 3NaF · AlF₃–

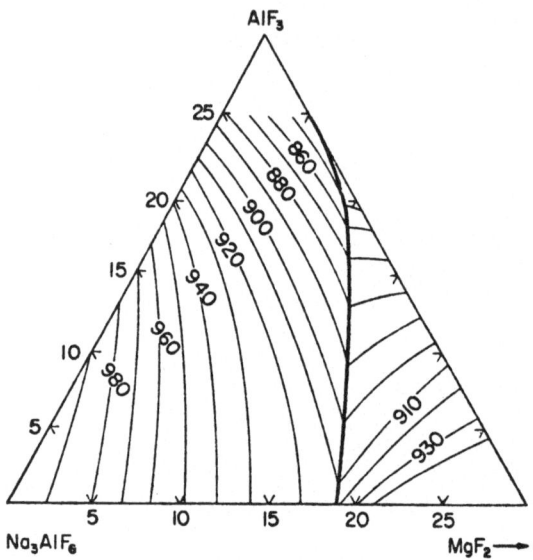

Fig. 211. The system NaF–MgF₂–AlF₃.

MgF₂, while not discussed in the source reference, is in fair agreement with previous work reported by A. M. Romanovskii and Ya. K. Berent, *Legkie Metal.* **6**:34 (1935). See Fig. 3449, *Phase Diagrams for Ceramists* (1969).

The System NaF–CaF₂–AlF₃. (See also the listing for this system immediately below.) See Fig. 212. Source: P. P. Fediotieff and W. P. Iljinskii, *Z. Anorg. Allgem. Chem.* **129**:106 (1923). The authors did not list the exact composition of the invariant points in the source reference. Later investigations by other workers established the phase relationships along the section 3NaF · AlF₃–CaF₂ (Fig. 207). The results of this work indicate that the NaF–CaF₂–AlF₃ ternary system has not yet been defined accurately.

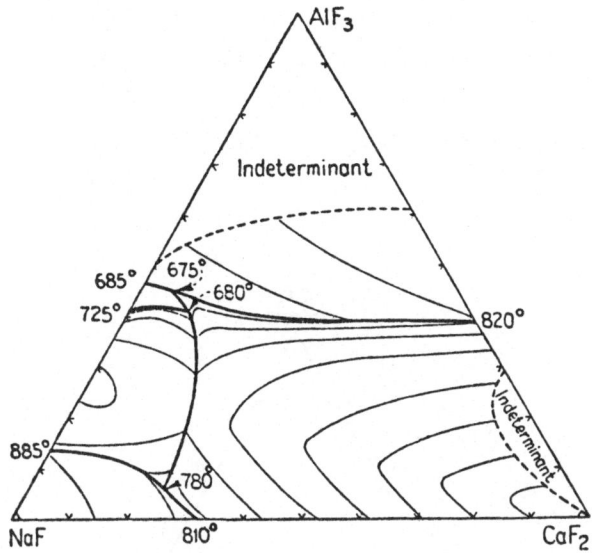

Fig. 212. The system NaF–CaF₂–AlF₃.

The System NaF–CaF₂–AlF₃. (See also the listing for this system immediately above.) See Fig. 213. Source: M. Verdan and R. Monnier,

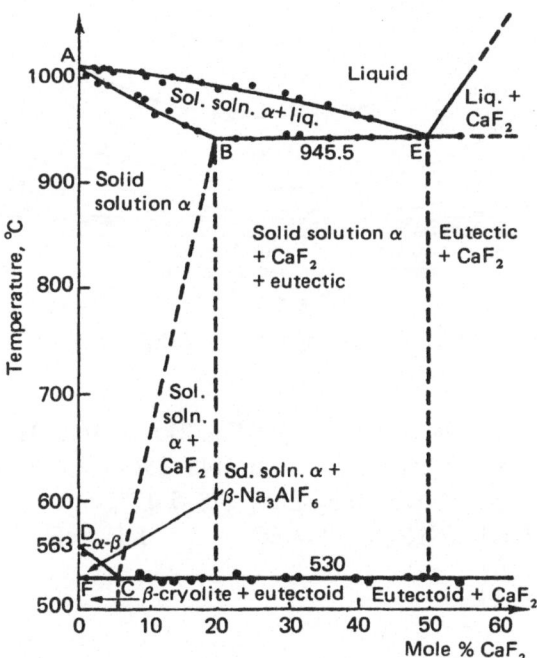

Fig. 213. The system NaF–CaF₂–AlF₃.

Rev. Intern. Hautes Temp. Refract. **9**:205 (1972). Invariant equilibrium compositions and temperatures are shown in the phase diagram.

The System LiF–BeF$_2$–CeF$_3$. See Fig. 214. Source: L. O. Gilpatrick, H. Insley, and C. J. Barton, MSR Program Semiannual Progress Report for Period Ending Aug. 30, 1970, ORNL-4622, p. 90.

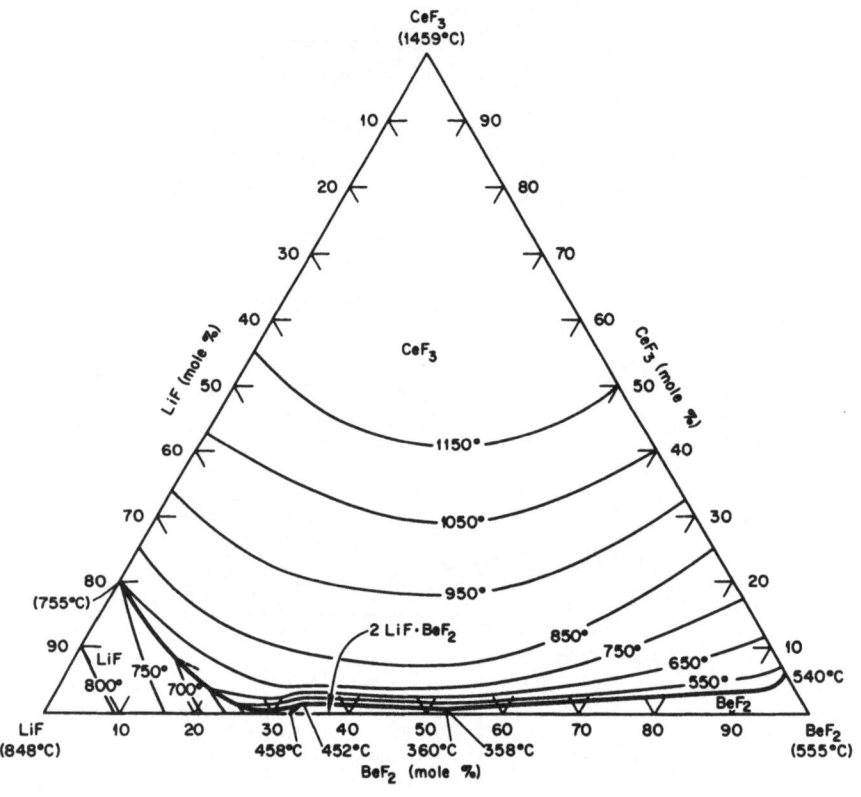

Fig. 214. The system LiF–BeF$_2$–CeF$_3$.

The System KF–BeF$_2$–LaF$_3$. See Fig. 215. Source: Hu Ch'ih-tsu and A. V. Novoselova, *Zh. Neorg. Khim.* **6**:2148 (1961). Invariant equilibrium points: $E = 610 \pm 5°$, KF–BeF$_2$–LaF$_3$ (78–5–17%); $P = 722 \pm 5°$, KF–BeF$_2$–LaF$_3$ (67.5–16.5–16%); $P = 667 \pm 5°$, KF–BeF$_2$–LaF$_3$ (71–12.5–16.5%).

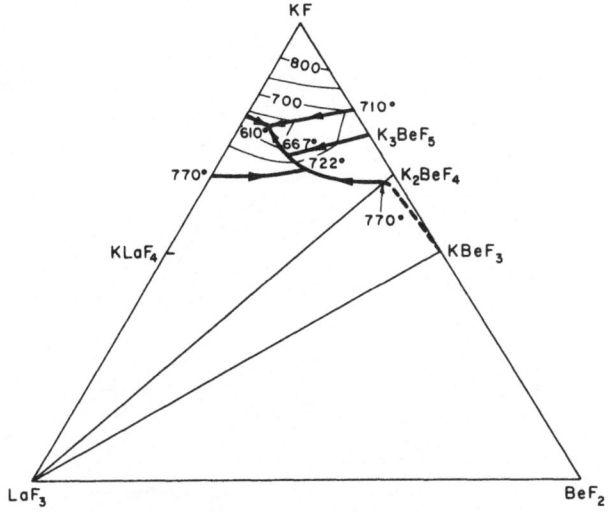

Fig. 215. The system KF–BeF$_2$–LaF$_3$.

The System LiF–BeF$_2$–ZrF$_4$. See Fig. 216. Source: R. E. Thoma, H. Insley, H. A. Friedman, and G. M. Hebert, *J. Nucl. Mater.* **27**:166 (1968). Invariant equilibrium points are given in the table.

Mole %			Temperature	Type of equilibrium
LiF	BeF$_2$	ZrF$_4$		
75	5	20	480	Peritectic
73	13	14	470	Peritectic
67	29.5	3.5	445	Peritectic
64.5	30.5	5	428	Peritectic
48	50	2	355	Eutectic
47.5	10	42.5	466	Peritectic
44	18	38	460	Eutectic
27	46	27	532	Eutectic
2	88	10	532	—
27	60	13	540	—
25	54	21	955	Critical solution point
73	13.5	13.5	475	Boundary curve max.
44	34	22	470	Quasi-binary eutectic
17	79	4	~545	Boundary curve max.

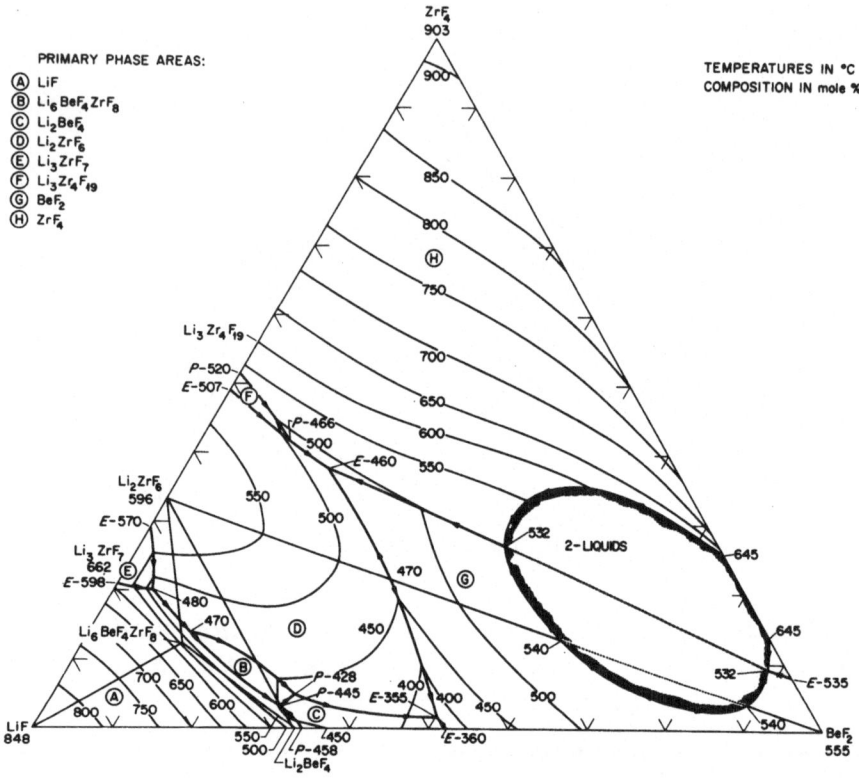

Fig. 216. The system LiF–BeF$_2$–ZrF$_4$.

Structural data: 6LiF · BeF$_2$ · ZrF$_4$: tetragonal, $I4_1/amd$, $a = 6.57$, $c = 18.62$ [D. R. Sears and J. H. Burns, *J. Chem. Phys.* **41**:3478 (1964)].

The System NaF–BeF$_2$–ZrF$_4$. See Fig. 217. Source: R. E. Thoma, H. Insley, T. N. McVay, H. A. Friedman, and C. F. Weaver, ORNL, unpublished work (1960–61). Invariant and singular equilibrium points are given in the table.

Mole %			Temperature	Type of equilibrium
NaF	BeF$_2$	ZrF$_4$		
70	26	4	535	Eutectic
57	40	3	335	Eutectic
55.5	39	5.5	340	Peritectic

(Continued)

	Mole %		Temperature	Type of equilibrium
NaF	BeF$_2$	ZrF$_4$		
50	42.5	7.5	345	Peritectic
44	52	4	350	Peritectic
37	25	37	458	Eutectic
28	36	36	480	Peritectic

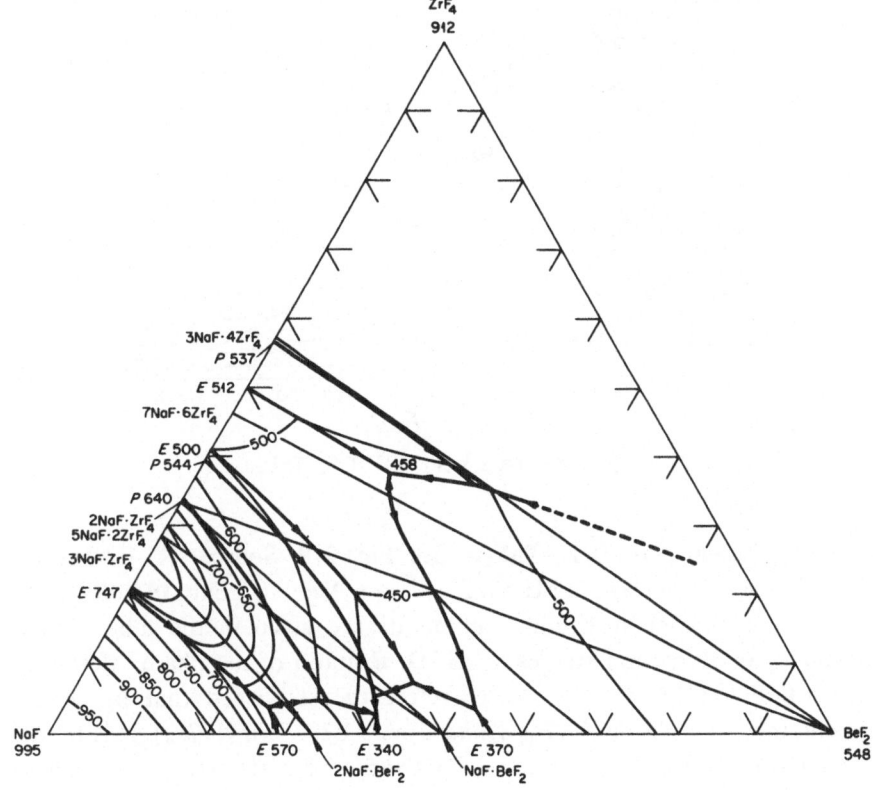

Fig. 217. The system NaF–BeF$_2$–ZrF$_4$.

The System LiF–BeF$_2$–ThF$_4$. See Fig. 218. Source: R. E. Thoma, H. Insley, H. A. Friedman, and C. F. Weaver, *J. Phys. Chem.* **64**:865 (1960). A review of the crystallization reactions occurring in the system was reported by R. E. Thoma and J. E. Ricci, Fractional Crystallization Reactions in the System LiF–BeF$_2$–ThF$_4$, ORNL-TM-2596 (1969).

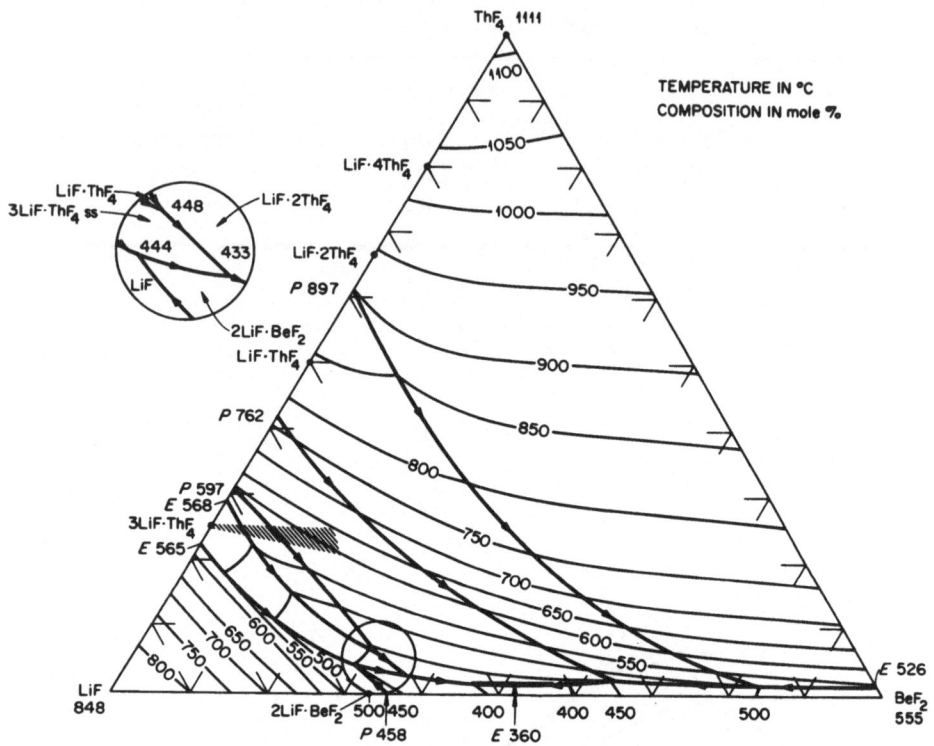

Fig. 218. The system LiF–BeF$_2$–ThF$_4$.

The System NaF–BeF$_2$–ThF$_4$. See Fig. 219. Source: R. E. Thoma, C. F. Weaver, H. Insley, and H. A. Friedman, *Nucl. Sci. Eng.* **19**:406 (1964). Invariant and singular equilibrium points are given in the table. Single crystal x-ray diffraction studies by G. D. Brunton of the phase inferred as NaF · BeF$_2$ · 3ThF$_4$ in the source reference established that its correct stoichiometry is Na$_3$BeTh$_{10}$F$_{45}$ [*Acta Cryst.* **B29**:2976 (1973)]. It is tetragonal, *P*4$_2$/*ncm*, *a* = 11.803(3), *c* = 23.420(5). The compound is isostructural with the phase observed in the NaF–BeF$_2$–UF$_4$ system (Fig. 73) and inferred to have the formula NaF · BeF$_2$ · 3UF$_4$. The results obtained by Brunton indicate that its correct stoichiometry is Na$_3$BeU$_{10}$F$_{45}$.

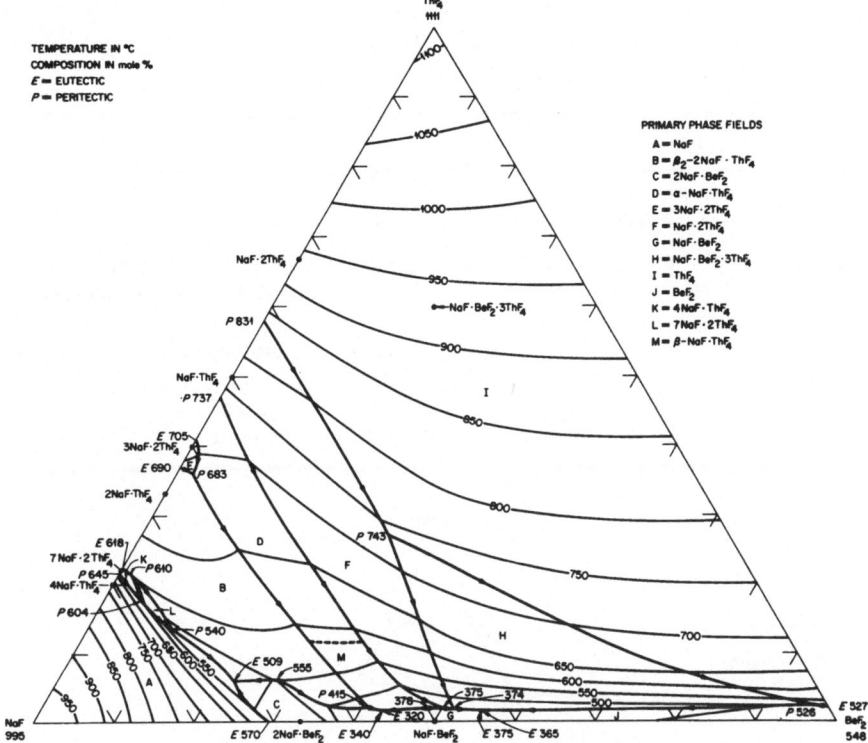

Fig. 219. The system NaF–BeF$_2$–ThF$_4$.

	Mole %		Temperature	Type of equilibrium
NaF	BeF$_2$	ThF$_4$		
78	4	18	604	Peritectic
77	2.5	20.5	610	Peritectic
76	10	14	540	Peritectic
72	22	6	509	Eutectic
66.7	27	6.3	555	Eutectic
62	2	36	683	Peritectic
57	41	2	415	Peritectic
55	43	2	320	Eutectic
50	48	2	378	Maximum on boundary curve
48	50	2	375	Peritectic
47	51	2	374	Peritectic
43	55	2	365	Eutectic
42	31	27	743	Peritectic
2	96	2	526	Peritectic

The System LiF–BeF$_2$–UF$_4$. See Fig. 220. Source: L. V. Jones *et al.*, *J. Am. Ceram. Soc.* **45**:79 (1962). Invariant and singular equilibrium points are given in the table.

Mole %			Temperature	Type of equilibrium
LiF	BeF$_2$	UF$_4$		
72	6	22	480	Peritectic (decomposition of 4LiF · UF$_4$ in the ternary system)
69	23	8	426	Eutectic
48	51.5	0.5	350	Eutectic
45.5	54	0.5	381	Peritectic
29.5	70	0.5	483	Peritectic

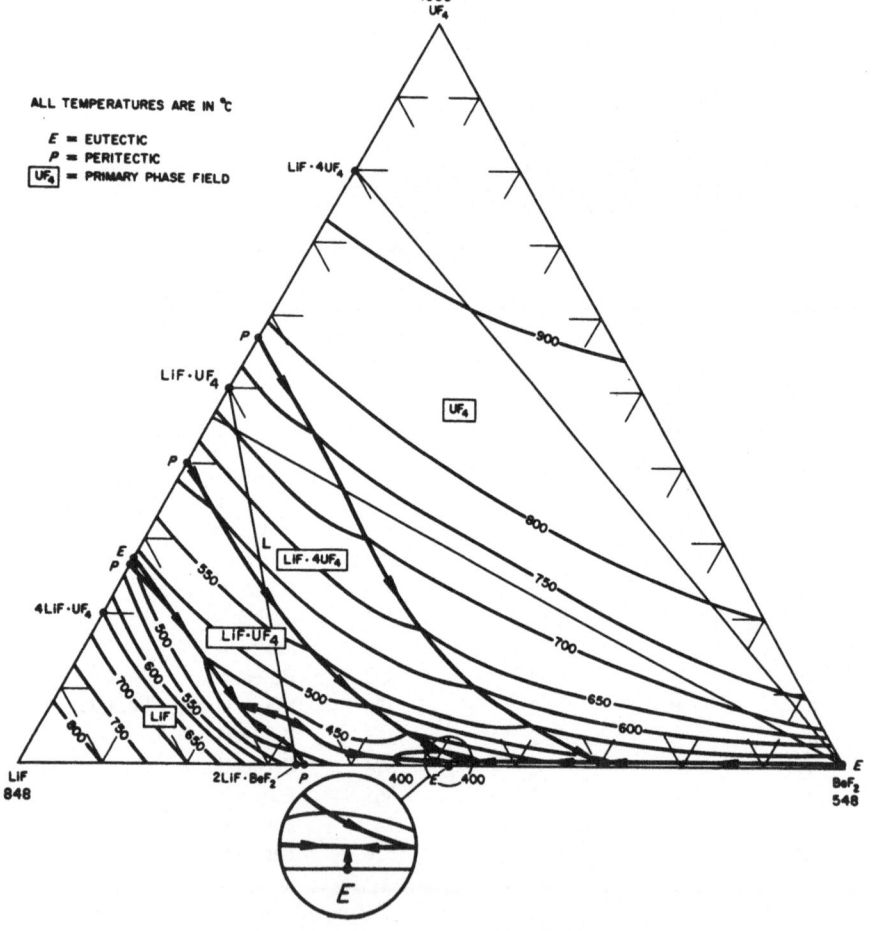

Fig. 220. The system LiF–BeF$_2$–UF$_4$.

The System NaF–BeF$_2$–UF$_4$. See Fig. 221. Source: J. F. Eichelberger, C. R. Hudgens, L. V. Jones, G. Pish, T. B. Rhinehammer, and L. J.

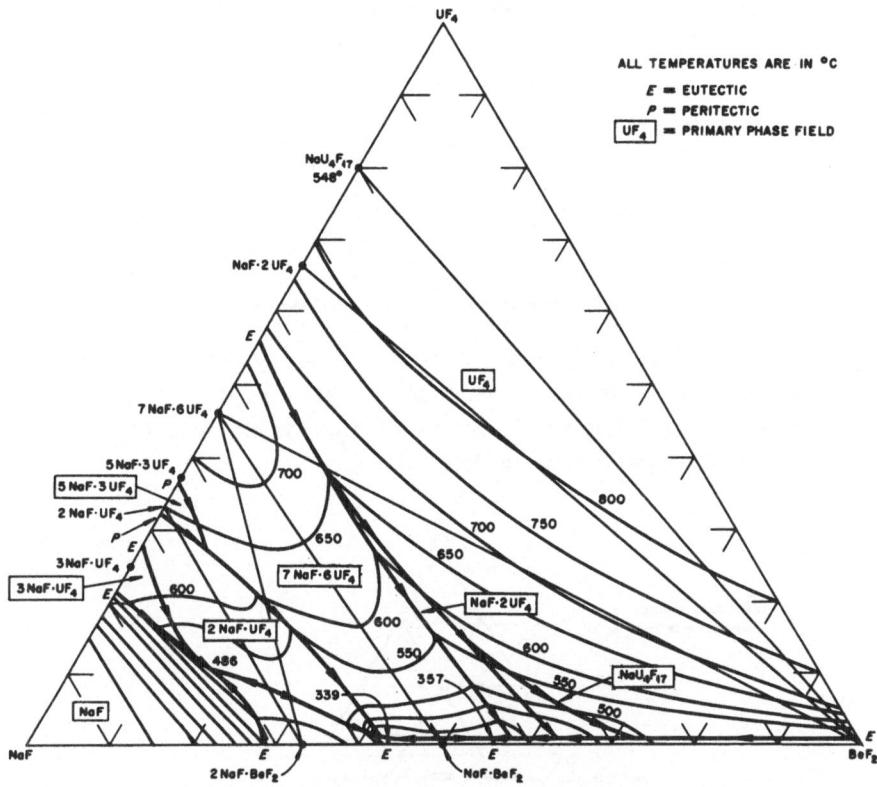

Fig. 221. The system NaF–BeF$_2$–UF$_4$.

Wittenberg, *J. Am. Ceram. Soc.* **46**:282 (1963). The minimum temperature in the quasibinary system NaF · BeF$_2$–7NaF · 6UF$_4$ occurs at 50.5 NaF–48.5–BeF$_2$–1.0 UF$_4$ (mole %) at 367°C. Invariant and singular equilibrium points are given in the table.

Mole %			Temperature	Type of equilibrium
NaF	BeF$_2$	UF$_4$		
74	12	14	500	Peritectic (decomposition of 3NaF · UF$_4$ in ternary system)
72.5	17	19.5	486	Eutectic
64.5	9	26.5	630	Peritectic (decomposition of 5NaF · 3UF$_4$ in ternary system)

(*Continued*)

Mole %			Temperature	Type of equilibrium
NaF	BeF_2	UF_4		
57	42	1	378	Peritectic
56	43.5	0.5	339	Eutectic
43.5	55.5	1	357	Eutectic
41	58	1	375	Peritectic
26	63	1	409	Peritectic
40	47	13	548	Peritectic
27	72	1	498	Peritectic

The System NaF–PbF_2–UF_4. See Fig. 222. Source: C. J. Barton, J. P. Blakely, G. J. Nessle, L. M. Bratcher, and W. R. Grimes, ORNL, unpublished work (1950–1951). Study of the system was discontinued before invariant equilibria were determined.

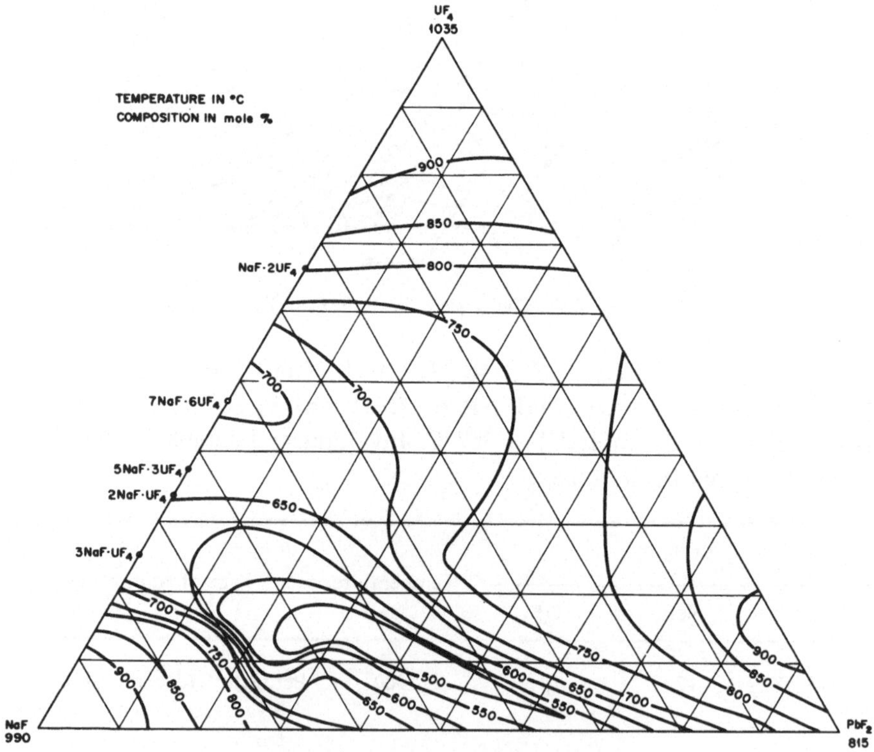

Fig. 222. The system NaF–PbF_2–UF_4.

The System KF–BeF_2–ZrF_4. See Fig. 223. Source: Chan Ngok Mai, Yu. M. Korenev, and A. V. Novoselova, *Russ. J. Inorg. Chem.* (*English*

Transl.) **10**:921 (1965). Composition–temperature sections are shown graphically in the source reference. In a subsequent report by the authors [*ibid.*, **10**:1683 (1965)], the phase relations along the limiting system 2KF · BeF₂–3KF · ZrF₄ are shown. In that work the eutectic is listed at 720° and at 55 mole % 2KF · BeF₂.

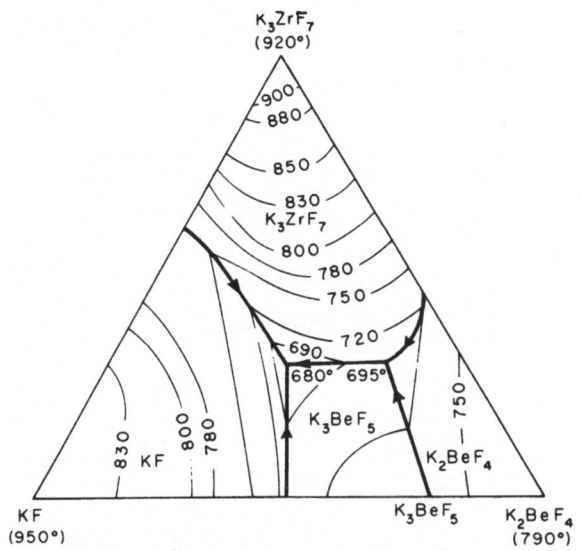

Fig. 223. The system KF–BeF₂–ZrF₄.

The System NaF–CrF₂–ZrF₄. See Fig. 224. Source: R. E. Thoma, B. S. Landau, and W. R. Grimes, ORNL, unpublished work (1957). The

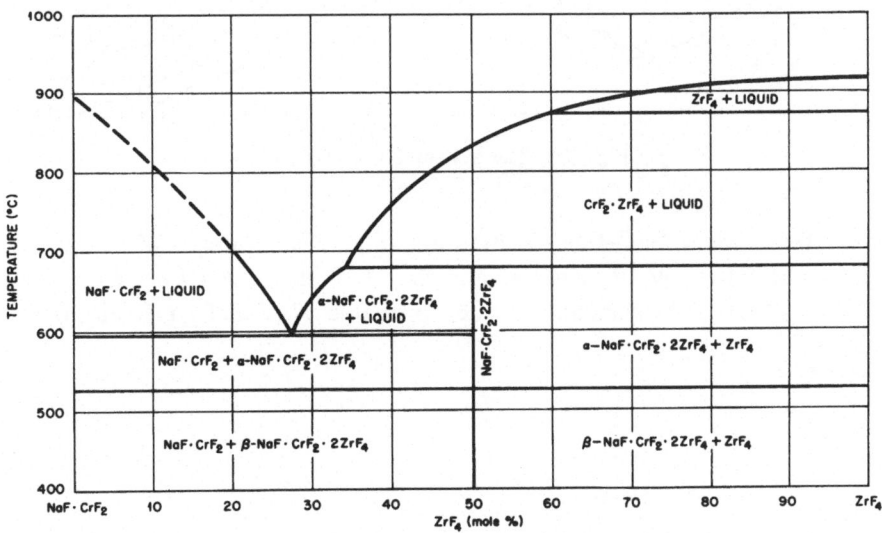

Fig. 224. The system NaF–CrF₂–ZrF₄.

ternary phase diagram of the system NaF–CrF$_2$–ZrF$_4$ has not been described except for the section NaF · CrF$_2$–ZrF$_4$ shown here.

The System KF–PbF$_2$–UF$_4$. See Fig. 225. Source: C. J. Barton, J. P. Blakely, G. J. Nessle, L. M. Bratcher, and W. R. Grimes, ORNL, unpublished work (1950–1951). The investigation conducted by the group cited above did not progress to the point where liquid–solid phase relationships are established.

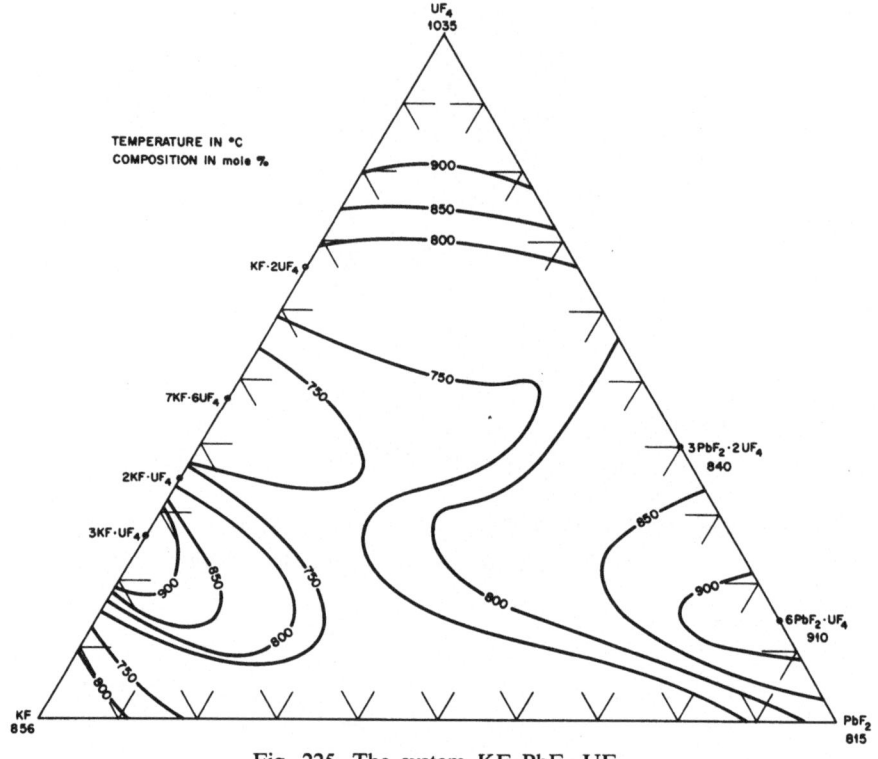

Fig. 225. The system KF–PbF$_2$–UF$_4$.

The System NaF–CeF$_3$–ZrF$_4$. See Fig. 226. Source: W. T. Ward, R. A. Strehlow, W. R. Grimes, and G. M. Watson, *J. Chem. Eng. Data* **5**:137 (1960). Enlargement indicates probable NaF–ZrF$_4$ primary phase fields.

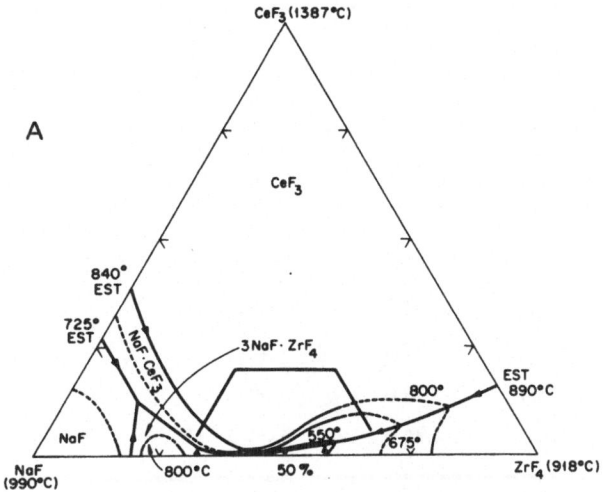

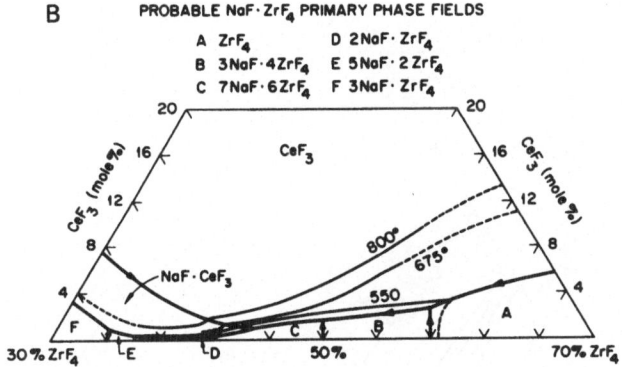

Fig. 226. The system NaF–CeF₃–ZrF₄. (B) gives detail of the region indicated in (A).

The System KF–AlF₃–ZrF₄. See Fig. 227. Source: R. E. Thoma, B. J. Sturm, and E. H. Guinn, Molten-Salt Solvents for Fluoride Volatility Processing of Aluminum-Matrix Nuclear Fuel Elements, ORNL-3594 (1964). Invariant and singular equilibrium points are given in the table.

Mole %			Temperature	Type of equilibrium
KF	AlF$_3$	ZrF$_4$		
86.0		14.0	765	Eutectic
63.0		36.0	590	Peritectic
60.0		40.0	445	Peritectic
58.0		42.0	430	Eutectic
45.0		55.0	440	Eutectic
63.0	15.0	22.0	490	Eutectic
~55.0	~5.0	~40.0	400	Eutectic

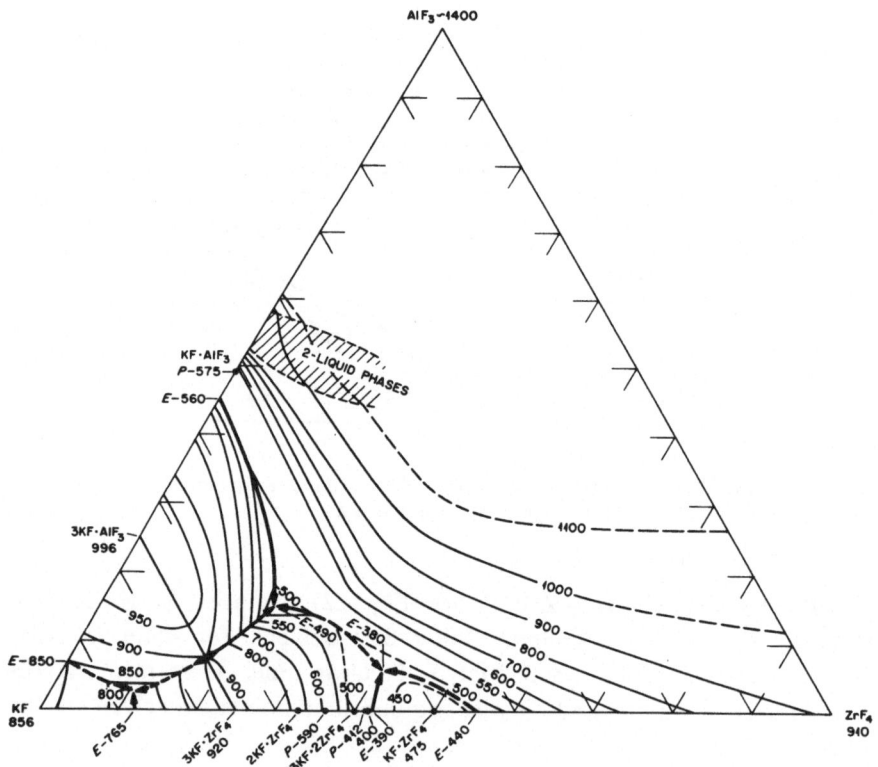

Fig. 227. The system KF–AlF$_3$–ZrF$_4$.

The System NaF–ZrF$_4$–ThF$_4$. See Fig. 228. Source: R. E. Thoma. H. A. Friedman, and H. Insley, ORNL, unpublished work (1955–1958), Invariant and singular equilibrium points are given in the table.

	Mole %		Temperature	Type of equilibrium
NaF	ZrF₄	ThF₄		
77	3	20	645	Peritectic
75.5	2.5	22	618	Eutectic
63	6	31[a]	683	Decomposition point of 3NaF · 2ThF₄ in ternary system
65.5	21.5	13	622	Peritectic
65.5	24.5	10	615	Peritectic
64.5	25.5	10	593	Peritectic
58	38	4	538	Peritectic
58	40	2	495	Eutectic
53	42	4	510	Peritectic
49.2	48	2.5	505	Eutectic

[a] Approximate composition.

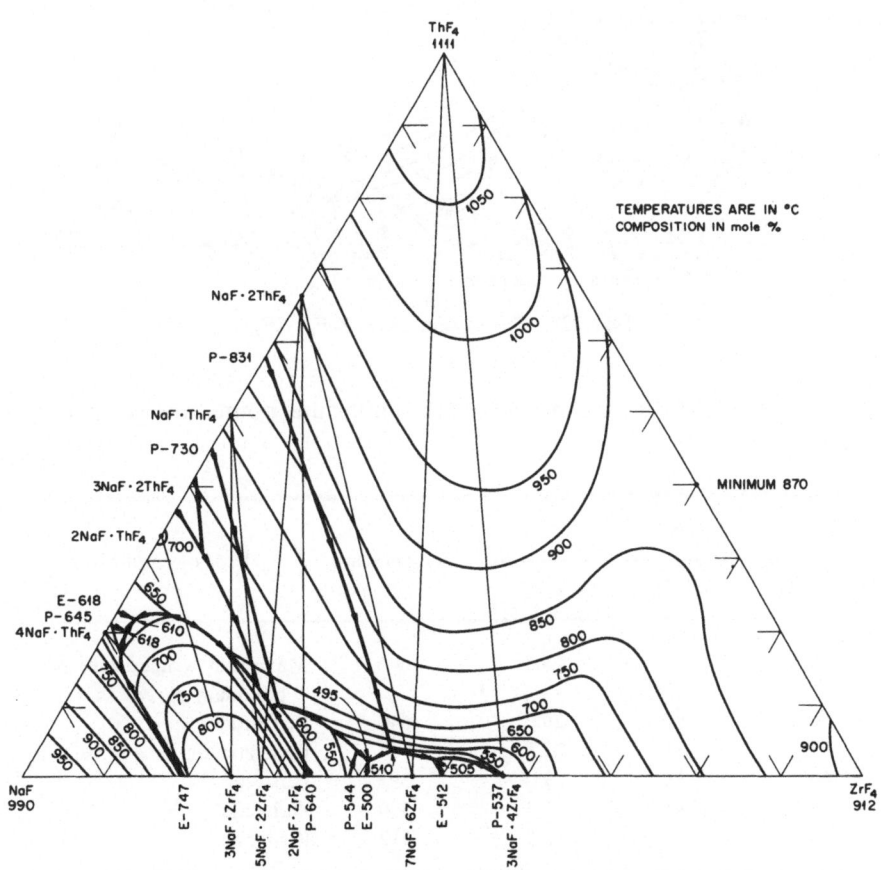

Fig. 228. The system NaF–ZrF₄–ThF₄.

The System NaF–ZrF₄–UF₄. See Fig. 229. Source: C. J. Barton, W. R. Grimes, H. Insley, R. E. Moore, and R. E. Thoma, *J. Phys. Chem.*

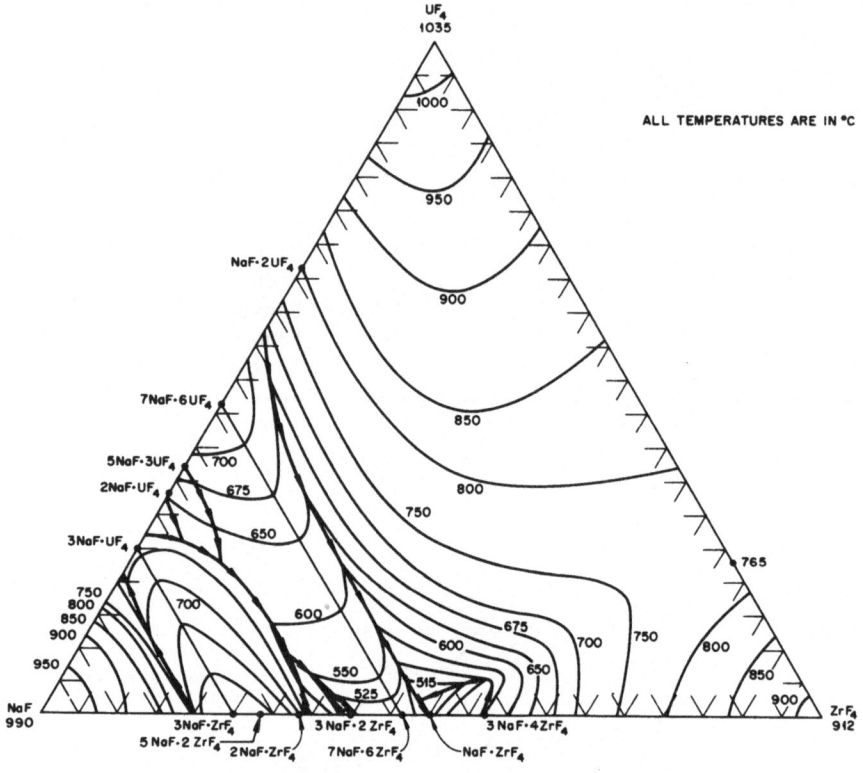

Fig. 229. The system NaF–ZrF₄–UF₄.

62:671 (1958). Invariant and singular equilibrium points are given in the table.

NaF	Mole % ZrF₄	UF₄	Temperature	Type of equilibrium
69.5	4.0	26.5	646	Maximum temperature of boundary curve
68.5	5.5	26.0	640	Peritectic
65.5	12.0	22.5	613	Peritectic or decomposition
64.0	27.0	9.0	592	Peritectic
61.5	34.5	4.0	540	Peritectic
50.5	47	2.5	513	Peritectic

The System LiF–ThF$_4$–UF$_4$. See Fig. 230. Source: C. F. Weaver, R. E. Thoma, H. Insley, and H. A. Friedman, *J. Am. Ceram. Soc.* **43**:214 (1960). Complete tables of data for the system were reported in: C. F.

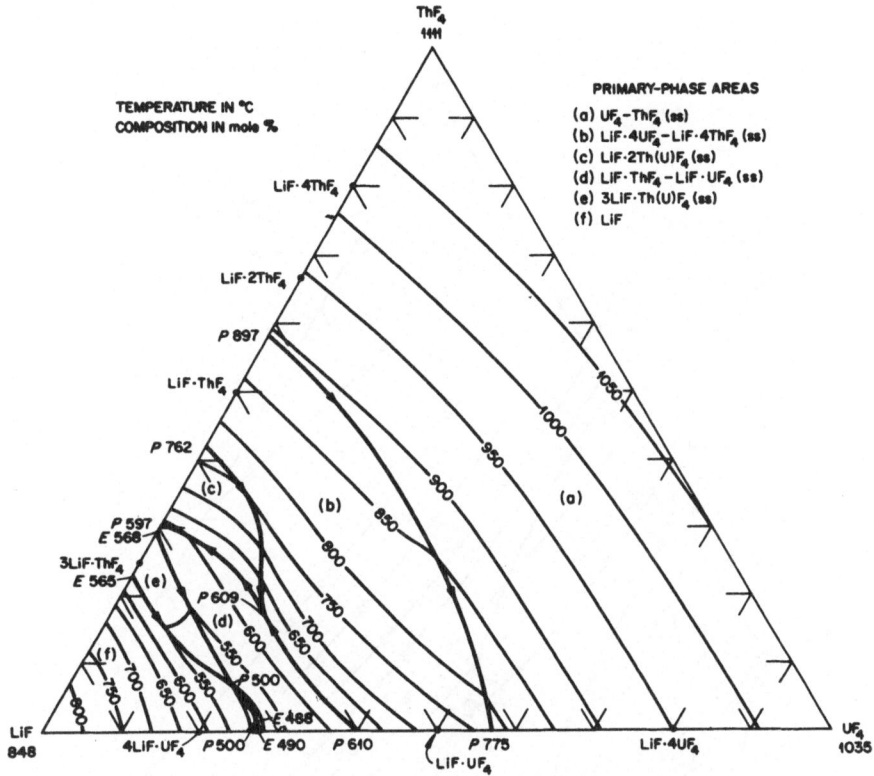

Fig. 230. The system LiF–ThF$_4$–UF$_4$.

Weaver *et al.*, Phase Equilibria in the Systems UF$_4$–ThF$_4$ and LiF–UF$_4$–ThF$_4$, ORNL-2719 (1959). Invariant and singular equilibrium points are given in the table.

	Mole %		Temperature	Type of equilibrium
LiF	ThF$_4$	UF$_4$		
72.5	7.0	20.5	500	Peritectic
72.0	1.5	26.5	488	Eutectic
53	18	19	609	Peritectic

The System NaF–ThF$_4$–UF$_4$. See Fig. 231. Source: R. E. Thoma, H. Insley, G. M. Hebert, H. A. Friedman, and C. F. Weaver, *J. Am. Ceram. Soc.* **46**:37 (1963). The report by R. E. Thoma *et al.*, Phase Equilibria in the System NaF–ThF$_4$–UF$_4$, ORNL-3304 (1963), contains all of the

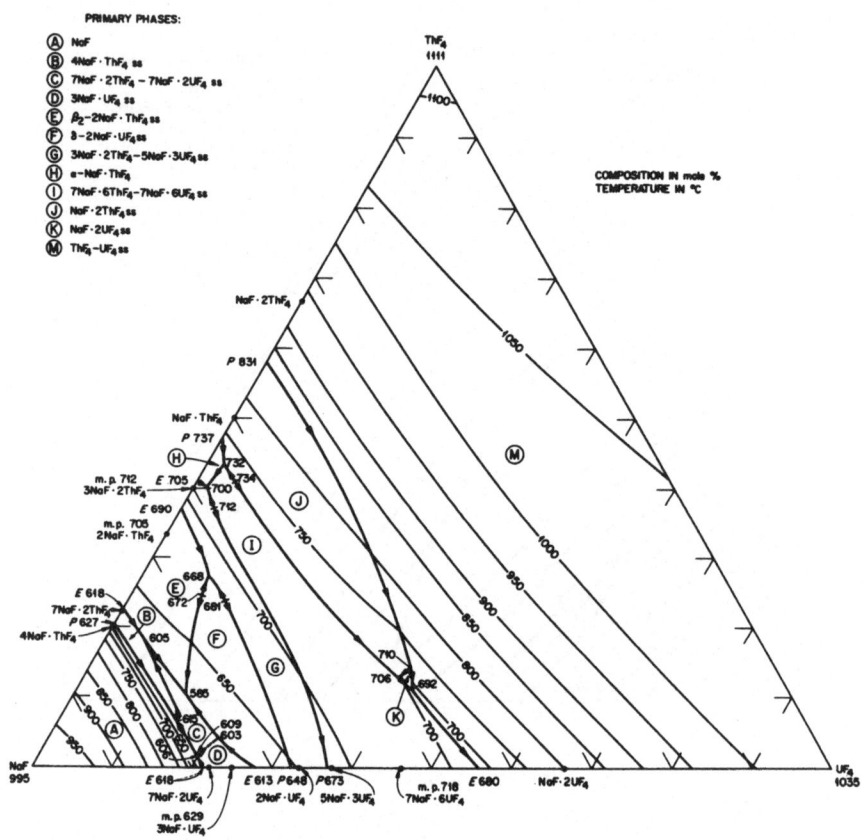

Fig. 231. The system NaF–ThF$_4$–UF$_4$.

experimental data used to construct the phase diagram. The table lists the composition and temperature of the singular and invariant equilibrium points in the system and, for clarity, the coexistent phases at each of these points.

Mole %			Temperature	Type of equilibrium	Phases present
NaF	ThF$_4$	UF$_4$			
78.5	6.5	15	615	Eutectic	NaF, 4NaF · (Th, U)F$_4$ ss, 7NaF · 2(Th, U)F$_4$ ss, liquid
78.5	5.5	16	616	Maximum on boundary curve	NaF, 7NaF · 2(Th, U)F$_4$ ss, liquid
78	7	15	616	Maximum on boundary curve	4NaF · (Th, U)F$_4$ ss, 7NaF · 2(Th, U)F$_4$ ss, liquid
77	2.5	20.5	609	Maximum on boundary curve	3NaF · (U, Th)F$_4$ ss, 7NaF · 2(Th, U)F$_4$ ss, liquid
77	18.5	4.5	605	Peritectic	4NaF · (Th, U)F$_4$ ss, 7NaF · 2(Th, U)F$_4$ ss, β_2-2NaF · (Th, U)F$_4$ ss, liquid
75.5	10.5	14	585	Eutectic	7NaF · 2(Th, U)F$_4$ ss, β_2-2NaF · (Th, U)F$_4$ ss, δ-2NaF · (U, Th)F$_4$ ss, liquid
74.5	4	21.5	603	Peritectic	3NaF · (U, Th)F$_4$ ss, 7NaF · 2(Th, U)F$_4$ ss, δ-2NaF · (U, Th)F$_4$ ss, liquid
66.5	24.5	9	672	Maximum on boundary curve	β_2-2NaF · (Th, U)F$_4$ ss, δ-2NaF · (U, Th)F$_4$ ss, liquid
64.5	27	8.5	668	Eutectic	β_2-2NaF · (Th, U)F$_4$ ss, δ-2NaF · (U, Th)F$_4$ ss, 3NaF · 2ThF$_4$–5NaF · 3UF$_4$ ss, liquid
64	24	12	681	Maximum on boundary curve	δ-2NaF · (U, Th)F$_4$ ss, 3NaF · 2ThF$_4$–5NaF · 3UF$_4$ ss
59	37	4	712	Maximum on boundary curve	3NaF · 2ThF$_4$–5NaF · 3UF$_4$ ss, 7NaF · 6(Th, U)F$_4$ ss, liquid
58.5	40	1.5	700	Eutectic	3NaF · 2ThF$_4$–5NaF · 3UF$_4$ ss, α-NaF · ThF$_4$, 7NaF · 6(Th, U)F$_4$ ss, liquid
54.5	43.5	2	732	Peritectic	α-NaF · ThF$_4$, 7NaF · 6(Th, U)F$_4$ ss, NaF · 2Th$_{1-x}$U$_x$F$_4$ ss, liquid
54.5	41	4.5	734	Maximum on boundary curve	7NaF · 6(Th, U)F$_4$ ss, NaF · 2Th$_{1-x}$U$_x$F$_4$ ss, liquid

(Continued)

Mole %			Temperature	Type of equilibrium	Phases present
NaF	ThF$_4$	UF$_4$			
48	13	39	706	Peritectic	7NaF · 6(Th, U)F$_4$ ss, NaF · 2Th$_{1-x}$U$_x$F$_4$ ss, NaF · 2U$_{1-x}$ThF$_4$ ss, liquid
47	11	42	692	Peritectic	7NaF · 6(Th, U)F$_4$ ss, NaF · 2U$_{1-x}$Th$_x$F$_4$ ss, (Th, U)F$_4$ ss, liquid
46	14	40	710	Peritectic	NaF · 2Th$_{1-x}$U$_x$F$_4$ ss, NaF · 2U$_{1-x}$Th$_x$F$_4$ ss, (Th, U)F$_4$ ss, liquid

The System BeF$_2$–ThF$_4$–UF$_4$. See Fig. 232. Source: C. F. Weaver, R. E. Thoma, H. A. Friedman, and G. M. Hebert, *J. Am. Ceram. Soc.*

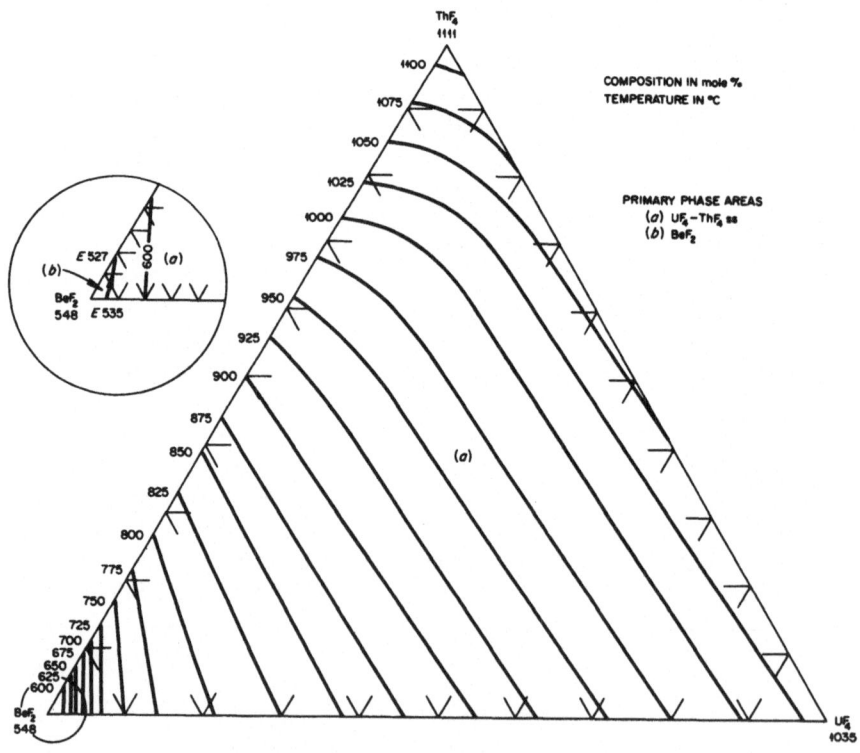

Fig. 232. The system BeF$_2$–ThF$_4$–UF$_4$.

44:146 (1961). No invariant equilibrium reactions occur in the system as a result of the formation of a continuous solid solution in the ThF_4–UF_4 binary system.

The System MgF_2–CaF_2–BaF_2. See Fig. 233. Source: V. T. Berezhnaya and G. A. Bukhalova, *Russ. J. Inorg. Chem. (English Transl.)* **6**:1091 (1961). Invariant equilibrium points: $E = 777°$, MgF_2–CaF_2–BaF_2 (27–

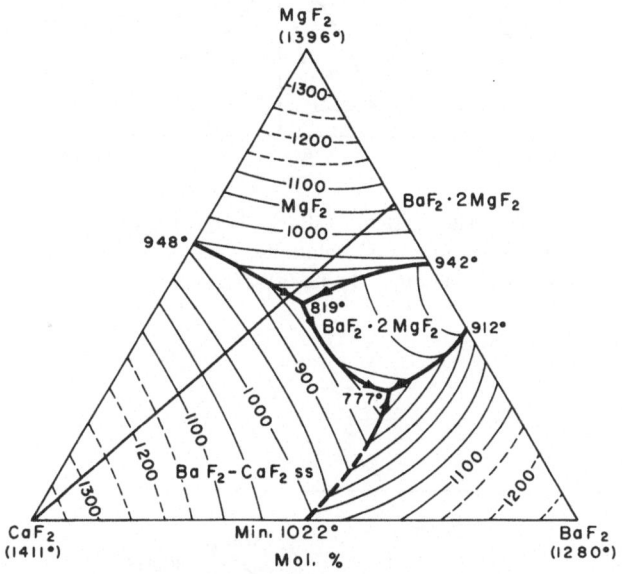

Fig. 233. The system MgF_2–CaF_2–BaF_2.

21–52%); $P = 819°$, MgF_2–CaF_2–BaF_2 (46–28–26%). A more recent version of the system was reported (in wt. %) by E. E. Lukashenko and E. A. Rentova, *Russ. J. Inorg. Chem. (English Transl.)* **15**:1766 (1970).

The $LiBaF_3$–KMF_3 Systems. See Figs. 234–236. Source: I. N. Belaev, S. A. Shilov, and S. Kh. Kulaeva, *Russ. J. Inorg. Chem. (English Transl.)* **15**:575 (1970). The systems with $KCdF_3$ and $KCaF_3$ are of the eutectic type, with the eutectics at $BaLiF_3$ 59.0 mole %, 695°, and 67.5 mole %, 703°, respectively. In the system with $KCdF_3$, there is restricted solid solubility in potassium fluorocadmate. The maximum solubility of $BaLiF_3$ at the eutectic temperature is 18 mole %. Continuous series of solid solutions are formed in systems in which the ratio of the radii of the interchanging ions does not exceed 1.34.

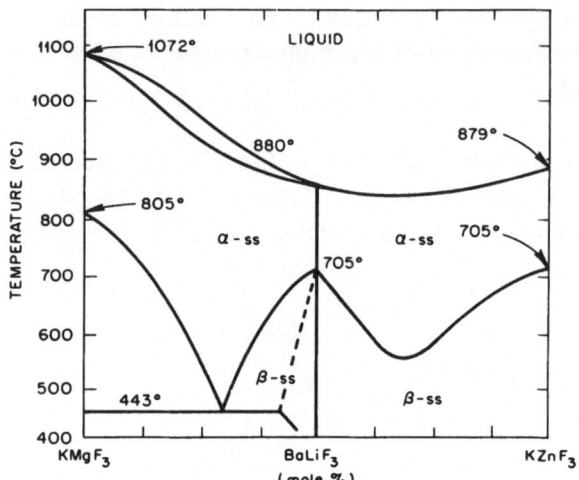

Fig. 234. The system LiBaF₃–KMg(Zn)F₃. α-s.s. = high-temperature solid solutions; β-s.s. = low-temperature solid solutions. (I) KMgF₃; (II) BaF₂.

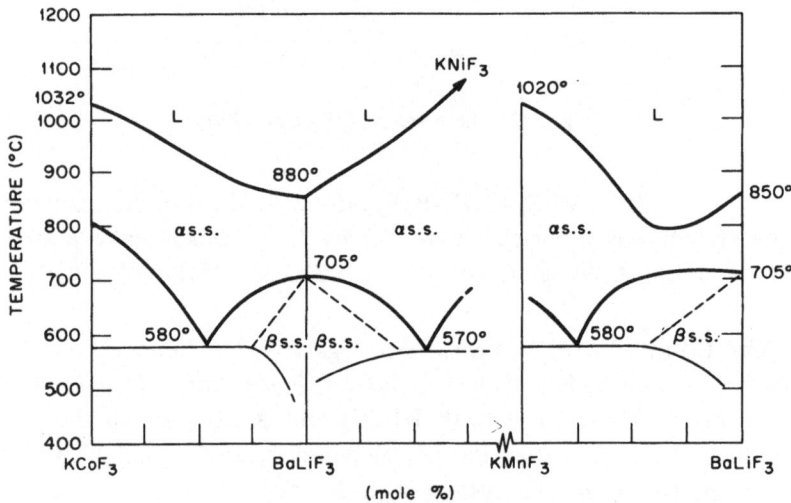

Fig. 235. The system LiBaF₃–KNi(Co,Mn)F₃. (I) KCoF₃; (II) KNiF₃; (III) KMnF₃; (IV) BaF₂.

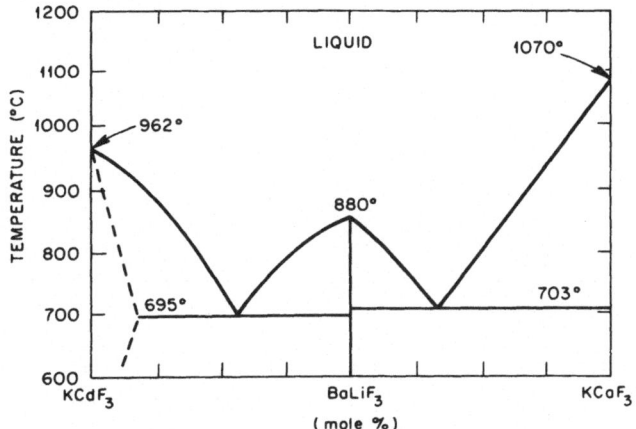

Fig. 236. The systems LiBaF₃–KCdF₃ and LiBaF₃–KCaF₃.
(I) LiBaF₃; (II) KCaF₃; (III) BaF₂.

The System Li₂TiF₆–Na₂TiF₆–K₂TiF₆. See Fig. 237. Source: R. V. Chernov and N. M. Ermolenko, *Russ. J. Inorg. Chem.* (*English Transl.*) **17**:1340 (1972). The minimum crystallization temperature in this partial

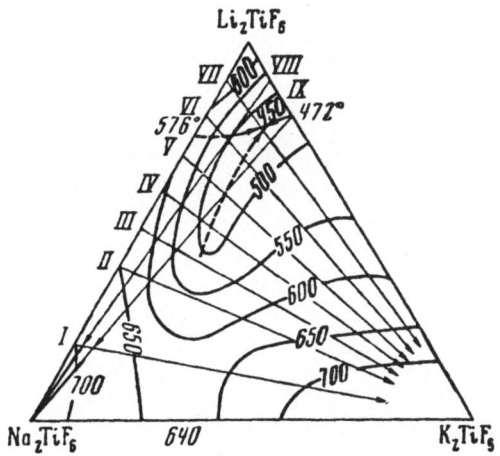

Fig. 237. The system Li₂TiF₆–Na₂TiF₆–K₂TiF₆.

section of the quaternary system was found to be 450° at the composition Li₂TiF₆–K₂TiF₆–Na₂TiF₆ (85–13–2 mole %). The authors noted that the eutectic temperature for the quaternary system is 420°.

INDEX